Zeta Functions in Algebra and Geometry

CONTEMPORARY MATHEMATICS

566

Zeta Functions in Algebra and Geometry

Second International Workshop
May 3–7, 2010
Universitat de les Illes Balears,
Palma de Mallorca, Spain

Antonio Campillo
Gabriel Cardona
Alejandro Melle-Hernández
Wim Veys
Wilson A. Zúñiga-Galindo
Editors

American Mathematical Society
Real Sociedad Matemática Española

American Mathematical Society
Providence, Rhode Island

2000 *Mathematics Subject Classification.* Primary 11F66, 11G25, 11L07, 11G50, 14D10, 14E18, 14G40, 22E50, 32S40, 57M27.

Library of Congress Cataloging-in-Publication Data

International Workshop on Zeta Functions in Algebra and Geometry (2nd : 2010 : Universitat de Les Illes Balears)

Zeta functions in algebra and geometry : second International Workshop on Zeta Functions in Algebra and Geometry, May 3–7, 2010, Universitat de Les Illes Balears, Palma de Mallorca, Spain / Antonio Campillo ... [et al.], editors.

p. cm. — (Contemporary Mathematics ; v. 566)

Includes bibliographical references.

ISBN 978-0-8218-6900-0 (alk. paper)

1. Functions, Zeta–Congresses. 2. Geometry, Algebraic–Congresses. 3. Algebraic varieties–Congresses. I. Campillo, Antonio, 1953– II. Title.

QA351.I58 2010
515′.56–dc23 2011050434

∞ The paper used in this book is acid-free and falls within the guidelines
established to ensure permanence and durability.
Visit the AMS home page at http://www.ams.org/

10 9 8 7 6 5 4 3 2 1 17 16 15 14 13 12

This volume is dedicated to Fritz Grunewald.

Fritz Grunewald 1949–2010

Contents

Preface

The present volume reflects the contents of the talks and some additional contributions given at the "*Second International Workshop on Zeta Functions in Algebra and Geometry*" held at Universitat de les Illes Balears, Palma de Mallorca, Spain, from May 3rd to May 7th, 2010.

Zeta functions can be attached to several mathematical objects like fields, groups, algebras, functions, and dynamical systems. Typically, zeta functions encode relevant arithmetic, algebraic, geometric or topological information about the original object. The conference was focused on the following topics:

(1) Arithmetic and geometric aspects of local, topological and motivic zeta functions,
(2) Poincaré series of valuations,
(3) Zeta functions of groups, rings and representations,
(4) Prehomogeneous vector spaces and their zeta functions,
(5) Height zeta functions.

Local zeta functions were introduced by A. Weil in the sixties and have been extensively studied by J.-I. Igusa, J. Denef and F. Loeser, among others. More recently, using ideas of motivic integration due to M. Kontsevich, a generalization of these functions, called motivic zeta functions, was introduced by Denef and Loeser. All these functions contain geometric, topological, and arithmetic information about mappings defined over local (and other) fields. In close terms, recently T. Hales discovered a motivic nature on integrals which play a central role in the Langlands program.

Using integration over spaces of functions in a spirit similar to motivic integration, A. Campillo, F. Delgado and S. M. Gusein-Zade study Poincaré series of some filtrations on the ring of germs of holomorphic functions of a singularity and its geometric and topological applications. In particular, unexpected connections relating valuation theory with zeta functions have been obtained.

M. du Sautoy and F. Grunewald, among others, have studied extensively zeta functions of groups which were introduced originally as potentially new invariants in attempts to understand the difficult problem of classifying infinite nilpotent groups. Recently du Sautoy has found that these zeta functions are an important tool in trying to understand the problem of classifying the wild class of finite p-groups.

Prehomogeneous vector spaces and their zeta functions were introduced by M. Sato, and have been studied extensively by T. Shintani, M. Kashiwara, F. Sato, T. Kimura, and A. Gyoja, among others. These spaces play a central role in the stunning generalization of Gauss's composition laws obtained by M. Bhargava.

The distribution of rational points of bounded height of smooth varieties over global fields is related to convergence properties of height zeta functions and estimated by the Batyrev-Manin conjecture and refinements. Current work by E. Peyre, Y. Tschinkel and A. Chambert-Loir, among others, provides extensive study and progress on the subject.

We organized the contributed papers into four parts. Part I, "*L-functions of varieties over finite fields and Artin L-functions*", contains the contributions of Pilar Bayer, Anne Frühbis-Krüger and Shabnam Kadir, and Antonio Rojas-León. Part II, "*Height zeta functions and arithmetic*", contains the contributions of Driss Essouabri, Yuri I. Manin and Sho Tanimoto and Yuri Tschinkel. Part III, "*Motivic zeta functions, Poincaré series, complex monodromy and knots*", contains the contributions of Nero Budur, Carlos Galindo and Francisco Monserrat, Evgeny Gorsky, Lars Halvard Halle and Johannes Nicaise, and András Némethi and Fernando Román. Part IV, "*Zeta functions for groups and representations*", contains the contributions of Nir Avni, Benjamin Klopsch, Uri Onn, and Christopher Voll, and Aner Shalev.

We now describe briefly the content of the articles forming this volume. There are contributions which are expository papers in each of the parts. Pilar Bayer's article discusses Artin L-functions of Galois representations of dimension 2 which is perfectly inserted in a complete historical context of the Artin conjecture.

The article of Tanimoto and Tschinkel surveys recent partial progress towards a proof of the Manin conjecture for equivariant compactifications of solvable algebraic groups. They use height zeta functions to study the asymptotic distribution of rational points of bounded height on projective equivariant compactifcations of semi-direct products.

The article of Budur is an excellent survey of some analytic invariants of singularities, mostly those ones which are related with the log canonical threshold, spectra associated with the mixed Hodge structure on the vanishing cohomology of Milnor fibers, multiplier ideals and jumping numbers, different zeta functions (monodromy zeta function, topological zeta function, Denef-Loeser motivic zeta function), and different versions of the Bernstein-Sato b-polynomial.

Related with this problem are the *monodromy conjectures*. What are *monodromy conjectures*? The answer to this question is the heart of the survey article of Halle and Nicaise. They give a very readily guide on new directions opened by a still mysterious conjecture formulated by Jun-Ichi Igusa (on p-adic integrals), that however seems quite natural from the point of view of the developments of algebraic geometry (conjecture of Borevich-Shafarevich, Weil conjectures, ...). They also give a new definition for motivic zeta functions of Calabi-Yau varieties over a complete discretely valued field in terms of analytic rigid geometry and base changes and proved that they verify a very precise global version of the monodromy conjecture.

Valuations are considered here in the context of singularity theory, which is one of the main sources of valuation theory as well as a research area in which valuations are an essential tool. The article of Galindo and Montserrat provides a concise survey of some aspects of the theory plane valuations offering a valuable view of the whole set and the current status of some of the top research problems.

Shalev's survey article gives an overview over a number of results on applications of certain zeta functions associated with groups to several topics including random generation, random walks on groups and commutator width.

The aim of the article of Frühbis-Krüger and Kadir is to give numerical examples to the conjectured change in degree of the zeta function for singular members of families of Calabi-Yau varieties over finite fields which are deformations of Fermat varieties.

Rojas-Leon's article contains some interesting and significant improvements to the classical Weil estimates for trigonometric sums associated to polynomials in one variable by utilizing Deligne-Katz-Laumon methods based on the local Fourier transform.

The main goal of Essouabri's article is to understand the asymptotic behavior of the number of rational points on Zariski open subsets of toric varieties in $\mathbb{P}^n(\mathbb{Q})$.

Manin's article lies at the interface of Diophantine geometry and model theory. The Manin's goal is: given certain combinatorial data about the set of K-rational points on a projective cubic surface defined over K, is to reconstruct the definition field K and the equation of the surface. The approach of the paper is based on Zilber's well-known reconstructions of algebraic geometry using model theory, but here one is not working over algebraically closed fields.

Heegard–Floer homology was introduced by Ozsváth and Szabó as a tool to understand Seiberg-Witten invariants of 3-manifold. Némethi and Román present a computation of the lattice cohomology of a special, but very important for singularity theory applications, class of 3-manifolds: they are obtained by surgery on an algebraic link in the 3-dimensional sphere. Lattice cohomology is a combinatorial construction starting from the plumbing graph of a manifold, which leads to certain cohomology groups. The connections with Seiberg–Witten invariants and Heegard–Floer theory are also presented.

Gorsky's article offers several very interesting conjectures related with homologies of torus knots $T_{n;m}$ using the combinatorics of $q;t$-Catalan numbers and their (several) generalizations.

The article of Avni, Klopsch, Onn and Voll is focused on the study of zeta functions associated to representations of some compact p-adic analytic groups by means of the Kirillov's orbit method, Clifford theory and p-adic integration.

The sponsors of Palma de Mallorca's Workshop include the Fundation for Scientific Research - Flanders (FWO), the Spanish Ministerio de Ciencia e Innovación, the local Govern de les Illes Balears, the Junta de Castilla y León, the Consell de Mallorca, the Ajuntament de Palma, the Caixa de Balears, the program Ingenio Mathematica, the Unversities Complutense de Madrid (UCM), Illes Balears (UIB) and Valladolid (UVA) and the Departament de Ciències Matemàtiques i Informàtica (UIB). We thank all of them and we also thank the American Mathematical Society (AMS) and the Real Sociedad Matemática Española (RSME) for agreeing to publish this volume as one of their common publications.

We finally want to thank all organizations and people that helped in organizing the conference and editing the proceedings, among others, the members of the Local Organizing Committee: Ll. Huguet (Chair), A. Campillo, G. Cardona, M. González–Hidalgo, A. Mir (Spain), the members of the Organizing Committee A. Melle-Hernández (Spain), W. Veys (Belgium), W. A. Zúñiga-Galindo (México) and the members of the Scientific Committee: A. Campillo (Spain), J. Denef (Belgium), F. Grunewald (Germany), S. M. Gusein-Zade (Russia), M. Larsen (USA), I. Luengo (Spain), Y. Tshinkel (USA), A. Yukie (Japan).

A short time before our workshop in Palma de Mallorca started, we heared the unexpected and sad news that Fritz Grunewald passed away. As leading specialist in the study of zeta functions in algebra, he was a distinguished speaker at the first edition of our "*International Workshop on Zeta Functions in Algebra and Geometry*" held in Segovia, Spain, in June 2007. Actually, he was very enthusiastic about that initiative and he immediately accepted to be a member of the scientific committee for the second edition in Palma. In that role, he was a great help for us. In fact, we still exchanged mails about the organization of the workshop few days before his decease.

During the first day of the workshop, the lectures of Dan Segal and Alex Lubotzky were in honour of Fritz: outstanding mathematician, extraordinary person and fantastic friend. This is indeed how we will remember him.

Antonio Campillo
Gabriel Cardona
Alejandro Melle-Hernández
Wim Veys
Wilson A. Zúñiga-Galindo

List of Participants

Sargis Aleksanyan
Institute of Mathematics of NAS
Armenia

Theofanis Alexoudas
Royal Holloway University of London
UK

Pilar Bayer Isant
Universitat de Barcelona
Spain

Iván Blanco-Chacón
Universitat de Barcelona
Spain

Bart Bories
Katholieke Universiteit Leuven
Belgium

Nero Budur
University of Notre Dame
USA

Antonio Campillo
Universidad de Valladolid
Spain

Gabriel Cardona Juanals
Universitat de les Illes Balears
Spain

Pierrette Cassou-Noguès
Université Bordeaux 1
France

Wouter Castryck
Katholieke Universiteit Leuven
Belgium

Helena Cobo
Katholieke Universiteit Leuven
Belgium

Félix Delgado de la Mata
Universidad de Valladolid
Spain

Josep Domingo-Ferrer
Universitat Rovira i Virgili
Spain

Wolfgang Ebeling
Leibniz Universität Hannover
Germany

Jordan S. Ellenberg
University of Wisconsin
USA

Driss Essouabri
Université Saint-Etienne
France

Alexander Esterov
Independent University of Moscow
Russia

Francesc Fitè
Universitat Politècnica de Catalunya
Spain

Anne Frühbis-Krüger
Leibniz Universität Hannover
Germany

Jeanneth Galeano Peñaloza
Cinvestav
Mexico

Carlos Galindo
Universitat Jaume I
Spain

Dorian Goldfeld
Columbia University
USA

Jon González Sánchez
Universidad de Cantabria
Spain

Manuel González Hidalgo
Universitat de les Illes Balears
Spain

Pedro Daniel González Pérez
Universidad Complutense de Madrid
Spain

Josep González Rovira
Universitat Politécnica de Catalunya
Spain

Evgeny Gorsky
Moscow State University
Russia

Sabir M. Gusein-Zade
Moscow State University
Russia

Gleb Gusev
Moscow State University
Russia

Yeni Hernández
Universidad Nacional Abierta
Mexico

Llorenç Huguet Rotger
Universitat de les Illes Balears
Spain

Benjamin Klopsch
Heinrich-Heine-Universität Düsseldorf
Germany

Takeyoshi Kogiso
Josai University
Japan

Pankaj Kumar
IGIDR, Mumbai
India

Joan-C. Lario
Universitat Politécnica de Catalunya
Spain

Michael Larsen
Indiana University
USA

Michael Lönne
Universität Bayreuth
Germany

Edwin León Cardenal
Cinvestav
Mexico

Wen-Wei Li
Institut de Mathématiques de Jussieu
France

François Loeser
Ecole Normale Supérieure Paris
France

Elisa Lorenzo
Universitat Politécnica de Catalunya
Spain

Alex Lubotzky
Einstein Institute of Mathematics
Jerusalem, Israel

Ignacio Luengo Velasco
Universidad Complutense de Madrid
Spain

Yuri Manin
Northwestern University
USA

Alejandro Melle Hernández
Universidad Complutense de Madrid
Spain

Arnau Mir Torres
Universitat de les Illes Balears
Spain

Francisco J. Monserrat
Universidad Politécnica de Valencia
Spain

Julio José Moyano Fernández
Universität Osnabrück
Germany

Mircea Mustata
University of Michigan
USA

András Némethi
Rényi Mathematical Institute
Budapest, Hungary

Johannes Nicaise
Katholieke Universiteit Leuven
Belgium

Uri Onn
Ben Gurion University of the Negev
Israel

Jorge Ortigas Galindo
Universidad de Zaragoza
Spain

Evija Ribnere
Heinrich-Heine-Universität Düsseldorf
Germany

Antonio Rojas León
Universidad de Sevilla
Spain

Dan Segal
University of Oxford
UK

Jan Schepers
Katholieke Universiteit Leuven
Belgium

Dirk Segers
Katholieke Universiteit Leuven
Belgium

Aner Shalev
Einstein Institute of Mathematics
Jerusalem, Israel

Rob Snocken
University of Southampton
UK

Alexander Stasinski
University of Southampton
UK

George Stoica
University of New Brunswick
Canada

Kiyoshi Takeuchi
University of Tsukuba
Japan

Takashi Taniguchi
Kobe University
Japan

Tomás Sánchez Giralda
Universidad de Valladolid
Spain

Javier Tordable
Google
Spain

Yuri Tschinkel
New York University
USA

Jan Tuitman
Katholieke Universiteit Leuven
Belgium

Wim Veys
Katholieke Universiteit Leuven
Belgium

David Villa
UNAM, campus Morelia-UMSNH
Mexico

Christopher Voll
University of Southampton
UK

Akihiko Yukie
Tohoku University
Japan

Wilson A. Zúñiga-Galindo
Cinvestav
Mexico

Shou-Wu Zhang
Columbia University New York
USA

Part I:
L-functions of varieties over finite fields and Artin L-functions

Contemporary Mathematics
Volume **566**, 2012
http://dx.doi.org/10.1090/conm/566/11212

Computational aspects of Artin L-functions

Pilar Bayer

ABSTRACT. Galois representations are a special type of algebraic arithmetical objects to which one can associate L-functions. The aim of this paper is the description of a procedure to calculate as many coefficients as needed of exotic Artin L-functions from the explicit resolution of some Galois embedding problems.

CONTENTS

Introduction

Artin L-functions $L(\rho, s)$ associated to complex Galois representations ρ are defined by Euler products, similar to those defining Dirichlet L-functions $L(\chi, s)$ associated to Dirichlet characters χ. It has been known for a long time that they can be extended to meromorphic functions on the whole complex plane. A wellknown conjecture, formulated by Artin in 1924, predicts that the functions $L(\rho, s)$ are, in fact, entire, with a possible exception of a pole at $s = 1$ when ρ contains the trivial representation.

By Kronecker-Weber's theorem, Artin L-functions for one-dimensional complex representations of the absolute Galois group $G_{\mathbb{Q}}$ of the field $\mathbb{Q}$ of rational numbers are Dirichlet L-functions and the holomorphic extension of these functions is wellknown. Old and recent advances in the modularity of Galois representations have shown that Artin's conjecture is true for those L-functions associated to two-dimensional irreducible complex representations of $G_{\mathbb{Q}}$ except possibly for those of even icosahedral type (cf. section 6), in which case the conjecture has not been

1991 *Mathematics Subject Classification.* Primary 11F66, 11F11, 11R32, 12F12.
Partially supported by MCYT, MTM2009-07024.

proven so far. Recent contributions towards the proof of the Artin's conjecture in the two-dimensional odd icosahedral case can be found in [**12**], [**13**], and [**29**].

ACKNOWLEDGEMENTS. I express my special thanks to the Scientific Committee of the *Second International Workshop on Zeta Functions in Algebra and Geometry* for having invited me to deliver a lecture, and to the Organizing Committee of the event for their work and commitment.

1. Modular forms and Maass forms

The modular group $\mathrm{SL}(2,\mathbb{Z})$ acts on the upper half-plane completed with cusps, $\mathcal{H}^* := \{z \in \mathbb{C} : \Im(z) > 0\} \cup \mathbb{P}^1(\mathbb{Q})$, in the usual manner:

$$\begin{array}{ccc} \mathrm{SL}(2,\mathbb{Z}) \times \mathcal{H}^* & \longrightarrow & \mathcal{H}^* \\ (\gamma,\ z) & \mapsto & \gamma(z) = \dfrac{az+b}{cz+d}, \end{array}$$

where $\gamma = \begin{bmatrix} a & b \\ c & d \end{bmatrix}$. Given an integer $N \geq 1$, let $\Gamma_0(N)$ denote the congruence group of level N defined by

$$\Gamma_0(N) = \left\{ \begin{bmatrix} a & b \\ c & d \end{bmatrix} \in \mathrm{SL}(2,\mathbb{Z}) : c \equiv 0 \pmod{N} \right\}.$$

Modular forms are complex analytic, or real analytic functions, defined on $\mathcal{H}^*$ satisfying certain functional equations and growth conditions that we will now recall.

DEFINITION 1.1. For an integer $k \geq 1$ and a Dirichlet character

$$\chi : (\mathbb{Z}/N\mathbb{Z})^* \to \mathbb{C}^*,$$

such that $\chi(-1) = (-1)^k$, a complex analytic modular form of type (N,k,χ) is a holomorphic function $f : \mathcal{H}^* \longrightarrow \mathbb{C}$ such that

$$f(\gamma(z)) = \chi(d)(cz+d)^k f(z), \text{ for any } \gamma = \begin{bmatrix} a & b \\ c & d \end{bmatrix} \in \Gamma_0(N).$$

If f is zero at the cusps, then f is said to be a cusp form.

The complex vector space of cusp forms of type (N,k,χ) will be denoted by $S(N,k,\chi)$. It is finite dimensional. In a neighborhood of the cusp at infinity, $i\infty$, any $f \in S(N,k,\chi)$ admits a Fourier expansion

$$f(z) = \sum_{n \geq 1} a_n q^n, \quad \text{where } q(z) = e^{2i\pi z}.$$

We denote by $S^{\mathrm{new}}(N,k,\chi)$ the new subspace of $S(N,k,\chi)$ in accordance with the terminology of Atkin-Lehner [**1**]. The Hecke operators act on the space $S(N,k,\chi)$ preserving $S^{\mathrm{new}}(N,k,\chi)$.

A holomorphic modular form f is uniquely determined by a suitable finite set of its Fourier coefficients. In [**39**], Murty refined this statement by showing that the first

$$(k/12)N \prod_{p|N} \left(1 + \frac{1}{p}\right)$$

Fourier coefficients suffice to determine any form $f \in S(N,k,\chi)$.

We shall also consider real analytic modular forms (usually called Maass forms, because they appeared for the first time in Maass' paper [**37**]). Let

$$\Delta_k = -y^2\left(\frac{\partial^2}{\partial^2 x}+\frac{\partial^2}{\partial^2 y}\right)+iky\frac{\partial}{\partial x}$$

be the Laplace operator of weight k.

DEFINITION 1.2. A Maass form of weight k, level N, and character χ is a real analytic complex-valued function f of $z = x+iy \in \mathcal{H}^*$ satisfying the following conditions:

(1) $\Delta_k(F) = \lambda F$, for some $\lambda \in \mathbb{C}$.

(2) For any $\gamma = \begin{bmatrix} a & b \\ c & d \end{bmatrix} \in \Gamma_0(N)$,

$$F(\gamma(z)) = \chi(d) j_\gamma(z)^k F(z),$$

where $j_\gamma(z) := \dfrac{cz+d}{|cz+d|} = e^{i\arg(cz+d)}$.

(3) F is of at most polynomial growth at the cusps of $\Gamma_0(N)$.

If, moreover, F is zero at the cusps then F is said to be a Maass cusp form. A weak Maass wave form is defined similarly but without the growth condition at the cusps.

We denote by $\mathcal{M}(N,k,\chi;\lambda)$ the space of Maass forms of Laplace eigenvalue λ. It is useful to write λ in the form $\lambda(s) = s(1-s)$, where $s = \dfrac{1}{2}+iR$ is a complex number, and so is R. Thus,

$$\lambda = \frac{1}{4}+R^2.$$

In [**41**], Selberg conjectured that if $\lambda > 0$, then $\lambda \geq \frac{1}{4}$; equivalently, that R is real. Since, as we shall see, there exist Maass forms with eigenvalue exactly equal to 1/4, if Selberg's conjecture is true, then it is sharp. The continuous spectrum of the Laplacian is well understood and it consists of the segment $[1/4, \infty)$.

At the neighborhood of $i\infty$, Maass cusp forms also admit Fourier expansions given by

$$F(x+iy) = \sum_{|n|>1} b(n) W_{\frac{k}{2}\mathrm{sgn}(n), iR}(4\pi|n|y) e^{2\pi i n x},$$

where W stands for the usual Whittaker function, normalized so that

$$W_{\alpha,\beta}(y) \sim e^{-\frac{y}{2}} y^\alpha, \quad \text{as } y \to \infty.$$

The following proposition relates holomorphic modular forms and Maass forms of the lowest Laplace eigenvalue; its proof can be found in [**22**] (cf. also [**25**]).

PROPOSITION 1.3. *Let F be a Maass cusp form of level N, weight k, character χ, and the lowest Laplace eigenvalue $\lambda\left(\frac{k}{2}\right) = \frac{k}{2}\left(1-\frac{k}{2}\right)$. Then*

$$f(z) := y^{-\frac{k}{2}} F(z)$$

is a holomorphic modular form of level N, weight k, and character χ and all such forms arise in this way.

If F is a Maass cusp form of Laplace eigenvalue $\frac{1}{4}$, then its Fourier expansion becomes

$$F(z) = y^{\frac{1}{2}} \sum_{n \geq 1} (4\pi n)^{\frac{1}{2}} a_F(n) q^n,$$

so that $f(z) = y^{-\frac{1}{2}} F(z)$, which is a holomorphic cusp form of weight one, has a Fourier expansion

$$f(z) = \sum_{n \geq 1} a_f(n) q^n,$$

with $a_f(n) = (4\pi n)^{\frac{1}{2}} a_F(n)$.

Our knowledge of the dimension of these spaces is quite different depending on whether $k \geq 2$ or $k = 1$. For weight $k \geq 2$, either from Riemann-Roch theorem or from Selberg trace formula, it is proven that

$$\dim S(N, k, \chi) = \frac{k-1}{12} \psi(N) + O(N^{1/2} d(N)),$$

$$\dim S^{new}(N, k, \chi) = \frac{k-1}{12} \varphi(N) + O((kN)^{2/3}),$$

uniform in k and N, where $d(N)$ denotes the number of divisors of N, $\varphi(N)$ stands for the Euler function

$$\varphi(N) = N \prod_{p|N} (1 - p^{-1}),$$

and

$$\psi(N) = N \prod_{p|N} (1 + p^{-1}).$$

For weight $k = 1$, it is expected that the following conjecture is true.

CONJECTURE 1.4 (cf. [**21**]). For N varying among the squarefree integers,

$$\dim S^{new}(N, 1, \chi) = \frac{h(K_N)}{2} + O_\varepsilon(N^\varepsilon),$$

with an O-constant independent of N and χ, and $h(K_N)$ denoting the number of elements of the ideal class group $Cl(K_N)$ of the imaginary quadratic field $K_N = \mathbb{Q}(\sqrt{-N})$.

2. Weight one holomorphic modular forms

Suppose that N is a prime and let $\chi_N = (./N)$ denote the Legendre symbol. Since no nonzero holomorphic cusps forms of weight one may exist unless $\chi_N(-1) = -1$, we are going to assume that $N \equiv 3 \pmod 4$.

Consider a nontrivial class character $\psi : Cl(K_N) \to \mathbb{C}^*$, for K_N denoting the imaginary quadratic field of discriminant $-N$. Since by pointwise multiplication the (irreducible) characters of any abelian group G form a dual group $\widehat{G}$ which is, non-canonically, isomorphic to G, we have $h(K_N) - 1$ choices for ψ. By Hecke, the theta function

$$\theta_\psi(z) := \sum_{\mathfrak{a} \subset \mathcal{O}_K} \psi(\mathfrak{a}) q^{N(\mathfrak{a})}$$

belongs to $S(N, 1, \chi_N)$. By class field theory, these theta functions correspond to cusps forms of dihedral type, in the sense that they are attached to Galois representation with dihedral image (cf. section 6). Let $S^{dih}(N, 1, \chi_N)$ be the vector space that they generate. Since we have $(h(K_N) - 1)/2$ independent forms of dihedral type and, by Siegel's theorem, is $h(K_N) > c(\varepsilon)N^{\frac{1}{2}-\varepsilon}$, we have the ineffective lower bound:

$$N^{1/2-\varepsilon} \ll_\varepsilon \dim S^{dih}(N, 1, \chi_N) = \frac{h(K_N) - 1}{2},$$

for any $0 < \varepsilon < 1/2$. The lower bound is ineffective since this happens to be the case in Siegel's theorem.

Most weight one holomorphic modular forms should be given by theta functions. The missing forms in this construction are called exotic and they are divided into the following types by a standard classification: dihedral, tetrahedral, octahedral, or icosahedral (cf. section 6). Thus we may write,

$$\dim S^{new}(N, 1, \chi_N) = \dim S^{dih}(N, 1, \chi_N) + \dim S^{ex}(N, 1, \chi_N),$$

and, conjecturally,

$$\dim S^{ex}(N, 1, \chi_N) \ll_\varepsilon N^\varepsilon, \text{ for any } \varepsilon > 0.$$

Important results in this direction have been obtained in recent decades. We mention some of them here.

THEOREM 2.1 (Duke, [**21**]). *For any prime N,*

$$\dim S(N, 1, \chi_N) \ll N^{11/12} \log^4 N,$$

with an absolute implied constant.

Note that the truth of conjecture 1.4 implies that

$$\dim S(N, 1, \chi_N) \ll N^{1/2} \log N.$$

THEOREM 2.2 (Michel-Venkatesh, [**38**]). *Fix a central character χ. The number of* GL(2)*-automorphic forms π of Galois type, with central character χ, and conductor N is $\ll_\varepsilon N^{e(G)+\varepsilon}$, where e is a real function on types defined by e(dihedral) $=$ $1/2$, e(tetahedral) $= 2/3$, e(octahedral) $= 4/5$, e(icosahedral) $= 6/7$.*

THEOREM 2.3 (Klüners, [**34**]). *Assume that all primes which exactly divide N are congruent to* 2 (mod 3)*. Then the dimension of the space of octahedral forms of weight* 1 *and conductor N is bounded above by $O_\varepsilon(N^{1/2+\varepsilon})$.*

THEOREM 2.4 (Bhargava-Ghate, [**6**]). *For any positive number X, let $\pi(X)$ denote the number of primes smaller than X and $N^{oct}_{prime}(X)$, the number of independent octahedral cuspidal newforms having prime level $< X$. Then*

(i) $N^{prime}_{oct}(X) = O(X/\log X)$.

(ii) $\dfrac{N^{prime}_{oct}(X)}{\pi(X)} = O(X^\varepsilon)$, *for any $\varepsilon > 0$.*

3. Some character tables

We now compile some basic facts on representation theory of finite groups, which go back to Frobenius, and which we are going to need in the next sections.

A linear representations of a finite group G is given by a vector space V, over some field k, and a homomorphism

$$\rho : G \longrightarrow \mathrm{GL}(V).$$

C_3	$1A$	$3A$	$3B$
order	1	1	1
χ_1	1	1	1
χ_2	1	ζ	ζ^2
χ_3	1	ζ^2	ζ

TABLE 1. Character table for the cyclic group C_3, where $\zeta = e^{\frac{2\pi i}{3}}$.

S_3	$1A$	$2A$	$3A$
order	1	3	2
χ_1	1	1	1
χ_2	1	-1	1
χ_3	2	0	-1

TABLE 2. Character table for the symmetric group S_3

We shall be mainly interested in the case where $k = \mathbb{C}$ and V is of finite dimension. The character χ_ρ of the representation is defined by

$$\chi_\rho : G \longrightarrow \mathbb{C}; \quad g \mapsto \operatorname{Tr}(\rho(g)).$$

Clearly,

$$\chi_\rho(1) = \dim V, \quad \chi_\rho(g^{-1}) = \overline{\chi_\rho(g)},$$

where the bar denotes complex conjugate, and χ_ρ is a class function; i.e., if $[g] = \{ugu^{-1}; u \in G\}$ stands for the conjugacy class of an element $g \in G$, then

$$\chi_\rho([g]) = \chi_\rho(g).$$

Valuable information about the complex representations of a finite group is collected in its character table. The rows of a character table are indexed by the irreducible representations of the group; the columns, by their conjugacy classes. The entries of the character table correspond to the values of the irreducible characters on those classes. Some examples are displayed in tables 1, 2, 3, 4. The necessary background to compute them can be found, for instance, in [**43**].

The character table of S_4 shows that this group does not admit faithful representations of dimension 2. Since we will be mainly interested in representations of this dimension, we also consider certain double covers of this group that support them. Although we have $H^2(S_4, C_2) \simeq C_2 \times C_2$, and, thus, S_4 admits 4 nonisomorphic double covers, there is a unique central extension

$$1 \to C_2 \to \widetilde{S_4} \to S_4 \to 1 \tag{3.1}$$

in which the central group admits a complex irreducible representation of dimension 2 with odd determinant. It is characterized by the fact that the transpositions of S_4 lift to involutions of $\widetilde{S_4}$. Moreover, $\widetilde{S_4} \simeq \mathrm{GL}(2, \mathbb{F}_3)$ (see e.g. [**8**]). The lifting of the conjugacy classes of S_4 to those of $\widetilde{S_4}$ is given in table 6.

4. Artin L-functions

Let $G_\mathbb{Q} = \mathrm{Gal}(\overline{\mathbb{Q}}|\mathbb{Q})$ be the absolute Galois group of the rational field, endowed with the pro-finite (Krull) topology. We take $\mathrm{GL}(V) \simeq \mathrm{GL}(n, \mathbb{C})$, endowed with

S_4	$1A$	$2A$	$2B$	$3A$	$4A$
order	1	6	3	8	6
χ_1	1	1	1	1	1
χ_2	1	-1	1	1	-1
χ_3	2	0	2	-1	0
χ_4	3	1	-1	0	-1
χ_5	3	-1	-1	0	1

TABLE 3. Character table for the symmetric group S_4

$\widetilde{S_4}$	$1\widetilde{A}$	$2\widetilde{A}$	$4\widetilde{A}$	$3\widetilde{A}$	$6\widetilde{A}$	$2\widetilde{B}$	$8\widetilde{A}$	$8\widetilde{B}$
order	1	1	12	6	8	8	6	6
χ_1	1	1	1	1	1	1	1	1
χ_2	1	1	1	1	1	-1	-1	-1
χ_3	2	2	2	-1	-1	0	0	0
χ_4	3	3	-1	0	0	1	-1	-1
χ_5	3	3	-1	0	0	-1	1	1
χ_6	2	-2	0	-1	1	0	$i\sqrt{2}$	$-i\sqrt{2}$
χ_7	2	-2	0	-1	1	0	$-i\sqrt{2}$	$i\sqrt{2}$
χ_8	4	-4	0	1	-1	0	0	0

TABLE 4. Character table for the group $\widetilde{S_4}$

the discrete topology. A Galois representation ρ is a continuous homomorphism

$$\rho : G_{\mathbb{Q}} = \mathrm{Gal}(\overline{\mathbb{Q}}|\mathbb{Q}) \longrightarrow \mathrm{GL}(n, \mathbb{C}).$$

Continuity means that ρ factorizes through the Galois group of a finite normal extension $K|\mathbb{Q}$, being $K := \overline{\mathbb{Q}}^{\ker(\rho)}$. Let $G_\rho = \rho(\mathrm{Gal}(K|\mathbb{Q}))$.

Let $\mathcal{O}_K$ denote the ring of integers of K, $p \in \mathbb{Z}$ a rational prime, and $\mathfrak{p}$ a prime ideal in $\mathcal{O}_K$ above (p). Associated to ρ and $\mathfrak{p}$, we may consider the descending chain of ramification subgroups of G_ρ

$$G_{-1,\rho}(\mathfrak{p}) \supseteq G_{0,\rho}(\mathfrak{p}) \supseteq \cdots \supseteq G_{s,\rho}(\mathfrak{p}) = (1).$$

Here $G_{-1,\rho}(\mathfrak{p})$ is the image of the decomposition group at $\mathfrak{p}$ in the extension $K|\mathbb{Q}$, and $G_{0,\rho}(\mathfrak{p})$ is the corresponding image of the inertia group. The representation ρ is unramified at $\mathfrak{p}$ if, and only if, $G_{0,\rho}(\mathfrak{p}) = (1)$. If this is the case, then the decomposition group at $\mathfrak{p}$ is cyclic and canonically generated by the Frobenius automorphism

$$G_{-1,\rho}(\mathfrak{p}) = \langle \mathrm{Frob}_{\rho,\mathfrak{p}} \rangle.$$

We shall write $\mathrm{Frob}_{\rho,p}$ to denote the conjugacy class $[\mathrm{Frob}_{\rho,\mathfrak{p}}]$ determined for those $\mathfrak{p}|p$; and $\mathrm{Frob}_{\rho,\infty}$, to denote that of $[\rho(c)]$, where $c \in G_{\mathbb{Q}}$ stands for the complex conjugation. The representation ρ is said to be odd if $\det(\mathrm{Frob}_{\rho,\infty}) = -1$; it is said to be even, otherwise. Explicit computation of higher ramification groups for $\widetilde{S_4}$-extensions can be found, for example, in [**5**].

The Artin conductor $N(\rho)$ is an important constant attached to ρ. It is defined by $N(\rho) = \prod_p p^{n(\rho,p)}$, where the exponents at each prime p are computed by taking into account the ramification groups for any prime divisor $\mathfrak{p}|p$:

$$n(\rho, p) = \sum_{i=0}^{\infty} \frac{1}{(G_{0,\rho}(\mathfrak{p}) : G_{i,\rho}(\mathfrak{p}))} \dim V/V^{G_{i,\rho}(\mathfrak{p})}.$$

They turn out to be positive integers.

The Artin L-function of ρ is defined by

$$L(\rho, s) = \prod_p \frac{1}{\det\left[\mathrm{I}_n - p^{-s}\mathrm{Frob}_{\rho,\mathfrak{p}}; V^{G_{0,\rho}(\mathfrak{p})}\right]} = \sum_{n=1}^{\infty} \frac{a_n}{n^s}, \quad \Re(s) > 1.$$

If p is unramified, then

$$a_p = \mathrm{Tr}(\mathrm{Frob}_{\rho,p}).$$

In order to extend Artin L-functions to the whole plane we need to include local gamma factors from the infinite places. The completed Artin L-function is defined by

$$\Lambda(\rho, s) := N(\rho)^{s/2}\Gamma(\rho, s)L(\rho, s), \quad \Re(s) > 1.$$

Let

$$\Gamma_{\mathbb{R}}(s) = \pi^{-s/2}\Gamma\left(\frac{s}{2}\right).$$

If, under the action of the complex conjugation, we have a decomposition

$$V = n_+(\infty)\chi_+ \oplus n_-(\infty)\chi_-,$$

with $n_+(\infty) = \dim V^{G_{0,\rho}(c)}$, and $n_-(\infty) = \mathrm{codim}\, V^{G_{0,\rho}(c)}$, then

$$\Gamma(\rho, s) = \Gamma_{\mathbb{R}}(s)^{n_+(\infty)}\Gamma_{\mathbb{R}}(s+1)^{n_-(\infty)}.$$

THEOREM 4.1 (Artin-Brauer, 1947). *The function $\Lambda(\rho, s)$ can be continued to the whole complex s-plane as a meromorphic function and satisfies a functional equation*

$$\Lambda(\rho, s) = W(\rho)\Lambda(\rho^*, 1 - s),$$

where ρ^* denotes the dual representation of ρ, and $|W(\rho)| = 1$.

CONJECTURE 4.2 (Artin, 1923). If ρ is an irreducible representation, aside from the trivial one, then $\Lambda(\rho, s)$ is an entire function.

For $n = 1$, Artin's conjecture is true by class-field theory. We are going to consider the state of the art of Artin's conjecture for $n = 2$ in the next section.

5. Modular forms of weight one and Artin L-functions

Assuming Artin's conjecture, every two-dimensional irreducible representation ρ of $G_{\mathbb{Q}}$ corresponds to a cusp form of weight 1, if $\det(\rho)$ is odd, and to a Maass cusp form of Laplace eigenvalue $\frac{1}{4}$, if $\det(\rho)$ is even (cf. [**28**], [**48**]).

Since Dirichlet characters (or Hecke characters) of $\mathbb{Q}$ can be viewed as automorphic forms on GL(1), it was conjectured by Langlands that any Artin L-function $L(\rho, s)$ of degree n should come from an automorphic cusp form $\pi(\rho)$ on GL(n). In particular, if $n = 2$, $L(\rho, s)$ should be the Dirichlet series of a cusp form; the form ought to be holomorphic, in the odd case; and a Maass form, in the even case. Thus, in dimension 2, Artin's conjecture follows from Langlands conjecture, due to the fact that the L-functions of cuspidal automorphic representations of GL(2) are holomorphic.

If $n = 2$ and the image of ρ is solvable, the existence of $\pi(\rho)$ was proven by Hecke, Langlands and Tunnell some years ago ([**28**], [**36**], [**47**]). In recent decades, and following Langlands' approach, significant progress has also been made in proving Artin's conjecture in the odd non-solvable case (cf. [**9**], [**12**], [**13**], [**23**], [**29**]).

One of the first results on Artin's conjecture on dimension 2 was obtained from Deligne-Serre's 1974 theorem concerning how to associate complex Galois representations to modular forms of weight one.

THEOREM 5.1 (Deligne-Serre, [**19**]). *Fix $N \geq 1$, and let χ (mod N) be an odd Dirichlet character; i. e. $\chi(-1) = -1$. Let $f(z) = \sum_{n>0} a_n q^n \in S(N,1,\chi)$ be a non-identically zero modular cusp form. Suppose that f is a normalized eigenfunction of the Hecke operators T_ℓ, $\ell \nmid N$, with eigenvalues a_ℓ. Then there exists an odd irreducible Galois representation*

$$\rho_f : G_{\mathbb{Q}} \longrightarrow \mathrm{GL}(2,\mathbb{C}),$$

unramified outside N, and such that

$$\mathrm{Tr}(\mathrm{Frob}_{\rho,p}) = a_p, \quad \det(\mathrm{Frob}_{\rho,p}) = \chi(p), \quad \textit{for } p \nmid N.$$

If, moreover, $f \in S^{new}(N,1,\chi)$, then the conductor of ρ is equal to N.

Let $f \in S^{new}(N,1,\chi)$ and consider the Mellin transform of f,

$$L(f,s) = \sum_{n>0} a_n n^{-s}.$$

Since, by Deligne-Serre, $L(\rho_f,s) = L(f,s)$, from Hecke's theory of Dirichlet series attached to cusps forms it follows that $L(\rho_f,s)$ is an entire function, so that Artin's conjecture is true for all two-dimensional representations ρ_f that arise in this way. Another consequence of the Deligne-Serre's theorem is the proof of the Ramanujan-Petersson conjecture for the Hecke eigenforms of weight one; i. e. $|a_p| \leq 2$ for any $p \nmid N$, since those a_p are seen as the sum of two roots of unity.

In 1989, Blasius and Ramakrishnan considered an analogous form of Deligne-Serre's theorem but in the context of Maass cusp forms. They showed in [**7**] that each Hecke-Maass cusp form F of Laplace eigenvalue $\frac{1}{4}$ defines an irreducible even representation $\rho_F : G_{\mathbb{Q}} \to \mathrm{GL}(2,\mathbb{C})$ such that $L(F,s) = L(\rho_F,s)$, modulo two hypotheses relative to the symplectic similitude group GSp(4). The hypotheses, which are necessary in order to translate the method of Deligne and Serre to the non-holomorphic context, concern the existence of compatible systems of 4-dimensional p-adic representations for Siegel modular forms of higher weight, and the structure of L-packets of automorphic cuspidal representations of GSp_2.

From the above considerations, it is reasonable to expect that tables of modular forms of weight 1, as well as those of Maass forms of Laplace eigenvalue 1/4, can be calculated from Artin L-functions of Galois representations of dimension 2. Nevertheless, two main difficulties arise. On the one hand, the computation of these Artin L-functions is by no means easy, since, as we shall see in the next sections, it involves the effective resolution of Galois inverse and Galois embedding problems. On the other hand, as we recalled in section 2, the dimension of the corresponding vector spaces of modular forms is still unknown.

G-Type	$\mathrm{Im}(\overline{\rho})$
dihedral	D_{2n}
tetrahedral	A_4
octahedral	S_4
icosahedral	A_5

TABLE 5. Two-dimensional irreducible complex Galois representation types

6. Linear and projective Galois representations

Let $V \simeq \mathbb{C}^2$ be a complex vector space of dimension 2. Any Galois linear representation $\rho : G_{\mathbb{Q}} \to \mathrm{GL}(V)$ determines a projective representation $\overline{\rho} : G_{\mathbb{Q}} \to \mathrm{PGL}(V)$ by composing with the projection $\pi : \mathrm{GL}(V) \to \mathrm{PGL}(V)$. We obtain a commutative diagram

$$\begin{array}{ccc} G_{\mathbb{Q}} & \xrightarrow{\rho} & \mathrm{GL}(2,\mathbb{C}) \\ & {\scriptstyle\overline{\rho}}\searrow & \downarrow{\scriptstyle\pi} \\ & & \mathrm{PGL}(2,\mathbb{C}) \end{array}$$

in which it is said that ρ is a lifting of $\overline{\rho}$. Both $\mathrm{Im}(\rho)$, $\mathrm{Im}(\overline{\rho})$, are finite groups and, moreover, the second one is a cyclic central extension of the first:

$$1 \to C_r \to \mathrm{Im}(\rho) \to \mathrm{Im}(\overline{\rho}) \to 1.$$

The order r of the kernel is called the index of ρ.

The representations ρ as above are classified in types, according to their images in $\mathrm{PGL}(2,\mathbb{C})$. We know after Klein [**33**] that any finite subgroup of $\mathrm{PGL}(2,\mathbb{C}) = \mathrm{PSL}(2,\mathbb{C})$ is either a cycle group C_n, a dihedral group D_n of order $2n$, $n \geq 2$, or the symmetry group of a Platonic solid: the tetrahedral group A_4 of order 12, the octahedral group S_4 of order 24, or the icosahedral group A_5 of order 60 (since dual solids have isomorphic symmetry groups). Accordingly, the irreducible representations of dimension two can be classified in the four types appearing in Table 5.

Two liftings of the same projective representation differ by a character $\chi : G_{\mathbb{Q}} \to \mathbb{C}^*$. By definition, a linear representation ρ is a minimal lifting of a projective representation $\overline{\rho}$ if it has minimum index among all the liftings of $\overline{\rho}$. The index of a minimal index is a power of 2.

Given a projective Galois representation $\overline{\rho} : G_{\mathbb{Q}} \to \mathrm{PGL}(2,\mathbb{C})$ and a central extension

$$1 \to C_r \to G \to \mathrm{Im}(\overline{\rho}) \to 1, \tag{6.1}$$

$\widetilde{S_4}$	$1\widetilde{A}, 2\widetilde{A}$	$4\widetilde{A}$	$2\widetilde{B}$	$3\widetilde{A}, 6\widetilde{A}$	$8\widetilde{A}, 8\widetilde{B}$
S_4	$1A$	$2A$	$2B$	$3A$	$4A$

TABLE 6. Lifting of conjugacy classes from S_4 to $\widetilde{S_4}$

the obstruction to the existence of a lifting ρ such that $\mathrm{Im}(\rho) = G$ is the element $\overline{\rho}^*(c) \in H^2(G_{\mathbb{Q}}, C_r)$, where $c \in H^2(\mathrm{Im}(\overline{\rho}), C_r)$ is the cohomology class defined by the exact sequence (6.1) and $\overline{\rho}^* : H^2(\mathrm{PGL}(2,\mathbb{C}), \mathbb{C}^*) \to H^2(G_{\mathbb{Q}}, C_r)$ is the morphism in cohomology defined by $\overline{\rho}$. Since, by a theorem of Tate, is $H^2(G_{\mathbb{Q}}, \mathbb{C}^*) = 0$ (cf. [**42**]), it turns out that any projective representation $\overline{\rho}$ has a lifting, provided that r is sufficiently large.

7. Computation of Artin L-functions

The computation of Artin L-functions of dihedral type offers no extra difficulties other than those involved in the effective determination of the class group of an imaginary quadratic field, since they are obtained from theta functions associated to Hecke ideal class characters.

As we have said, the other two-dimensional cases are considered exotic, since they conjecturally generate much smaller spaces. There are no odd tetrahedral or icosahedral representations of index equal to 2. Several facts concerning extensions of octahedral type and index $r \geq 2$, and tetrahedral type or icosahedral type and index $r > 2$ are discussed in [**2**], [**3**] [4], [9], [**10**], [**15**], [**16**], [**20**], [**30**], [**31**], [**32**], [**40**].

In what follows we are going to restrict ourselves to Galois embedding problems of octahedral type and index 2. Let $f(X) \in \mathbb{Q}[X]$ be an irreducible polynomial of degree 4 and denote by x_i $(1 \leq i \leq 4)$ its zeros in an algebraic closure $\overline{\mathbb{Q}}$ that we are going to fix. Let $K_1 = \mathbb{Q}(x_1)$ be a root field, $K = \mathbb{Q}(x_1, x_2, x_3, x_4)$ its algebraic closure, and suppose that $\mathrm{Gal}(K|\mathbb{Q}) \simeq S_4$. We consider the Galois embedding problem defined by diagram

(7.1)
$$\begin{array}{ccc} & & G_{\mathbb{Q}} \\ & \overset{?}{\swarrow} & \downarrow \varphi \\ \mathrm{Gal}(?|\mathbb{Q}) & \longrightarrow & \mathrm{Gal}(K|\mathbb{Q}) \\ \downarrow ?\simeq & & \downarrow \simeq \\ \widetilde{S_4} & \longrightarrow & S_4. \end{array}$$

For short, we shall denote this embedding problem by

$$(EP): \ \widetilde{S_4} \to S_4 \simeq \mathrm{Gal}(K|\mathbb{Q}).$$

By definition, the embedding problem (EP) is solvable if there exists an extension $\widetilde{K}|K$ such that $\mathrm{Gal}(\widetilde{K}|\mathbb{Q}) \simeq \widetilde{S_4}$ and such that the following diagram is commutative

$$\begin{array}{ccc} \mathrm{Gal}(\widetilde{K}|\mathbb{Q}) & \longrightarrow & \mathrm{Gal}(K|\mathbb{Q}) \\ \downarrow & & \downarrow \\ \widetilde{S_4} & \longrightarrow & S_4. \end{array}$$

The obstruction to the solvability of (EP) is given by the class of a 2-cocycle $\varphi^*(\varepsilon) \in H^2(G_{\mathbb{Q}}, C_2)$, where $\varepsilon \in H^2(S_4, C_2)$ is the cohomology class defined by the extension 3.1. As is wellknown from Galois cohomology, the cohomology group $H^2(G_{\mathbb{Q}}, C_2)$ is isomorphic to $\mathrm{Br}_2(\mathbb{Q})$, the 2-torsion subgroup of the Brauer group $\mathrm{Br}(\mathbb{Q})$. Since the elements of this subgroup are given by isomorphy classes of quaternion algebras and these are classified by Hilbert symbols, the obstruction to the solvability of (EP) can be expressed in terms of these symbols. The following theorem, due to Serre, provides the key point to effectively compute the obstruction.

THEOREM 7.1 (Serre, [**44**]). *The embedding problem (EP) is solvable if and only if*

$$w(\mathrm{Tr}_{K_1|\mathbb{Q}}) \otimes (2, d) = 1 \in \mathrm{Br}_2(\mathbb{Q}),$$

where d is the discriminant of $K_1|\mathbb{Q}$, $w(\mathrm{Tr})$ denotes the Hasse-Witt invariant of the quadratic form defined by the trace, and $(2, d)$ stands for Hilbert symbol.

A table of quartic fields was constructed by Godwin [**27**]. Let $f(X) = a_4X^4 + a_3X^3 + a_2X^2 + a_1X + a_0$, $K_1 \simeq K[X]/(f(X))$, and d denote the discriminant of K_1. By computing the entries appearing in Serre's formula, it can be seen that $\mathrm{Gal}(K|\mathbb{Q}) \simeq S_4$ and that the embedding problem (EP) is solvable for 37 values of d in the range $-3280 \leq d < 0$. For instance, this is the case for $d = -283, -331, -3267, -3271$ (cf. [**18**]).

Suppose that we have chosen a quartic field of discriminant $d < 0$ for which (EP) is solvable. Now we consider either of the faithful representations $\widetilde{S_4} \to \mathrm{GL}(2, \mathbb{C})$ (see Table 4) and the Galois representation which factorizes through $\mathrm{Gal}(\widetilde{K}|\mathbb{Q}) \simeq \widetilde{S_4}$:

$$\rho : \; G_{\mathbb{Q}} \to \mathrm{Gal}(\widetilde{K}|\mathbb{Q}) \hookrightarrow \mathrm{GL}(2, \mathbb{C}).$$

Since $\det\rho = \chi_d$ is the quadratic character attached to $\mathbb{Q}(\sqrt{d})$ and $d < 0$, we have that ρ is odd and, by construction, it is irreducible.

In Table 7, λ_f denotes a prime ideal in the ring of integers $\mathcal{O}_{K_1}$, over a rational prime p, of residue degree equal to f. Of course, we must have that $\sum_i f_i = 4$ at any prime p unramified in K_1.

A first step in the computation of $L(\rho, s)$ is provided by the following proposition.

PROPOSITION 7.2. *Let $p \nmid d$ be a prime. Let $\mathrm{Frob}_{\rho,p} \subseteq \widetilde{S_4}$ be the Frobenius substitution and $\mathrm{Frob}_{\overline{\rho},p} \subseteq S_4$ be its image under $\overline{\rho}$. Then $\mathrm{Frob}_{\overline{\rho},p}$ determines a_p^2 according to the decomposition types contained in Table 7.*

D – type	$p\mathcal{O}_1$	$\mathrm{Frob}_{\overline{\rho},p}$	$\mathrm{Frob}_{\rho,p}$	a_p^2	$\det(\rho)$
I	$\mathfrak{p}_1\mathfrak{p}_1'\mathfrak{p}_1''\mathfrak{p}_1'''$	$1A$	$1\widetilde{A}, 2\widetilde{A}$	4	1
II	$\mathfrak{p}_2\mathfrak{p}_2'$	$2A$	$4\widetilde{A}$	0	1
III	$\mathfrak{p}_1\mathfrak{p}_1'\mathfrak{p}_2$	$2B$	$2\widetilde{B}$	0	-1
IV	$\mathfrak{p}_1\mathfrak{p}_3$	$3A$	$3\widetilde{A}, 6\widetilde{A}$	1	1
V	$\mathfrak{p}_4$	$4A$	$8\widetilde{A}, 8\widetilde{B}$	-2	-1

TABLE 7. Values of a_p^2

In order to fix ideas, we are going to assume that the Galois representation ρ has been chosen so that

$\mathrm{Frob}_{\rho,p}$	$1\widetilde{A}$	$2\widetilde{A}$	$2\widetilde{B}$	$3\widetilde{A}$	$4\widetilde{A}$	$6\widetilde{A}$	$8\widetilde{A}$	$8\widetilde{B}$
a_p	2	-2	0	-1	0	1	$i\sqrt{2}$	$-i\sqrt{2}$

TABLE 8

The second row of the table is deduced from the character table of $\widetilde{S}_4$ and the sense of this election is to fix a name for those conjugacy classes of $\widetilde{S_4}$ that collapse to the same conjugacy class of S_4. From all the above, the following statement follows.

PROPOSITION 7.3 (D-Types I, IV). *Assume that (EP) is solvable and let* $\widetilde{K} = K(\sqrt{\gamma})$ *be a solution. Suppose that* $\mathrm{Frob}_{\overline{\rho},p} = 1A$ *or* $\mathrm{Frob}_{\overline{\rho},p} = 3A$, *so that* $\mathrm{Frob}_{\rho,p} \in \{1\widetilde{A}, 2\widetilde{A}\}$ *or* $\mathrm{Frob}_{\rho,p} \in \{3\widetilde{A}, 6\widetilde{A}\}$. *Then*

$$\mathrm{Frob}_{\rho,p} \in \{1\widetilde{A}, 3\widetilde{A}\} \quad \textit{if and only if} \quad \gamma \in K_{\mathfrak{p}}^{*2}$$

for $\mathfrak{p}|p$ *a prime in* $\mathcal{O}_K$ *and being* $K_{\mathfrak{p}}$ *the completion of* K *at the place* $\mathfrak{p}$.

A much more delicate question is to decide the conjugacy class for the Frobenius elements of decomposition type V.

THEOREM 7.4 (D-type V). *Assume that (EP) is solvable and let* $\widetilde{K}$ *be a solution. Suppose that* $p \neq 2$ *is a prime for which* $\mathrm{Frob}_{\overline{\rho},p} = 4A$, *so that* $\mathrm{Frob}_{\rho,p} \in \{8\widetilde{A}, 8\widetilde{B}\}$. *Then there exists an element* $\gamma \in K^*$ *for which* $\widetilde{K} = K(\sqrt{\gamma})$ *and*

$$\varepsilon\gamma^{\frac{p-1}{2}} \equiv -\frac{1}{2} + \frac{t^{-1}(\gamma) - t(\gamma)}{2\gamma} \pmod{\mathfrak{p}_4},$$

where $t = (1,2,3) \in S_4$, $\varepsilon = \pm 1$, *and* $a_p = \varepsilon i\sqrt{2}$. *Thus*

$$\mathrm{Frob}_{\rho,p} = 8\widetilde{A} \quad \textit{if and only if} \quad \varepsilon = 1.$$

Proof. Let $s = (1,2,3,4) \in 4A$ be a representative of the conjugacy class in S_4 of the 4-cycles, and $s_A \in 8\widetilde{A}$, $s_B \in 8\widetilde{A}$ be the two liftings of s to $\widetilde{S}_4$. Since $\widetilde{K}|\mathbb{Q}$ is a Galois extension, there exists an element $b_s \in K^*$ such that

$$s(\gamma) = b_s^2 \gamma.$$

We label the things in this way:

$$s_A(\gamma^{1/2}) = b_s\gamma^{1/2}, \; s_B(\gamma^{1/2}) = -b_s\gamma^{1/2}.$$

By taking into account the action of the Frobenius substitution, we obtain

$$\mathrm{Frob}_{\rho,p} = 8\widetilde{A} \quad \text{if and only if} \quad \gamma^{p/2} \equiv b_s\gamma^{1/2} \pmod{\mathfrak{p}_4},$$

or, equivalently,

$$\gamma^{(p-1)/2} \equiv b_s \pmod{\mathfrak{p}_4} \quad \text{if and only if} \quad a_p = i\sqrt{2}.$$

Now, by choosing the solution γ of the embedding problem as in Theorem 7.5, one can prove that

$$b_s = -\frac{1}{2} + \frac{t^{-1}(\gamma) - t(\gamma)}{2\gamma}. \quad \square$$

We have thus seen that the explicit calculation of $L(\rho, s)$ can be deduced from the explicit knowledge of the elements γ that solve the embedding problem (EP). The construction of explicit solutions of embedding problems is quite involved. The first results concerning A_n and S_n fields and central extensions of index 2 were solved in Crespo's thesis [**15**], [**16**] (cf. also [**3**]). Later, the author extended the method to many other central extensions.

In what follows, we are going to review the construction of the element γ in the particular case of the embedding problem (EP), since it is basic for the implementation of the algorithm that will compute $L(\rho, s)$ in this case. The moral is the following: although it is difficult to identify the element γ directly in K, there is a visible non-zero element in a certain Clifford algebra attached to the solvable embedding problem whose spinor norm is effectively computable and solves (EP).

THEOREM 7.5 (Crespo, [**15**],[**16**]). *Let* $f(X) = a_4X^4 + a_3X^3 + a_2X^2 + a_1X + a_0 \in \mathbb{Q}[X]$ *be an irreducible polynomial with Galois group* S_4. *Let* $K_1 = \mathbb{Q}(x_1)$, $d = \mathrm{disc}(K_1)$, *and* $K = \mathbb{Q}(x_1, x_2, x_3, x_4)$. *Let* $M = [x_j^i]_{0\le i\le 3; 1\le j\le 4}$, *so that* $T = M^tM = \mathrm{Tr}_{K_1|\mathbb{Q}}(XY)$. *Suppose that the embedding problem*

$$(EP): \; \widetilde{S_4} \to S_4 \simeq \mathrm{Gal}(K|\mathbb{Q})$$

is solvable. Then there exists a matrix $P \in \mathrm{GL}(4, \mathbb{Q})$ *such that* $P^tTP = \mathrm{diag}[1,1,2,2d]$ *and*

$$\gamma := \det\left[MPR + I_4\right] \neq 0, \qquad \textit{where } R = \begin{bmatrix} 1 & & & \\ & 1 & & \\ & & \frac{1}{2} & \frac{1}{2} \\ & & \frac{1}{2}\sqrt{d} & -\frac{1}{2}\sqrt{d} \end{bmatrix}.$$

The fields $\widetilde{K}_c := K(\sqrt{c\gamma})$, $c \in \mathbb{Q}^*/\mathbb{Q}^{*2}$, *are the solutions to* (EP).

Proof. (sketch) We consider the diagram of extensions given by

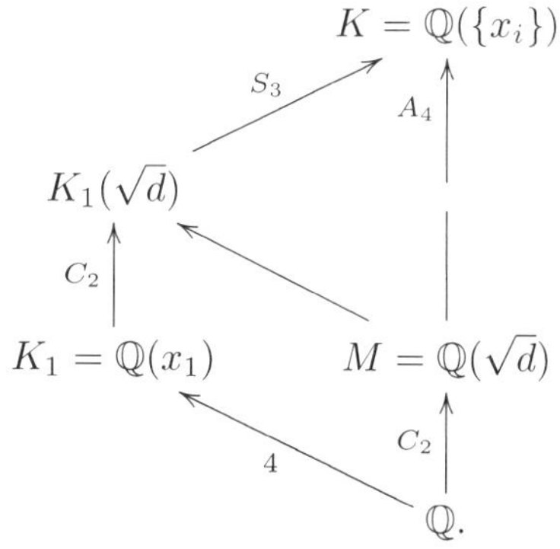

Let $Q_4 = X_1^2 + X_2^2 + X_3^2 + X_4^2$. Since we have an isomorphism of quadratic spaces

$$\mathrm{Tr}_{K_1(\sqrt{d})|M}(X^2) \otimes_M K \simeq Q_4 \otimes K,$$

we shall have an isomorphism of the Clifford algebras attached to them:

$$C(K_1(\sqrt{d})) \otimes_M K \simeq C(K^4).$$

Now, the solvability of the embedding problem (EP) implies that of the

$$\widetilde{A_4} \to A_4 \simeq \mathrm{Gal}(K|M). \tag{7.2}$$

By Serre's formula, this implies the existence of an isomorphism

$$\mathrm{Tr}_{K_1(\sqrt{d})|M}(X^2) \simeq Q_4 \otimes M.$$

Thus, the solvability condition can be translated in terms of the existence of an isomorphism of Clifford algebras

$$C(K_1(\sqrt{d})) \simeq C(M^4).$$

By taking the images of the canonical basis by the inverses of the above isomorphisms, we find elements $v_i \in C(K_1(\sqrt{d})) \otimes_M K$ and $w_i \in C(K_1(\sqrt{d}))$ such that:

$$v_i^2 = w_i^2 = 1, \quad v_i v_j = -v_j v_i, \quad w_i w_j = -w_j w_i, \quad 1 \le i, j \le 4, i \ne j,$$
$$v_i^s = v_{s(i)}, \quad w_i^s = w_i, \quad \text{for any } s \in A_4.$$

Moreover, they can be chosen so that

$$z = \sum_{\varepsilon_i \in \{0,1\}} v_1^{\varepsilon_1} \dots v_4^{\varepsilon_4} w_1^{\varepsilon_4} \dots w_1^{\varepsilon_1} \ne 0.$$

In this case, the spinor norm $\mathrm{SpinN}(z) = 2^4\gamma \in K^*$. and the field $K(\sqrt{\gamma})$ is a Galois extension of M with Galois group $\widetilde{A}_4$. The reason for this is the following: the central extension $\widetilde{A}_4$ obtained by restriction to the alternating group (see the diagram at the beginning of the proof) admits a spinor description

$$\begin{array}{ccccccccc}
1 & \longrightarrow & C_2 & \longrightarrow & \widetilde{A}_4 & \longrightarrow & A_4 & \longrightarrow & 1 \\
 & & \downarrow & & \downarrow & & \downarrow & & \\
1 & \longrightarrow & C_2 & \longrightarrow & \mathrm{Spin}_4(K) & \longrightarrow & \mathrm{SO}_4(K), & &
\end{array}$$

where $\mathrm{Spin}_4(K)$ is isomorphic to a subgroup of the multiplicative group $C(K^4)^*$. We can take

$$z = \sum_{\varepsilon_i \in \{0,1\}} v_1^{\varepsilon_1} \dots v_4^{\varepsilon_4} w_1^{\varepsilon_4} \dots w_1^{\varepsilon_1} \in C(K_1(\sqrt{d}) \otimes_M K)^* \simeq C(K^4)^*,$$

Now $\mathrm{SpinN}(z) = 2^4\gamma \in K^*$ and $K(\sqrt{\gamma})|M$ is a Galois extensions with Galois group $\widetilde{A}_4$. Since, moreover, the element γ can be chosen so that $r(\gamma) = \gamma$, where $r = (3, 4)$, then $\widetilde{K} := K(\sqrt{\gamma})$ is Galois over $\mathbb{Q}$ and yields a solution to (EP).

Example. Let $f(X) = X^4 + 5X^3 + 6X^2 - 3$. Then $d = -3^3 \cdot 11^2 = -3267$, $\mathrm{Gal}(K|\mathbb{Q}) \simeq S_4$ and (EP) is solvable. We can take

$$\begin{aligned} \gamma = \quad & -11(54x_1^3x_2^2 + 222x_1^2x_2^2 + 6x_1x_2^2 - 366x_2^2 \\ & +174x_1^3x_2 + 798x_1^2x_2 + 357x_1x_2 - 1152x_2 \\ & +78x_1^3 + 537x_1^2 + 517x_1 - 1300), \end{aligned}$$

where $f(x_1) = f(x_2) = 0$. Now, if we take into account the two complex representations of degree 2 of $\widetilde{S}_4$ to define $\rho, \rho^c : G_{\mathbb{Q}} \to \mathrm{GL}(2, \mathbb{C})$, the first coefficients a_p, a_p^c of the Artin L-functions $L(\rho, s)$ and $L(\rho^c, s)$ are given by

p	2	5	7	13	17	19	23	29	31	37	41	43	47	...
a_p	$i\sqrt{2}$	$i\sqrt{2}$	1	1	$i\sqrt{2}$	0	$-i\sqrt{2}$	0	1	0	$i\sqrt{2}$	0	$-i\sqrt{2}$	...
a_p^c	$-i\sqrt{2}$	$-i\sqrt{2}$	1	1	$-i\sqrt{2}$	0	$i\sqrt{2}$	0	1	0	$-i\sqrt{2}$	0	$i\sqrt{2}$	...

TABLE 9. Coefficients of two Artin L-functions of octahedral type

References

[1] Atkin, A. O. L.; Lehner, J.: Hecke operators on $\Gamma_0(m)$. *Math. Ann.* 185 (1970), 134–170. MR0268123 (42:3022)

[2] Basmaji, J.; Kiming, I.: A table of A_5-fields. *On Artin's conjecture for odd 2-dimensional representations*, 37–46, 122–141. Lecture Notes in Math., 1585. Springer, 1994. MR1322317 (96e:11141)

[3] Bayer, P.: Embedding problems with kernel of order two. *Séminaire de Théorie des Nombres, Paris* 1986, 1987, 27–34, Progr. Math., 75. Birkhäuser, 1988. MR990504 (90c:12004)

[4] Bayer, P.; Frey, G.: Galois representations of octahedral type and 2-coverings of elliptic curves. *Math. Z.* 207 (1991), no. 3, 395–408. MR1115172 (92d:11058)

[5] Bayer, P.; Rio, A.: Dyadic exercises for octahedral extensions. *J. reine u. angew. Math.* 517 (1999), 1–17. MR1728550 (2001a:11191)

[6] Bhargava, M.; Ghate, E.: On the average number of octahedral newforms of prime level. *Math. Ann.* 344 (2009), no. 4, 749–768. MR2507622 (2010c:11050)

[7] Blasius, D.; Ramakrishnan, D.: Maass forms and Galois representations. *Galois groups over* $\mathbf{Q}$ (Berkeley, CA, 1987), 33–77. Math. Sci. Res. Inst. Publ., 16. Springer, 1989. MR1012167 (90m:11078)

[8] Bonnafé, C.: *Representations of* $\mathrm{SL}_2(\mathbf{F}_q)$. Algebra and Applications, 13. Springer, 2011. xxii+186 pp. MR2732651 (2011m:20029)

[9] Buhler, J.: An icosahedral modular form of weight one. *Modular functions of one variable,* V (Proc. Second Internat. Conf., Univ. Bonn, Bonn, 1976), pp. 289–294. Lecture Notes in Math., Vol. 601, Springer, 1977. MR0472702 (57:12395)

[10] Buhler, J.: *Icosahedral Galois representations.* Lecture Notes in Mathematics, Vol. 654. Springer, 1978. ii+143 pp. MR0506171 (58:22019)

[11] Bump, D.: *Automorphic forms and representations.* Cambridge Studies in Advanced Mathematics, 55. Cambridge University Press, 1997. MR1431508 (97k:11080)

[12] Buzzard, K.; Dickinson, M.; Shepherd-Barron, N.; Taylor, R.: On icosahedral Artin representations. *Duke Math. J.* 109 (2001), no. 2, 283–318. MR1845181 (2002k:11078)

[13] Buzzard, K.; Stein, W.: A mod five approach to modularity of icosahedral Galois representations. *Pacific J. Math.* 203 (2002), no. 2, 265–282. MR1897901 (2003c:11052)

[14] Cohen, H.: *A course in computational algebraic number theory.* Graduate Texts in Mathematics, 138. Springer, 1993, xii+534 pp. MR1228206 (94i:11105)

[15] Crespo, T.: Explicit construction of $\widetilde{A}_n$ type fields. *J. Algebra* 127 (1989), 452-461. MR1028464 (91a:12006)

[16] Crespo, T.: Explicit construction of $2S_n$ Galois extensions. *J. Algebra* 129 (1990), no. 2, 312–319. MR1040941 (91d:11135)

[17] Crespo, T.: Central extensions of the alternating group as Galois groups. *Acta Arith.* 66 (1994), no. 3, 229–236. MR1276990 (95f:12008)

[18] Crespo, T.: Galois representations, embedding problems and modular forms. Journées Arithmétiques (Barcelona, 1995). *Collect. Math.* 48 (1997), no. 1–2, 63–83. MR1464017 (98j:11101)

[19] Deligne, P.; Serre, J-P.: Formes modulaires de poids 1. *Ann. Sci. École Norm. Sup.* (4), 7 (1974), 507–530 (1975). MR0379379 (52:284)

[20] Doud, D.: S_4 and $\tilde{S}_4$ extensions of $\mathbf{Q}$ ramified at only one prime. *J. Number Theory* 75 (1999), no. 2, 185–197. MR1681628 (2000b:11060)

[21] Duke, W.: The dimension of the space of cusp forms of weight one. *Internat. Math. Res. Notices* 1995, no. 2, 99–109. MR1317646 (95m:11042)

[22] Duke, W.; Friedlander, J. B.; Iwaniec, H.: The subconvexity problem for Artin L-functions. *Invent. Math.* 149 (2002), no. 3, 489–577. MR1923476 (2004e:11046)

[23] Frey, G.: Construction and arithmetical applications of modular forms of low weight. *Elliptic curves and related topics,* 1–21. CRM Proc. Lecture Notes, 4. Amer. Math. Soc., Providence, RI, 1994. MR1260951 (95b:11042)

[24] Frey, G. (ed.): *On Artin's conjecture for odd 2-dimensional representations.* Lecture Notes in Mathematics, 1585. Springer, 1994. MR1322315 (95i:11001)

[25] Ganguly, S.: On the dimension of the space of cusp forms of octahedral type. *Int. J. Number Theory* 6 (2010), no. 4, 767–783. MR2661279

[26] Gelbart, S.: *Automorphic forms on adele groups.* Ann. of Math. Studies 83. Princeton Univ. Press, Princeton, NJ, 1975. MR0379375 (52:280)

[27] Godwin, H. J. On quartic fields of signature one with small discriminant. II. *Math. Comp.* 42 (1984), no. 166, 707–711. Corrigenda: "On quartic fields of signature one with small discriminant. II". *Math. Comp.* 43 (1984), no. 168, 621. MR736462 (85i:11092a)

[28] Hecke, E.: Über die Bestimmung Dirichletscher Reihen durch ihre Funktionalgleichung. *Math. Ann.* 112 (1936), 664–699. MR1513069

[29] Khare, C.; Wintenberger, J-P.: Serre's modularity conjecture. I. *Invent. Math.* 178 (2009), no. 3, 485–504. MR2551763 (2010k:11087)

[30] Kiming, I.: On the liftings of 2-dimensional projective Galois representations over $\mathbf{Q}$. *J. Number Theory* 56 (1996), no. 1, 12–35. MR1370194 (96j:11162)

[31] Kiming, Ian; Verrill, Helena A. On modular mod l Galois representations with exceptional images. *J. Number Theory* 110 (2005), no. 2, 236–266, 11F80 MR2122608 (2006g:11104)

[32] Kiming, I.; Wang, X. D.: Examples of 2-dimensional, odd Galois representations of A_5-type over $\mathbf{Q}$ satisfying the Artin conjecture. *On Artin's conjecture for odd 2-dimensional representations*, 109–121. Lecture Notes in Math., 1585. Springer, 1994. MR1322321 (96a:11128)

[33] Klein, F.: *Vorlesungen über das Ikosaeder und die Auflösung der Gleichungen vom fünften Grade.* Reprint of the 1884 original. Edited, with an introduction and commentary by Peter Slodowy. Birkhäuser, 1993. MR1315530 (96g:01046)

[34] Klüners, J.: The number of S_4-fields with given discriminant. *Acta Arith.* 122 (2006), no. 2, 185–194. MR2234835 (2007b:11175)

[35] Langlands, R. P.: On the notion of an automorphic representation. A supplement, in *Automorphic Forms, Representations and L-functions*, ed. by A. Borel and W. Casselman. Proc. Sympos. Pure Math. 33, part 1, 203–207, A.M.S., Providence, 1979. MR546619 (83f:12010)

[36] Langlands, R. P.: *Base change for GL(2).* Ann. of Math. Studies 96. Princeton Univ. Press, Princeton, NJ, 1980. MR574808 (82a:10032)

[37] Maass, H.: Über eine neue Art von nichtanalytischen automorphen Funktionen und die Bestimmung Dirichletscher Reihen durch Funktionalgleichungen. *Math. Ann.* 121 (1949), 141–183, MR0031519 (11:163c)
[38] Michel, P.; Venkatesh, A.: On the dimension of the space of cusp forms associated to 2-dimensional complex Galois representations. *Int. Math. Res. Not.* 2002, no. 38, 2021–2027. MR1925874 (2003i:11064)
[39] Murty, M. Ram.: Congruences between modular forms. *Analytic number theory* (Kyoto, 1996), 309–320. London Math. Soc. Lecture Note Ser., 247, Cambridge Univ. Press, Cambridge, 1997. MR1694998 (2000c:11073)
[40] Quer, J.: Liftings of projective 2-dimensional Galois representations and embedding problems. *J. Algebra* 171 (1995), no. 2, 541–566. MR1315912 (96b:12009)
[41] Selberg, A.: On the estimation of Fourier coefficients of modular forms. Proc. Sympos. Pure Math., Vol. VIII, Amer. Math. Soc., Providence, RI, 1965, pp. 1–15. MR0182610 (32:93)
[42] Serre, J.-P.: Modular forms of weight one and Galois representations. *Algebraic number fields: L-functions and Galois properties.* A. Frölich, ed. Proc. Sympos., Univ. Durham, Durham, 1975, pp. 193–268. Academic Press, London, 1977. MR0450201 (56:8497)
[43] Serre, J-P.: *Linear representations of finite groups.* Translated from the second French edition by Leonard L. Scott. Graduate Texts in Mathematics, Vol. 42. Springer, 1977. x+170 pp. MR0450380 (56:8675)
[44] Serre, J-P.: L'invariant de Witt de la forme $\mathrm{Tr}(x^2)$. *Comment. Math. Helv.* 59 (1984), no. 4, 651–676. MR780081 (86k:11067)
[45] Shepherd-Barron, N. I.; Taylor, R.: mod 2 and mod 5 icosahedral representations. *J. Amer. Math. Soc.* 10 (1997), no. 2, 283–298. MR1415322 (97h:11060)
[46] Shimura, G.: *Introduction to the Arithmetic Theory of Automorphic Functions.* Iwanami Shoten and Princeton University Press, 1971. MR0314766 (47:3318)
[47] Tunnell, J.: Artin's conjecture for representations of octahedral type. *Bull. A.M.S.* 5 (1981), 173–175. MR621884 (82j:12015)
[48] Weil. A.: Über die Bestimmung Dirichletscher Reihen durch Funktionalgleichungen. *Math. Ann.* 168 (1967), 149–156. MR0207658 (34:7473)

Departament d'Àlgebra i Geometria, Facultat de Matemàtiques, Universitat de Barcelona, 08007 Barcelona, Spain

Contemporary Mathematics
Volume **566**, 2012
http://dx.doi.org/10.1090/conm/566/11213

Zeta functions for families of Calabi–Yau n-folds with singularities

Anne Frühbis-Krüger and Shabnam Kadir

ABSTRACT. We consider families of Calabi–Yau n-folds containing singular fibres and study relations between the occurring singularity structure and the decomposition of the local (Weil) zeta-function. For 1-parameter families, this provides new insights into the combinatorial structure of the strong equivalence classes arising in the Candelas–de la Ossa–Rodrigues-Villegas approach for computing the zeta-function. This can also be extended to families with more parameters as is explored in several examples, where the singularity analysis provides correct predictions for the changes of degree in the decomposition of the zeta-function when passing to singular fibres. These observations provide first evidence in higher dimensions for Lauder's conjectured analogue of the Clemens–Schmid exact sequence.

1. Introduction

After a decade and a half of string theorists studying Calabi–Yau manifolds over fields of characteristic zero, particularly in the context of mirror symmetry, Candelas, de la Ossa and Rodrigues-Villegas [**CdOV1**] began the exploration of arithmetic mirror symmetry. Calabi–Yau manifolds over finite characteristic thus became objects of interest to physicists as well as mathematicians. After the discovery that the moduli spaces of all known Calabi–Yau manifolds form a web linked via conifold transitions [**GH**], the interest on the part of physicists decreased significantly concerning more complicated singularities which occur at other interesting points in the complex structure moduli space. However, newer results such as [**KLS**] suggest that it might be worthwhile to reconsider this and ask questions such as: Is string theory viable on spaces with singularities with high Milnor numbers and even non-isolated singularities? Can the D-brane interpretation of conifold (i.e. ordinary double points) transitions by Greene, Strominger and Morrison [**S, GMS**] be extended to what would be more complicated phase transitions? Questions of this type have not been considered very deeply yet - in part, because the study of singularities with more structure requires different methods. In this article, we want to start an approach in this direction by specifically studying properties at the singular fibres of families of Calabi–Yau varieties. In [**KLS**] the first question

1991 *Mathematics Subject Classification.* Primary 11G25; Secondary 14Q15, 14D06.
Key words and phrases. Zeta function, Calabi-Yau, n-fold, singularities.
The first author was supported in part by DFG-Schwerpunktprogramm 1489 'Algorithmic and experimental methods in algebra, geometry and number theory'.

was addressed by finding points in the moduli space where the singular Calabi–Yau manifolds exhibited modularity, (i.e. their cohomological L-series were completely determined by certain modular cusp forms) as a consequence of the rank of certain motives decreasing in size at singularities. For an overview of Calabi–Yau modularity the reader may consult [**HKS**]. Our approach, which enables the specification of exactly how much the degree of the contribution to the zeta function associated to each strong orbit (which in turn is directly related to motive rank) decreases, would further aid such investigations.

The local Weil zeta-function for certain families of Calabi–Yau varieties of various dimensions decomposes into pieces parametrized by monomials which are related to the toric data of the Calabi–Yau varieties [**CdOV1, CdOV2, CdO, K04, K06**]. It was shown in these papers that this decomposition points to deeper structures, since these monomials can also be related to the periods which satisfy Picard-Fuchs equations. Away from the singular fibres, this phenomenon of a link to p-adic periods was explained for one-parameter families using Monsky–Washnitzer cohomology in [**Kl**]. The families considered there all have the property that one distinguished member of each family is a diagonal variety of Fermat type; these are very accessible to explicit computations and are known to possess decompositions in terms of Fermat motives [**GY, KY**].

At certain values of the parameter, the corresponding variety becomes singular, and it was observed in Conjecture 7.3, page 137 in [**K04**] and §7.2 of [**K06**] that the degree of the contribution to each piece decreases according to the types of singularities encountered in explicit examples. In order to test whether the observations in [**K04, K06**] concerning the degenerations of the zeta functions for singular Calabi–Yau varieties hold more generally, we analyse the discriminant locus and singularity structure for general 1-parameter and some explicit 2-parameter families of Calabi–Yau varieties with distinguished fibre of Fermat-type and compare the results to the structure of their zeta functions. In particular, this provides strong evidence for conjectures connecting the numbers and types of singularities in the discriminant locus with certain combinatorial arguments arising from motivic and zeta function considerations and proves them for the considered cases by a direct comparison. In all cases with isolated singularities the total Milnor number of the singularities is given precisely by the degeneration in the degree of the various parts of the zeta function. Observations on finer combinatorial properties of the decomposition are also possible; for the 1-parameter families, the decomposition of the singular locus and the Milnor numbers of the types of singularities occuring are reflected in the analysis of the structure of this degeneration. For these considerations, the choice of using Dwork's original approach for computing the zeta-function was influenced by two constraints: by the presence of isolated singularities in the cases of interest and by the goal to also study higher-dimensional examples, which basically rules out explicit resolution of singularities in many cases due to the intrinsic complexity of the algorithm.

After fixing notation and stating references for standard facts about the local zeta function at good primes in Section 2, we first analyze the occurring singularities in detail in Section 3. There we focus on combinatorial aspects in the calculations, which by themselves do not seem very exciting at first glance, but reoccur from a different perspective in the computation of the zeta functions for the corresponding singular fibres in the subsequent section. This correspondence is then explored further in Section 5 for explicit examples of 2-parameter families and leads to the conjectures at the end of the article linking the singularity structure and the decomposition of the zeta function. If these conjectures hold, then a singularity analysis in the singular fibres coupled with a calculation of the zeta function away from the singular fibres already provides a large amount of vital information on the zeta function at the singularities by using well-established standard methods of singularity theory and of point counting.

The authors would like to thank the members of the Institut für Algebraische Geometrie and the Graduiertenkolleg 'Analysis, Geometrie und Stringtheorie' for the good working atmosphere and the insightful discussions. All computations of the singularity analysis were done in SINGULAR [**DGPS**], for the zeta-function calculations, Mathematica [**Mth**] was used.

2. Facts about the zeta-function

A pair of reflexive polyhedra (Δ, Δ^*) is known to give rise to a pair of mirror Calabi–Yau families $(\hat{V}_{f,\Delta}, \hat{V}_{f,\Delta^*})$. In this setting, Batyrev proved that topological invariants such as the Hodge numbers could be written in terms of the toric combinatorial data given by the reflexive polytopes. For the case of families of Calabi–Yau varieties which are deformations of a Fermat variety, the data of the reflexive polytope is encoded in certain monomials. For a detailed treatment of toric constructions of mirror symmetric Calabi–Yau manifolds see [**Bat**] or §4.1, page 53 of [**CK**].

First we recall a few standard definitions: the arithmetic structure of Calabi–Yau varieties can be encoded in the congruent or local zeta function. The Weil Conjectures (proven by Deligne [**Del1**] in 1974) show that the local zeta function is a rational function determined by the cohomology of the variety.

DEFINITION 2.1 (Local zeta function). The local zeta function for a smooth projective variety X over $\mathbb{F}_p$ is defined as follows:

$$\zeta(X/\mathbb{F}_p, t) := \exp\left(\sum_{r\in\mathbb{N}} \#X(\mathbb{F}_{p^r})\frac{t^r}{r}\right), \tag{2.1}$$

where $\#X(\mathbb{F}_{p^r})$ is the number of rational points of the variety.

For families of Calabi–Yau manifolds in weighted projective space the local zeta function can be computed in various ways, we however shall utilise exclusively methods first developed by Dwork in his proof of the rationality part of the Weil conjectures [**Dw1, Dw2**]. We thus use Gauss sums composed of the additive Dwork character, Θ and the multiplicative Teichmüller character (see e.g. [**CdOV1**]), $\omega^n(x)$:

$$G_n = \sum_{x\in\mathbb{F}_p^*} \Theta(x)\omega^n(x). \tag{2.2}$$

When a variety is defined as the vanishing locus of a polynomial $f \in k[X_1, \ldots, X_n]$, where k is a finite field, a non-trivial additive character like Dwork's character can be exploited to count points over k. Since $\Theta(x)$ is a character:

$$\sum_{y\in k}\Theta(yf(x)) = \begin{cases} 0 & \text{if } P(x)\neq 0, \\ q := \mathrm{Card}(k) & \text{if } f(x)=0\ ; \end{cases} \tag{2.3}$$

hence

$$\sum_{x\in k^n}\sum_{y\in k}\Theta(yf(x)) = q\#X(\mathbb{F}_{p^r})\ , \tag{2.4}$$

The above equation can be expressed in terms of Gauss sums which are amenable to computation via the Gross-Koblitz formula [**GK**]. All zeta function computations in this paper use an implementation of this method on Mathematica developed in [**K04, K06**]. In our context, the choice of this method was mainly influenced by the fact that it is also suitable for treating singular Calabi–Yau varieties, whereas most other approaches are restricted to the non-singular case. Lauder's extension of the deformation method (cf. conjecture 4.11 of [**L2**]) to the singular case relies on the existence of an analogue of the Clemens-Schmid exact sequence in positive characteristic which is currently only conjectural.

Just as in the examples of [**CdOV1, CdOV2, K04, K06**], these methods enable us to show that the number of points and hence the zeta function decomposes into parts labelled by strong β-classes, $\mathcal{C}_\beta$.[1].

$$\zeta(t,a) = \zeta_{\mathbf{c}onst}(t)\prod_{\mathcal{C}_\beta}\zeta_{\mathcal{C}_\beta}(t,a)$$

where $\zeta_{\mathbf{c}onst}(t)$ is a simple term, independent of the parameter a, and the β-classes are defined as follows:

Definition 2.2 (Strong motivic β-equivalence classes). For a given set of weights, $\mathbf{w} = (w_1,\ldots,w_n)$, $d = \sum_i w_i$, $w_i|d\ \forall i$, identify the set of all monomials with the set of all exponents of the monomials. We now consider a subset thereof defined as

$$\mathcal{M} := \mathcal{M}(\mathbf{w}) := \left\{\mathbf{x} = (x_1,\ldots,x_i,\ldots,x_n)\in\prod_{i=1}^n w_i\mathbb{Z}/d\mathbb{Z}\ \mid \mathbf{x}\cdot\mathbf{w} = ld,\ l\in\mathbb{Z}\right\}.$$

It is easy to see that $0\leq l\leq n-1$.
Let $l(\mathbf{x}) := \mathbf{x}\cdot\mathbf{w}/d$. Given a $\beta\in\mathcal{M}$ with $l(\beta)=1$, we can quotient out the set $\mathcal{M}$ with the equivalence relation $\sim_\beta$ on monomials, where

$$\mathbf{x}\sim_\beta\mathbf{y}\Leftrightarrow\mathbf{y}=\mathbf{x}+t\beta,\ t\in\mathbb{Z},$$

From now on we shall assume (unless otherwise stated) that the ith exponent of each monomial is taken $mod\frac{d}{w_i}$. The equivalence classes, $\mathcal{C}_\beta$, thus obtained shall be referred to as the **strong β -equivalence classes**.

Remark 2.3. For families of Calabi–Yau varieties which are deformations of smooth varieties of Fermat type the toric data is equivalent to specifying the monomials $\mathbf{x}\in\mathcal{M}$ for which $l(\mathbf{x})=1$, see [**Sk**] for the general case and [**CdOK**] for the special case of three-folds in explicit detail.

Following from §1.3 of [**Kl**] we define:

[1] These papers do not explicitly refer to 'strong equivalence classes', the term was coined later by Kloosterman in section 1.2 of [**Kl**]

DEFINITION 2.4 (Weak β-equivalence classes). We call two monomials $\mathbf{x}$ and $\mathbf{y}$ weakly equivalent if there exists $j \in \mathbb{Z}/d\mathbb{Z}$ and invertible $s, t \in (\mathbb{Z}/d\mathbb{Z})^*$ such that

$$s\mathbf{x} + t\mathbf{y} = j\beta,$$

where β is the deformation vector.

Each weak equivalence class can be subdivided into a finite number of strong equivalence classes. Following directly from Theorem 6.4 and Corollary 6.10 [**Kl**] we have the following proposition:

PROPOSITION 2.5 ([**Kl**]). *Considering smooth fibres of a 1-parameter family of Calabi–Yau varieties, the factor of the local zeta function associated to a weak β-class and to the parameter value a is $\zeta_{\mathcal{C}_\beta}(t, a)$ which is an element of $\mathbb{Q}[t]$.*

This degree of the factor of the zeta function associated to each weak β-class, can be computed as the number of monomials in the class which do not contain $\left(\frac{d}{w_i} - 1\right)$ in its ith component (c.f. Definition 2.2 in [**Kl**]*).*

REMARK 2.6. In all cases computed it was found furthermore that the local zeta function associated to a strong β-class was at worst a fractional power

$$\zeta_{\mathcal{C}_\beta}(t, a) = \left(\frac{P(t, a)}{Q(t, a)}\right)^{\frac{r}{s}},$$

where $P(t, a)$ and $Q(t, a)$ are polynomials and $r, s \in \mathbf{Z}$.

Kloosterman's explanation of the above-stated relation using Monsky-Washnitzer cohomology breaks down when the variety in question is singular. A key aim of this article is to explore the degenerations of the various pieces of the zeta function for singular fibres. More sophisticated theoretical tools such as limiting mixed Frobenius structures in rigid cohomology will be needed to explain the degenerations. Lauder [**L2**] provides a preliminary exploration of this through the introduction of a conjectured analogue of the Clemens-Schmid exact sequence, but his testing ground for the conjecture mostly consists of families of curves.

In this article we are able to supplement Lauder's examples through looking at singularities of higher-dimensional varieties, not just low dimension ones, as e.g. curves are prone to oversimplification due to their low dimensionality and could thus be misleading. All our results for 1-parameter families are applicable in all dimensions. Moreover, all arguments are explicit and no step requires desingularization, which would effectively have blocked the simultaneous view in all dimensions. In this article we intentionally only provide phenomenological (and for 2-parameter families also experimental) data, but no theoretical explanation for the observed correspondences, because we see it merely as the first step in this direction. We wish to disseminate the observations as soon as possible and would prefer to devote another article to the theoretical side in due time.

3. Singularity analysis for some families of Fermat-type Calabi–Yau n-folds

In this section, we collect data about the discriminant and the singularities of the fibres. To this end, we first consider general 1-parameter families in detail and then proceed to general observations on 2-parameter families which establish the background for the explicit examples in Section 5.

3.1. Facts about numerical invariants of isolated hypersurface singularities. Let $f \in \mathbb{C}\{\underline{x}\}$ define an isolated hypersurface singularity of dimension n, i.e. a germ (X, x) which has an isolated singularity at x. Then the n-th Betti number of a non-singular nearby fibre is usually referred to as the Milnor number $\mu(X, x)$ of the singularity and the dimension of the base space of a miniversal deformation of (X, x) is called the Tjurina number $\tau(X, x)$. For hypersurfaces and complete intersections these two invariants are closely related as can be seen from the algebraic description of the two numbers:

LEMMA 3.1. [**GLS**]

$$\mu(X,x) = \dim_{\mathbb{C}} \mathbb{C}\{\underline{x}\} / \left\langle \frac{\partial f}{\partial x_1}, \ldots, \frac{\partial f}{\partial x_n} \right\rangle$$
$$\tau(X,x) = \dim_{\mathbb{C}} \mathbb{C}\{\underline{x}\} / \left\langle f, \frac{\partial f}{\partial x_1}, \ldots, \frac{\partial f}{\partial x_n} \right\rangle$$

The ideal $\left\langle f, \frac{\partial f}{\partial x_1}, \ldots, \frac{\partial f}{\partial x_n} \right\rangle$ *is often referred to as the Tjurina ideal.*

REMARK 3.2. Forming the analogous quotients for a polynomial f in the corresponding polynomial ring, one obtains the sum over all local Milnor numbers (or Tjurina numbers respectively) of the respective affine hypersurface defined by f.

3.2. 1-parameter families. For the 1-parameter families, we can explicitly specify Gröbner Bases[2] for the relative Tjurina ideal[3] w.r.t. a lexicographical ordering, where the parameter a of the family is considered smaller than any of the variables. As a consequence, we can specify the discriminant of the family, count the number of singularities in each fibre over the base space and determine the Milnor numbers of the occurring singularities. A priori this is not very interesting, but later on it will turn out that the same kind of combinatorial data which arise here also appear in the computation of the Weil zeta function at singular fibres of the family. Moreover, we shall consider 2-parameter families later on, which specialize to such 1-parameter families, if one parameter is set to zero. For these considerations, we shall make use of the explicit calculations of this subsection.

Before stating the result explicitly, we need to recall one small observation which will yield a key argument in the proof:

LEMMA 3.3. *Consider a polynomial ring* $R[x]$ *over some (noetherian commutative) ring* R *(with unit). Let* $f = Ax^\alpha - C$, $g = Bx^\beta - D$ *for some* $A, B, C, D \in R$, $\alpha, \beta \in \mathbb{Z}$. *Then the ideal* $\langle f, g \rangle$ *contains polynomials which we can symbolically write as*

$$\begin{array}{rcl} A^r D^s x^{\gcd(\alpha,\beta)} & - & C^r B^s, \\ C^{\frac{\beta}{\gcd(\alpha,\beta)}-r} B^{\frac{\alpha}{\gcd(\alpha,\beta)}-s} x^{\gcd(\alpha,\beta)} & - & A^{\frac{\beta}{\gcd(\alpha,\beta)}-r} D^{\frac{\alpha}{\gcd(\alpha,\beta)}-s} \\ A^{\frac{\beta}{\gcd(\alpha,\beta)}} D^{\frac{\alpha}{\gcd(\alpha,\beta)}} & - & C^{\frac{\beta}{\gcd(\alpha,\beta)}} B^{\frac{\alpha}{\gcd(\alpha,\beta)}} \end{array}$$

where r, s *are integers arising from the Bézout identity* $r\alpha - s\beta = \gcd(\alpha, \beta)$; *to avoid ambiguities, we choose precisely the ones arising from the extended Euclidean*

[2]For a precise definition of a Gröbner Basis we refer the reader to any standard textbook on computational commutative algebra as e.g. [**GP**], [**KR**], [**CLO**].

[3]The relative Tjurina ideal is the Tjurina ideal in which only differentiation w.r.t. to the variables, but not the parameters are taken.

algorithm as either r and s or as $\frac{\beta}{\gcd(\alpha,\beta)} - r$ and $\frac{\alpha}{\gcd(\alpha,\beta)} - s$ making sure that r and s are both positive integers.

Using this, we can now state the main lemma of this section:

LEMMA 3.4. *Let $\mathcal{X} \subset \mathbb{P}_{w_1,\ldots,w_n}$, $\gcd(w_1,\ldots,w_n) = 1$, be the 1-parameter family of Calabi–Yau varieties*[4] *given by the polynomial*

$$F = \left(\sum_{i=1}^{n} x_i^{\frac{d}{w_i}}\right) + a \cdot \prod_{i=1}^{k} x_i^{\beta_i}$$

of the weighted degree $d = \sum_{i=1}^{n} w_i = \sum_{i=1}^{k} \beta_i w_i$, with $\beta_i \neq 0 \forall 1 \leq i \leq k$ and $\beta_i = 0$ for $i > k$. Let $\gamma := \gcd(\beta_1 w_1, \ldots, \beta_k w_k)$. Then the discriminant of the family is

$$V\left(a^{\frac{d}{\gamma}} + (-1)^{\frac{d}{\gamma}-1} \frac{d^{\frac{d}{\gamma}}}{\prod_{i=1}^{k} \beta_i w_i^{\frac{\beta_i w_i}{\gamma}}}\right) \subset \mathbb{A}^1_{\mathbb{C}}.$$

In the respective fibre above each of the $\frac{d}{\gamma}$ points of the discriminant there are precisely

$$\frac{\gcd(w_1,\ldots,w_k)}{\prod_{i=1}^{k} w_i} \cdot d^{k-2} \cdot \gamma$$

singularities with local equation

$$\sum_{i=2}^{k} x_i^2 + \sum_{i=k+1}^{n} x_i^{\frac{d}{w_i}}$$

of Milnor number $\prod_{i=k+1}^{n}(\frac{d}{w_i} - 1)$ and no further singularities.

PROOF. Preparations:
As we are considering hypersurfaces here, the relative T^1 is of the form $(\mathbb{C}[a])[\underline{x}]/J$, where

$$J = \left\langle F, \frac{\partial F}{\partial x_1}, \ldots \frac{\partial F}{\partial x_n} \right\rangle$$

is the relative Tjurina ideal (due to the weighted homogeneity and the resulting Euler relation, we can drop one of the $n+1$ generators.). More precisely, this ideal actually describes the relative T^1 of the affine cone over our family and we therefore need to ignore all contributions for which the associated prime is the irrelevant ideal. This is not difficult here, since intersection with any of the k first coordinate hyperplanes immediately leads to an $\langle x_1, \ldots, x_n \rangle$ primary ideal, and hence passage to any of the first k affine charts immediately removes precisely the unwanted part, but nothing else. As we are in weighted projective space and want to count singularities, our choice of the appropriate affine charts needs a little bit of extra caution: a priori we count points before the identification and thus might obtain a multiple of the correct number. Hence the calculated number needs to be divided by the weight of the respective variable. To simplify the presentation of the subsequent steps, we choose the chart $x_1 \neq 0$.

[4]Note that up to permutation of variables any monomial 1-parameter family of Calabi–Yau varieties with given zero fibre of Fermat type and perturbation term of weighted degree d can be written in this form.

Gröbner Basis:
Our next step is to compute a Gröbner basis of the relative Tjurina ideal in this chart where α_i denotes $\frac{d}{w_i}$ to shorten notation. For the structure of the final result, it turns out to be most suitable to choose a lexicographical ordering with $x_2 > \cdots > x_n > a$.

$$\begin{aligned} f_0 &= \left(\sum_{j=2}^{n} x_j^{\alpha_j}\right) + 1 + a\prod_{j=2}^{k} x_j^{\beta_j} \\ f_i &= \frac{\partial f_0}{\partial x_i} = \alpha_i x_i^{\alpha_i - 1} + a\beta_i x^{\beta_i - 1}\prod_{\substack{j=1 \\ j\neq i}}^{k} x_j^{\beta_j} \quad \text{for} \quad 2 \le i \le k \\ f_i &= \frac{\partial f_0}{\partial x_i} = \alpha_i x_i^{\alpha_i - 1} \quad \text{for} \quad k+1 \le i \le n \end{aligned}$$

As $f_0 - \sum_{i=2}^{n} \frac{1}{\alpha_i} x_i f_i = a\frac{1}{\alpha_1}\left(\prod_{i=2}^{k} x_i^{\beta_i}\right) + 1$, we may safely set

$$h_0 = \frac{1}{\alpha_1}\left(a\prod_{i=2}^{k} x_i^{\beta_i}\right) + 1$$

instead of the original f_0.
Forming $f_2 x_2 - \beta_2\alpha_1 h_0$ and the s-polynomials of the pairs $(f_2, h_0), \ldots, (f_k, h_0)$, we obtain new polynomials

$$h_i = x_i^{\alpha_i} - \frac{\beta_i \alpha_1}{\alpha_i} \quad 2 \le i \le k.$$

The leading monomials of these h_i, $2 \le i \le k$ and of the f_i, $k < i \le n$, are obviously pure powers in the respective variables x_i. We shall use them later on when computing the discriminant.
Considering h_0 and h_2, we now apply Remark 3.3 (polynomial 1 or 2 respectively) and obtain a polynomial

$$g_2 = x_2^{\gcd(\beta_2,\alpha_2)} - \underbrace{\left(\frac{\beta_i\alpha_1}{\alpha_i}\right)^r}_{:=c_1} \cdot \left(a\prod_{i=3}^{k} x_i^{\beta_i}\right)^s$$

for suitable exponents $r, s \in \mathbb{N}$ as specified in the remark. Please note that the exponent of x_2, $\gcd(\beta_2, \alpha_2)$ can be written as $\frac{1}{w_2}\gcd(d, \beta_2 w_2)$. By polynomial 3 of the same remark

$$h_{0,new} = c_1^{\frac{\beta_2}{\gcd(\alpha_2,\beta_2)}} \cdot \left(\frac{1}{\alpha_1} a \prod_{i=3}^{k} x_i^{\beta_i}\right)^{\frac{\alpha_2}{\gcd(\alpha_2,\beta_2)}} - 1$$

In this expression, the use of properties of gcd shows that the exponent of x_3 is of the form $\frac{d}{\gcd(d,\beta_2 w_2)}$. Reducing all of the h_i by g_2, we obtain polynomials which no longer depend on x_2, because all occurrences of x_2 in the g_2 were of the form $x_2^{\beta_2}$. We are hence in the situation to apply Remark 3.3 again, this time to x_3 and can eventually iterate the process $k-2$ times. This leads to polynomials of the form

$$g_i = x_i^{\frac{d}{w_i}\frac{\gcd(d,\beta_2 w_2,\ldots,\beta_i w_i)}{\gcd(d,\beta_2 w_2,\ldots,\beta_{i-1} w_{i-1})}} - c_i \cdot p_i(x_{i+1}, \ldots, x_k)$$

for each $3 \leq i \leq k$.
To determine the discriminant we could now continue one step further, eliminating x_k, but here it is easier to observe (e.g. by explicit polynomial division) that for any polynomial $1 - p(\underline{x}, a)$, also every polynomial $1 - p(\underline{x}, a)^k$ is in the ideal. Applying this to h_0 and the $\frac{d}{\gamma}$-th power, where

$$\gamma = \gcd(\beta_1 w_1, \ldots, \beta_k w_k) = \gcd(d, \beta_2 w_2, \ldots, \beta_k w_k),$$

we obtain

$$h_{k+1} = 1 - \left(\frac{1}{\alpha_1} a \prod_{i=2}^{k} x_i^{\beta_i}\right)^{\frac{d}{\gamma}}.$$

But the exponents α_i of the leading monomials of the h_i all divide $\beta_i \gamma$ for $2 \leq i \leq k$ by construction which allows reduction of h_{k+1} by these and leads to the claimed expression

$$g_{n+1} = a^{\frac{d}{\gamma}} \frac{\prod_{i=1}^{k} (\beta_i w_i)^{\frac{\beta_i w_o}{\gamma}}}{d^{\frac{d}{\gamma}}} + (-1)^{\frac{d}{\gamma}-1}.$$

To finish the Gröbner basis calculation, let us first consider the set of polynomials $S = \{h_2, \ldots, h_n, g_2, \ldots, g_k, g_{n+1}\}$. For $2 \leq i \leq k$ we drop h_i from it, if the x_i-degree of g_i is strictly smaller than the one of h_i, otherwise we drop g_i. The resulting set then contains n polynomials of which each of the first $n-1$ has a pure power of the respective variable x_i as leading monomial, and the last element g_{k+1} which has a leading monomial not involving any of the x_i. Hence this set obviously forms a Gröbner basis of some ideal, because all s-polynomials vanish by the product criterion. It then remains to show that the original polynomials $f_0, \ldots, f_n$ reduce to zero w.r.t. this set which can be checked by a straight forward but lengthy calculation.

Reading off the data:
It is clear that a takes precisely the $\frac{d}{\gamma}$ values

$$\sqrt[d/\gamma]{\frac{d^{\frac{d}{\gamma}}}{\prod_{i=1}^{k} (\beta_i w_i)^{\frac{\beta_i w_i}{\gamma}}}} \cdot \zeta$$

where ζ runs through all the $\frac{d}{\gamma}$-th roots of unity. At each of these points in the base, we can obtain the number of singularities by plugging in the value for a into g_k and counting solutions, followed by the values for a and x_k into g_{k-1} and so on, where $x_{k+1} = \cdots = x_n = 0$. This leads to the expression

$$\frac{1}{w_2} \gcd(d, \beta_2 w_2) \frac{d}{w_3} \frac{\gcd(d, \beta_2 w_2, \beta_3 w_3)}{\gcd(d, \beta_2 w_2)} \cdots \frac{d}{w_k} \frac{\gcd(d, \beta_2 w_2, \ldots, \beta_k w_k)}{\gcd(d, \beta_2 w_2, \ldots, \beta_{k-1} w_{k-1})}$$

for the number of singular points, which after simplification of the expression and multiplication by $\frac{\gcd(w_1, \ldots, w_k)}{w_1}$ (to take account of the identification of points in weighted projective space) leads to the claimed number.
The multiplicity of each of these points is then given by the product of the powers of the variables x_i in the polynomials h_i, $k+1 \leq i \leq n$. As the Gröbner basis generates the global Tjurina ideal of the fibre for each fixed value of a, the corresponding support describes the singular locus and the local multiplicity at each of the finitely many points is precisely the Tjurina number. By considering the corresponding

local equations, we can then check that the Tjurina number and Milnor number coincide for each arising singularity. □

Considering the extreme cases of the families with the highest and lowest numbers of singularities, we obtain:

COROLLARY 3.5. *Let $\mathcal{X} \subset \mathbb{P}_{w_1,\ldots,w_n}$ be the 1-parameter family of Calabi–Yau varieties given by the polynomial*

$$F = \left(\sum_{i=1}^{n} x_i^{\frac{d}{w_i}}\right) + a \cdot \prod_{i=1}^{n} x_i$$

where $d = \sum_{i=1}^{n} w_i$ and $\gcd(w_1, \ldots, w_n) = 1$. Then the discriminant of the family is

$$V\left(a^d + (-1)^{d-1}\frac{d^d}{\prod_{i=1}^{n} w_i^{w_i}}\right) \subset \mathbb{A}^1_{\mathbb{C}}.$$

In the respective fibre, above each of the d points of the discriminant, there are precisely

$$\frac{d^{n-2}}{\prod_{i=1}^{n} w_i}$$

ordinary double points (with Milnor number $\mu = 1$ and Tjurina number $\tau = 1$) and no further singularities.

COROLLARY 3.6. *Let $\mathcal{X} \subset \mathbb{P}_{w_1,\ldots,w_n}$ be the 1-parameter family of Calabi–Yau varieties given by the polynomial*

$$F = \left(\sum_{i=1}^{n} x_i^{\frac{d}{w_i}}\right) + a x_1^{\frac{d-w_n}{w_1}} x_n$$

where $d = \sum_{i=1}^{5} w_i$ and $w_1 | w_n$. Then the discriminant of the family is

$$V\left(a^{\frac{d}{w_n}} + (-1)^{\frac{d}{w_n}-1}\frac{d^{\frac{d}{w_n}}}{w_n(d-w_n)^{\frac{d}{w_n}-1}}\right) \subset \mathbb{A}^1_{\mathbb{C}}.$$

In the respective fibre above each of the $\frac{d}{w_n}$ points of the discriminant there is precisely 1 isolated singularity of which the local normal form (after moving to the coordinate origin) is

$$\left(\sum_{i=2}^{n-1} x_i^{\frac{d}{w_i}}\right) + x_n^2$$

with Milnor number $\mu = \prod_{i=2}^{n-1}\left(\frac{d}{w_i} - 1\right)$.

3.3. Some particular 2-parameter families. In this case, the Gröbner basis of the relative Tjurina ideal is far too complicated to write down in general. Nevertheless, it is possible to follow the lines of some of the calculations of the previous subsection to specify and study the discriminant of some families. By analysis of the discriminant it is then possible to precisely classify the arising singularities in explicit families.

LEMMA 3.7. *Let $\mathcal{X} \subset \mathbb{P}_{w_1,\ldots,w_n}$ be the 2-parameter family of Calabi–Yau (n-2)-folds given by the polynomial*

$$F = \sum_{i=1}^{n} x_i^{\frac{d}{w_i}} + a \prod_{i=1}^{n} x_i + b x_1^{\beta_1} x_2^{\beta_2}$$

where $d = \sum_{i=1}^n w_i$ and $\beta_1 w_1 + \beta_2 w_2 = d$. Then the discriminant of this 2-parameter family is reducible and its irreducible components can be sorted into two different kinds:

- *Lines L_i parallel to the a-axis, which are determined by the discriminant of*

$$x^{\frac{d}{w_2}} + b x^{\beta_2} + 1$$

- *A (possibly reducible) curve C which can be specified as the resultant of*

$$a^d x_2^d - \frac{d^{d-2}}{\prod_{i=3}^{n} w_i^{w_i}} \left(\beta_1 b x_2 + \frac{d}{w_1} \right)$$

and

$$\frac{d}{w_2} x_2^{\frac{d}{w_2}} + (\beta_2 - \beta_1) b x_2^{\beta_2} - \frac{d}{w_1}.$$

PROOF. As before, we choose a suitable affine chart, say $x_1 \neq 0$, and fix a lexicographical monomial ordering $x_n > \cdots > x_2 > a > b$. But here an explicit computation of a Gröbner basis of the Tjurina ideal cannot be performed in all generality. Instead, we can proceed analogously to the steps of the proof of Lemma 3.4 and obtain the following elements of the ideal:

$$\begin{aligned} h_i &= \frac{d}{w_i} x_i^{\frac{d}{w_i}} - \beta_1 b x_2^{\beta_2} - \frac{d}{w_1} \forall 3 \leq i \leq n \\ h_2 &= \frac{d}{w_2} x_2^{\frac{d}{w_2}} + (\beta_2 - \beta_1) b x_2^{\beta_2} - \frac{d}{w_1} \\ h_0 &= a \prod_{i=2}^{n} x_i + \left(\beta_1 b x_2^{\beta_2} + \frac{d}{w_1} \right) \end{aligned}$$

As before, we can again conclude that also

$$\left(a \prod_{i=2}^{n} x_i \right)^d - \left(\beta_1 b x_2^{\beta_2} + \frac{d}{w_1} \right)^d$$

is in the ideal and forming a normal form w.r.t. $h_3, \ldots, h_n$ then yields

$$\left(\beta_1 b x_2^{\beta_2} + \frac{d}{w_1} \right)^{\sum_{i=3}^n w_i} \cdot \left(a^d x_2^d \prod_{i=3}^{n} w_i^{w_i} - d^{d-2} \left(\beta_1 b x_2^{\beta_2} + \frac{d}{w_1} \right)^{w_1 + w_2} \right)$$

At this point, we can branch our computation and consider each factor separately.

<u>$g_1 = \left(\beta_1 b x_2^{\beta_2} + \frac{d}{w_1} \right)$</u>: Here we directly obtain

$$g_2 = g_1 + h_2 = \frac{d}{w_2} x_2^{\frac{d}{w_2}} + \beta_2 b x_2^{\beta_2}$$

and

$$g_3 = \frac{w_1}{d} g_1 + \frac{w_2}{d} g_2 = x_2^{\frac{d}{w_2}} + b x_2^{\beta_2} + 1.$$

Therefore the resultant of g_2 and g_3 is also contained in the ideal. On the other hand, $g_2 = x_2 \frac{\partial g_3}{\partial x_2}$ and hence the above resultant is just the discriminant of g_3 by the rules for computing resultants and the fact that $Res_{x_2}(g_3, x_2) = 1$.

$\underline{g_4 = a^d x_2^d \prod_{i=3}^n w_i^{w_i} - d^{d-2}\left(\beta_1 b x_2^{\beta_2} + \frac{d}{w_1}\right)^{w_1+w_2}}$: As g_4 and h_2 are both in the ideal so is their resultant w.r.t. x_2 which describes the desired curve. □

On the basis of this lemma, it is now easy to treat interesting special cases, which we want to consider in a later section of this article, by a straight-forward computation. In order to treat such examples by the combinatorial algorithm for determining the zeta-function, the two perturbation monomials need to be in the same strong β-orbit in the sense that the orbit structure w.r.t. the second monomial refines the one w.r.t. the first monomial. As this is a rather restrictive condition on the possible choices of monomials, we only state a choice of three explicit examples in Section 5.

4. The influence of singularity data on strong β-classes

In the previous section, we analysed the singularity structure of some 1- and 2-parameter families of Calabi–Yau varieties and, in particular, the structure of the Milnor algebra which encodes cohomological information about the singularities. Now we shift our focus to the computation of the local zeta-function for these families and re-encounter combinatorial data which we already saw in the previous section, in particular the combinatorics of the strong β-classes.

REMARK 4.1. Recalling definition 2.2 of strong motivic β-classes in $\mathcal{M}$, it is easy to show that each strong β-class, $\mathcal{C}_\beta$, is a set with cardinality d_β, where

$$d_\beta = \operatorname*{lcm}_{\beta_i \neq 0}(\operatorname{ord}(\beta_i)) = \operatorname*{lcm}_{\beta_i \neq 0}\left(\frac{d}{\gcd(\beta_i w_i, d)}\right) = \frac{d}{\gcd_{\beta_i \neq 0}(\beta_i w_i)}.$$

Hence the total number of strong β-classes, $\mathcal{O}_\beta$ is

$$\mathcal{O}_\beta = |\mathcal{M}| \frac{\gcd_{\beta_i \neq 0}(\beta_i w_i)}{d}.$$

LEMMA 4.2. *Let $w_1, \ldots, w_n$ be a set of weights satisfying the conditions of Definition 2.2. The total number of elements in $\mathcal{M}$ is*

$$|\mathcal{M}| = \left(\prod_{i=1}^n \frac{d}{w_i}\right) \frac{1}{d_{\mathbf{c}}},$$

where d_c denotes the cardinality of a strong $\mathbf{c} = (1, \ldots, 1)$-class.

PROOF. The total number of monomials in

$$\mathcal{W} := \prod_{i=1}^n w_i \mathbb{Z}/d\mathbb{Z}$$

is given by the product of the number of possible entries in each position, i.e. $\prod_{i=1}^n \frac{d}{w_i}$. Modulo d, the weighted degree of an element of $\mathcal{W}$ can take any value in $\{0, \dots, d-1\}$ and the number of elements of $\mathcal{W}$ mapping to the same class of weighted degree modulo d is precisely $\frac{1}{d}|\mathcal{W}|$. Hence, this is the number of elements of weighted degree 0 modulo d, i.e.

$$|\mathcal{M}| = \frac{1}{d}\prod_{i=1}^{n}\frac{d}{w_i}.$$

For later considerations, it will be convenient to modify this formula slightly using that $\gcd(w_1, \dots, w_n) = 1$ implies $d = d_c$, which proves the claimed formula. □

REMARK 4.3. For any given $k \in \{1, \dots, n\}$, we can partition $\mathcal{M}$ into subsets for which the last $(n-k)$ entries coincide. A priori, there are $\prod_{i=k+1}^{n}\frac{d}{w_i}$ possibilities for the last $(n-k)$ entries. As the front part of any element of $\mathcal{M}$, i.e. the first k entries of the element, can only provide weighted degrees which are multiples of $\gcd(w_1, \dots, w_k)$ and as the weighted degree of any element of $\mathcal{M}$ is a multiple of d, not all combinations of the last $(n-k)$ entries can actually occur, but only those which themselves also provide multiples of $\gcd(w_1, \dots, w_k)$ as weighted degree. Hence the total number of these subsets of $\mathcal{M}$ is

$$\frac{1}{\gcd(w_1, \dots, w_k)}\prod_{i=k+1}^{n}\frac{d}{w_i}.$$

Combining these observations and the lemma, we obtain the following result for the number of strong β-classes which share the same last $(n-k)$ entries:

COROLLARY 4.4. *Let $\beta \in \mathcal{M}$ satisfy $l(\beta) = 1$ and $\beta_{k+1} = \dots = \beta_n = 0$. Then the number of elements of $\mathcal{M}$ which share the same last $(n-k)$ entries is precisely*

$$\frac{\gcd(w_1, \dots, w_k)}{d}\prod_{i=1}^{k}\frac{d}{w_i}$$

and the number of strong β-classes with these last $(n-k)$ entries is

$$T_\beta = \frac{\gcd(w_1, \dots, w_k)}{d}\frac{\gcd(\beta_1 w_1, \dots, \beta_k w_k)}{d}\prod_{i=1}^{k}\frac{d}{w_i},$$

which coincides with the total number of singularities in the singular fibre of a 1-parameter family of Fermat-type Calabi–Yau varieties with perturbation term $\underline{x}^\beta$ as considered in section 3.

Applying this corollary to the two special cases of 1-parameter families considered in 3, we find precisely the number of A_1-singularities in the case $\beta = (1, \dots, 1)$ and 1 for the completion of the square. This establishes the first of the two correspondences, which we discuss here. The second one is more subtle and links the Milnor number to the contributions of each β-class to the zeta-function. It is known that among the monomials in $\mathcal{M}$ only those that do not contain any entry of the form $\left(\frac{d}{w_i} - 1\right)$ in the i-th position should be counted when computing the degree of the associated piece of the zeta-function. Therefore counting the number of possible ways of constructing such monomials seems a natural question to consider and leads to the following observation:

LEMMA 4.5. *Let $\beta \in \mathcal{M}$ satisfy $l(\beta) = 1$, $\beta_{k+1} = \cdots = \beta_n = 0$ and $\gcd(w_1, \ldots, w_k) = 1$. Then the number of tuples which appear as the last $(n-k)$ entries in an element of $\mathcal{M}$ and do not involve any entry of the form $\left(\frac{d}{w_i} - 1\right)$, is precisely*

$$\prod_{i=k+1}^{n} \left(\frac{d}{w_i} - 1\right).$$

This coincides with the Milnor number of the appearing singularities according to 3.

PROOF. As $\gcd(w_1, \ldots, w_k) = 1$, any weighted degree $\sum_{i=k+1}^{n} \alpha_i w_i$ can be completed to a multiple of d by some contribution of the first k entries. Of these only the ones with $\alpha_i \neq \left(\frac{d}{w_i} - 1\right)$ need to be counted which after a direct application of the inclusion-exclusion formula yields the desired expression. □

Combining the result of this lemma and the preceding corollary, we see that in the case of $\gcd(w_1, \ldots, w_k) = 1$ the total number of strong β-classes is precisely the total Milnor number. On the other hand, explicit computation showed that for all families of Calabi–Yau 3-folds with one perturbation considered, the degree of the zeta-function drops by exactly the total Milnor number, e.g. for the case of the canonical perturbation, this is the total number of conifold singularities, when passing to a singular fibre. We will see further occurrences of these coincidences in explicit examples for 2-parameter families in the next section.

The correspondence between the findings of the singularity analysis and the intermediate results of the calculation of the zeta-function via the combinatorics of the strong β-classes can be shown to further illuminate the internal structure of the combinatorial objects involved. As the calculations in the general case are rather technical and might block the view for the key observation, we only state this for the case $\beta = (1, \ldots, 1)$:

REMARK 4.6. By using standard facts about the gcd, the cardinality of the set $\mathcal{M}$ can also be stated as

$$|\mathcal{M}| = \prod_{i=2}^{n} \gcd\left(\frac{d}{w_i}, \frac{d}{\gcd(w_1, \ldots, w_{i-1})}\right),$$

which better reflects the combinatorial structure of $\mathcal{M}$. [5]. Consider the first two weights w_1 and w_2. The **c**-subclasses associated to each weight have lengths $L_1 = \frac{d}{w_1}$, $L_2 = \frac{d}{w_2}$ respectively. The i-th coordinates of the ordered monomials in every **c**-class take values in the range $0, 1, 2, \ldots, \left(\frac{d}{w_i} - 1\right)$ going up by 1 cyclically. The greatest common divisor of these two **c**-subclass lengths, $g_{1,2} = \gcd\left(\frac{d}{w_1}, \frac{d}{w_2}\right)$, can be used to divide the ranges $0, 1, 2, \ldots, \left(\frac{d}{w_i} - 1\right)$ into $g_{1,2}$ disjoint partitioning sets given by:

$$S_{i_k} = \left\{k, k + g_{1,2}, k + 2g_{1,2}, \ldots, k + \left(\frac{L_i}{g_{1,2}} - 1\right) g_{1,2}\right\}, \ 0 \leq k \leq (g_{1,2} - 1).$$

[5] Note that this decomposition into a product holds for any ordering of the weights.

We can now divide the monomials in $\mathcal{M}$ with ith coordinate in S_{i_k} $(i = 1, 2)$ into $g_{1,2}$ distinct sets. Hence we have established that

$$g_{1,2} = \gcd\left(\frac{d}{w_1}, \frac{d}{w_2}\right) \mid |\mathcal{M}|,$$

thus accounting for the first factor in the formula. Iterating this process, we next compute the **c**-subclass length associated to the pair of weights (w_1, w_2), which we shall label $L_{1,2} = \frac{d}{\gcd(w_1,w_2)}$. Then we find analogously to the previous step:

$$g_{(1,2),3} = \gcd\left(\frac{d}{w_3}, \frac{d}{\gcd(w_1, w_2)}\right),$$

which again leads to a further partitioning. Eventually, this leads to a sequence of refinements of the partitioning which reflects the claimed expression for the number of elements in $\mathcal{M}$.

5. Examples of 2-parameter families

The observations for the 1-parameter families might still be a combinatorial coincidence, but passing to 2-parameter families where the singularity analysis is no longer purely combinatorial we still see the same phenomena: The total Milnor number of a singular fibre matches the change of the degree of the zeta-function when moving from a smooth to a singular fibre. Our observations provide evidence that Lauder's conjecture of an analogue to the Clemens-Schmid exact sequence in [**L2**] for projective hypersurfaces could be extended to the case of hypersurfaces in weighted projective space. In particular, the proof of Thoerem 2.15 (semi-stable reduction for algebraic de Rham cohomology) in [**L2**] Section 2.3 of [**L2**] is based on the Griffiths-Dwork method; the latter readily generalizes to the case of toric hypersurfaces (see [**CK**], section 5.3.2).

The three considered examples are:

5.1. A family in $\mathbb{P}_{(1,1,2,2,2)}$. Considering the family in $\mathbb{P}_{(1,1,2,2,2)}$ given by

$$F = x^8 + y^8 + z^4 + u^4 + v^4 + a \cdot xyzuv + b \cdot x^4y^4,$$

the discriminant consists of two lines $L_1 = V(b-2)$ and $L_2 = V(b+2)$ (denote $L = L_1 \cup L_2$) and the curve C which possesses the two components $C_1 = V(a^4 - 256b + 512)$ and $C_2 = V(a^4 - 256b - 512)$. For the singular fibres of the family the following singularity types occur:

$(a, b) \in L \setminus (L \cap C)$: 4 singularities of type $T_{4,4,4}$ $(\mu = 11)$
$(a, b) \in C \setminus (C \cap L)$: 64 ordinary double points
$(a, b) \in L \cap C$, $a \neq 0$: 4 singularities of type $T_{4,4,4}$ $(\mu = 11)$ and
64 ordinary double points
(transversal intersections of the components of the discriminant)
$(0, b) \in L \cap C$: 4 singularities with local normal form $x^2 + z^4 + u^4 + v^4$
$(\mu = 27)$
(higher order contact of the components of the discriminant)

When computing the zeta function, we see the following degrees of the contributions depending on the considered fibre of the family. The contributions are labeled by the respective $(1,1,1,1,1)$-classes; classes only differing by a permutation of entries are collected in one line.[6] Computations were made for good primes $p \in 3,5,7,11,13,17,19,23$ with orders between 8 for the smaller primes and 3 for the larger ones:

Degree of Contribution $R_{\mathbf{v}}(t)$ According to Singularity					
Monomial $\mathbf{v}$	Perm.	Smooth	64 A_1 with $\mu_{X,x} = 1$	4 $T_{4,4,4}$ with $\mu_{X,x} = 11$	Both or 4 $\mu_{X,x} = 27$
		$\mu_X = 0$	$\mu_X = 64$	$\mu_X = 44$	$\mu_X = 108$
(0,0,0,0,0)	1	6	5	4	3
(7,1,0,0,0)	2	4	3	2	1
(6,2,0,0,0)	1	4	3	2	1
(2,2,2,0,0)	3	4	3	3	2
(3,1,2,0,0)	6	2	1	1	0
(4,0,2,0,0)	3	4	3	3	2
(7,7,1,0,0)	3	3	2	2	1
(6,0,1,0,0)	6	3	2	2	1
(5,1,1,0,0)	3	4	3	3	2
(7,3,3,0,0)	3	4	3	3	2
(2,0,3,0,0)	6	3	2	2	1
(1,1,3,0,0)	3	3	2	2	1
(1,1,0,1,2)	6	2	1	2	1
(2,0,0,1,2)	12	2	1	2	1
(7,3,0,1,2)	6	0	-1	0	-1
degree:		168	104	124	60
degree change:			64	44	108

The coincidence of the total Milnor number with the total drop in degree as evident in this table, provides experimental evidence for Lauder's conjecture.

For this first example of a particular family, we also provide the explicit zeta-function in one case, to justify the omission of this data in the later examples. Zeta function data is too richly detailed for the chosen focus of the article. For $p = 7$ and a fibre ($b = -2$, $a = 3$ or 4) of the family with 4 $T_{4,4,4}$ singularities, the zeta-function of our family has the following contributions:

Monomial $\mathbf{v}$	Contribution	Power $\lambda_{\mathbf{v}}$
(0,0,0,0,0)	$(1 + 18t + 2 \cdot 41pt^2 + 18p^3t^3 + p^6t^4)$	1
(0,2,1,1,1)	$(1 - pt)(1 + pt)$	2
(6,2,0,0,0)	$(1 - 2pt + p^3t^2)$	1
(0,0,0,2,2)	$(1 + pt)(1 + 2pt + p^3t^2)$	3
(2,0,1,3,3)	$[(1 - pt)(1 + pt)]^{\frac{1}{2}}$	6
(4,0,2,0,0)	$(1 + pt)(1 + 2pt + p^3t^2)$	3

[6]We list the number of permutations in the column labeled 'Perm.'. Pairs of strong monomial classes in $(5,1,1,0,0) and (7,3,3,0,0)$, $(6,0,1,0,0) and (2,0,3,0,0)$, and $(7,7,1,0,0) and (1,1,3,0,0)$ are weakly equivalent.

(0,0,2,1,1)	$[(1+p^3t^2)(1-pt)(1+pt)]^{\frac{1}{2}}$	3
(6,0,1,0,0)	$[(1-2pt+p^3t^2)(1+2pt+p^3t^2)]^{\frac{1}{2}}$	6
(0,4,0,3,3)	$[(1+p^3t^2)^2(1-pt)(1+pt)]^{\frac{1}{2}}$	3
(4,0,1,1,0)	$[(1+p^3t^2)^2(1-pt)(1+pt)]^{\frac{1}{2}}$	3
(2,0,3,0,0)	$[(1-2pt+p^3t^2)(1+2pt+p^3t^2)]^{\frac{1}{2}}$	6
(2,2,1,1,0)	$[(1+p^3t^2)(1-pt)(1+pt)]^{\frac{1}{2}}$	3
(0,0,3,1,0)	$(1-pt)(1+pt)$	6
(2,0,2,1,0)	$[(1-2pt+p^3t^2)(1+2pt+p^3t^2)]^{\frac{1}{2}}$	12
(4,0,2,3,1)	1	6

Note that the square roots arise from the algorithmic computation of the zeta function, but never occur in the final result, because the corresponding contributions always arise in pairs.

5.2. A family in $\mathbb{P}_{(1,1,2,2,6)}$. Considering the family in $\mathbb{P}_{(1,1,2,2,6)}$ given by

$$F = x^{12} + y^{12} + z^6 + u^6 + v^2 + a \cdot xyzuv + b \cdot x^6y^6,$$

the discriminant consists of two lines $L_1 = V(b-2)$ and $L_2 = V(b+2)$ (denote $L = L_1 \cup L_2$) and the curve C which possesses the two components $C_1 = V(a^6 - 1728b + 3456)$ and $C_2 = V(a^6 - 1728b - 3456)$. For the singular fibres of the family the following singularity types occur:

$(a,b) \in L \setminus (L \cap C)$: 6 singularities of type $T_{2,6,6} = Y^1_{2,2}$ ($\mu = 13$)
$(a,b) \in C \setminus (C \cap L)$: 72 ordinary double points
$(a,b) \in L \cap C$, $a \neq 0$: 6 singularities of type $T_{2,6,6}$ ($\mu = 13$) and 72 ordinary double points
(transversal intersections of the components of the discriminant)
$(0,b) \in L \cap C$: 6 singularities with local normal form $x^2 + z^6 + u^6 + v^2$ ($\mu = 25$)
(higher order contact of the components of the discriminant)

Here the contributions to the factors of the zeta-function are displayed in the following table. Computations were made for good primes $p \in 5, 7, 11, 13, 17$ up to order 6 for the smaller primes and order 4 for the larger ones.

Numerically Computed deg $R_{\mathbf{v}}(t)$ According to Singularity					
Monomial $\mathbf{v}$	Perm.	Smooth	72 A_1 with $\mu_{X,x} = 1$	6 $T_{2,6,6}$ with $\mu_{X,x} = 13$	Both or 6 $\mu_{X,x} = 25$
		$\mu_X = 0$	$\mu_X = 72$	$\mu_X = 78$	$\mu_X = 150$
(0,0,0,0,0)	1	6	5	4	3
(11,1,0,0,0)	2	4	3	2	1
(10,2,0,0,0)	2	6	5	4	3
(9,3,0,0,0)	1	4	3	2	1
(10,0,0,1,0)	4	4	3	3	2
(9,1,0,1,0)	4	3	2	2	1
(8,2,0,1,0)	2	4	3	3	2
(5,5,0,1,0)	2	3	2	2	1

(8,0,2,0,0)	4	6	5	4	3
(7,1,2,0,0)	2	4	3	2	1
(5,3,2,0,0)	4	4	3	2	1
(4,4,2,0,0)	2	6	5	4	3
(6,0,3,0,0)	2	4	3	3	2
(5,1,3,0,0)	4	2	1	1	0
(4,2,3,0,0)	4	4	3	3	2
(3,3,3,0,0)	2	4	3	3	2
(6,0,0,0,1)	1	4	3	3	2
(5,1,0,0,1)	2	4	3	3	2
(4,2,0,0,1)	2	2	1	1	0
(3,3,0,0,1)	1	4	3	3	2
(2,2,0,1,1)	2	3	2	2	1
(3,1,0,1,1)	4	4	3	3	2
(10,6,0,1,1)	4	3	2	2	1
(11,5,0,1,1)	2	4	3	3	2
(1,1,2,0,1)	2	2	1	2	1
(2,0,2,0,1)	4	2	1	2	1
(9,5,2,0,1)	4	2	1	2	1
(10,4,2,0,1)	2	0	-1	0	-1
degree:		254	182	176	104
degree change:			72	78	150

5.3. A family in $\mathbb{P}_{(1,1,3,3,4)}$. Considering the family in $\mathbb{P}_{(1,1,3,3,4)}$ given by

$$F = x^{12} + y^{12} + z^4 + u^4 + v^3 + a \cdot xyzuv + b \cdot x^4y^4v,$$

the discriminant consists of three lines $L = V(b^3 + 27)$ and the curve $C = V(a^{12} - a^8b^4 - 576a^8b + 512a^4b^5 + 96768a^4b^2 - 65536b^6 - 3538944b^3 - 47775744)$. For the singular fibres of the family the following singularity types occur:

$(a, b) \in L \setminus (L \cap C)$**:** 12 ordinary double points
$(a, b) \in C \setminus ((L \cap C) \cup C_{sing})$**:** 48 ordinary double points
$(0, b) \in L \cap C$**:** 12 singularities of type X_9
(higher order contact of components of the dicriminant)
$(a, b) \in L \cap C, a \neq 0$**:** 60 ordinary double points
(transversal intersections of the components of the discriminant)
$(a, b) \in V(9a^4 - 16b^4, b^3 - 108) \subset C_{sing}$**:** 48 A_2 singularities
$(a, b) \in V(a^4 - 288b, b^3 - 216) \subset C_{sing}$**:** 96 A_1 singularities

Numerically Computed deg $R_{\mathbf{v}}(t)$ According to Singularity							
Monomial $\mathbf{v}$	Perm.	Smooth	12 A_1	48 A_1	60 A_1	48 A_2	12 X_9
(0,0,0,0,0)	1	6	5	5	4	4	3
(11,1,0,0,0)	2	4	3	3	2	2	1
(10,2,0,0,0)	2	4	3	3	2	2	1
(9,3,0,0,0)	2	6	5	5	4	4	3
(8,4,0,0,0)	2	6	5	5	4	4	3
(7,5,0,0,0)	2	4	3	3	2	2	1
(6,6,0,0,0)	1	6	5	5	4	4	3

(9,0,1,0,0)	4	4	4	3	3	2	2
(8,1,1,0,0)	4	4	4	3	3	2	2
(7,2,1,0,0)	4	2	2	1	1	0	0
(6,3,1,0,0)	4	4	4	3	3	2	2
(5,4,1,0,0)	4	4	4	3	3	2	2
(11,10,1,0,0)	4	2	2	1	1	0	0
(6,0,2,0,0)	2	4	4	3	3	2	2
(5,1,2,0,0)	4	2	2	1	1	0	0
(4,2,2,0,0)	4	4	4	3	3	2	2
(3,3,2,0,0)	2	4	4	3	3	2	2
degree:		180	168	132	120	84	72
degree change:			12	48	60	96	108

Computations were made for good primes $p \in 5, 7, 11, 13, 17$ up to order 6 for the smaller primes and order 4 for the larger ones.

6. Conclusion

For one-parameter families it has been shown that the combinatorics of the monomial equivalence classes, which split up the zeta function, is intimately related to the singularity structure of the varieties. Moreover, all computed examples[7] have also shown that the change of degree of the contribution by each labeled part of the zeta function follows patterns of the set of strong β-classes. The total change of degree of the zeta function upon passing to a singular fibre has been observed as coinciding with the total Milnor number of the singular fibre.

For the more involved case of two-parameter families, it is also apparent from the finite number of cases computed, that the combinatorics of the strong equivalence classes once again seem to be reflected in the singularity structure. From this arises the following conjecture, which strongly refines the conjectures 7.3, page 137 in [**K04**] and in §7.2 of [**K06**]:

CONJECTURE 6.1 (Singularity -geometric/combinatorial duality). *Given a family of Calabi–Yau varieties with special fibre of Fermat type, the total Milnor number of each arising singular fibre is expressible in terms of the change of the degree of the zeta function when passing to the singular fibre.*
The singularity structure as reflected in the relative Milnor (and Tjurina) algebra of the family encodes information on the degree changes of factors of the zeta function labeled by β-classes.

The degenerative properties of the zeta functions at singular points studied here (and the global L-series they give rise to) were recently exploited in [**KLS**] in order to investigate the phenomenon of 'string modularity'. The main result was that for several families (all containing a Fermat member as a special fibre), the modular

[7] In addition to the examples stated in this article, all Calabi–Yau 3-folds of Fermat-type have been systematically studied combinatorially from our point of view. For a number of interesting cases, which did not pose too many computational difficulties for the Mathematica programs, the explicit zeta-functions have been determined for low primes – all showing the same behaviour. We choose to include only 3 explicit examples of 2-parameter families which already cover most of our observations, because adding further examples would not show new phenomena.

form associated to part of the global zeta function or L-series found at a degenerate, non-Fermat point in the moduli space agreed with that of the motivic L-series of a different weighted Fermat variety. These pairs are called L-correlated and provide evidence that the conformal field theory at deformed fibres (currently difficult to define) are related to those of the well-defined rational conformal field theories of Fermat-type manifolds (Gepner models) with a completely different geometry. Our singularity-theoretic and combinatorial results would aid exploration of both finding more examples of singular members of Calabi–Yau families exhibiting modularity, and perhaps more L-correlated 'string-modular' pairs.

References

[Bat] V. Batyrev, *Dual Polyhedra and the Mirror Symmetry for Calabi–Yau Hypersurfaces in Toric Varieties* Duke math J.69, No. 2, (1993) 349. MR1269718 (95c:14046)

[CdO] P. Candelas and X. de la Ossa, *The zeta-function of a p-aadic manifolds, Dwork theory for physicists*, arXiv: hep-th/0705.2056

[CdOK] P. Candelas, X. de la Ossa, S. Katz, *Mirror Symmetry for Calabi–Yau Hypersurfaces in Weighted Projected $\mathbb{P}_4$ and Extensions of Landau-Ginzburg Theory*, arXiv: hep-th/9412117.

[CdOV1] P.Candelas, X. de la Ossa, F.Rodriguez Villegas, *Calabi–Yau Manifolds over Finite Fields I*, arXiv:hep-th/0012233.

[CdOV2] P.Candelas, X. de la Ossa, and F.Rodriguez Villegas, *Calabi–Yau Manifolds over Finite Fields II*, Fields Institute Communications Volume **38**,(2003).

[CLO] D. Cox, J. Little, D. O'Shea: *Ideals, Varieties and Algorithms*, 3rd edition, Springer (2007) MR2290010 (2007h:13036)

[CK] D. Cox and S. Katz, *Mirror Symmetry and Algebraic Geometry*, Math. Surveys and Monographs, **68**, Amer. Math. Soc. (1999) MR1677117 (2000d:14048)

[DGPS] Decker, W.; Greuel, G.-M.; Pfister, G.; Schönemann, H.: SINGULAR 3-1-2 — *A computer algebra system for polynomial computations.* http://www.singular.uni-kl.de (2010).

[Del1] P. Deligne, *La Conjecture de Weil I*, Publ.Math.IHES **43** (1974) 273–307. MR0340258 (49:5013)

[Del2] P. Deligne, *La Conjecture de Weil II*, Publ.Math.IHES **52** (1980) 137–252. MR601520 (83c:14017)

[Dw1] B.M.Dwork, *On the rationality of the Zeta Function of an Algebraic Variety*, Amer.J.Math. **82** (1960) 631. MR0140494 (25:3914)

[Dw2] B. Dwork, *On the zeta-function of a hypersurface*, Publ. Math. I.H.É.S., 12 (1962) 5–68. MR0159823 (28:3039)

[Gr] A. Grothendieck, *Formulé de Lefschetz ét rationalité de fontion de L*, Séminaire Bourbaki **279**, 1964/1965, 1–15.

[GH] P. Green and T. Hübsch, *Connecting moduli spaces of Calabi–Yau threefolds*, Commun. Math. Phys. B298 (1988) 493–525. MR969210 (90a:14050)

[GK] B. H. Gross and N. Koblitz, *Gauss Sums and p-adic Γ-function*, Annals of Math. 109 569 (1979). MR534763 (80g:12015)

[GMS] B.R. Greene, D. Morrison and A. Strominger, *Black hole condensation and the unification of string vacua*, Nucl. Phys. B451 (1995) 109–120, arXiv: hep-th/9504145. MR1352415 (96m:83085)

[GLS] G.-M. Greuel, C. Lossen, E. Shustin: *Introduction to Singularities and Deformation*, Springer (2007) MR2290112 (2008b:32013)

[GP] G.-M. Greuel, G. Pfister:*A* SINGULAR *Introduction to Commutative Algebra*, 2nd edition, Springer (2008) MR2363237 (2008j:13001)

[GY] F. Gouvêa and N. Yui, *Arithmetic of Diagonal Hypersurfaces over Finite Fields*, London Math. Soc. Lecture Notes Series **209**, Cambridge University Press (1995) MR1340424 (97k:11095)

[HKS] K. Hulek, R. Kloosterman, M. Schuett, *Modularity of Calabi–Yau Manifolds*, Global aspects of complex geometry, pp. 271–309, Springer, Berlin (2006)

[K04] S. N. Kadir, *The Arithmetic of Calabi–Yau Manifolds and Mirror Symmetry*, University of Oxford (2004) arXiv: hep-th/0409202.

[K06] S. Kadir, *Arithmetic mirror symmetry for a two-parameter family of Calabi–Yau manifolds.* Mirror symmetry. V, 35–86, AMS/IP Stud. Adv. Math., 38, Amer. Math. Soc., Providence, RI, (2006) MR2282954 (2008e:14026)

[KLS] S. Kadir, M. Lynker and R. Schimmrigk, *String Modular Phases in Calabi–Yau Families*, Journal of Geometry and Physics (2011), doi: 10.1016/j.geomphys.2011.04.010, arXiv: 1012.5807 (hep-th).

[KY] S. Kadir and N. Yui, *Motives and Mirror Symmetry for Calabi–Yau Orbifolds* , Fields Institute Communications Volume **54**,(2008). MR2454318 (2009j:14050)

[Kl] R. Kloosterman, *The zeta function of monomial deformations of Fermat hypersurfaces*, Algebra Number Theory 1 (2007), no. 4, 421–450. MR2368956 (2008j:14044)

[KR] M. Kreuzer, L. Robbiano: *Computational Commutative Algebra*, 2nd edition, Springer (2008). MR2723052 (2011h:13041)

[L1] A. Lauder, *Counting solutions to equations in many variables over finite fields*, Foundations of Computational Mathematics 4 No. 3 (2004) 221–267 MR2078663 (2005f:14048)

[L2] A. Lauder, *Degenerations and limit Frobenius structures in rigid cohomology*, to appear in LMS J. Comp. Math.. arXiv: math.NT0912.5185. MR2777002

[Mth] Wolfram Research, Inc., Mathematica, Version 5.1, Champaign, IL (2004).

[Sk] H. Skarke, *Weight systems for toric Calabi–Yau varieties and reflexivity of Newton polyhedra*", Modern Phys. A 11 (1996), no. 20, 1637–1652, arXiv: alg-geom/9603007. MR1397227 (97e:14054)

[S] A. Strominger, *Massless black holes and conifolds in string theory*, Nucl. Phys. B451 (1995) 96–108. MR1352414 (96m:83084)

Institut f. Algebraische Geometrie, Leibniz Universität Hannover, Welfengarten 1, 30167 Hannover, Germany

E-mail address: anne@math.uni-hannover.de

Institut f. Algebraische Geometrie, Leibniz Universität Hannover, Welfengarten 1, 30167 Hannover, Germany

E-mail address: kadir@math.uni-hannover.de

Contemporary Mathematics
Volume **566**, 2012
http://dx.doi.org/10.1090/conm/566/11214

Estimates for exponential sums with a large automorphism group

Antonio Rojas-León

ABSTRACT. We prove some improvements of the classical Weil bound for one variable additive and multiplicative character sums associated to a polynomial over a finite field $k = \mathbb{F}_q$ for two classes of polynomials which are invariant under a large abelian group of automorphisms of the affine line $\mathbb{A}^1_k$: those invariant under translation by elements of k and those invariant under homotheties with ratios in a large subgroup of the multiplicative group of k. In both cases, we are able to improve the bound by a factor of $\sqrt{q}$ over an extension of k of cardinality sufficiently large compared to the degree of f.

1. Introduction

Let $k = \mathbb{F}_q$ be a finite field with q elements. As a consequence Weil's bound for the number of rational points on a curve over k, one can obtain estimates for character sums defined on the affine line $\mathbb{A}^1_k$ (cf. [**6**],[**17**]). Let us describe the precise results.

Let $f \in k[x]$ be a polynomial of degree d and $\psi : k \to \mathbb{C}^\star$ a non-trivial additive character. Consider the sum $\sum_{x\in k} \psi(f(x))$ (and, more generally,

$$\sum_{x\in k_r} \psi(\mathrm{Tr}_{k_r/k}(f(x)))$$

for a finite extension k_r of k of degree r). Then, if d is prime to p, we have the estimate

$$\left|\sum_{x\in k_r} \psi(\mathrm{Tr}_{k_r/k}(f(x)))\right| \leq (d-1)q^{\frac{r}{2}}.$$

If d is divisible by p, we can reduce to the previous case using the following trick. Since $t \mapsto \psi(t^p)$ is a non-trivial additive character, there must be some $a \in k$ such that $\psi(t^p) = \psi(at)$ for every $t \in k$. If $f(x) = a_d x^d + a_{d-1}x^{d-1} + \cdots$ with

2010 *Mathematics Subject Classification.* Primary 11L07,11T23,11G15.

Partially supported by P08-FQM-03894 (Junta de Andalucía), MTM2007-66929 and FEDER.

$d = ep$, let $b_d \in k$ be such that $b_d^p = a_d$, then

$$\begin{aligned}\psi(\mathrm{Tr}_{k_r/k}(f(x))) &= \psi(\mathrm{Tr}_{k_r/k}((b_d x^e)^p))\psi(\mathrm{Tr}_{k_r/k}(f(x) - a_d x^d)) = \\ &= \psi(\mathrm{Tr}_{k_r/k}(b_d x^e)^p)\psi(\mathrm{Tr}_{k_r/k}(f(x) - a_d x^d)) = \\ &= \psi(a \cdot \mathrm{Tr}_{k_r/k}(b_d x^e))\psi(\mathrm{Tr}_{k_r/k}(f(x) - a_d x^d)) = \\ &= \psi(\mathrm{Tr}_{k_r/k}(f(x) - a_d x^d + a b_d x^e)).\end{aligned}$$

We keep reducing the polynomial in this way until we get a polynomial with degree d' prime to p. Then we apply the prime to p case and obtain an estimate

$$\left|\sum_{x \in k_r} \psi(\mathrm{Tr}_{k_r/k}(f(x)))\right| \leq (d'-1)q^{\frac{r}{2}}.$$

except when d' is zero (that is, when $f = c + g^p - ag$ for some constant c and some $g \in k[x]$). If the character ψ is obtained from a character of the prime subfield $\mathbb{F}_p$ by pulling back via the trace map, then $a = 1$.

Similarly, if $\chi : k^\star \to \mathbb{C}^\star$ is a multiplicative character of order $m > 1$ and $f \in k[x]$ is not an m-th power, we have an estimate

$$\left|\sum_{x \in k_r} \chi(\mathrm{N}_{k_r/k}(f(x)))\right| \leq (e-1)q^{\frac{r}{2}} \leq (d-1)q^{\frac{r}{2}}$$

where e is the number of distinct roots of f.

In this article we will improve these estimates for a special class of polynomials: those which are either translation invariant or homothety invariant, that is, either $f(x+\lambda) = f(x)$ for every $\lambda \in k$ or $f(\lambda x) = f(x)$ for every $\lambda \in k^\star$ (or every λ in a large subgroup of $k^\star$). For such polynomials, there is a large abelian group G of automorphisms of $\mathbb{A}^1_{\bar{k}}$ such that $f \circ \sigma = f$ for every $\sigma \in G$.

On the level of ℓ-adic cohomology, this gives an action of G on the pull-back by f of the Artin-Schreier and Kummer sheaves associated to ψ and χ respectively [**1**, 1.7], so they induce an action on their cohomology. The character sums can be expressed as the trace of the geometric k_r-Frobenius action on this cohomology, by Grothendieck's trace formula. The above estimates are a consequence of the fact that this action has all eigenvalues of archimedean absolute value $\leq q^{\frac{r}{2}}$. Precisely, if $S_r = \sum_{x \in k_r} \psi(\mathrm{Tr}_{k_r/k}(f(x)))$ (respectively $U_r = \sum_{x \in k_r} \chi(\mathrm{N}_{k_r/k}(f(x)))$) the L-functions

$$L(\psi, f; T) := \exp \sum_{r \geq 1} S_r \frac{T^r}{r}$$

and

$$L(\chi, f; T) := \exp \sum_{r \geq 1} U_r \frac{T^r}{r}$$

are the polynomials $\det(1 - T \cdot \mathrm{Frob}_k | \mathrm{H}^1_c(\mathbb{A}^1_{\bar{k}}, f^\star \mathcal{L}_\psi))$ of degree $d'-1$ and $\det(1 - T \cdot \mathrm{Frob}_k | \mathrm{H}^1_c(\mathbb{A}^1_{\bar{k}}, f^\star \mathcal{L}_\chi))$ of degree $e-1$ respectively.

Now under the action of the abelian group G, this cohomology splits as a direct sum of eigenspaces for the different characters of G. Under certain generic conditions, it is natural to expect some cancellation among the traces of the Frobenius actions on these eigenspaces, thus giving a substantial improvement of Weil's estimate if G is large (namely by a $\sqrt{\#G}$ factor). Compare [**15**], where an improvement for the Weil estimate for the number of rational points on Artin-Schreier curves was obtained using the same arguments we apply in this article.

For the translation invariant case (sections 2 ans 3), we obtain this improvement using the local theory of ℓ-adic Fourier transform [**14**] and Katz' computation of the geometric monodromy groups for some families of exponential sums [**7**], [**9**]. The argument is similar to that in [**15**]. For the homothety invariant case (sections 4 and 5), we use Weil descent together with certain properties of the convolution of sheaves on $\mathbb{G}_{m,k}$.

Throughout this article, $k = \mathbb{F}_q$ will be a finite field of characteristic p, $\bar{k} = \bar{\mathbb{F}}_q$ a fixed algebraic closure and $k_r = \mathbb{F}_{q^r}$ the unique extension of k of degree r in $\bar{k}$. We will fix a prime $\ell \neq p$, and work with ℓ-adic cohomology. In order to speak about weights without ambiguity, we will fix a field isomorphism $\iota : \bar{\mathbb{Q}}_\ell \to \mathbb{C}$. We will use this isomorphism to identify $\bar{\mathbb{Q}}_\ell$ and $\mathbb{C}$ without making any further mention to it. When we speak about weights, we will mean weights with respect to the chosen isomorphism ι.

2. Additive character sums for translation invariant polynomials

Let $f \in k[x]$ be a polynomial. f is said to be *translation invariant* if $f(x+a) = f(x)$ for every $a \in k$.

LEMMA 2.1. *Let $f \in k[x]$. The following conditions are equivalent:*

(a) *f is translation invariant.*
(b) *There exists $g \in k[x]$ such that $f(x) = g(x^q - x)$.*

PROOF. *(b)* $\Rightarrow$ *(a)* is clear. Suppose that f is translation invariant. If the degree of f is $< q$, the polynomial $f(x) - f(0)$ has at least q roots (all elements of k) and degree $< q$, so it is identically zero. So f is the constant polynomial $f(0)$. Otherwise, we can write $f(x) = (x^q - x)h(x) + r(x)$ with $\deg(r) < q$. For every $a \in k$ we have then $f(x+a) = (x^q - x)h(x+a) + r(x+a) = (x^q - x)h(x) + r(x)$, so $(x^q - x)(h(x+a) - h(x)) = r(x) - r(x+a)$. Since the right hand side has degree $< q$, we conclude that $h(x+a) - h(x) = r(x+a) - r(x) = 0$. $r(x)$ is then translation invariant and therefore constant, for its degree is less than q, and h is also translation invariant of degree $\deg(f) - q$. By induction, there is $t \in k[x]$ such that $h(x) = t(x^q - x)$. So we take $g(x) = xt(x) + r$. $\square$

Let $f \in k[x]$ be translation invariant, and $g \in k[x]$ of degree d such that $f(x) = g(x^q - x)$. Let $\psi : k \to \bar{\mathbb{Q}}_\ell^\star$ be a non-trivial additive character. The Artin-Scheier-reduced degree of f (i.e. the lowest degree of a polynomial which is Artin-Schreier equivalent to f) is $q(d-1)+1$ (since $g(x^q - x) = a_d x^{qd} + d a_d x^{q(d-1)+1} +$ (terms of degree $\leq q(d-1)$)). Therefore the Weil bound for exponential sums gives

$$\left| \sum_{x \in k_r} \psi(\mathrm{Tr}_{k_r/k}(f(x))) \right| \leq q(d-1)q^{\frac{r}{2}} = (d-1)q^{\frac{r}{2}+1}$$

On the other hand, since $f(x)=g(x^q-x)$ we get, for every $r\geq 1$,

$$\sum_{x\in k_r}\psi(\mathrm{Tr}_{k_r/k}(f(x)))=\sum_{x\in k_r}\psi(\mathrm{Tr}_{k_r/k}(g(x^q-x)))=$$
$$=\sum_{t\in k_r}\#\{x\in k_r|x^q-x=t\}\psi(\mathrm{Tr}_{k_r/k}(g(t)))=$$
$$=\sum_{t\in k_r}\sum_{u\in k}\psi(u\mathrm{Tr}_{k_r/k}(t))\psi(\mathrm{Tr}_{k_r/k}(g(t)))=\sum_{u\in k}\sum_{t\in k_r}\psi(\mathrm{Tr}_{k_r/k}(g(t)+ut)).$$

Consider the $\bar{\mathbb{Q}}_\ell$-sheaf $\mathcal{L}_{\psi(g)}:=g^\star\mathcal{L}_\psi$ on $\mathbb{A}^1_k$, where $\mathcal{L}_\psi$ is the Artin-Schreier sheaf associated to ψ. The Fourier transform of the object $\mathcal{L}_{\psi(g)}[1]$ with respect to ψ [**13**] is a single sheaf $\mathcal{F}_g$ placed in degree -1. The sheaf $\mathcal{F}_g$ is irreducible and smooth of rank $d-1$ on $\mathbb{A}^1_k$, and totally wild at infinity with a single slope $\frac{d}{d-1}$ and Swan conductor d [**7**, Theorem 17]. We have

$$\sum_{x\in k_r}\psi(\mathrm{Tr}_{k_r/k}(f(x)))=\sum_{x\in k_r}\psi(\mathrm{Tr}_{k_r/k}(g(x^q-x)))=\sum_{u\in k}\sum_{t\in k_r}\psi(\mathrm{Tr}_{k_r/k}(g(t)+ut))=$$
$$=-\sum_{u\in k}\mathrm{Tr}(\mathrm{Frob}^r_{k,u}|(\mathcal{F}_g)_u)=-\sum_{u\in k}\mathrm{Tr}(\mathrm{Frob}_{k,u}|[\mathcal{F}_g]^r_u)\tag{1}$$

where $[\mathcal{F}_g]^r$ is the r-th Adams power of $\mathcal{F}_g$ [**4**].

Let $g(x)=\sum_{i=0}^d a_ix^i$. Recall the following facts about the local and global monodromies of the sheaf $\mathcal{F}_g$:

(1) Suppose that $p>d$ and k contains all $2(d-1)$-th roots of $-da_d$. Let $u(t)=\sum_{i\geq 0}r_it^{1-i}\in tk[[t^{-1}]]$ be a power series such that $f'(t)+u(t)^{d-1}=0$ and let $v(t)=\sum_{i\geq 0}s_it^{1-i}$ be the inverse image of t under the automorphism $k((t^{-1}))\to k((t^{-1}))$ defined by $t^{-1}\mapsto u(t)^{-1}$ (cf. [**3**, Proposition 3.1]). Let $h(t)=\sum_{i=0}^d b_it^i$ be the polynomial obtained from $f(v(t))+v(t)t^{d-1}\in t^dk[[t^{-1}]]$ by removing the terms with negative exponent. Then, as a representation of the decomposition group D_∞ at infinity, we have

$$\mathcal{F}_g\cong[d-1]_\star(\mathcal{L}_{\psi(h(t))}\otimes\mathcal{L}_{\rho^d(s_0t)})\otimes\rho(d(d-1)a_d/2)^{deg}\otimes g(\rho,\psi)^{deg}$$

where $\rho:k^\star\to\bar{\mathbb{Q}}_\ell^\star$ is the quadratic character, $g(\rho,\psi)=-\sum_{t\in k}\rho(t)\psi(t)$ the corresponding Gauss sum and $[d-1]_\star:\mathbb{G}_{m,k}\to\mathbb{G}_{m,k}$ the $(d-1)$-th power map [**5**, Equation 4]. Notice that $s_0^{d-1}=-1/da_d$.

(2) Suppose that $p>2$, and let $G\subseteq\mathrm{GL}(V)$ be the geometric monodromy group of $\mathcal{F}_g$, where V is its stalk at a geometric generic point. Then by [**16**, Propositions 11.1 and 11.6], either G is finite or G_0 (the unit connected component of G) is $\mathrm{SL}(V)$ or $\mathrm{Sp}(V)$ in its standard representation. By [**7**, proof of Theorem 19], for $p>d$ the Sp case occurs if and only if $g(x+c)+d$ is odd for some $c,d\in k$. Moreover for $p>2d-1$ G is never finite by [**7**, Theorem 19]. See [**5**, Section 2] for some other criterions that rule out the finite monodromy case in the $p\leq 2d-1$ case.

The determinant of $\mathcal{F}_g$ is computed over $\bar{k}$ in [**7**, Theorem 17]. In order to obtain a good estimate in the exceptional case below, we need to find its value over k.

LEMMA 2.2. *Suppose that $p > d$ and k contains all $2(d-1)$-th roots of $-da_d$. Then*

$$\det \mathcal{F}_g \cong \mathcal{L}_{\psi((d-1)b_{d-1}t+(d-1)b_0)} \otimes \rho^d(-1)^{deg} \otimes \rho^{d-1}(d(d-1)a_d/2)^{deg} \otimes (g(\rho,\psi)^{d-1})^{deg}$$

PROOF. Note that the result is compatible with [**7**, Theorem 17], since $b_{d-1} = a_{d-1}s_0^{d-1} = a_{d-1}/r_0^{d-1} = -a_{d-1}/da_d$ as one can easily check.

Let $D_\infty^{d-1} \subseteq D_\infty$ be the closed subgroup of index $d-1$ which fixes $1/t^{d-1}$. Since k contains all $(d-1)$-th roots of unity, D_∞^{d-1} is normal in D_∞ and the quotient D_∞/D_∞^{d-1} is generated by $t \mapsto \zeta t$, where $\zeta \in k$ is a primitive $(d-1)$-th root of unity. Using the previous description of the representation of D_∞ given by $\mathcal{F}_g$, we get an isomorphism of D_∞^{d-1}-representations

$$[d-1]^\star \mathcal{F}_g \cong$$

$$\cong \left(\bigoplus_{i=0}^{d-2} (t \mapsto \zeta^i t)^\star \mathcal{L}_{\psi(h(t))} \otimes \mathcal{L}_{\rho^d(s_0 t)} \right) \otimes \rho(d(d-1)a_d/2)^{deg} \otimes g(\rho,\psi)^{deg} \cong$$

$$\cong \left(\bigoplus_{i=0}^{d-2} \mathcal{L}_{\psi(h(\zeta^i t))} \otimes \mathcal{L}_{\rho^d(s_0 \zeta^i t)} \right) \otimes \rho(d(d-1)a_d/2)^{deg} \otimes g(\rho,\psi)^{deg}$$

so

$$[d-1]^\star \det \mathcal{F}_g \cong \det[d-1]^\star \mathcal{F}_g \cong$$

$$\cong \left(\bigotimes_{i=0}^{d-2} \mathcal{L}_{\psi(h(\zeta^i t))} \otimes \mathcal{L}_{\rho^d(s_0 \zeta^i t)} \right) \otimes \rho^{d-1}(d(d-1)a_d/2)^{deg} \otimes (g(\rho,\psi)^{d-1})^{deg} \cong$$

$$\cong \mathcal{L}_{\psi(\sum_{i=0}^{d-2} h(\zeta^i t))} \otimes \mathcal{L}_{\rho^d(\prod_{i=0}^{d-2}(s_0\zeta^i t))} \otimes \rho^{d-1}(d(d-1)a_d/2)^{deg} \otimes (g(\rho,\psi)^{d-1})^{deg} \cong$$

$$\cong \mathcal{L}_{\psi((d-1)b_{d-1}t^{d-1}+(d-1)b_0)} \otimes \mathcal{L}_{\rho^d((-1)^d(s_0 t)^{d-1})} \otimes$$
$$\otimes \rho^{d-1}(d(d-1)a_d/2)^{deg} \otimes (g(\rho,\psi)^{d-1})^{deg} \cong$$
$$\cong \mathcal{L}_{\psi((d-1)b_{d-1}t^{d-1}+(d-1)b_0)} \otimes \mathcal{L}_{\rho^{d(d-1)}(-s_0 t)} \otimes$$
$$\otimes \rho^d(-1)^{deg} \otimes \rho^{d-1}(d(d-1)a_d/2)^{deg} \otimes (g(\rho,\psi)^{d-1})^{deg} \cong$$
$$\cong \mathcal{L}_{\psi((d-1)b_{d-1}t^{d-1}+(d-1)b_0)} \otimes \rho^d(-1)^{deg} \otimes \rho^{d-1}(d(d-1)a_d/2)^{deg} \otimes (g(\rho,\psi)^{d-1})^{deg}$$

since $\sum_{i=0}^{d-2}(\zeta^j)^i = 0$ for $(d-1) \not| j$, $d(d-1)$ is even and $\prod_{i=0}^{d-2} \zeta^i = (-1)^d$.

In particular, $[d-1]^\star(\det \mathcal{F}_g)$ and

$$[d-1]^\star \mathcal{L}_{\psi((d-1)b_{d-1}t+(d-1)b_0)} \otimes \rho^d(-1)^{deg} \otimes \rho^{d-1}(d(d-1)a_d/2)^{deg} \otimes (g(\rho,\psi)^{d-1})^{deg}$$

are isomorphic characters of D_∞^{d-1}, so there is some character $\chi : k^\star \to \bar{\mathbb{Q}}_\ell^\star$ with $\chi^{d-1} = \mathbf{1}$ such that

$$\det \mathcal{F}_g \cong \mathcal{L}_\chi \otimes \mathcal{L}_{\psi((d-1)b_{d-1}t+(d-1)b_0)} \otimes$$
$$\otimes \rho^d(-1)^{deg} \otimes \rho^{d-1}(d(d-1)a_d/2)^{deg} \otimes (g(\rho,\psi)^{d-1})^{deg}$$

as representations of D_∞. But then

$$\widehat{(\det \mathcal{F}_g)} \otimes \mathcal{L}_\chi \otimes \mathcal{L}_{\psi((d-1)b_{d-1}t+(d-1)b_0)} \otimes$$
$$\otimes \rho^d(-1)^{deg} \otimes \rho^{d-1}(d(d-1)a_d/2)^{deg} \otimes (g(\rho,\psi)^{d-1})^{deg}$$

is a rank 1 smooth sheaf on $\mathbb{G}_{m,k}$, tamely ramified at 0 and unramified at infinity, so it must be geometrically trivial, that is, χ is trivial (since everything else is unramified at 0). Moreover, since the Frobenius action is trivial at infinity it must be the trivial sheaf. Therefore

$$\det \mathcal{F}_g \cong \mathcal{L}_{\psi((d-1)b_{d-1}t+(d-1)b_0)} \otimes \rho^d(-1)^{deg} \otimes \rho^{d-1}(d(d-1)a_d/2)^{deg} \otimes (g(\rho,\psi)^{d-1})^{deg}$$

as sheaves on $\mathbb{A}^1_{\bar{k}}$. □

PROPOSITION 2.3. *Suppose that $p > d$, the sheaf $\mathcal{F}_g$ does not have finite monodromy (e.g. $p > 2d-1$) and there do not exist $c, d \in k$ such that $g(x+c)+d$ is odd. Then we have an estimate*

$$\left| \sum_{x \in k_r} \psi(\mathrm{Tr}_{k_r/k}(f(x))) \right| \leq C_{d,r} q^{\frac{r+1}{2}}$$

where

$$C_{d,r} = \frac{1}{d-1} \sum_{i=0}^{d-1} |i-1| \binom{d-2+r-i}{r-i} \binom{d-1}{i}$$

unless $a_{d-1} = 0$ and $r = d-1$, in which case there is an estimate

$$\left| \sum_{x \in k_r} \psi(\mathrm{Tr}_{k_r/k}(f(x))) - A \right| \leq C_{d,r} q^{\frac{r+1}{2}}.$$

where $A = (-1)^{d-1} q \cdot \rho^d(-1)(\psi(b_0)\rho(d(d-1)a_d/2)g(\rho,\psi))^{d-1}$.

PROOF. By [**4**, Section 1], we have

$$\sum_{x \in k_r} \psi(\mathrm{Tr}_{k_r/k}(f(x))) = -\sum_{u \in k} \mathrm{Tr}(\mathrm{Frob}_{k,u} | [\mathcal{F}_g]^r_u) =$$

$$= \sum_{i=0}^{r} (-1)^{i-1}(i-1) \mathrm{Tr}(\mathrm{Frob}_k, \mathrm{H}^1_c(\mathbb{A}^1_{\bar{k}}, \mathrm{Sym}^{r-i}\mathcal{F}_g \otimes \wedge^i \mathcal{F}_g)) -$$

$$- \sum_{i=0}^{r} (-1)^{i-1}(i-1) \mathrm{Tr}(\mathrm{Frob}_k, \mathrm{H}^2_c(\mathbb{A}^1_{\bar{k}}, \mathrm{Sym}^{r-i}\mathcal{F}_g \otimes \wedge^i \mathcal{F}_g)).$$

Let $G \subseteq \mathrm{GL}(V)$ be the geometric monodromy group of $\mathcal{F}_g$. Under the hypotheses of the proposition, the unit connected component of G is $\mathrm{SL}(V)$, so G is the inverse image of its image by the determinant. By lemma 2.2, G is $\mathrm{SL}(V)$ if $b_{d-1} = 0$ (if and only if $a_{d-1} = 0$) and $\mathrm{GL}_p(V) = \mu_p \cdot \mathrm{SL}(V)$ (since $p > d$, so p does not divide $d-1$) if $b_{d-1} \neq 0$.

For every i, the dimension of $\mathrm{H}^2_c(\mathbb{A}^1_{\bar{k}}, \mathrm{Sym}^{r-i}\mathcal{F}_g \otimes \wedge^i \mathcal{F}_g)$ is the dimension of the coinvariant (or the invariant) space of the action of G on $\mathrm{Sym}^{r-i}V \otimes \wedge^i V$. By [**15**, Corollary 5], the action of $\mathrm{SL}(V) \subseteq G$ on $\mathrm{Sym}^{r-i}V \otimes \wedge^i V$ has no invariants unless $r = d-1$ and $i = r, r-1$, in which case the invariant space W_i is one-dimensional. If $a_{d-1} \neq 0$, a generator ζ_p of the quotient $G/\mathrm{SL}(V) \cong \mu_p$ acts on W_i via multiplication by ζ_p^{d-1}, which can not be trivial since $p > d$. So the action of G has no invariants on $\mathrm{Sym}^{r-i}V \otimes \wedge^i V$ for any i if $a_{d-1} \neq 0$.

In that case, since $\mathrm{H}^1_c(\mathbb{A}^1_{\bar{k}}, \mathrm{Sym}^{r-i}\mathcal{F}_g \otimes \wedge^i \mathcal{F}_g)$ is mixed of weights $\leq r+1$ we get

$$\left| \sum_{x \in k_r} \psi(\mathrm{Tr}_{k_r/k}(f(x))) \right| \leq \sum_{i=0}^{r} |i-1| \dim \mathrm{H}^1_c(\mathbb{A}^1_{\bar{k}}, \mathrm{Sym}^{r-i}\mathcal{F}_g \otimes \wedge^i \mathcal{F}_g) \cdot q^{\frac{r+1}{2}}.$$

Moreover, by the Ogg-Shafarevic formula we have

$$\begin{aligned}\dim \mathrm{H}^1_c(\mathbb{A}^1_{\bar{k}}, \mathrm{Sym}^{r-i}\mathcal{F}_g \otimes \wedge^i \mathcal{F}_g) &= -\chi(\mathbb{A}^1_{\bar{k}}, \mathrm{Sym}^{r-i}\mathcal{F}_g \otimes \wedge^i \mathcal{F}_g) = \\ &= \mathrm{Swan}_\infty(\mathrm{Sym}^{r-i}\mathcal{F}_g \otimes \wedge^i \mathcal{F}_g) - \mathrm{rank}(\mathrm{Sym}^{r-i}\mathcal{F}_g \otimes \wedge^i \mathcal{F}_g) \leq \\ &\leq \frac{1}{d-1}\mathrm{rank}(\mathrm{Sym}^{r-i}\mathcal{F}_g \otimes \wedge^i \mathcal{F}_g) = \frac{1}{d-1}\binom{d-2+r-i}{r-i}\binom{d-1}{i}\end{aligned}$$

since all slopes at infinity of $\mathcal{F}_g$ (and a fortiori of $\mathrm{Sym}^{r-i}\mathcal{F}_g \otimes \wedge^i \mathcal{F}_g$) are $\leq \frac{d}{d-1}$.

Suppose now that $a_{d-1} = 0$ and $r = d-1$. As in [**15**, Corollary 5], we have

$$\begin{aligned}&\sum_{i=0}^{r}(-1)^{i-1}(i-1)\mathrm{Tr}(\mathrm{Frob}_k, \mathrm{H}^2_c(\mathbb{A}^1_{\bar{k}}, \mathrm{Sym}^{r-i}\mathcal{F}_g \otimes \wedge^i \mathcal{F}_g)) = \\ &= (-1)^r(r-2)\mathrm{Tr}(\mathrm{Frob}_k, \mathrm{H}^2_c(\mathbb{A}^1_{\bar{k}}, \mathrm{Sym}^1\mathcal{F}_g \otimes \wedge^{r-1}\mathcal{F}_g)) + \\ &+(-1)^{r-1}(r-1)\mathrm{Tr}(\mathrm{Frob}_k, \mathrm{H}^2_c(\mathbb{A}^1_{\bar{k}}, \wedge^r \mathcal{F}_g)) = \\ &= (-1)^{r-1}\mathrm{Tr}(\mathrm{Frob}_k, \mathrm{H}^2_c(\mathbb{A}^1_{\bar{k}}, \det \mathcal{F}_g)) = \\ &= (-1)^d q \cdot \psi((d-1)b_0)\rho^d(-1)\rho^{d-1}(d(d-1)a_d/2)g(\rho,\psi)^{d-1} = \\ &= (-1)^d q \cdot \rho^d(-1)(\psi(b_0)\rho(d(d-1)a_d/2)g(\rho,\psi))^{d-1}\end{aligned}$$

by lemma 2.2. We conclude as above using the fact that, for the two values of i for which $\mathrm{H}^2_c(\mathbb{A}^1_{\bar{k}}, \mathrm{Sym}^{r-i}\mathcal{F}_g \otimes \wedge^i \mathcal{F}_g)$ is one-dimensional, the sheaf $\mathrm{Sym}^{r-i}\mathcal{F}_g \otimes \wedge^i \mathcal{F}_g$ has at least one slope equal to 0 at infinity, so

$$\begin{aligned}\dim \mathrm{H}^1_c(\mathbb{A}^1_{\bar{k}}, \mathrm{Sym}^{r-i}\mathcal{F}_g \otimes \wedge^i \mathcal{F}_g) &= 1 - \chi(\mathbb{A}^1_{\bar{k}}, \mathrm{Sym}^{r-i}\mathcal{F}_g \otimes \wedge^i \mathcal{F}_g) = \\ &= 1 + \mathrm{Swan}_\infty(\mathrm{Sym}^{r-i}\mathcal{F}_g \otimes \wedge^i \mathcal{F}_g) - \mathrm{rank}(\mathrm{Sym}^{r-i}\mathcal{F}_g \otimes \wedge^i \mathcal{F}_g) \leq \\ &\leq 1 + \frac{d}{d-1}(\mathrm{rank}(\mathrm{Sym}^{r-i}\mathcal{F}_g \otimes \wedge^i \mathcal{F}_g) - 1) - \mathrm{rank}(\mathrm{Sym}^{r-i}\mathcal{F}_g \otimes \wedge^i \mathcal{F}_g) < \\ &< \frac{1}{d-1}\mathrm{rank}(\mathrm{Sym}^{r-i}\mathcal{F}_g \otimes \wedge^i \mathcal{F}_g) = \frac{1}{d-1}\binom{d-2+r-i}{r-i}\binom{d-1}{i}.\end{aligned}$$

□

PROPOSITION 2.4. *Suppose that $p > d$, the sheaf $\mathcal{F}_g$ does not have finite monodromy (e.g. $p > 2d-1$) and there exist $\alpha, \beta \in k$ such that $g(x+\alpha)+\beta$ is odd (so d is odd). Then we have an estimate*

$$\left|\sum_{x\in k_r}\psi(\mathrm{Tr}_{k_r/k}(f(x)))\right| \leq C_{d,r}q^{\frac{r+1}{2}}$$

where

$$C_{d,r} = \frac{1}{d-1}\sum_{i=0}^{d-1}|i-1|\binom{d-2+r-i}{r-i}\binom{d-1}{i}$$

unless $a_{d-1} = 0$ and $r \leq d-1$ is even, in which case there is an estimate

$$\left|\sum_{x\in k_r}\psi(\mathrm{Tr}_{k_r/k}(f(x))) - (-1)^r\psi(-\beta)^r q^{\frac{r}{2}+1}\right| < C_{d,r}q^{\frac{r+1}{2}}.$$

PROOF. The proof is similar to the previous one. In this case, the unit connected component of G is $\mathrm{Sp}(V)$, so by lemma 2.2 G is $\mathrm{Sp}(V)$ if $b_{d-1} = 0$ (if and only if $a_{d-1} = 0$) and $\mu_p \cdot \mathrm{SL}(V)$ (since $p > d$, so p does not divide $d-1$) if $b_{d-1} \neq 0$.

By [**10**, lemma on p.62], the action of $\mathrm{Sp}(V) \subseteq G$ on $\mathrm{Sym}^{r-i}V \otimes \wedge^i V$ has no invariants unless $r \leq d-1$ is even and $i = r, r-1$, in which case the invariant space W_i is one-dimensional. If $a_{d-1} \neq 0$, a generator ζ_p of the quotient $G/\mathrm{Sp}(V) \cong \mu_p$ acts on W_i via multiplication by ζ_p^{d-1}, which can not be trivial since $p > d$. So the action of G has no invariants on $\mathrm{Sym}^{r-i}V \otimes \wedge^i V$ for any i if $a_{d-1} \neq 0$. We conclude this case as in the previous proposition.

Suppose now that $a_{d-1} = 0$, $r \leq d-1$ is even and $i = r$ or $r-1$. Since the coefficient of x^{d-1} in $g(x)$ is 0, the coefficient in $g(x+\alpha)+\beta$ is $da_d\alpha$, so it can only be an odd polynomial if $\alpha = 0$. That is, $g(x)+\beta$ is odd, or equivalently, $g(-x) = -2\beta - g(x)$. Then the sheaf $\psi(\beta)^{deg} \otimes \mathcal{F}_g(1/2)$ is self-dual: since the dual of $\mathcal{L}_{\psi(g)}$ is $\mathcal{L}_{\psi(-g)}(1)$, using that $D \circ FT_\psi = [-1]^\star FT_\psi \circ D(1)$ [**13**, Corollaire 2.1.5] we get that the dual of $\mathcal{F}_g = \mathcal{H}^{-1}(FT_\psi(\mathcal{L}_{\psi(g)}[1]))$ is

$$[-1]^\star \mathcal{H}^{-1}(FT_\psi \mathcal{L}_{\psi(-g)}(1)) = [-1]^\star \mathcal{F}_{-g}(1) = [-1]^\star \mathcal{F}_{2\beta+g(-x)}(1) = \psi(2\beta)^{deg} \otimes \mathcal{F}_g(1)$$

so $\psi(\beta)^{deg} \otimes \mathcal{F}_g(1/2)$ is self-dual (symplectically, since it is so geometrically by [**7**, Theorem 19]). In particular, the one-dimensional $\mathrm{Sp}(V)$-invariant subspace of $(\mathrm{Sym}^{r-i}\mathcal{F}_g \otimes \wedge^i \mathcal{F}_g) \otimes \psi(\beta)^{r \cdot deg}(r/2)$ is also invariant under all Frobenii. So W_i is in fact the geometrically constant sheaf $\psi(-\beta)^{r \cdot deg}(-r/2)$. In particular

$$\begin{aligned}
&\sum_{i=0}^{r}(-1)^{i-1}(i-1)\mathrm{Tr}(\mathrm{Frob}_k, \mathrm{H}^2_c(\mathbb{A}^1_{\bar{k}}, \mathrm{Sym}^{r-i}\mathcal{F}_g \otimes \wedge^i \mathcal{F}_g)) = \\
&= (-1)^r(r-2)\mathrm{Tr}(\mathrm{Frob}_k, \mathrm{H}^2_c(\mathbb{A}^1_{\bar{k}}, \mathrm{Sym}^1\mathcal{F}_g \otimes \wedge^{r-1}\mathcal{F}_g)) + \\
&\quad + (-1)^{r-1}(r-1)\mathrm{Tr}(\mathrm{Frob}_k, \mathrm{H}^2_c(\mathbb{A}^1_{\bar{k}}, \wedge^r \mathcal{F}_g)) = \\
&= (-1)^{r-1}\mathrm{Tr}(\mathrm{Frob}_k, \mathrm{H}^2_c(\mathbb{A}^1_{\bar{k}}, \psi(-\beta)^{r \cdot deg}(-r/2))) = (-1)^{r-1}\psi(-\beta)^r q^{\frac{r}{2}+1}.
\end{aligned}$$

We conclude as in the previous proposition. □

3. Multiplicative character sums for translation invariant polynomials

Let $f \in k[x]$ be translation invariant, and $g \in k[x]$ of degree d such that $f(x) = g(x^q - x)$. Let $\chi : k^\star \to \bar{\mathbb{Q}}_\ell^\star$ a non-trivial multiplicative character of order m, extended by zero to all of k. Since f has degree qd, Weil's bound gives in this case

$$\left| \sum_{x \in k_r} \chi(\mathrm{N}_{k_r/k}(f(x))) \right| \leq (qd-1)q^{\frac{r}{2}}.$$

On the other hand we have, for every $r \geq 1$,

$$\begin{aligned}
\sum_{x \in k_r} \chi(\mathrm{N}_{k_r/k}(f(x))) &= \sum_{x \in k_r} \chi(\mathrm{N}_{k_r/k}(g(x^q - x))) = \\
&= \sum_{t \in k_r} \#\{x \in k_r | x^q - x = t\} \chi(\mathrm{N}_{k_r/k}(g(t))) = \\
= \sum_{t \in k_r}\sum_{u \in k} \psi(u\mathrm{Tr}_{k_r/k}(t))\chi(\mathrm{N}_{k_r/k}(g(t))) &= \sum_{u \in k}\sum_{t \in k_r} \psi(u\mathrm{Tr}_{k_r/k}(t))\chi(\mathrm{N}_{k_r/k}(g(t))).
\end{aligned}$$

Consider the $\bar{\mathbb{Q}}_\ell$-sheaf $\mathcal{L}_{\chi(g)} := g^\star \mathcal{L}_\chi$ on $\mathbb{A}^1_k$, where $\mathcal{L}_\chi$ is the Kummer sheaf on $\mathbb{G}_{m,k}$ associated to χ [**1**, 1.7], extended by zero to $\mathbb{A}^1_k$. Suppose that g is square-free and its degree d is prime to p. Then $\mathcal{L}_{\chi(g)}$ is an irreducible middle extension sheaf, smooth on the complement of the subscheme $Z \subseteq \mathbb{A}^1_k$ defined by $g = 0$. Since there is at least one point where it is not smooth, it is not isomorphic to an Artin-Schreier sheaf and therefore the Fourier transform of $\mathcal{L}_{\chi(g)}[1]$ is a single irreducible middle extension sheaf $\mathcal{F}_g$ placed in degree -1 [**8**, 8.2]. We have

$$\sum_{x\in k_r} \chi(\mathrm{N}_{k_r/k}(f(x))) = \sum_{x\in k_r} \chi(\mathrm{N}_{k_r/k}(g(x^q - x))) =$$

$$= \sum_{u\in k} \sum_{t\in k_r} \psi(u\mathrm{Tr}_{k_r/k}(t))\chi(\mathrm{N}_{k_r/k}(g(t))) =$$

$$= -\sum_{u\in k} \mathrm{Tr}(\mathrm{Frob}^r_{k,u}|(\mathcal{F}_g)_u) = -\sum_{u\in k} \mathrm{Tr}(\mathrm{Frob}_{k,u}|[\mathcal{F}_g]^r_u) \tag{2}$$

where $[\mathcal{F}_g]^r$ is the r-th Adams power of $\mathcal{F}_g$.

PROPOSITION 3.1. *The sheaf $\mathcal{F}_g$ has generic rank d, it is smooth on $\mathbb{G}_{m,k}$ and tamely ramified at 0. Its rank at 0 is $d-1$. If all roots of g are in k, the action of the decomposition group D_∞ on the generic stalk of $\mathcal{F}_g$ splits as a direct sum $\bigoplus_a \chi(g'(a))^{deg} \otimes g(\chi,\psi)^{deg} \otimes \mathcal{L}_{\bar{\chi}} \otimes \mathcal{L}_{\psi_a}$ where the sum is taken over the roots of f, $\mathcal{L}_{\psi_a}$ is the Artin-Schreier sheaf corresponding to the character $t \mapsto \psi(at)$ and $g(\chi,\psi) = -\sum_t \chi(t)\psi(t)$ if the Gauss sum.*

PROOF. The generic rank of $\mathcal{F}_g$ is the dimension of $\mathrm{H}^1_c(\mathbb{A}^1_{\bar{k}}, \mathcal{L}_{\chi(g)} \otimes \mathcal{L}_{\psi_z})$ for generic z. Since $\mathcal{L}_{\chi(g)}$ is tamely ramified everywhere and has rank one, for any $z \neq 0$ $\mathcal{L}_{\chi(g)} \otimes \mathcal{L}_{\psi_z}$ is tamely ramified at every point of $\mathbb{A}^1_{\bar{k}}$ and totally wild at infinity with Swan conductor 1. In particular its H^i_c vanish for $i \neq 1$. By the Ogg-Shafarevic formula, its Euler characteristic is then $1-d-1=-d$, since there are d points in $\mathbb{A}^1_{\bar{k}}$ where the stalk is zero. Therefore $\dim \mathrm{H}^1_c(\mathbb{A}^1_{\bar{k}}, \mathcal{L}_{\chi(g)} \otimes \mathcal{L}_{\psi_z}) = d$ for every $z \neq 0$. Similarly, it is $d-1$ for $z=0$. Since $\mathcal{F}_g$ is a middle extension, it is smooth exactly on the open set where the rank is maximal, so it is smooth on $\mathbb{G}_{m,k}$. It is tamely ramified at zero, since $\mathcal{L}_{\chi(g)}$ is tamely ramified at infinity [**14**, Théorème 2.4.3].

Suppose now that all roots of g are in k, and let a be one such root. In an étale neighborhood of a, the sheaf $\mathcal{L}_{\chi(g)}$ is isomorphic to $\mathcal{L}_{\chi(g'(a)(x-a))} = \chi(g'(a))^{deg} \otimes \mathcal{L}_{\chi(x-a)}$, since $g(x) = g'(a)(x-a)\frac{g(x)}{g'(a)(x-a)}$ and $\frac{g(x)}{g'(a)(x-a)}$ is an m-th power in the henselization of $\mathbb{A}^1_{\bar{k}}$ at a (since its image in the residue field is 1). Applying Laumon's local Fourier transform [**14**, Proposition 2.5.3.1] and using that Fourier transform commutes with tensoring by unramified sheaves, we deduce that the D_∞-representation $\mathcal{F}_g$ contains $(LFT^{(0,\infty)}_\psi \chi(g'(a))^{deg} \otimes \mathcal{L}_\chi) \otimes \mathcal{L}_{\psi_a} = \chi(g'(a))^{deg} \otimes g(\chi,\psi)^{deg} \otimes \mathcal{L}_{\bar{\chi}} \otimes \mathcal{L}_{\psi_a}$ as a direct summand. Since g has d distinct roots we obtain d different terms this way, which is the rank of $\mathcal{F}_g$, so its monodromy at ∞ is the direct sum of these terms. □

Define by induction the sequence of polynomials $g_n(x) \in k[x]$ for $n \geq 1$ by: $g_1(x) = g(x)$, and for $n \geq 1$ $g_{n+1}(x)$ is the resultant in t of $g_n(t)$ and $g(x-t)$.

COROLLARY 3.2. *Suppose that either m does not divide r or $g_r(0) \neq 0$. Then we have an estimate*

$$\left| \sum_{x\in k_r} \chi(\mathrm{N}_{k_r/k}(f(x))) \right| \leq C_{d,r} q^{\frac{r+1}{2}}$$

where

$$C_{d,r} = \sum_{i=0}^{r} |i-1| \left(\binom{d-1+r-i}{r-i}\binom{d}{i} - \binom{d-2+r-i}{r-i}\binom{d-1}{i} \right).$$

PROOF. By the previous proposition, the action of the inertia group I_∞ on $\mathcal{F}_g^{\otimes r}$ splits as a direct sum over the r-uples of roots of f

$$\bigoplus_{(a_1,\ldots,a_r)} \mathcal{L}_{\bar\chi}^{\otimes r} \otimes \mathcal{L}_{\psi_{a_1}} \otimes \cdots \otimes \mathcal{L}_{\psi_{a_r}} = \bigoplus_{(a_1,\ldots,a_r)} \mathcal{L}_{\bar\chi}^{\otimes r} \otimes \mathcal{L}_{\psi_{a_1+\cdots+a_r}}.$$

For each $(a_1,\ldots,a_r)$, the character $\mathcal{L}_{\bar\chi}^{\otimes r} \otimes \mathcal{L}_{\psi_{a_1+\cdots+a_r}}$ is trivial if and only if both $\mathcal{L}_{\bar\chi}^{\otimes r}$ and $\mathcal{L}_{\psi_{a_1+\cdots+a_r}}$ are trivial, that is, if and only if m divides r and $a_1+\cdots+a_r = 0$. Under the hypotheses of the corollary, at least one of these conditions does not hold (since the sums $a_1+\cdots+a_r$ are the roots of g_r). So $\mathcal{F}_g^{\otimes r}$ has no invariants under the action of I_∞ and, a fortiori, under the action of the larger group $\pi_1(\mathbb{G}_{m,\bar k},\bar\eta)$. Since $\mathrm{Sym}^{r-i}\mathcal{F}_g \otimes \wedge^i \mathcal{F}_g$ is a subsheaf of $\mathcal{F}_g^{\otimes r}$ for every i, we conclude that $\mathrm{H}^2_c(\mathbb{A}^1_{\bar k}, \mathrm{Sym}^{r-i}\mathcal{F}_g \otimes \wedge^i\mathcal{F}_g) = 0$ for every $i = 0,\ldots,r$. Therefore

$$\sum_{x\in k_r} \chi(\mathrm{N}_{k_r/k}(f(x))) = -\sum_{u\in k} \mathrm{Tr}(\mathrm{Frob}_{k,u}|[\mathcal{F}_g]^r_u) =$$
$$= \sum_{i=0}^{r} (-1)^{i-1}(i-1)\mathrm{Tr}(\mathrm{Frob}_k, \mathrm{H}^1_c(\mathbb{A}^1_{\bar k}, \mathrm{Sym}^{r-i}\mathcal{F}_g \otimes \wedge^i\mathcal{F}_g)).$$

Since $\mathrm{H}^1_c(\mathbb{A}^1_{\bar k}, \mathrm{Sym}^{r-i}\mathcal{F}_g \otimes \wedge^i\mathcal{F}_g)$ is mixed of weights $\leq r+1$, we get

$$\left|\sum_{x\in k_r} \chi(\mathrm{N}_{k_r/k}(f(x)))\right| \leq \sum_{i=0}^{r} |i-1| \dim \mathrm{H}^1_c(\mathbb{A}^1_{\bar k}, \mathrm{Sym}^{r-i}\mathcal{F}_g \otimes \wedge^i\mathcal{F}_g) \cdot q^{\frac{r+1}{2}}.$$

And by the Ogg-Shafarevic formula, we have

$$\dim \mathrm{H}^1_c(\mathbb{A}^1_{\bar k}, \mathrm{Sym}^{r-i}\mathcal{F}_g \otimes \wedge^i\mathcal{F}_g) = -\chi(\mathbb{A}^1_{\bar k}, \mathrm{Sym}^{r-i}\mathcal{F}_g \otimes \wedge^i\mathcal{F}_g) =$$
$$= \mathrm{Swan}_\infty(\mathrm{Sym}^{r-i}\mathcal{F}_g \otimes \wedge^i\mathcal{F}_g) - \mathrm{rank}_0(\mathrm{Sym}^{r-i}\mathcal{F}_g \otimes \wedge^i\mathcal{F}_g) \leq$$
$$\leq \binom{d-1+r-i}{r-i}\binom{d}{i} - \binom{d-2+r-i}{r-i}\binom{d-1}{i}$$

by the previous proposition, since $\mathcal{F}_g$ is smooth on $\mathbb{G}_{m,k}$, tamely ramified at 0 and all its slopes at infinity (and thus all slopes of of $\mathrm{Sym}^{r-i}\mathcal{F}_g \otimes \wedge^i\mathcal{F}_g$) are ≤ 1. □

COROLLARY 3.3. *If all roots of* $g(x) = \sum_{i=0}^d a_i x^i$ *are in* k, *the determinant of* $\mathcal{F}_g$ *is* $\chi((-1)^{d(d-1)/2}a_d^{-(d-2)}\mathrm{disc}(g))^{deg} \otimes (g(\chi,\psi)^d)^{deg} \otimes \mathcal{L}_{\bar\chi^d} \otimes \mathcal{L}_{\psi_{-a_{d-1}/a_d}}$.

PROOF. By proposition 3.1, the action of D_∞ on the determinant of $\mathcal{F}_g$ is given by

$$\mathcal{G} := \bigotimes_a \chi(g'(a))^{deg} \otimes g(\chi,\psi)^{deg} \otimes \mathcal{L}_{\bar\chi} \otimes \mathcal{L}_{\psi_a} =$$
$$= \chi(\prod_a g'(a))^{deg} \otimes (g(\chi,\psi)^d)^{deg} \otimes \mathcal{L}_{\bar\chi^d} \otimes \mathcal{L}_{\psi_{\sum a}}$$

where the product is taken over the roots of g. Now $\sum a = -a_{d-1}/a_d$, and

$$\prod_a g'(a) = \prod_a a_d \prod_{g(b)=0, b\neq a} (b-a) =$$

$$= a_d^d \prod_{g(a)=g(b)=0, a\neq b} (a-b) = (-1)^{d(d-1)/2} a_d^{-(d-2)} \mathrm{disc}(g).$$

Therefore $\det(\mathcal{F}_g)\otimes\hat{\mathcal{G}}$ is smooth on $\mathbb{G}_{m,k}$, tamely ramified at zero and unramified at infinity, so it is geometrically constant. Looking at the Frobenius action at 0, it must be the constant sheaf $\bar{\mathbb{Q}}_\ell$. We conclude that $\det(\mathcal{F}_g)\cong\mathcal{G}$. □

PROPOSITION 3.4. *Let $h(x) = g(x - \frac{a_{d-1}}{da_d})$. Suppose that $p > 2d+1$ and h is not odd (for d odd) or even (for d even). Then the geometric monodromy group G of $\mathcal{F}_g$ is $\mathrm{GL}_{sp}(V)$ if $a_{d-1}\neq 0$ and $\mathrm{GL}_s(V)$ if $a_{d-1}=0$, where V is the geometric generic stalk of $\mathcal{F}_g$ and s is the order of χ^d.*

PROOF. Since $\mathcal{L}_{\chi(g)}$ is the translate of $\mathcal{L}_{\chi(h)}$ by $a := \frac{a_{d-1}}{da_d}$, we have $\mathcal{F}_g = \mathcal{F}_h\otimes\mathcal{L}_{\psi_a}$. If G (respectively G') is the geometric monodromy group of $\mathcal{F}_g$ (resp. $\mathcal{F}_h$), we have then $G\subseteq\mu_p\cdot G'$ and $G'\subseteq\mu_p\cdot G$. In particular, the unit connected components G_0 and G'_0 are the same. Since $\mathcal{F}_g$ is pure, G_0 is a semisimple group [**2**, Corollaire 1.3.9], so by [**9**, Theorem 7.6.3.1], $\mathcal{F}_g$ is Lie-irreducible and G_0 is one of $\mathrm{SL}(V)$, $\mathrm{Sp}(V)$ (only possible if $\chi^d=\mathbf{1}$) or $\mathrm{SO}(V)$ (only possible if χ^d has order 2). We will see that, under the given hypotheses, the last two options are not possible.

By corollary 3.3, the determinant of $\mathcal{F}_h$ is geometrically isomorphic to $\mathcal{L}_{\bar{\chi}^d}$. By [**7**, Proposition 6], the factor group G'/G'_0 is cyclic of finite prime to p order. In particular, there exists some prime to p integer e such that the geometric monodromy group of the pull-back $[e]^\star\mathcal{F}_h$ is in G'_0, where $[e]:\mathbb{G}_{m,k}\to\mathbb{G}_{m,k}$ is the e-th power map. If $G'_0=\mathrm{Sp}(V)$ or $\mathrm{SO}(V)$, $[e]^\star\mathcal{F}_h$ would then be geometrically self-dual. By proposition 3.1, its restriction to the inertia group I_∞ is the direct sum of $[e]^\star\mathcal{L}_{\psi_b}\otimes\mathcal{L}_{\bar{\chi}^e}$ taken over the roots b of h. Its dual is then the direct sum on $[e]^\star\mathcal{L}_{\psi_{-b}}\otimes\mathcal{L}_{\chi^e}$. Given that the dual of $[e]^\star\mathcal{L}_{\psi_b}$ is $[e]^\star\mathcal{L}_{\psi_{-b}}$, in order for this to be self-dual as a representation of I_∞ a necessary condition is that the set of roots of h is symmetric with respect to 0, that is, that h is either even or odd (since it is a priori square-free).

So, if h is neither even nor odd, G_0 is $\mathrm{SL}(V)$. Then G is $\mathrm{GL}_n(V)$, where n is the geometric order of the determinant of $\mathcal{F}_g$. By corollary 3.3, this order is sp if $a_{d-1}\neq 0$ and s if $a_{d-1}=0$. □

COROLLARY 3.5. *Let $h(x) = g(x-\frac{a_{d-1}}{da_d})$. Suppose that $p>2d+1$ and h is not odd (for d odd) or even (for d even). Then we have an estimate*

$$\left|\sum_{x\in k_r}\chi(\mathrm{N}_{k_r/k}(f(x)))\right| \leq C_{d,r}q^{\frac{r+1}{2}}$$

where

$$C_{d,r} = \sum_{i=0}^{r}|i-1|\left(\binom{d-1+r-i}{r-i}\binom{d}{i} - \binom{d-2+r-i}{r-i}\binom{d-1}{i}\right)$$

unless $r = d$, χ^d is trivial and $a_{d-1} = 0$, in which case there exists an ℓ-adic unit $\beta \in \bar{\mathbb{Q}}_\ell$ with $|\beta| = q^{\frac{d}{2}}$ such that

$$\left| \sum_{x \in k_r} \chi(\mathrm{N}_{k_r/k}(f(x))) - (-1)^d q\beta \right| \leq C_{d,r} q^{\frac{r+1}{2}}.$$

If k contains all roots of g, then $\beta = \chi((-1)^{d(d-1)/2} a_d^{-(d-2)} \mathrm{disc}(g)) g(\chi, \psi)^d$.

Proof. By the previous proposition, the monodromy group G of $\mathcal{F}_g$ is $\mathrm{GL}_{sp}(V)$ if $a_{d-1} \neq 0$ and $\mathrm{GL}_s(V)$ if $a_{d-1} = 0$. We proceed as in the proof of proposition 2.3: G_0 has no invariants on $\mathrm{Sym}^{r-i} V \otimes \wedge^i V$ unless $r = d$ and $i = r, r-1$, in which case the invariant space is one-dimensional and G acts on it via multiplication by the determinant. So the action of G does not have invariants unless $a_{d-1} = 0$ and χ^d is trivial (i.e. $m|d$) by corollary 3.3. In that case we obtain the estimate as in 2.3, using the value for $C_{d,r}$ computed in corollary 3.2.

In the exceptional case, we have again

$$\sum_{i=0}^{r} (-1)^{i-1} (i-1) \mathrm{Tr}(\mathrm{Frob}_k, \mathrm{H}^2_c(\mathbb{A}^1_{\bar{k}}, \mathrm{Sym}^{r-i} \mathcal{F}_g \otimes \wedge^i \mathcal{F}_g)) =$$
$$= (-1)^{r-1} \mathrm{Tr}(\mathrm{Frob}_k, \mathrm{H}^2_c(\mathbb{A}^1_{\bar{k}}, \det \mathcal{F}_g)).$$

Now $\det \mathcal{F}_g$ is geometrically constant of weight d, so there exists an ℓ-adic unit β with $|\beta| = 1$ such that $\det \mathcal{F}_g = (\beta q^{\frac{d}{2}})^{deg}$. Then $\mathrm{Tr}(\mathrm{Frob}_k, \mathrm{H}^2_c(\mathbb{A}^1_{\bar{k}}, \det \mathcal{F}_g)) = \beta q^{\frac{d}{2}+1}$. If k contains all roots of g, the value of β is given in corollary 3.3. We conclude as in proposition 2.3 using that, for the two values of i for which $\mathrm{H}^2_c(\mathbb{A}^1_{\bar{k}}, \mathrm{Sym}^{r-i} \mathcal{F}_g \otimes \wedge^i \mathcal{F}_g)$ is one-dimensional, the sheaf $\mathrm{Sym}^{r-i} \mathcal{F}_g \otimes \wedge^i \mathcal{F}_g$ has at least one slope equal to 0 at infinity, so

$$\dim \mathrm{H}^1_c(\mathbb{A}^1_{\bar{k}}, \mathrm{Sym}^{r-i} \mathcal{F}_g \otimes \wedge^i \mathcal{F}_g) = 1 - \chi(\mathbb{A}^1_{\bar{k}}, \mathrm{Sym}^{r-i} \mathcal{F}_g \otimes \wedge^i \mathcal{F}_g) =$$
$$= 1 + \mathrm{Swan}_\infty(\mathrm{Sym}^{r-i} \mathcal{F}_g \otimes \wedge^i \mathcal{F}_g) - \mathrm{rank}_0(\mathrm{Sym}^{r-i} \mathcal{F}_g \otimes \wedge^i \mathcal{F}_g) \leq$$
$$\leq \mathrm{gen.rank}(\mathrm{Sym}^{r-i} \mathcal{F}_g \otimes \wedge^i \mathcal{F}_g) - \mathrm{rank}_0(\mathrm{Sym}^{r-i} \mathcal{F}_g \otimes \wedge^i \mathcal{F}_g) =$$
$$= \binom{d-1+r-i}{r-i} \binom{d}{i} - \binom{d-2+r-i}{r-i} \binom{d-1}{i}.$$

□

4. Additive character sums for homothety invariant polynomials

Let $f \in k_r[x]$ be a polynomial and $e|q-1$ an integer. Let $\Gamma_e \subseteq k^\star$ be the unique subgroup of $k^\star$ of index e. We say that f is *Γ_e-homothety invariant* if $f(\lambda x) = f(x)$ for every $\lambda \in \Gamma_e$. Equivalently, if $f(\lambda^e x) = f(x)$ for every $\lambda \in k^\star$. An argument similar to that in lemma 2.1 shows

Lemma 4.1. *Let $f \in k_r[x]$ and $e|q-1$. The following conditions are equivalent:*

(a) *f is Γ_e-homothety invariant.*
(b) *There exists $g \in k_r[x]$ such that $f(x) = g(x^{\frac{q-1}{e}})$.*

Let $f \in k_r[x]$ be Γ_e-homothety invariant, $g \in k_r[x]$ of degree d such that $f(x) = g(x^{\frac{q-1}{e}})$ and $\psi : k \to \bar{\mathbb{Q}}_\ell^\star$ a non-trivial additive character. Weil's bound gives in this case

$$\left|\sum_{x\in k_r}\psi(\mathrm{Tr}_{k_r/k}(f(x)))\right|\leq\left(\frac{d(q-1)}{e}-1\right)q^{\frac{r}{2}}.$$

On the other hand,

$$\sum_{x\in k_r}\psi(\mathrm{Tr}_{k_r/k}(f(x)))=\psi(\mathrm{Tr}_{k_r/k}(f(0)))+\sum_{x\in k_r^\star}\psi(\mathrm{Tr}_{k_r/k}(g(x^{\frac{q-1}{e}})))=$$

$$=\psi(\mathrm{Tr}_{k_r/k}(f(0)))+\frac{q-1}{e}\sum_{\mathrm{N}_{k_r/k}(x)^e=1}\psi(\mathrm{Tr}_{k_r/k}(g(x)))=$$

$$=\psi(\mathrm{Tr}_{k_r/k}(f(0)))+\frac{q-1}{e}\sum_{\mu^e=1}\sum_{\mathrm{N}_{k_r/k}(x)=\mu}\psi(\mathrm{Tr}_{k_r/k}(g(x))). \tag{3}$$

For each μ, we will estimate the sum $\sum_{\mathrm{N}_{k_r/k}(x)=\mu}\psi(\mathrm{Tr}_{k_r/k}(g(x)))$ using Weil descent. Fix a basis $\{\alpha_1,\dots,\alpha_r\}$ of k_r over k, and let

$$P(x_1,\dots,x_r)=\prod_\sigma(\sigma(\alpha_1)x_1+\cdots+\sigma(\alpha_r)x_r),$$

where the product is taken over all $\sigma\in\mathrm{Gal}(k_r/k)$. Since P is $\mathrm{Gal}(k_r/k)$-invariant, its coefficients are in k. By construction, for every $(x_1,\dots,x_r)\in k^r$ we have $P(x_1,\dots,x_r)=\mathrm{N}_{k_r/k}(\alpha_1x_1+\cdots+\alpha_rx_r)$. Therefore

$$\sum_{\mathrm{N}_{k_r/k}(x)=\mu}\psi(\mathrm{Tr}_{k_r/k}(g(x)))=\sum_{P(x_1,\dots,x_r)=\mu}\psi(\mathrm{Tr}_{k_r/k}(g(\alpha_1x_1+\cdots+\alpha_rx_r))=$$

$$=\sum_{P(x_1,\dots,x_r)=\mu}\psi\left(\sum_\sigma g^\sigma(\sigma(\alpha_1)x_1+\cdots+\sigma(\alpha_r)x_r)\right)$$

where g^σ is the polynomial obtained by applying σ to the coefficients of g, and the sum is taken over all r-tuples $(x_1,\dots,x_r)\in k^r$ such that $P(x_1,\dots,x_r)=\mu$. By Grothendieck's trace formula, we get

$$\sum_{\mathrm{N}_{k_r/k}(x)=\mu}\psi(\mathrm{Tr}_{k_r/k}(g(x)))=\sum_{i=0}^{2r-2}\mathrm{Tr}(\mathrm{Frob}_k|\mathrm{H}^i_c(V_\mu\otimes\bar{k},\mathcal{L}_{\psi(G)})) \tag{4}$$

where V_μ is the hypersurface defined in $\mathbb{A}^r_k$ by the equation $P(x_1,\dots,x_r)=\mu$ and $G=\sum_\sigma g^\sigma(\sigma(\alpha_1)x_1+\cdots+\sigma(\alpha_r)x_r)\in k[x]$ (since it is $\mathrm{Gal}(k_r/k)$-invariant).

PROPOSITION 4.2. *Suppose that g has degree d prime to p. For any $\mu\in k^\star$, $\mathrm{H}^i_c(V_\mu\otimes\bar{k},\mathcal{L}_{\psi(G)})=0$ for $i\neq r-1$ and $\dim\mathrm{H}^{r-1}_c(V_\mu\otimes\bar{k},\mathcal{L}_{\psi(G)})=rd^{r-1}$.*

PROOF. Over k_r, the map $(x_1,\dots,x_r)\mapsto(\sigma(\alpha_1)x_1+\cdots+\sigma(\alpha_r)x_r)_{\sigma\in\mathrm{Gal}(k_r/k)}$ is a (linear) isomorphism between $\mathbb{A}^r_{k_r}$ and $\mathbb{A}^{\mathrm{Gal}(k_r/k)}_{k_r}$. The pull-back of P under this automorphism is just $x_1\cdots x_r$. So $V_\mu\otimes\bar{k}$ is isomorphic to the hypersurface $x_1\cdots x_r=\mu$, and the sheaf $\mathcal{L}_{\psi(G)}$ corresponds under this isomorphism to the sheaf $\mathcal{L}_{\psi(\sum_\sigma g^\sigma(x_\sigma))}=\boxtimes_\sigma\mathcal{L}_{\psi(g^\sigma)}$ where $\mathcal{L}_{\psi(g^\sigma)}$ is the pull-back of the Artin-Schreier sheaf $\mathcal{L}_\psi$ by g^σ.

For every $\sigma\in\mathrm{Gal}(k_r/k)$, the sheaf $\mathcal{L}_{\psi(g^\sigma)}$ is smooth on $\mathbb{A}^1_{\bar{k}}$ of rank one, with slope d at infinity. [**8**, Theorem 5.1] shows that the class of objects of the form $\mathcal{G}[1]$ where $\mathcal{G}$ is a smooth $\bar{\mathbb{Q}}_\ell$-sheaf on $\mathbb{G}_{m,\bar{k}}$, tamely ramified at 0 and totally wild at infinity is invariant under convolution. In particular, if $m : \mathbb{G}_{m,\bar{k}}^{\mathrm{Gal}(k_r/k)} \to \mathbb{G}_{m,\bar{k}}$ is the multiplication map, $\mathrm{R}^i m_!(\boxtimes_\sigma \mathcal{L}_{\psi(g^\sigma)}) = 0$ for $i \neq r-1$ and $\mathrm{R}^{r-1} m_!(\boxtimes_\sigma \mathcal{L}_{\psi(g^\sigma)})$ is smooth on $\mathbb{G}_{m,\bar{k}}$ of rank rd^{r-1}, tamely ramified at 0 and totally wild at infinity with Swan conductor d^r [**8**, Theorem 5.1(4,5)]. Taking the fibre at μ proves the proposition using proper base change. □

COROLLARY 4.3. *Suppose that g has degree d prime to p. Then*

$$\left| \sum_{x \in k_r^\star} \psi(\mathrm{Tr}_{k_r/k}(f(x))) \right| \leq rd^{r-1}(q-1)q^{\frac{r-1}{2}}$$

PROOF. Since $\mathcal{L}_{\psi(G)}$ is pure of weight 0, $\mathrm{H}_c^{r-1}(V_\mu \otimes \bar{k}, \mathcal{L}_{\psi(G)})$ is mixed of weights $\leq r-1$ for every μ (in fact it is pure of weight $r-1$ by [**8**, Theorem 5.1(7)]). So the previous proposition together with (4) implies

$$\left| \sum_{\mathrm{N}_{k_r/k}(x) = \mu} \psi(\mathrm{Tr}_{k_r/k}(g(x))) \right| \leq rd^{r-1}q^{\frac{r-1}{2}}$$

for every $\mu \in k^\star$. We conclude by using (3). □

5. Multiplicative character sums for homothety invariant polynomials

Let $e|q-1$ an integer and $f(x) = g(x^{\frac{q-1}{e}}) \in k_r[x]$ Γ_e-homothety invariant as in the previous section. Let $d = \deg(g)$ and $\chi : k^\star \to \bar{\mathbb{Q}}_\ell^\star$ a non-trivial multiplicative characer of order m. Weil's bound gives

$$\left| \sum_{x \in k_r} \chi(\mathrm{N}_{k_r/k}(f(x))) \right| \leq \left(\frac{d(q-1)}{e} - 1 \right) q^{\frac{r}{2}}$$

if g is not an m-th power. On the other hand, we have

$$\sum_{x \in k_r} \chi(\mathrm{N}_{k_r/k}(f(x))) = \chi(\mathrm{N}_{k_r/k}(f(0))) + \sum_{x \in k_r^\star} \chi(\mathrm{N}_{k_r/k}(g(x^{\frac{q-1}{e}}))) =$$

$$= \chi(\mathrm{N}_{k_r/k}(f(0))) + \frac{q-1}{e} \sum_{\mathrm{N}_{k_r/k}(x)^e = 1} \chi(\mathrm{N}_{k_r/k}(g(x))) =$$

$$= \chi(\mathrm{N}_{k_r/k}(f(0))) + \frac{q-1}{e} \sum_{\mu^e = 1} \sum_{\mathrm{N}_{k_r/k}(x) = \mu} \chi(\mathrm{N}_{k_r/k}(g(x))). \tag{5}$$

In order to estimate the sum $\sum_{\mathrm{N}_{k_r/k}(x)=\mu} \chi(\mathrm{N}_{k_r/k}(g(x)))$, we may and will assume without loss of generality that $g(0) \neq 0$: otherwise, writing $g(x) = x^a g_0(x)$ with $g_0(0) \neq 0$,

$$\sum_{\mathrm{N}_{k_r/k}(x)=\mu} \chi(\mathrm{N}_{k_r/k}(g(x))) = \sum_{\mathrm{N}_{k_r/k}(x)=\mu} \chi(\mathrm{N}_{k_r/k}(x^a g_0(x))) =$$

$$= \sum_{\mathrm{N}_{k_r/k}(x)=\mu} \chi(\mathrm{N}_{k_r/k}(x^a))\chi(\mathrm{N}_{k_r/k}(g_0(x))) = \chi(\mu)^a \sum_{\mathrm{N}_{k_r/k}(x)=\mu} \chi(\mathrm{N}_{k_r/k}(g_0(x))),$$

with $|\chi(\mu)^a| = 1$.

Let $P = \prod_\sigma (\sigma(\alpha_1)x_1 + \cdots + \sigma(\alpha_r)x_r)$ be as in the previous section, then

$$\sum_{\mathrm{N}_{k_r/k}(x)=\mu} \chi(\mathrm{N}_{k_r/k}(g(x))) = \sum_{P(x_1,\ldots,x_r)=\mu} \chi(\mathrm{N}_{k_r/k}(g(\alpha_1 x_1 + \cdots + \alpha_r x_r))) =$$

$$= \sum_{P(x_1,\ldots,x_r)=\mu} \chi\left(\prod_\sigma g^\sigma(\sigma(\alpha_1)x_1 + \cdots + \sigma(\alpha_r)x_r)\right)$$

so, by Grothendieck's trace formula,

$$\sum_{\mathrm{N}_{k_r/k}(x)=\mu} \chi(\mathrm{N}_{k_r/k}(g(x))) = \sum_{i=0}^{2r-2} \mathrm{Tr}(\mathrm{Frob}_k|\mathrm{H}^i_c(V_\mu \otimes \bar{k}, \mathcal{L}_{\chi(H)})) \tag{6}$$

where V_μ is the same as in the previous section and

$$H(x_1,\ldots,x_r) = \prod_\sigma g^\sigma(\sigma(\alpha_1)x_1 + \cdots + \sigma(\alpha_r)x_r),$$

the product taken over the elements of $\mathrm{Gal}(k_r/k)$.

Over k_r, the map

$$(x_1,\ldots,x_r) \mapsto (\sigma(\alpha_1)x_1 + \cdots + \sigma(\alpha_r)x_r)_{\sigma\in\mathrm{Gal}(k_r/k)}$$

is an isomorphism betweem $\mathbb{A}^r_{k_r}$ and $\mathbb{A}^{\mathrm{Gal}(k_r/k)}_{k_r}$, and the pull-back of P under this automorphism is $x_1\cdots x_r$. So $V_\mu \otimes \bar{k}$ is isomorphic to the hypersurface $x_1\cdots x_r = \mu$, and the sheaf $\mathcal{L}_{\chi(H)}$ corresponds under this isomorphism to the sheaf $\mathcal{L}_{\chi(\prod_\sigma g^\sigma(x_\sigma))} = \boxtimes_\sigma \mathcal{L}_{\chi(g^\sigma)}$ where $\mathcal{L}_{\chi(g^\sigma)}$ is the pull-back of the Kummer sheaf $\mathcal{L}_\chi$ by g^σ. Thus $\dim \mathrm{H}^i_c(V_\mu \otimes \bar{k}, \mathcal{L}_{\chi(H)}) = \dim \mathrm{H}^i_c(\{x_1\cdots x_r = \mu\}, \boxtimes_\sigma \mathcal{L}_{\chi(g^\sigma)})$. By proper base change, $\mathrm{H}^i_c(\{x_1\cdots x_r = \mu\}, \boxtimes_\sigma \mathcal{L}_{\chi(g^\sigma)})$ is the fibre at μ of the sheaf $\mathrm{R}^i m_!(\boxtimes_\sigma \mathcal{L}_{\chi(g^\sigma)})$, where $m : \mathbb{A}^{\mathrm{Gal}(k_r/k)}_{\bar{k}} \to \mathbb{A}^1_{\bar{k}}$ is the multiplication map.

PROPOSITION 5.1. *Let $g_1,\ldots,g_r \in k_r[x]$ be square-free of degree d with $g_i(0) \neq 0$, $m : \mathbb{A}^r_{k_r} \to \mathbb{A}^1_{k_r}$ the multiplication map and $K_r := \mathrm{R}m_!(\mathcal{L}_{\chi(g_1)} \boxtimes \cdots \boxtimes \mathcal{L}_{\chi(g_r)})$. Suppose that χ^d is not trivial. Then $K_r = \mathcal{L}_r[1-r]$ for a middle extension sheaf $\mathcal{L}_r$ of generic rank rd^{r-1} and pure of weight $r-1$ (on the open set where it is smooth), which is totally ramified at infinity and unipotent at 0, with $\mathrm{H}^1_c(\mathbb{A}^1_{\bar{k}}, \mathcal{L}_r)$ pure of weight r and dimension $(d-1)^r$.*

PROOF. We will proceed by induction, as in [**1**, Théorème 7.8]. For $r = 1$, $\mathcal{L}_r = \mathcal{L}_{\chi(g_1)}$ and all results are well known (see e.g. [**11**]). The sheaf is smooth of rank 1 on the complement of the set of roots of g_1, and the monodromy group at a root α acts via the non-trivial character χ, so $\mathcal{L}_{\chi(g_1)}$ is a middle extension at α.

Suppose everything has been proven for $r-1$. Then

$$K_r = \mathrm{R}m_!(\mathcal{L}_{\chi(g_1)} \boxtimes \cdots \boxtimes \mathcal{L}_{\chi(g_r)}) = \mathrm{R}m_{2!}(\mathrm{R}m_{1!}(\mathcal{L}_{\chi(g_1)} \boxtimes \cdots \boxtimes \mathcal{L}_{\chi(g_{r-1})}) \boxtimes \mathcal{L}_{\chi(g_r)}) =$$

$$= \mathrm{R}m_{2!}(K_{r-1} \boxtimes \mathcal{L}_{\chi(g_r)}) = \mathrm{R}m_{2!}(\mathcal{L}_{r-1}[2-r] \boxtimes \mathcal{L}_{\chi(g_r)})$$

where $m_1 : \mathbb{A}^{r-1}_{k_r} \to \mathbb{A}^1_{k_r}$ and $m_2 : \mathbb{A}^2_{k_r} \to \mathbb{A}^1_{k_r}$ are the multiplication maps.

The fibre of K_r at $t \in \bar{k}$ is then $\mathrm{R}\Gamma_c(\{xy = t\} \subseteq \mathbb{A}^2_{\bar{k}}, \mathcal{L}_{r-1} \boxtimes \mathcal{L}_{\chi(g_r)})[2-r]$. If $t \neq 0$, $\{xy = t\}$ is isomorphic to $\mathbb{G}_m$ via the projection on x, so the fibre is $\mathrm{R}\Gamma_c(\mathbb{G}_{m,\bar{k}}, \mathcal{L}_{r-1} \boxtimes \sigma_t^\star \mathcal{L}_{\chi(g_r)})[2-r]$, where $\sigma_t : \mathbb{G}_{m,\bar{k}} \to \mathbb{G}_{m,\bar{k}}$ is the involution $x \mapsto t/x$. Since $\mathcal{L}_{r-1}$ is totally ramified at 0 (and unramified at infinity) and $\sigma_t^\star \mathcal{L}_{\chi(g_r)}$ is unramified at 0 (and totally ramified at infinity), their tensor product is totally ramified at both 0 and infinity. In particular, its H_c^2 is vanishes. On the other hand, $\mathcal{L}_{r-1}$ and $\mathcal{L}_{\chi(g_r)}$ do not have punctual sections [**12**, Corollary 6 and Proposition 9], so neither does $\mathcal{L}_{r-1}\otimes\sigma_t^\star\mathcal{L}_{\chi(g_r)}$ and thus its H_c^0 vanishes. We conclude that the restriction of K_r to $\mathbb{G}_m$ is a single sheaf placed in degree $1+(r-2)=r-1$.

The fibre of K_r at 0 is $\mathrm{R}\Gamma_c(\{xy=0\}\subseteq\mathbb{A}^2_{\bar{k}},\mathcal{L}_{r-1}\boxtimes\mathcal{L}_{\chi(g_r)})[2-r]$. The group $\mathrm{H}_c^2(\{xy=0\},\mathcal{L}_{r-1}\boxtimes\mathcal{L}_{\chi(g_r)})$ vanishes, because so does H_c^2 of its restriction to $x=0$ (which is a constant times $\mathcal{L}_{\chi(g_r)}$, totally ramified at infinity) and to $y=0$ (which is a constant times $\mathcal{L}_{r-1}$, also totally ramified at infinity). The group H_c^0 also vanishes, because neither the restiction of $\mathcal{L}_{r-1}\boxtimes\mathcal{L}_{\chi(g_r)}$ to $x=0$ nor its restriction to $y=0$ have punctual sections. So the stalk of K_r at 0 is also concentrated in degree $r-1$.

Once we know K_r is a single sheaf $\mathcal{L}_r=\mathrm{R}^{r-1}m_!(\mathcal{L}_{\chi(g_1)}\boxtimes\cdots\boxtimes\mathcal{L}_{\chi(g_r)})$, since $\mathrm{H}_c^i(\mathbb{A}^1_{\bar{k}},\mathcal{L}_{\chi(g_i)})=0$ for $i\neq 1$ and has dimension $d-1$ and is pure of weight 1 for $i=1$ we get, by Künneth, that $\mathrm{H}_c^i(\mathbb{A}^1_{\bar{k}},\mathcal{L}_r)=0$ for $i\neq 1$ and it has dimension $(d-1)^r$ and is pure of weight r for $i=1$. Similarly, since the inverse image of $\mathbb{G}_{m,\bar{k}}$ under the multiplication map is $\mathbb{G}^r_{m,\bar{k}}$, $\mathrm{H}_c^1(\mathbb{G}_{m,\bar{k}},\mathcal{L}_r)=0$ for $i\neq 1$ and it has dimension d^r for $i=1$. In particular, the rank of $\mathcal{L}_r$ at 0 is $\chi(\mathbb{A}^1_{\bar{k}},\mathcal{L}_r)-\chi(\mathbb{G}_{m,\bar{k}},\mathcal{L}_r)=d^r-(d-1)^r$.

Let $t\in\bar{k}$ be a point which is not the product of a ramification point of $\mathcal{L}_r$ and a ramification point of $\mathcal{L}_{\chi(g_r)}$. Then at every point of $\mathbb{G}_{m,\bar{k}}$ at least one of $\mathcal{L}_{r-1}$, $\sigma_t^\star\mathcal{L}_{\chi(g_r)}$ is smooth. Since $\mathcal{L}_{r-1}$ has unipotent monodromy at 0 and $\sigma_t^\star\mathcal{L}_{\chi(g_r)}$ is unramified at ∞, by the Ogg-Shafarevic formula we have

$$-\chi(\mathbb{G}_{m,\bar{k}},\mathcal{L}_{r-1})=\mathrm{Swan}_\infty\mathcal{L}_{r-1}+\sum_{s\in\bar{k}^\star}(\mathrm{Swan}_s\mathcal{L}_{r-1}+\mathrm{drop}_s\mathcal{L}_{r-1})$$

and

$$-\chi(\mathbb{G}_{m,\bar{k}},\sigma_t^\star\mathcal{L}_{\chi(g_r)})=\mathrm{Swan}_0\mathcal{L}_{\chi(g_r)}+\sum_{s\in\bar{k}^\star}(\mathrm{Swan}_{t/s}\mathcal{L}_{\chi(g_r)}+\mathrm{drop}_{t/s}\mathcal{L}_{\chi(g_r)})$$

The local term at $u\in\bar{k}^\star$ (sum of the Swan conductor and the drop of the rank) gets multiplied by e upon tensoring with un unramified sheaf of rank e. The local term at 0 or ∞ (the Swan conductor) gets multiplied by e upon tensoring with a sheaf of rank e with unipotent monodromy. We conclude that

$$-\chi(\mathbb{G}_{m,\bar{k}},\mathcal{L}_{r-1}\otimes\sigma_t^\star\mathcal{L}_{\chi(g_r)})=-(\mathrm{rank}\,\mathcal{L}_{\chi(g_r)})\chi(\mathbb{G}_{m,\bar{k}},\mathcal{L}_{r-1})-$$
$$-(\mathrm{rank}\,\mathcal{L}_{r-1})\chi(\mathbb{G}_{m,\bar{k}},\sigma_t^\star\mathcal{L}_{\chi(g_r)})=d^{r-1}+d(r-1)d^{r-2}=rd^{r-1}.$$

This is the generic rank of $\mathcal{L}_r$.

Being a middle extension is a local property which is invariant under tensoring by unramified sheaves. Since, at every point of $\mathbb{G}_{m,\bar{k}}$, at least one of $\mathcal{L}_{r-1}$, $\sigma_t^\star\mathcal{L}_{\chi(g_r)}$ is unramified and they are both middle extensions (by the induction hypothesis), their tensor product is a middle extension on $\mathbb{G}_{m,\bar{k}}$. Since it is totally ramified at both 0 and ∞, we conclude that $\mathrm{H}_c^1(\mathbb{G}_{m,\bar{k}},\mathcal{L}_{r-1}\otimes\sigma_t^\star\mathcal{L}_{\chi(g_r)})$ is pure of weight $(r-2)+1=r-1$ [**2**, Théorème 3.2.3]. So $\mathcal{L}_r$ is pure of weight $r-1$ on the open set where it is smooth.

Now let $j_W:W\hookrightarrow\mathbb{A}^1_{\bar{k}}$ be the inclusion of the largest open sen on which $\mathcal{L}_r$ is smooth. Since $\mathcal{L}_r$ has no punctual sections, there is an injection $0\to\mathcal{L}_r\to j_{W\star}j_W^\star\mathcal{L}_r$, let $\mathcal{Q}$ be its punctual cokernel. We have an exact sequence

$$0\to\mathrm{H}_c^0(\mathbb{A}^1_{\bar{k}},\mathcal{Q})\to\mathrm{H}_c^1(\mathbb{A}^1_{\bar{k}},\mathcal{L}_r)\to\mathrm{H}_c^1(\mathbb{A}^1_{\bar{k}},j_{W\star}j_W^\star\mathcal{L}_r)\to 0$$

where $\mathrm{H}^0_c(\mathbb{A}^1_{\bar{k}}, \mathcal{Q})$ has weight $\leq r-1$. Since $\mathrm{H}^1_c(\mathbb{A}^1_{\bar{k}}, \mathcal{L}_r)$ is pure of weight r, we conclude that $\mathrm{H}^0_c(\mathbb{A}^1_{\bar{k}}, \mathcal{Q})$ and therefore $\mathcal{Q}$ are zero, so $\mathcal{L}_r$ is a middle extension. Now let $j : \mathbb{A}^1_{\bar{k}} \hookrightarrow \mathbb{P}^1_{\bar{k}}$ be the inclusion, again we get an exact sequence

$$0 \to \mathcal{L}_r^{I_\infty} \to \mathrm{H}^1_c(\mathbb{A}^1_{\bar{k}}, \mathcal{L}_r) \to \mathrm{H}^1_c(\mathbb{P}^1_{\bar{k}}, j_\star \mathcal{L}_r) \to 0$$

with $\mathcal{L}_r^{I_\infty}$ of weight $\leq r-1$, since $\mathrm{H}^1_c(\mathbb{A}^1_{\bar{k}}, \mathcal{L}_r)$ is pure of weight r we conclude that $\mathcal{L}_r^{I_\infty} = 0$, that is, $\mathcal{L}_r$ is totally ramified at infinity.

It remains to prove that $\mathcal{L}_r$ has unipotent monodromy at zero. Consider the exact sequence

$$0 \to \mathcal{L}_r^{I_0} \to \mathrm{H}^1_c(\mathbb{G}_{m,\bar{k}}, \mathcal{L}_r) \to \mathrm{H}^1_c(\mathbb{A}^1_{\bar{k}}, \mathcal{L}_r) \to 0$$

which identifies $\mathcal{L}_r^{I_0}$ with the weight $< r$ part of $\mathrm{H}^1_c(\mathbb{G}_{m,\bar{k}}, \mathcal{L}_r)$. Since $\mathrm{H}^1_c(\mathbb{G}_{m,\bar{k}}, \mathcal{L}_r) = \bigotimes_{i=1}^r \mathrm{H}^1_c(\mathbb{G}_{m,\bar{k}}, \mathcal{L}_{\chi(g_i)})$ and $\mathrm{H}^1_c(\mathbb{G}_{m,\bar{k}}, \mathcal{L}_{\chi(g_i)})$ has $d-1$ Frobenius eigenvalues of weight 1 and one of weight 0, we conclude that $\mathrm{H}^1_c(\mathbb{G}_{m,\bar{k}}, \mathcal{L}_r)$ has $\binom{r}{i}(d-1)^i$ eigenvalues of weight i for every $i = 0, \ldots, r$. By [**8**, Theorem 7.0.7], an eigenvalue of weight $i < r$ on $\mathcal{L}_r^{I_0}$ corresponds to a unipotent Jordan block of size $r-i$ for the action of I_0. So the sum of the sizes of the unipotent Jordan blocks for the monodromy of $\mathcal{L}_r$ at 0 is

$$\sum_{i=0}^{r-1} \binom{r}{i}(d-1)^i(r-i) = r\sum_{i=0}^{r-1}\binom{r}{i}(d-1)^i - r\sum_{i=0}^{r-1}\binom{r-1}{i-1}(d-1)^i =$$
$$= r\sum_{i=0}^{r-1}\binom{r-1}{i}(d-1)^i = r(1+d-1)^{r-1} = rd^{r-1}$$

which is the generic rank of $\mathcal{L}_r$. So the unipotent Jordan blocks fill out the entire monodromy at 0. □

COROLLARY 5.2. *Suppose that g is square-free of degree d prime to p and χ^d is not trivial. For any $\mu \in k^\star$, $\mathrm{H}^i_c(V_\mu \otimes \bar{k}, \mathcal{L}_{\chi(H)}) = 0$ for $i \neq r-1$ and $\dim \mathrm{H}^{r-1}_c(V_\mu \otimes \bar{k}, \mathcal{L}_{\chi(H)}) = rd^{r-1}$.*

PROOF. Apply the previous proposition with $(g_1, \ldots, g_r) = (g^\sigma)_{\sigma \in \mathrm{Gal}(k_r/k)}$, and proper base change. □

COROLLARY 5.3. *Suppose that g is square-free of degree d prime to p and χ^d is not trivial. Then*

$$\left|\sum_{x \in k_r^\star} \chi(\mathrm{N}_{k_r/k}(f(x)))\right| \leq rd^{r-1}(q-1)q^{\frac{r-1}{2}}$$

PROOF. Since $\mathcal{L}_{\chi(H)}$ is pure of weight 0, $\mathrm{H}^{r-1}_c(V_\mu \otimes \bar{k}, \mathcal{L}_{\chi(H)})$ has weights $\leq r-1$ for every μ. So the previous corollary together with (6) implies

$$\left|\sum_{\mathrm{N}_{k_r/k}(x)=\mu} \chi(\mathrm{N}_{k_r/k}(g(x)))\right| \leq rd^{r-1}q^{\frac{r-1}{2}}$$

for every $\mu \in k^\star$. We conclude by using (5). □

REMARK 5.4. *The following example shows that the hypothesis χ^d non-trivial is necessary. Let p be odd, $r = 2$, $g(x) = x^2 + 1$ and $\rho : k^\star \to \bar{\mathbb{Q}}_\ell^\star$ the quadratic character. Then*

$$\sum_{\mathrm{N}_{k_r/k}(x)=1} \rho(\mathrm{N}_{k_r/k}(x^2+1)) = \sum_{x^{q+1}=1} \rho((x^2+1)(x^{2q}+1)) =$$

$$= \sum_{x^{q+1}=1} \rho(x^2 + x^{2q} + 2) = \sum_{x^{q+1}=1} \rho((x+x^q)^2) \geq q - 1$$

since $x + x^q = \mathrm{Tr}_{k_r/k}(x) \in k$ and therefore $\rho((x+x^q)^2) = \rho(x+x^q)^2 = 1$ unless $x + x^q = 0$, which only happens for $x^2 = -1$, that is, for at most two values of x. So we can never have an estimate of the form

$$\left| \sum_{\mathrm{N}_{k_r/k}(x)=1} \rho(\mathrm{N}_{k_r/k}(x^2+1)) \right| \leq C \cdot q^{\frac{1}{2}}$$

which is valid for all q.

References

1. P. Deligne, *Application de la formule des traces aux sommes trigonométriques, dans Cohomologie Etale, Séminaire de Géométrie Algébrique du Bois-Marie, SGA 4 1/2*, Lecture Notes in Math **569**, 168–232.
2. ______, *La conjecture de Weil. II*, Publications Mathématiques de l'IHÉS **52** (1980), no. 1, 137–252. MR601520 (83c:14017)
3. L. Fu, *Calculation of l-adic local Fourier transformations*, Manuscripta mathematica **133** (2010), no. 3, 409–464. MR2729262
4. L. Fu and D. Wan, *Moment L-functions, partial L-functions and partial exponential sums*, Mathematische Annalen **328** (2004), no. 1, 193–228. MR2030375 (2004k:11099)
5. C. Douglas Haessig and Antonio Rojas-Leon, *L-functions of symmetric powers of the generalized Airy family of exponential sums*, International Journal of Number Theory, **7** No. 8 (2011), 2019–2064.
6. H. Hasse, *Theorie der relativ-zyklischen algebraischen Funktionenkörper, insbesondere bei endlichem Konstantenkörper.*, Journal für die reine und angewandte Mathematik (1935), no. 172, 37–54.
7. N.M. Katz, *On the Monodromy Groups Attached to Certain Families of Exponential Sums*, Duke Mathematical Journal **54** (1987). MR885774 (88i:11053)
8. ______, *Gauss Sums, Kloosterman Sums, and Monodromy Groups*, Annals of Mathematics Studies, vol. 116, Princeton University Press, 1988. MR955052 (91a:11028)
9. ______, *Exponential Sums and Differential Equations*, Annals of Mathematics Studies, vol. 124, Princeton University Press, 1990. MR1081536 (93a:14009)
10. ______, *Frobenius-Schur indicator and the ubiquity of Brock-Granville quadratic excess*, Finite Fields and Their Applications **7** (2001), no. 1, 45–69. MR1803935 (2002d:11069)
11. ______, *Estimates for nonsingular multiplicative character sums*, International Mathematics Research Notices **2002** (2002), no. 7, 333–349. MR1883179 (2003a:11106)
12. ______, *A semicontinuity result for monodromy under degeneration*, Forum Mathematicum **15** (2003), no. 2, 191–200. MR1956963 (2004c:14039)
13. N.M. Katz and G. Laumon, *Transformation de Fourier et majoration de sommes exponentielles*, Publications Mathématiques de l'IHÉS **62** (1985), no. 1, 145–202. MR823177 (87i:14017)
14. G. Laumon, *Transformation de Fourier, constantes d'équations fonctionnelles et conjecture de Weil*, Publications Mathématiques de l'IHÉS **65** (1987), no. 1, 131–210. MR908218 (88g:14019)
15. A. Rojas-León and D. Wan, *Improvements of the Weil bound for Artin-Schreier curves*, Mathematische Annalen **351**, No. 2 (2011), 417–442. MR2836667

16. O. Šuch, *Monodromy of Airy and Kloosterman sheaves*, Duke Mathematical Journal **103** (2000), no. 3, 397–444. MR1763654 (2001g:11132)
17. A. Weil, *On some exponential sums*, Proceedings of the National Academy of Sciences of the United States of America **34** (1948), no. 5, 204. MR0027006 (10:234e)

DEPARTAMANTO DE ÁLGEBRA, UNIVERSIDAD DE SEVILLA, APDO 1160, 41080 SEVILLA, SPAIN
E-mail address: `arojas@us.es`

Part II:
Height zeta functions and arithmetic

Contemporary Mathematics
Volume **566**, 2012
http://dx.doi.org/10.1090/conm/566/11216

Height zeta functions on generalized projective toric varieties

Driss Essouabri

ABSTRACT. In this paper we study the analytic properties of height zeta functions associated to generalized projective toric varieties. As an application, we obtain asymptotic expansions of the counting functions of rational points of generalized projective toric varieties provided with a large class of heights.

1. Introduction and detailed description of the problem

Let V be a projective algebraic variety over $\mathbb{Q}$. We can study the density of rational points of V by first choosing a line bundle $\mathcal{L}$ such that its class in the Picard group of V is contained in the interior of the cone of effective divisors. In this case, for a suitable Zariski open subset U of V, some positive tensor power of $\mathcal{L}$ defines a projective embedding $\phi : U(\mathbb{Q}) \hookrightarrow \mathbb{P}^{n-1}(\mathbb{Q})$ for some n. There are then several ways to measure the density of $\phi\, U(\mathbb{Q})$. First, we choose a family of local norms $v = (v_p)_{p\leq\infty}$ whose Euler product $H_v = \prod_{p\leq\infty} v_p$ specifies a height H_v on $\phi\, U(\mathbb{Q})$ in the evident way. It suffices to set $H_{\mathcal{L},v}(M) := H_v\left(\phi(M)\right)$ for any point $M \in U(\mathbb{Q})$.
By definition, the density of the rational points of $U(\mathbb{Q})$ with respect to $H_{\mathcal{L},v}$ is the function $B \to N_{\mathcal{L},v}(U,B) := \{M \in U(\mathbb{Q}) \mid H_{\mathcal{L},v}(M) \leq B\}$.

Manin's conjecture (see [**Bro**], [**Ch**], [**P2**] for more details) concerns the asymptotic behavior of the density for large B. It asserts the existence of constants $a = a(L)$, $b = b(L)$ and $C = C(U, L, v) > 0$, such that:

$$N_{\mathcal{L},v}(U,B) = C\ B^a (\log B)^{b-1}(1+o(1)) \quad (B \to +\infty) \tag{1}$$

A tauberian theorem derive the asymptotic (1) above from the analytic properties of the height-zeta functions defined by the following:

$$Z_{\mathcal{L},v}(U;s) := \sum_{\mathbf{x}\in U(\mathbb{Q})} H_{\mathcal{L},v}(x)^{-s}. \tag{2}$$

For nonsingular toric varieties, a suitable refinement of the conjecture was first proved by Batyrev-Tschinkel [**BT**]. Improvements in the error term were then given by Salberger [**Sal**], and a bit later by de la Bretèche [**Bre3**].

1991 *Mathematics Subject Classification.* Primary 11M41 ,14G10, 14G05.

Key words and phrases. Manin's conjecture, heights, rational points, zeta functions, meromorphic continuation, Newton polyhedron.

The author wishes to express his thanks to Ben Lichtin for his many helpful suggestions and his careful reading which improved the presentation and also the English of this paper. He also expresses his thanks to the referee for his report and his several relevant remarks.

Each of these works used a *particular* height function H_∞. The most important feature is its choice of v_∞, which was defined as follows:

$$\forall \mathbf{y} = (y_1, \dots, y_n) \in \mathbb{Q}_\infty^n (= \mathbb{R}^n), \qquad v_\infty(\mathbf{y}) = \max_i |y_i|.$$

Although this is quite convenient to use for torics, there is no fundamental reason why one should a priori limit the effort to prove the conjecture to this particular height function. In addition, since the existence of a precise asymptotic for one height function need not imply anything equally precise for some other height function, a proof of Manin's conjecture for toric varieties and heights other than H_∞ is not a consequence of the work cited above.

In this paper we will study zeta function and the density of rational points on generalized projective toric varieties equipped with a large class of heights.
Following B. Sturmfels (see §1 and lemma 1.1 of [**St**]), we define here a **generalized projective toric variety** (also called binomial variety) as follows:

$$V(\mathbf{A}) := \{(x_1 : \dots : x_n) \in \mathbb{P}^{n-1}(\mathbb{Q}) \mid \prod_{\substack{j=1,\dots,n \\ a_{i,j}>0}} x_j^{a_{i,j}} = \prod_{\substack{j=1,\dots,n \\ a_{i,j}<0}} x_j^{-a_{i,j}} \, \forall i = 1, \dots, l\}$$

where $\mathbf{A}$ is an $l \times n$ matrix with entries in $\mathbb{Z}$, whose rows $\mathbf{a}_i = (a_{i,1}, \dots, a_{i,n})$ each satisfy the property that $\sum_{j=1}^n a_{i,j} = 0$.
This stands in contrast to common practise in algebraic geometry (see [**Co**]*), where toric varieties are assumed to be normal (this follows from the construction of toric varieties from fans). Normality (and therefore smoothness) imposes strong combinatorial restrictions on the matrix* $\mathbf{A}$ *(see §2 of* [**St**] *for more details).* **We don't assume such conditions here** *and the reader should consider this fact when comparing our results with those obtained in earlier works cited above.*

The heights or (generalized heights) we consider here are defined as follows:
Let $P = P(X_1, \dots, X_n)$ be a generalized polynomial (i.e. exponents of monomials can be arbitrary nonnegative real numbers) with positive coefficients, elliptic on $[0, \infty)^n$ [1] and homogeneous of degree $d > 0$. To such P, we associate the family of local norms v_p on $\mathbb{Q}_p^n$ ($p \le \infty$) by setting, $v_p(\mathbf{y}) = \max |y_j|_p$ for any $p < \infty$ and $\mathbf{y} = (y_1, \dots, y_n)$ and[2]

$$v_\infty(\mathbf{y}) = (P(|y_1|, \dots, |y_n|))^{1/d} \qquad \forall \mathbf{y} = (y_1, \dots, y_n) \in \mathbb{Q}_\infty^n.$$

A simple exercise involving heights then shows that the height H_P associated[3] to this family of local norms is given as follows. For any $\mathbf{x} = (x_1 : \dots : x_n) \in \mathbb{P}^{n-1}(\mathbb{Q})$:

$$H_P(\mathbf{x}) := (P(|x_1|, \dots, |x_n|))^{1/d} \text{ if } x_1, \dots, x_n \text{ are coprime integers.} \tag{3}$$

Very little appears to be known about the asymptotic density of projective varieties for such heights H_P. A few earlier works are [**P1**], [**E2**] or [**Sw**], but these are limited to very special choices of P. In particular, the reader should appreciate

---

[1] P is elliptic on $[0, \infty)^n$ if its restriction to this set vanishes only at $(0, \dots, 0)$. In the following, the term "elliptic" means "elliptic on $[0, \infty)^n$ ".

[2] If P is a quadratic form or is of the form $P = X_1^d + \dots + X_n^d$, then clearly the triangle inegality is verified. However, in order to preserve height under embedding, the definitions of metric used to build heights (see for example ([**P2**], chap. 2, §2.2) or [**P1**]) do not assume in general that the triangle inequality must hold.

[3] Precisely, this is the height associated to the pair $(\iota^*\mathcal{O}(1), v)$ where $\iota : V(\mathbf{A}) \hookrightarrow \mathbb{P}^{n-1}(\mathbb{Q})$ is the canonical embedding, $\mathcal{O}(1)$ is the standard line bundle on $\mathbb{P}^{n-1}(\mathbb{Q})$, and $v = (v_p)_{p\le\infty}$ is the family of local norms. For simplicity this height is denoted H_P.

the fact that none of the general methods developed to study the asymptotic density with respect to H_∞ can be expected to apply to any other height H_P. To be convinced of this basic fact, it is enough to look at the simple case where $V = \mathbb{P}^{n-1}(\mathbb{Q})$ and $U = \{(x_1, \ldots, x_n) \in \mathbb{P}^{n-1}(\mathbb{Q}) \mid x_1 \ldots x_n \neq 0\}$. In this case, an easy calculation shows that

$$(4) \quad N_{H_\infty}(U;t) = 2^{n-1}\#\{\mathbf{m} \in \mathbb{N}^n;\ \max m_i \leq t \text{ and } gcd\,((m_i)) = 1\} \sim \frac{2^{n-1}\ t^n}{\zeta(n)}.$$

On the other hand, if $P = P(X_1, \ldots, X_n)$ is a homogeneous elliptic polynomial of degree $d > 0$ (for example $P = X_1^d + \cdots + X_n^d$), we have

$$\begin{aligned} N_{H_P}(U;t) &= 2^{n-1}\#\{\mathbf{m} \in \mathbb{N}^n \mid P(\mathbf{m}) \leq t^d \text{ and } gcd(m_i) = 1\} \\ (5) \qquad &\sim \frac{2^{n-1}}{\zeta(n)}\#\{\mathbf{m} \in \mathbb{N}^n \mid P(\mathbf{m}) \leq t^d\}. \end{aligned}$$

The asymptotic for the second factor in (5) is given by

$$\#\{\mathbf{m} \in \mathbb{N}^n \mid P(\mathbf{m}) \leq t^d\} \sim C(P;n)t^n, \text{ where } C(P;n) = \frac{1}{n}\int_{\mathbb{S}^{n-1}\cap\mathbb{R}_+^n} P_d^{-n/d}(\mathbf{v})d\sigma(\mathbf{v}),$$

$\mathbb{S}^{n-1}$ denotes the unit sphere of $\mathbb{R}^n$, and $d\sigma$ its Lebesgue measure. This is a classical result of Mahler [**Ma**]. It should be evident to the reader that this cannot be determined from the asymptotic for the counting function of (4).

Our results are formulated in terms of a polyhedron in $[0,\infty)^n$, which can be associated to $V(\mathbf{A})$ in a natural way, and, in addition, the maximal torus $U(\mathbf{A}) := \{\mathbf{x} \in V(\mathbf{A}) \mid x_1 \ldots x_n \neq 0\}$.

The first result, Theorem 1 (see §3), establishes meromorphic continuation of the height zeta function $Z_{H_P}(U(\mathbf{A});s) := \sum_{\mathbf{x}\in U(\mathbf{A})} H_P(\mathbf{x})^{-s}$ beyond its domain of convergence and determines explicitly the principal part at its largest pole (which is very important for applications).

In the case of projective space $\mathbb{P}^{n-1}(\mathbb{Q})$ the height zeta function reduces to the study of Dirichlet's series of the form $Z(Q;s) = \sum_{\mathbf{m}\in\mathbb{N}^n} Q(m_1, \ldots, m_n)^{-s}$ where Q is a suitable polynomial and the analytic properties of this last series are closely related to the nature of the **singularity at infinity of the polynomial** Q (see [**Me**], [**Ma**], [**Ca2**], [**Sar1**], [**L**], [**E1**],..).

An important feature of Theorem 1 is the precise description for **the top order polar term** *of $Z_{H_P}(U(\mathbf{A});s)$, in which one sees very clearly the joint dependance upon the polynomial P that defines the height H_P and the geometry of the variety $V(\mathbf{A})$.* This joint dependence upon the nature of singularity of P at infinity and the geometry of the variety deserves to be completed in an even more general setting.

As application, Corollary 1 (see §3), shows that if the diagonal intersects the polyhedron in a compact face, then there exist constants $a = a(\mathbf{A}), b = b(\mathbf{A})$ and $C = C(\mathbf{A}, H_P)$ such that as $t \to \infty$:

$$N_{H_P}(U(\mathbf{A});t) := \#\{\mathbf{x} \in U(\mathbf{A}) : H_P(\mathbf{x}) \leq t\} = C\ t^a \log(t)^{b-1}\left(1 + O\left((\log t)^{-1}\right)\right).$$

The constants a and b are also characterized quite simply in terms of this polyhedron. The second part of Corollary 1 refines this conclusion by asserting that if the dimension of this face *equals* the dimension of $V(\mathbf{A})$, then $C > 0$. In this event, we are also able to give an explicit expression for C, the form of which is a reasonable generalization of that given in (5).

Our second main result uses the fact that we are able to give a very precise description of the polyhedron for the class of hypersurfaces $\{\mathbf{x} \in \mathbb{P}^{n-1}(\mathbb{Q}) \mid x_1^{a_1} \dots x_{n-1}^{a_{n-1}} = x_n^{|\mathbf{a}|}\}$. The particular case of the singular hypersurface $\{\mathbf{x} \in \mathbb{P}^{n-1}(\mathbb{Q}) \mid x_1 \dots x_{n-1} = x_n^n\}$, provided with the height H_∞, was studied in several papers ([**BT**], [**Bre1**], [**F**], [**HBM**], [**BEL**]). However, besides a result of Swinnerton-Dyer [**Sw**] for the singular cubic $X_1X_2X_3 = X_4^3$ with a single height function *not* H_∞, nothing comparable to our Theorem 2 appears to exist in the literature.

All our results above follow from our *fundamental* theorem (i.e Theorem 3 in §3.2) which is the main ingredient of this paper. This allows one to study analytic properties for "mixed zeta functions" $Z(f; P; s)$ (see §3.2.2), and in particular, to determine explicitly the principal part at its largest pole. Such zeta functions combine together in one function the multiplicative features of the variety and the additive nature of the polynomial that defines the height H_P.

2. Notations and preliminaries

2.1. Notations.

(1) $\mathbb{N} = \{1, 2, \dots\}$, $\mathbb{N}_0 = \mathbb{N} \cup \{0\}$;

(2) The expression: $f(\mathbf{y}, \mathbf{x}) \ll_{\mathbf{y}} g(\mathbf{x})$ uniformly in $\mathbf{x} \in X$ means there exists $A = A(\mathbf{y}) > 0$, such that, $\forall \mathbf{x} \in X \quad |f(\mathbf{y}, \mathbf{x})| \le A\, g(\mathbf{x})$;

(3) For any $\mathbf{x} = (x_1, .., x_n) \in \mathbb{R}^n$, we set $\|\mathbf{x}\| = \|\mathbf{x}\|_2 = \sqrt{x_1^2 + .. + x_n^2}$ and $|\mathbf{x}| = |x_1| + .. + |x_n|$. We denote the canonical basis of $\mathbb{R}^n$ by $(\mathbf{e}_1, \dots, \mathbf{e}_n)$. The standard inner product on $\mathbb{R}^n$ is denoted by $\langle ., . \rangle$.
We set also $\mathbf{0} = (0, \dots, 0)$ and $\mathbf{1} = (1, \dots, 1)$;

(4) We denote a vector in $\mathbb{C}^n$ $\mathbf{s} = (s_1, \dots, s_n)$, and write $\mathbf{s} = \boldsymbol{\sigma} + i\boldsymbol{\tau}$, where $\boldsymbol{\sigma} = (\sigma_1, \dots, \sigma_n)$ and $\boldsymbol{\tau} = (\tau_1, \dots, \tau_n)$ are the real resp. imaginary components of $\mathbf{s}$ (i.e. $\sigma_i = \Re(s_i)$ and $\tau_i = \Im(s_i)$ for all i). We also write $\langle \mathbf{x}, \mathbf{s} \rangle$ for $\sum_i x_i s_i$ if $\mathbf{x} \in \mathbb{R}^n, \mathbf{s} \in \mathbb{C}^n$;

(5) Given $\boldsymbol{\alpha} \in \mathbb{N}_0^n$, we write $\mathbf{X}^{\boldsymbol{\alpha}}$ for the monomial $X_1^{\alpha_1} \cdots X_n^{\alpha_n}$. For an analytic function $h(\mathbf{X}) = \sum_{\boldsymbol{\alpha}} a_{\boldsymbol{\alpha}} \mathbf{X}^{\boldsymbol{\alpha}}$, the set $supp(h) := \{\boldsymbol{\alpha} \mid a_{\boldsymbol{\alpha}} \neq 0\}$ is called the support of h;

(6) $f : \mathbb{N}^n \to \mathbb{C}$ is said to be multiplicative if for all $\mathbf{m} = (m_1, \dots, m_n)$, $\mathbf{m}' = (m_1', \dots, m_n') \in \mathbb{N}^n$ satisfying $gcd\,(lcm\,(m_i), lcm\,(m_i')) = 1$ we have $f\,(m_1 m_1', \dots, m_n m_n') = f(\mathbf{m}).f(\mathbf{m}')$;

(7) Let F be a meromorphic function on a domain $\mathcal{D}$ of $\mathbb{C}^n$ and let $\mathcal{S}$ be the support of its polar divisor. F is said to be of moderate growth if there exists $a, b > 0$ such that $\forall \delta > 0$, $F(\mathbf{s}) \ll_{\boldsymbol{\sigma}, \delta} 1 + |\boldsymbol{\tau}|^{a|\boldsymbol{\sigma}| + b}$ uniformly in $\mathbf{s} = \boldsymbol{\sigma} + i\boldsymbol{\tau} \in \mathcal{D}$ verifying $d(\mathbf{s}, \mathcal{S}) \ge \delta$.

2.2. Preliminaries from convex analysis.

2.2.1. *Standard constructions.* For the reader's convenience, some classical notions from convex analysis that will be used throughout the article are assembled here. For more details see for example the book [**R**].

• Let $A = \{\boldsymbol{\alpha}^1, \dots, \boldsymbol{\alpha}^q\}$ be a finite subset of $\mathbb{R}^n$.
The **convex hull** of A is $conv(A) := \{\sum_{i=1}^q \lambda_i \boldsymbol{\alpha}^i \mid (\lambda_1, .., \lambda_q) \in \mathbb{R}_+^q$ and $\sum_{i=1}^q \lambda_i = 1\}$ and its interior is $conv^*(A) := \{\sum_{i=1}^q \lambda_i \boldsymbol{\alpha}^i \mid (\lambda_1, .., \lambda_q) \in \mathbb{R}_+^{*q}$ and $\sum_{i=1}^q \lambda_i = 1\}$. The **convex cone** of A is $con(A) := \{\sum_{i=1}^q \lambda_i \boldsymbol{\alpha}^i \mid (\lambda_1, \dots, \lambda_q) \in \mathbb{R}_+^q\}$ and its (relative) interior is $con^*(A) := \{\sum_{i=1}^q \lambda_i \boldsymbol{\alpha}^i \mid (\lambda_1, \dots, \lambda_q) \in \mathbb{R}_+^{*q}\}$.

• Let $\Sigma = \{\mathbf{x} \in \mathbb{R}_+^n \mid \langle \boldsymbol{\beta}, \mathbf{x}\rangle \geq 1\ \forall \boldsymbol{\beta} \in I\}$ where I is a finite (nonempty) subset of $\mathbb{R}_+^n \setminus \{\mathbf{0}\}$. Σ is a convex polyhedron of $\mathbb{R}_+^n \setminus \{\mathbf{0}\}$.

(1) Let $\mathbf{a} \in \mathbb{R}_+^n \setminus \{\mathbf{0}\}$, we define $m(\mathbf{a}) := \inf_{\mathbf{x}\in\Sigma}\langle \mathbf{a}, \mathbf{x}\rangle$ and the face of Σ with polar vector $\mathbf{a}$ (or the first meet locus of $\mathbf{a}$) as $\mathcal{F}_\Sigma(\mathbf{a}) = \{\mathbf{x} \in \Sigma \mid \langle \mathbf{a}, \mathbf{x}\rangle = m(\mathbf{a})\}$;

(2) The faces of Σ are the sets $\mathcal{F}_\Sigma(\mathbf{a})$ ($\mathbf{a} \in \mathbb{R}_+^n \setminus \{\mathbf{0}\}$). A facet of Σ is a face of maximal dimension;

(3) Let F be a face of Σ. The cone $pol(F) := \big\{\mathbf{a} \in \mathbb{R}_+^n \setminus \{\mathbf{0}\} \mid F = \mathcal{F}_\Sigma(\mathbf{a})\big\}$ is called the polar cone associated to F and its elements are called polar vectors of F.
A polar vector $\mathbf{a} \in pol(F)$ is said to be a normalized polar vector of F if $m(\mathbf{a}) = 1$ (i.e $F := \{\mathbf{x} \in \Sigma \mid \langle \mathbf{a}, \mathbf{x}\rangle = 1\}$). We denote by $Pol_0(F)$ the set of normalized polar vector of F;

(4) We define the *index* of Σ by $\iota(\Sigma) := \min\{|\boldsymbol{\alpha}|;\ \boldsymbol{\alpha} \in \Sigma\}$. It is clear that $\mathcal{F}_\Sigma(\mathbf{1}) = \{\mathbf{x} \in \Sigma;\ |\mathbf{x}| = \iota(\Sigma)\}$.

• Let J be a subset of $[0,\infty)^n \setminus \{\mathbf{0}\}$, the set $\mathcal{E}(J) = convex\ hull\big(J + \mathbb{R}_+^n\big)$ is the Newton polyhedron of J. We denote also by $\mathcal{E}^\infty(J) = convex\ hull\big(J - \mathbb{R}_+^n\big)$ its Newton polyhedron at infinity.

2.3. Construction of the mixed volume constant $A_0(\mathcal{T}; P)$.

2.3.1. *The volume constant $A_0(P)$.* Let $P(\mathbf{X}) = \sum_{\boldsymbol{\alpha}\in supp(P)} a_{\boldsymbol{\alpha}} \mathbf{X}^{\boldsymbol{\alpha}}$ be a generalized polynomial with positive coefficients that depends upon all the variables $X_1, \ldots, X_n$. We apply the discussion in [**Sar1**] (see also [**Sar2**]) to define a "volume constant" for P.

By definition, the Newton polyhedron of P (at infinity) is the set $\mathcal{E}^\infty(P) := \big(conv(supp(P)) - \mathbb{R}_+^n\big)$. All the definitions introduced in §2.2.1 apply to such a polyhedron once we change the definition of $m(\mathbf{a})$ to equal $\sup_{\mathbf{x}\in\mathcal{E}^\infty(P)} < \mathbf{a}, \mathbf{x} >$ for any $\mathbf{a} \neq \mathbf{0} \in \mathbb{R}_+^n$. In particular, the notions of face, facet, (normalized) polar vector, etc. all extend in a straightforward way.

Let G_0 be the smallest face of $\mathcal{E}^\infty(P)$ which meets the diagonal $\Delta = \mathbb{R}_+\mathbf{1}$. We denote by σ_0 the unique positive real number t that satisfies $t^{-1}\mathbf{1} \in G_0$. Thus, there exists a unique vector subspace $\overrightarrow{G}_0$ of largest codimension ρ_0 such that $G_0 \subset \sigma_0^{-1}\mathbf{1} + \overrightarrow{G}_0$. Both ρ_0, σ_0 evidently depend upon P, but it is not necessary to indicate this in the notation. We also set $P_{G_0}(X) = \sum_{\boldsymbol{\alpha}\in G_0} a_{\boldsymbol{\alpha}} \mathbf{X}^{\boldsymbol{\alpha}}$.

There exist finitely many facets of $\mathcal{E}^\infty(P)$ that intersect in G_0. We denote their normalized polar vectors by $\boldsymbol{\lambda}_1, \ldots, \boldsymbol{\lambda}_N$.

By a permutation of the coordinates X_i one can suppose that $\oplus_{i=1}^{\rho_0} \mathbb{R}\mathbf{e}_i \oplus \overrightarrow{G_0} = \mathbb{R}^n$, and that $\{\mathbf{e}_{m+1}, \ldots, \mathbf{e}_n\}$ is the set of standard basis vectors to which G_0 is parallel (i.e. for which $G_0 = G_0 - \mathbb{R}_+\mathbf{e}_i$). If G_0 is compact then $m = n$.

Set $\Lambda = Conv\{\mathbf{0}, \boldsymbol{\lambda}_1, \ldots, \boldsymbol{\lambda}_N, \mathbf{e}_{\rho_0+1}, \ldots, \mathbf{e}_n\}$. It follows that $dim\Lambda = n$.

DEFINITION 1. The **volume constant** associated to P is:
$A_0(P) := n!\ Vol(\Lambda)\ \int_{[1,+\infty[^{n-m}} \left(\int_{\mathbb{R}_+^{m-\rho_0}} P_{G_0}^{-\sigma_0}(\mathbf{1}, \mathbf{x}, \mathbf{y})\ d\mathbf{x}\right) d\mathbf{y}\,.$

In ([**Sar1**], chap 3, th. 1.6) (also see [**Sar2**]), P. Sargos proved the following important result about the function $Y(P; s) := \int_{[1,+\infty[^n} P(\mathbf{x})^{-s}\, d\mathbf{x}$. This generalized earlier work of Cassou-Noguès [**Ca2**].

Theorem. (P. Sargos [Sar1] chap. 3, see also [Sar2]).
Let P be a generalized polynomial with positive coefficients. Then $Y(P; s)$ converges

absolutely in $\{\sigma - \Re s > \upsilon_0\}$, *and has a meromorphic continuation to* $\mathbb{C}$ *with largest pole at* $s = \sigma_0$ *of order* ρ_0. *In addition,* $Y(P;s) \sim_{s\to\sigma_0} A_0(P)\ (s-\sigma_0)^{-\rho_0}$. *Thus,* $A_0(P) > 0$.

Remark. The previous Theorem implies (see [**Sar2**]), that $\frac{A_0(P)}{\sigma_0(\rho_0-1)!}t^{\sigma_0}\log^{\rho_0-1} t$ equals the dominant term for the volume of the set $\{\mathbf{x} \in [1,\infty)^n \mid P(\mathbf{x}) \le t\}$ as $t \to \infty$.

For general results on volume constant near the origin, see also ([**DNS**], §5).

When P is elliptic, we recover Mahler's result.

PROPOSITION 1 ([**Ma**]). *If* $P = P(X_1,\ldots,X_n)$ *is an elliptic polynomial of degree* $d \ge 1$. *Then,* $\sigma_0 = \frac{n}{d}$, $\rho_0 = 1$ *and* $A_0(P) = \dfrac{1}{d}\displaystyle\int_{\mathbb{S}^{n-1}\cap\mathbb{R}_+^n} P_d^{-n/d}(\mathbf{v})\, d\sigma(\mathbf{v})$, *where* P_d *is the homogeneous part of* P *of degree* d, *and* $d\sigma$ *is induced Lebesgue measure on the unit sphere* $\mathbb{S}^{n-1}$.

2.3.2. *Construction of the volume constant* $A_0(I,\mathbf{u},\mathbf{b})$. Let $I = \{\boldsymbol{\beta}_k\}$ be a finite sequence of distinct elements of $\mathbb{R}_+^r \setminus \{\mathbf{0}\}$, $\mathbf{u} = \{u(\boldsymbol{\beta})\}_{\boldsymbol{\beta}\in I}$ a set of positive integers, and $\mathbf{b} = (b_1,\ldots,b_r) \in \mathbb{R}_+^{*r}$. To the triple $(I;\mathbf{u};\mathbf{b})$, we associate the generalized polynomial $P_{(I;\mathbf{u};\mathbf{b})}$ in $q := \sum_{\boldsymbol{\beta}\in I} u(\boldsymbol{\beta})$ variables as follows:

First, we form a sequence $\boldsymbol{\alpha}^1, \boldsymbol{\alpha}^2, \ldots$ whose elements are the $\boldsymbol{\beta}_k \in I$ but with each element repeated exactly $u(\boldsymbol{\beta}_k)$ times. The indexing of the $\boldsymbol{\alpha}^i$ is specified by the ordering of the $\boldsymbol{\beta}_j$ as follows. We set
$\boldsymbol{\alpha}^1,\ldots,\boldsymbol{\alpha}^{u(\boldsymbol{\beta}_1)} = \boldsymbol{\beta}_1$, $\boldsymbol{\alpha}^{u(\boldsymbol{\beta}_1)+1},\ldots,\boldsymbol{\alpha}^{u(\boldsymbol{\beta}_1)+u(\boldsymbol{\beta}_2)} = \boldsymbol{\beta}_2$, etc.

We next form a $q\times r$ matrix whose row vectors are the $\boldsymbol{\alpha}^i$. Denoting its column vectors by $\gamma^1,\ldots,\gamma^r$, we obtain r vectors in $\mathbb{R}_+^q$, with which we now define the generalized polynomial $P_{(I;\mathbf{u};\mathbf{b})}(\mathbf{Y}) := \sum_{i=1}^r b_i \mathbf{Y}^{\gamma^i}$ on $\mathbb{R}_+^q$. **We then denote the volume constant of** $P_{(I;\mathbf{u};\mathbf{b})}$ by $A_0(I;\mathbf{u};\mathbf{b})$.

2.3.3. *The mixed volume constant* $A_0(\mathcal{T};P)$. We now apply the preceding construction by starting with a pair $\mathcal{T} = (I,\mathbf{u})$ of finite sets and a generalized polynomial $P(\mathbf{X}) = b_1\mathbf{X}^{\gamma^1} + \cdots + b_r\mathbf{X}^{\gamma^r}$ on $\mathbb{R}_+^n$ with positive coefficients. We assume that $I \subset \mathbb{R}_+^n \setminus \{0\}$. The elements of $\mathbf{u}$ are positive integers that depend upon the elements of I. Thus, $\mathbf{u} = \{u(\boldsymbol{\eta})\}_{\boldsymbol{\eta}\in I}$. We also set $\mathbf{b} = (b_1,\ldots,b_r)$.

Given the $r\times n$ matrix $\boldsymbol{\Gamma}$ whose row vectors are $\gamma^1,\ldots,\gamma^r$, we associate to $\mathcal{T}$ and P the following "mixed" objects $I^*, \mathbf{u}^*$. These will play an important role.

(1) $I^* = \boldsymbol{\Gamma}(I) = \{\boldsymbol{\Gamma}(\boldsymbol{\eta});\ \boldsymbol{\eta}\in I\} \subset \mathbb{R}_+^r \setminus \{\mathbf{0}\}$;

(2) $\mathbf{u}^* = (u^*(\boldsymbol{\beta}))_{\boldsymbol{\beta}\in I^*}$ where $u^*(\boldsymbol{\beta}) = \sum_{\{\boldsymbol{\Gamma}(\boldsymbol{\eta})=\boldsymbol{\beta}\}} u(\boldsymbol{\eta})$ for each $\boldsymbol{\beta}\in I^*$.

We define finally the **mixed volume** constant by: $A_0(\mathcal{T};P) = A_0(I^*;\mathbf{u}^*;\mathbf{b})$, the **volume constant** of the generalized polynomial $P_{(I^*;\mathbf{u}^*,\mathbf{b})}$ on $\mathbb{R}_+^n$ (see §2.3.2)

3. Statements of main results

3.1. Main results about height zeta functions on generalized toric varieties. Let $\mathbf{A}$ a $l\times n$ matrix with entries in $\mathbb{Z}$, whose rows $\mathbf{a}_i = (a_{i,1},\ldots,a_{i,n})$ each satisfy the property that $\sum_{j=1}^n a_{i,j} = 0$. We consider the generalized projective toric varieties defined by:

(6)

$$V(\mathbf{A}) := \{(x_1:\cdots:x_n)\in\mathbb{P}^{n-1}(\mathbb{Q}) \mid \prod_{\substack{j=1,\ldots,n\\ a_{i,j}>0}} x_j^{a_{i,j}} = \prod_{\substack{j=1,\ldots,n\\ a_{i,j}<0}} x_j^{-a_{i,j}}\ \forall i=1,\ldots,l\}$$

Denote by $U(\mathbf{A}) := \{(x_1 : \cdots : x_n) \in V(\mathbf{A}) : x_1 \ldots x_n \neq 0\}$ its maximal torus. We assume, without loss of generality, that the rows $\mathbf{a}_i$ ($i = 1, \ldots, l$) are linearly independent over $\mathbb{Q}$. It follows that:

$$(7) \qquad rank(\mathbf{A}) = l \text{ and } dim V(\mathbf{A}) = n - 1 - rank(\mathbf{A}) = n - 1 - l.$$

Define also

$$T(\mathbf{A}) := \left\{\nu \in \mathbb{N}_0^n \mid \mathbf{A}(\nu) = \mathbf{0} \text{ and } \prod_i \nu_i = 0\right\} \text{ and } T^*(\mathbf{A}) = T(\mathbf{A}) \setminus \{\mathbf{0}\};$$

$$(8)\ c(\mathbf{A}) := \frac{1}{2}\#\{\epsilon \in \{-1, +1\}^n \mid \prod_{j=1}^n \epsilon_j^{a_{i,j}} = 1\ \forall i = 1, \ldots, l\}$$

Define $\mathcal{E}(\mathbf{A}) := \mathcal{E}(T^*(\mathbf{A})) = convex\ hull(T^*(\mathbf{A}) + \mathbb{R}_+^n)$ the Newton polyhedron of $T^*(\mathbf{A})$ and set $\mathcal{F}_0(\mathbf{A})$ to denote the smallest face of $\mathcal{E}(\mathbf{A})$ that meets the diagonal. We then introduce the following:

(1) $\rho(\mathbf{A}) := \#(\mathcal{F}_0(\mathbf{A}) \cap T^*(\mathbf{A})) - dim(\mathcal{F}_0(\mathbf{A}))$;
(2) $\mathcal{E}^0(\mathbf{A}) := \{\mathbf{x} \in \mathbb{R}^n;\ \langle \mathbf{x}, \boldsymbol{\nu}\rangle \geq 1\ \forall \boldsymbol{\nu} \in \mathcal{E}(\mathbf{A})\}$ (the dual of $\mathcal{E}(\mathbf{A})$);
(3) $\iota(\mathbf{A}) := \min\{|\mathbf{c}|;\ \mathbf{c} \in \mathcal{E}^0(\mathbf{A}) \cap \mathbb{R}_+^n\}$ (the index of $\mathcal{E}(\mathbf{A})$).

Fix now a (generalized) polynomial $P = P(X_1, \ldots, X_n)$ with positive coefficients and assume that P is elliptic and homogeneous of degree $d > 0$. Denote by H_P the height of $\mathbb{P}^{n-1}(\mathbb{Q})$ associated to P (see (3) in §1). We introduce also the following notations:

(1) Writing P as a sum of monomials $P(X) = b_1 X^{\boldsymbol{\gamma}^1} + \cdots + b_r X^{\boldsymbol{\gamma}^r}$, we set $\mathbf{b} = (b_1, \ldots, b_r) \in \mathbb{R}_+^{*r}$
(2) Defining $\boldsymbol{\alpha}^1, \ldots, \boldsymbol{\alpha}^n \in \mathbb{R}_+^r$ to be the row vectors of the matrix that equals the transpose of the matrix with rows $\boldsymbol{\gamma}^1, \ldots, \boldsymbol{\gamma}^r$, we set $I^*(\mathbf{A}) = \{\sum_{i=1}^n \beta_i \boldsymbol{\alpha}^i \mid \boldsymbol{\beta} \in \mathcal{F}_0(\mathbf{A}) \cap T^*(\mathbf{A})\}$.

We can now state the first result as follows.

THEOREM 1. *If the Newton polyhedron $\mathcal{E}(\mathbf{A})$ has a compact face which meets the diagonal, then the height zeta function* $\mathbf{s} \mapsto Z_{H_P}(U(\mathbf{A}); s) := \sum_{M \in U(\mathbf{A})} H_P^{-s}(M)$ *is holomorphic in the half-plane $\{s \in \mathbb{C} \mid \sigma > \iota(\mathbf{A})\}$, and there exists $\eta > 0$ such that $s \mapsto Z_{H_P}(U(\mathbf{A}); s)$ has meromorphic continuation with moderate growth to the half-plane $\{\sigma > \iota(\mathbf{A}) - \eta\}$ with only one possible pole at $s = \iota(\mathbf{A})$ of order at most $\rho(\mathbf{A})$.*

If we assume in addition that $dim(\mathcal{F}_0(\mathbf{A})) = dim V(\mathbf{A})$, then $s = \iota(\mathbf{A})$ is indeed a pole of order $\rho(\mathbf{A})$ and $Z_{H_P}(U(\mathbf{A}); s) \sim_{s \to \iota(\mathbf{A})} \dfrac{C_0(\mathbf{A}; H_P)}{(s - \iota(\mathbf{A}))^{\rho(\mathbf{A})}}$, where

$$C_0(\mathbf{A}; H_P) := c(\mathbf{A})\ d^{\rho(\mathbf{A})} A_0(I^*(\mathbf{A}); \mathbf{1}; \mathbf{b}) \prod_p [(1 - \frac{1}{p})^{\#\mathcal{F}_0(\mathbf{A}) \cap T^*(\mathbf{A})} (\sum_{\boldsymbol{\nu} \in T(\mathbf{A})} \frac{1}{p^{\langle \boldsymbol{\nu}, \mathbf{c}\rangle}})]$$
$$> 0,$$

$c(\mathbf{A})$ is defined by (8), $A_0(I^*(\mathbf{A}); \mathbf{1}; \mathbf{b})$ is the volume constant associated to the polynomial $P_{(I^*(\mathbf{A}), \mathbf{1}, \mathbf{b})}$ (see §2.3.2), and where $\mathbf{c}$ is any[4] *normalized polar vector of the face $\mathcal{F}_0(\mathbf{A})$.*

[4]The constant $C_0(\mathbf{A}; H_P)$ does not depend on this choice.

By a simple adaptation of a standard tauberian argument of Landau (see for example [**E2**], Prop. 3.1)), we deduce from Theorem 1 the following arithmetical consequence:

COROLLARY 1. *If the Newton polyhedron $\mathcal{E}(\mathbf{A})$ has a compact face which meets the diagonal, then there exists a polynomial Q of degree at most $\rho(\mathbf{A})-1$ and $\theta>0$ such that as $t\to+\infty$:*

$$N_{H_P}(U(\mathbf{A});t):=\#\{M\in U(\mathbf{A})\mid H_P(M)\le t\}=t^{\iota(\mathbf{A})}Q(\log(t))+O\left(t^{\iota(\mathbf{A})-\theta}\right).$$

If we assume in addition that $dim\,(\mathcal{F}_0(\mathbf{A}))=dimV(\mathbf{A})=n-1-l$, then $Q\neq 0$, $degQ=\rho(\mathbf{A})-1$ and:

$$N_{H_P}(U(\mathbf{A});t)=C\,(\mathbf{A};H_P)\ t^{\iota(\mathbf{A})}(\log t)^{\rho(\mathbf{A})-1}\ \left(1+O\left((\log t)^{-1}\right)\right)\ \textit{where}$$

$C\,(\mathbf{A};H_P)\ :=\ \dfrac{C_0\,(\mathbf{A};H_P)}{\iota(\mathbf{A})\ (\rho(\mathbf{A})-1)!}$, *($C_0\,(\mathbf{A};H_P)$ is the constant volume defined in Theorem 1 above).*

Theorem 1 and its corollary 1 are general results that apply to any generalized projective toric variety. Our second result applies Corollary 1 to a particular class of toric hypersurfaces that correspond to a class of problems from multiplicative number theory.

Let $n\in\mathbb{N}$ $(n\ge 3)$ and $\mathbf{a}=(a_1,\dots,a_{n-1})\in\mathbb{N}^{n-1}$. Set $q=|\mathbf{a}|=a_1+\dots+a_{n-1}$.

Consider the hypersurface: $X_{n-1}(\mathbf{a})=\{\mathbf{x}\in\mathbb{P}^{n-1}(\mathbb{Q})\mid x_1^{a_1}\dots x_{n-1}^{a_{n-1}}=x_n^q\}$ with torus: $U_{n-1}(\mathbf{a})=\{\mathbf{x}\in X_{n-1}(\mathbf{a})\mid x_1\dots x_{n-1}\neq 0\}$.

Let $P=P(X_1,\dots,X_n)$ be a generalized polynomial as in Theorem 1. Define:

$$\begin{aligned}
L_n(\mathbf{a}) &:= \left\{\boldsymbol{\nu}=(\nu_1,\dots,\nu_{n-1})\in\mathbb{N}_0^{n-1};\ q|\langle\mathbf{a},\boldsymbol{\nu}\rangle \text{ and } \nu_1\dots\nu_{n-1}=0\right\}\setminus\{\mathbf{0}\};\\
\mathcal{E}(\mathbf{a}) &:= \mathcal{E}(L_n(\mathbf{a}))=\textit{convex hull}\big(L_n(\mathbf{a})+\mathbb{R}_+^{n-1}\big) \text{ its Newton polyhedron;}\\
\mathcal{F}_0(\mathbf{a}) &= \text{the smallest face of } \mathcal{E}(\mathbf{a}) \text{ which meets the diagonal } \Delta=\mathbb{R}_+\mathbf{1};\\
J_n(\mathbf{a}) &:= L_n(\mathbf{a})\cap\mathcal{F}_0(\mathbf{a}) \quad\text{and}\quad \rho(\mathbf{a})=\#\left(J_n(\mathbf{a})\right)-n+2;\\
c(\mathbf{a}) &:= \frac{1}{2}\#\left\{(\varepsilon_1,\dots,\varepsilon_n)\in\{-1,+1\}^n\mid \varepsilon_1^{a_1}\dots\varepsilon_{n-1}^{a_{n-1}}=\varepsilon_n^q\right\}.
\end{aligned}$$

Then we have:

THEOREM 2. *Let $\mathbf{c}\in\mathbb{R}_+^{*(n-1)}$ be a normalized polar vector of the face $\mathcal{F}_0(\mathbf{a})$. There exists a polynomial Q of degree at most $\rho(\mathbf{a})-1$ and $\theta>0$ such that:*

$$N_{H_P}(U_{n-1}(\mathbf{a});t)=\#\{M\in U_n(\mathbf{a})\mid H_P(M)\le t\}=t^{|\mathbf{c}|}Q(\log t)+O\left(t^{1-\theta}\right).$$

If we assume in addition that $\mathcal{F}_0(\mathbf{a})$ is a facet of the polyhedron $\mathcal{E}(\mathbf{a})$, then $Q\neq 0$, $degQ=\rho(\mathbf{a})-1$ and

$$N_{H_P}(U_{n-1}(\mathbf{a});t)=C\,(\mathbf{a};H_P)\ t^{|\mathbf{c}|}(\log t)^{\rho(\mathbf{a})-1}\times\Big(1+O\big((\log t)^{-1}\big)\Big)$$

where:

$$C\,(\mathbf{a};H_P){:=}\frac{c(\mathbf{a})\ d^{\rho(\mathbf{a})}\ A_0\left(\mathcal{T}_{\mathbf{c}};\tilde{P}\right)}{|\mathbf{c}|\cdot(\rho(\mathbf{a})-1)!}\cdot\prod_p\left(1-\frac{1}{p}\right)^{\rho(\mathbf{a})+n-2}\sum_{\substack{\boldsymbol{\nu}\in\mathbb{N}_0^{n-1};\ q\ |\ \langle\mathbf{a},\boldsymbol{\nu}\rangle\\ \nu_1\cdots\nu_{n-1}=0}}p^{-\langle\mathbf{c},\boldsymbol{\nu}\rangle}>0,$$

$A_0\left(\mathcal{T}_{\mathbf{c}};\tilde{P}\right) > 0$ is the mixed volume constant (see §2.3.3) associated to the polynomial $\tilde{P}(X_1,\ldots,X_{n-1}) := P(X_1,\cdots,X_{n-1},\prod_{j=1}^{n-1} X_j^{a_j/q})$ and to the pair $\mathcal{T}_{\mathbf{c}} = \left(J_n(\mathbf{a}),(u(\boldsymbol{\beta})_{\boldsymbol{\beta}\in J_n(\mathbf{a})}\right)$ with $u(\boldsymbol{\beta}) = 1\ \forall \boldsymbol{\beta} \in J_n(\mathbf{a})$.

Remark 1: An interesting question is to determine the precise set of exponents $a_1,\ldots,a_{n-1} \geq 1$ (for given q and n) such that $\mathcal{F}_0(\mathbf{a})$ is a facet of $\mathcal{E}(\mathbf{a})$. It seems reasonable to believe that the complement of this set is thin in a suitable sense (when q is allowed to be arbitrary).

Remark 2: *If $\mathbf{a} = (a_1,\ldots,a_{n-1}) \in \mathbb{N}^{n-1}$ satisfies the property that each a_i divides $q = a_1 + \cdots + a_{n-1}$, then $\mathcal{F}_0(\mathbf{a}) = conv\left(\{\frac{q}{a_i}\mathbf{e}_i \mid i = 1,\ldots,n-1\}\right)$. Thus, $\mathcal{F}_0(\mathbf{a})$ is a facet of $\mathcal{E}(\mathbf{a})$ and the more precise second part of Theorem 2 applies.*

Remark 3: The analogue of Theorem 2 had been proved for the height H_∞ and the particular surface $X_3(\mathbf{1})$ in several earlier works (see [**F**], [**Bre1**], [**Sal**], [**HBM**]). More recently, the article [**BEL**] extended these earlier results to $X_{n-1}(\mathbf{1})$ for any $n \geq 3$ (but only used H_∞). Theorem 2 should therefore be understood as a natural generalization of all these earlier results.

To illustrate concretely what the theorem is saying, we write out the details when $d = 2$, $n = 4$, and $P = \sum_{i=1}^4 X_i^2$. In this event, to apply the discussion in §2.3.3 to find $A_0(\mathcal{T},\tilde{P})$, we must compute the volume constant $A_0(P_3)$ (see definition 1 in §2.3.1) for the generalized polynomial $P_3 := \tilde{P}_{(J_3^*,\mathbf{u}_3^*,\mathbf{1})}$. An exercise left to the reader will show the following:

$$P_3{=}X_1^6X_4^4X_5^4X_6^2X_8^2{+}X_2^6X_4^2X_6^4X_7^4X_9^2{+}X_3^6X_5^2X_7^2X_8^4X_9^4{+}X_1^2X_2^2X_3^2X_4^2X_5^2X_6^2X_7^2X_8^2X_9^2.$$

By comparing our result with the result obtained by La Bretèche [Bre1] for the height H_∞, we get

$$N_{H_P}(U_3(\mathbb{Q});t) \sim 2^{11}\ A_0(P_3)\ N_{H_\infty}(U_3(\mathbb{Q});t) \textbf{ as } t \to \infty.$$

Therefore, the constant volume $A_0(P_3)$ associated to the polynomial P_3 defined above, measures the dependency of the counting function on the height.

3.2. Main results about mixed zeta functions. Theorem 1, Corollary 1 and Theorem 2 are simple consequences of our *fundamental* theorem (see Theorem 3 in §3.2.2 below) which is the main ingredient of this paper.

3.2.1. *Functions of finite type: Definition and examples.* For any arithmetic function $f : \mathbb{N}^n \to \mathbb{C}$, we can define, at least formally, the Dirichlet series

$$\mathcal{M}(f;\mathbf{s}) := \sum_{\mathbf{m}\in\mathbb{N}^n} \frac{f(m_1,\ldots,m_n)}{m^{s_1}\ldots m_n^{s_n}}. \tag{9}$$

Several works (see for example [**BEL**], [**Bre2**], [**K**], [**Mo**]) indicate that the following property should be satisfied by classes of f that are typically encountered in arithmetic problems.

DEFINITION 2. An arithmetic function $f : \mathbb{N}^n \to \mathbb{C}$ is said to be of **finite type** at a point $\mathbf{c} \in \mathbb{R}_+^{*n}$ if $\mathcal{M}(f;\mathbf{s})$ converges absolutely in $\{\mathbf{s} \in \mathbb{C}^n \mid \sigma_i = \Re(s_i) > c_i\ \forall i\}$ and can be continued as a meromorphic function to a neighborhood of $\mathbf{c}$, as follows: *There exists a pair $\mathcal{T}_{\mathbf{c}} = (I_{\mathbf{c}},\mathbf{u})$, where $I_{\mathbf{c}}$ is a non empty subset of $\mathbb{R}_+^n \setminus \{\mathbf{0}\}$ and*

$u = \big(u(\boldsymbol{\beta})\big)_{\boldsymbol{\beta}\in I_{\mathbf{c}}}$ *is a vector of positive integers, such that*

$$\text{(10)}\qquad \mathbf{s} \mapsto H(f;\mathcal{T}_{\mathbf{c}};\mathbf{s}) := \Big(\prod_{\boldsymbol{\beta}\in I_{\mathbf{c}}} \left(\langle \mathbf{s},\boldsymbol{\beta}\rangle\right)^{u(\boldsymbol{\beta})}\Big)\ \mathcal{M}(f;\mathbf{c}+\mathbf{s})$$

has a holomorphic continuation with moderate growth (see (7) §2.1) to the set $\{\mathbf{s}\in\mathbb{C}^n \mid \sigma_i > -\varepsilon_0\ \forall i = 1,\dots,n\}$ *for some* $\varepsilon_0 > 0$.

We can assume[5] that $\langle \boldsymbol{\beta},\mathbf{c}\rangle = 1$ for each $\boldsymbol{\beta}\in I_{\mathbf{c}}$. In this case we call $\mathcal{T}_{\mathbf{c}} = (I_{\mathbf{c}},\mathbf{u})$ a "**regularizing pair**" of $\mathcal{M}(f;\mathbf{s})$ at $\mathbf{c}$.
If in addition $H(f;\mathcal{T}_{\mathbf{c}};\mathbf{0})\neq 0$, we call the pair $\mathcal{T}_{\mathbf{c}}$ the **polar type** of $\mathcal{M}(f;.)$ at $\mathbf{c}$.

3.2.2. *Main result in the case of arithmetic functions of finite type.* . Let $P = P(X_1,\dots,X_n)$ be a generalized polynomial with positive coefficients and of degree $d>0$. Let $f:\mathbb{N}^n\mapsto\mathbb{C}$ be an arithmetic function of finite type at a point $\mathbf{c}\in\mathbb{R}_+^{*n}$. Let $\mathcal{T}_{\mathbf{c}} = (I_{\mathbf{c}},\mathbf{u})$ be a regularizing pair of $\mathcal{M}(f;.)$ at $\mathbf{c}$ as in Definition 2.

$$\text{(11)}\qquad \text{Set}\qquad Z(f;P;s) := \sum_{(m_1,\dots,m_n)\in\mathbb{N}^n} \frac{f(m_1,\dots,m_n)}{P(m_1,\dots,m_n)^{s/d}}.$$

Our results above follow from the following *fundamental* theorem:

THEOREM 3. *If P is elliptic and homogeneous, then $s\mapsto Z(f;P;s)$ is holomorphic in $\{s:\sigma>|\mathbf{c}|\}$ and there exists $\eta>0$ such that $s\mapsto Z(f;P;s)$ has a meromorphic continuation[6] with moderate growth to $\{\sigma>|\mathbf{c}|-\eta\}$ with at most one pole at $s=|\mathbf{c}|$ of order at most $\rho_0(\mathcal{T}_{\mathbf{c}}) := \sum_{\boldsymbol{\beta}\in I_{\mathbf{c}}} u(\boldsymbol{\beta}) - rank\,(I_{\mathbf{c}})+1$.*
Assume in addition that the following two properties are satisfied:

(1) $\mathbf{1}\in con^*(I_{\mathbf{c}})$;
(2) *there exists a function K holomorphic in a tubular neighborhood[7] of $\mathbf{0}$ such that: $H(f;\mathcal{T}_{\mathbf{c}};\mathbf{s}) = K\left((\langle\boldsymbol{\beta},\mathbf{s}\rangle)_{\boldsymbol{\beta}\in I_{\mathbf{c}}}\right)$, where $H(f;\mathcal{T}_{\mathbf{c}};\mathbf{s})$ is the function defined in (10).*

Then,

$$Z(f;P;s) = \frac{C_0(f;P)}{(s-|\mathbf{c}|)^{\rho_0(\mathcal{T}_{\mathbf{c}})}} + O\left(\frac{1}{(s-|\mathbf{c}|)^{\rho_0(\mathcal{T}_{\mathbf{c}})-1}}\right)\quad \text{as } s\to|\mathbf{c}|,$$

where $C_0(f;P) := H(f;\mathcal{T}_{\mathbf{c}};\mathbf{0})d^{\rho_0(\mathcal{T}_{\mathbf{c}})}A_0(\mathcal{T}_{\mathbf{c}},P)$ and where $A_0(\mathcal{T}_{\mathbf{c}},P)>0$ is the mixed volume constant associated to P and $\mathcal{T}_{\mathbf{c}}$ (see §2.3.3).
In particular, $s=|\mathbf{c}|$ is a pole of order $\rho_0(\mathcal{T}_{\mathbf{c}})$ if and only if $H(f;\mathcal{T}_{\mathbf{c}};\mathbf{0})\neq 0$.

By a simple adaptation of a standard tauberian argument of Landau (see for example [**E2**], Prop. 3.1)), we deduce from Theorem 3 the following arithmetical consequence:

---

[5] by replacing each vector $\boldsymbol{\beta}$ by the vector $\frac{1}{\langle\boldsymbol{\beta},\mathbf{c}\rangle}\boldsymbol{\beta}$

[6] Set $\mathcal{F}_f = \{\mathbf{c}\in\mathbb{R}_+^{*n}\mid f \text{ is of finite type at } \mathbf{c}\}$, $\sigma_f = \inf\{|\mathbf{c}|;\ \mathbf{c}\in\mathcal{F}_f\}$ and $\rho_f = \inf\{\rho_0(\mathcal{T}_{\mathbf{c}})\mid \mathbf{c}\in\mathcal{F}_f \text{ and } |\mathbf{c}|=\sigma_f\}$. The first part of Theorem 3 actually says that $Z(f;P;s)$ is *holomorphic* in $\{s:\sigma>\sigma_f\}$ and there exists $\eta>0$ such that $s\mapsto Z(f;P;s)$ has a meromorphic continuation with moderate growth to $\{\sigma>\sigma_f-\eta\}$ with at most one pole at $s=\sigma_f$ of order at most ρ_f. In particular, if $\mathbf{c}$ satisfies all assumptions of second part of Theorem 3, then $|\mathbf{c}|=\sigma_f$.

[7] By tubular neighborhood of $\mathbf{0}$, we mean a neighborhood of the form $\{\mathbf{s}\in\mathbb{C}^n\mid\ |\sigma_i|<\varepsilon\ \forall i\}$ where $\varepsilon>0$.

COROLLARY 2. *If P is as above and $f \geq 0$. There exist a polynomial Q of degree at most $\rho_0(\mathcal{T}_{\mathbf{c}}) - 1$ and $\theta > 0$ such that:*

$$N(f; P^{1/d}; t) := \sum_{\{\mathbf{m} \in \mathbb{N}^n;\ P^{1/d}(\mathbf{m}) \leq t\}} f(m_1, \ldots, m_n) = t^{|\mathbf{c}|} Q(\log t) + O\left(t^{|\mathbf{c}|-\theta}\right).$$

In particular, there exists a nonnegative constant $C(f; P)$ such that

$$N(f; P^{1/d}; t) = C(f; P)\ t^{|\mathbf{c}|} (\log t)^{\rho_0(\mathcal{T}_{\mathbf{c}})-1} + O\left(t^{|\mathbf{c}|} (\log t)^{\rho_0(\mathcal{T}_{\mathbf{c}})-2}\right) \quad \text{as } t \to \infty,$$

Assume in addition that the properties 1 and 2 of theorem 3 are satisfied. Then,

$$C(f; P) := \frac{C_0(f; P)}{|\mathbf{c}|\ (\rho_0(\mathcal{T}_{\mathbf{c}}) - 1)!} = \frac{H(f; \mathcal{T}_{\mathbf{c}}; \mathbf{0})\ d^{\rho_0(\mathcal{T}_{\mathbf{c}})}\ A_0(\mathcal{T}_{\mathbf{c}}, P)}{|\mathbf{c}|\ (\rho_0(\mathcal{T}_{\mathbf{c}}) - 1)!} \geq 0,$$

In particular, $C(f; P) > 0$ if and only if $H(f; \mathcal{T}_{\mathbf{c}}; \mathbf{0}) \neq 0$.

Remark 1. To our knowledge, the only result comparable to Corollary 2 is due to La Bretèche [**Bre2**] who proved estimates for densities $N(f; \|.\|_\infty; t)$ using the max norm $\|\mathbf{x}\|_\infty = \max_i |x_i|$. Corollary 2 extends his results to a large class of norms or generalized norms. ◇

Remark 2. Up to normalization factors, the volume constant $C_0(f; P)$ in Theorem 3 is the product of two terms, one arithmetic the other geometric. The arithmetic factor equals $H(f; \mathcal{T}_{\mathbf{c}}; \mathbf{0})$. In the examples that have been worked out, this is typically an eulerian product and depends only on the arithmetic function f. The second part is the mixed volume constant $A_0(\mathcal{T}_{\mathbf{c}}; P)$ which reflects a joint dependence upon both f and P.
An interesting point to be observed here, is that $A_0(\mathcal{T}_{\mathbf{c}}; P)$ is, by definition, the volume constant of a polynomial constructed explicitly from both f and P (see §2.3.3). This polynomial is not necessarily elliptic and has in general more variables than the original polynomial P.

Remark 3. Assumption (2) of Theorem 3 is always satisfied if *rank* $I_{\mathbf{c}} = n$.

Remark 4. It is clear that any monomial $f(\mathbf{m}) = m_1^{\mu_1} \ldots m_n^{\mu_n}$ is of finite type since $\mathcal{M}(f; \mathbf{s}) = \prod_{i=1}^n \zeta(s_i - \mu_i)$. Thus, the polar type of $\mathcal{M}(f; .)$ at $\mathbf{c} = (1 + \mu_1, \ldots, 1 + \mu_n)$ is $\mathcal{T}_{\mathbf{c}} = (I; \mathbf{u})$ where $I = \{\frac{1}{\mu_i + 1} \mathbf{e}_i : i = 1, \ldots, n\}$ and $u(\boldsymbol{\beta}) = 1$ for all $\boldsymbol{\beta} \in I$. It's easy to see that all assumptions of Corollary 2 are satisfied. In particular, for $\boldsymbol{\mu} = \mathbf{0}$ (i.e $f \equiv 1$), Corollary 2 implies that:

$$\#\{\mathbf{m} \in \mathbb{N}^n \mid \mathcal{N}_P(\mathbf{m}) \leq x\} = \frac{1}{n} \left(\int_{\mathbb{S}^{n-1} \cap \mathbb{R}_+^n} P^{-n/d}(\mathbf{u})\ d\sigma(\mathbf{u}) \right) x^n + O(x^{n-\delta}). \tag{12}$$

Therefore conclusion of Corollary 2 agree with that obtained by Mahler [**Ma**] (see Proposition 1) and Sargos (see [**Sar1**] chap 3 or [**Sar2**]).

3.2.3. *Uniform multiplicative functions.* The Theorem 3 (and its corollary 2) above applies to a large class of arithmetic functions f. However, to use this theorem we need to choose a suitable singular point $\mathbf{c}$ in the boundary of the domain of convergence of $\mathcal{M}(f; \mathbf{s})$ to determine its polar type $\mathcal{T}_{\mathbf{c}}$ and to verify all the required assumptions. In general, It is not easy to choose such suitable point $\mathbf{c}$ and this question was not treated by La Bretèche in [**Bre2**]. In proposition 2 below we will address this problem for the class of uniform multiplicative functions f which are sufficient to prove our main results about rational points on generalized projective toric varieties.

DEFINITION 3. A multiplicative function $f : \mathbb{N}^n \mapsto \mathbb{N}_0$ is said to be **uniform** if there exists a function $g = g_f : \mathbb{N}_0^n \mapsto \mathbb{N}_0$ and two constants $M, C > 0$ such that for all prime numbers p and all $\boldsymbol{\nu} \in \mathbb{N}_0^n$, $f(p^{\nu_1}, \ldots, p^{\nu_n}) = g(\boldsymbol{\nu}) \leq C\,(1 + |\boldsymbol{\nu}|)^M$.

We fix a uniform multiplicative function $f : \mathbb{N}^n \mapsto \mathbb{N}_0$ throughout the rest of §3.2.3. We then define:

(1) $S^*(g) = \{\boldsymbol{\nu} \in \mathbb{N}_0^n \setminus \{\mathbf{0}\} \mid g(\boldsymbol{\nu}) \neq 0\}$ and assume that $S^*(g) \neq \emptyset$;
(2) $\mathcal{E}(f) := \mathcal{E}(S^*(g)) = \mathit{convex\ hull}\big(S^*(g) + \mathbb{R}_+^n\big)$ the Newton polyhedron determined by $S^*(g)$;
(3) $\mathcal{E}(f)^o := \{\mathbf{x} \in \mathbb{R}^n \mid \langle \mathbf{x}, \boldsymbol{\nu}\rangle \geq 1\ \forall \boldsymbol{\nu} \in \mathcal{E}(f)\}$ the dual of $\mathcal{E}(f)$;
(4) $\iota(f) := \min\left\{|\mathbf{c}| \mid \mathbf{c} \in \mathcal{E}(f)^o \cap \mathbb{R}_+^n\right\}$ (the "index" of f);
(5) $\mathcal{F}_0(f) :=$ the smallest face of $\mathcal{E}(f)$ which meets the diagonal. We denote its set of normalized polar vectors by $pol_0\left(\mathcal{F}_0(f)\right)$;
(6) $I_f := \mathcal{F}_0(f) \cap S^*(g)$ and $\mathbf{u}_f := (g(\boldsymbol{\nu}))_{\boldsymbol{\nu} \in I_f}$.

With theses notations and those of definition 2 in §3.2.1, we have:

PROPOSITION 2. *Assume f is a uniform multiplicative function and that the face $\mathcal{F}_0(f)$ is compact. Let $\mathbf{c} \in pol_0(\mathcal{F}_0(f))$. Then f is of finite type at $\mathbf{c}$, $\mathbf{1} \in con^*(I_f)$, and the pair $\mathcal{T}_\mathbf{c} = (I_f; \mathbf{u}_f)$ is the polar type of $\mathcal{M}(f;.)$ at $\mathbf{c}$.*

Moreover, we have

$$H(f;\mathcal{T}_\mathbf{c};\mathbf{0}) = \prod_p \left(1 - \frac{1}{p}\right)^{\sum_{\boldsymbol{\nu}\in I_f} g(\boldsymbol{\nu})} \left(\sum_{\boldsymbol{\nu}\in\mathbb{N}_0^n} \frac{g(\boldsymbol{\nu})}{p^{\langle \boldsymbol{\nu}, \mathbf{c}\rangle}}\right) > 0.$$

If we assume in addition that $dim\mathcal{F}_0(f) = rank(S^*(g)) - 1$, there exists a function K holomorphic in a tubular neighborhood of $\mathbf{0}$ such that: $H(f;\mathcal{T}_\mathbf{c};\mathbf{s}) = K\left(\left(\langle \boldsymbol{\nu}, \mathbf{s}\rangle\right)_{\boldsymbol{\nu}\in I_f}\right)$.

Remark: The assumption $dim\,\mathcal{F}_0(f) = rank(S^*(g)) - 1$ is automatically satisfied if for example the face $\mathcal{F}_0(f)$ is a facet of $\mathcal{E}(f)$.

Remark: The polyhedron $\mathcal{E}(f)$ of interest here is *not* in the $\mathbf{s}$ space that one might think would normally be associated to the multiple zeta function $\mathcal{M}(f;\mathbf{s})$, but rather is in the exponent space of $\boldsymbol{\nu}$; i.e. the domain of g.This implies in particular that the $\mathbf{s}$ space is, in some sense, playing the role of the polar (or dual) space.

4. Proof of Theorem 3

The starting point of our method is the remarkable formula of Mellin:

(13)
$$\frac{\Gamma(s)}{\left(\sum_{k=0}^r w_k\right)^s} = \frac{1}{(2\pi i)^r} \int_{\rho_1 - i\infty}^{\rho_1 + i\infty} \cdots \int_{\rho_r - i\infty}^{\rho_r + i\infty} \frac{\Gamma(s - z_1 - \cdots - z_r)\prod_{i=1}^r \Gamma(z_i)\, d\mathbf{z}}{w_0^{s - z_1 - \cdots - z_r}\left(\prod_{k=1}^r w_k^{z_k}\right)}$$

valid if $\forall i = 0, \ldots, r$, $\Re(w_i) > 0$, $\forall i = 1, \ldots, r$ $\rho_i > 0$ and $\Re(s) > \rho_1 + \cdots + \rho_r$.

Methods that use Mellin's formula in the classical case do not adapt easily to prove Theorem 3. The main reason this appears to be the case is that the inductive procedure, in which one inducts on the number of monomials in an expression for P, is incapable of the precision we need to prove our main result, that is, an explicit description of the top order term in the principal part of $Z(f;P;s)$ at its first pole. So, in order to prove theorem 3, our strategy is the following: First, we use Mellin's formula (13) (with a suitable integer r) to write $Z(f;P;s)$ as an integral of a twist of $\mathcal{M}(f;\mathbf{s})$ over a chain of $\mathbb{C}^r$ (see section §4.2 bellow). This is needed because the arithmetic information needed to derive information about the possible first pole σ_0 is contained in the polar divisor of the series $\mathcal{M}(f;\mathbf{s})$.

In §4.1.2 the important ingredient for this section (i.e lemma 3) is proved. This lemma gives (under suitable assumptions) the analytic continuation of integrals over a chain in $\mathbb{C}^r$, identifies for each of them a possible first pole σ_0, and gives a *very precise* bound for its order ρ_0. This precise bound is *crucial* in the proof of the main result of this paper that is the second part of theorem 3. To obtain this information, we must argue more carefully than in classical proofs that use Mellin's Formula. For this reason our proof is more technical, and uses in particular a double induction argument.

In §4.1.3 we will prove our second important ingredient (i.e lemma 4). This lemma gives for a class of integrals over chains in $\mathbb{C}^r$, the top order term in the principal part at the first pole.

In §4.1.1 we will prove two elementary but useful lemmas (lemmas 1 and 2) which will help us justify the convergence of integrals over chains of $\mathbb{C}^r$.

4.1. Four lemmas and their proofs.

4.1.1. *Two elementary lemmas (i.e lemmas 1 and 2).*

LEMMA 1. *Let $\boldsymbol{\mu}^1,\dots,\boldsymbol{\mu}^k$ be k vectors of $\mathbb{R}^r$ and let $l\in\mathbb{R}$. Set for all $\tau\in\mathbb{R}$:*

$$F_r(\tau)=\int_{\mathbb{R}^r}\prod_{i=1}^{r}(1+|y_i|)^l e^{\frac{\pi}{2}\left(|\tau|-\sum_{i=1}^r|y_i|-\sum_{i=1}^k|\langle\boldsymbol{\mu}^i,\mathbf{y}\rangle|-\left|\tau-\sum_{i=1}^r y_i-\sum_{i=1}^k\langle\boldsymbol{\mu}^i,\mathbf{y}\rangle\right|\right)}dy_1\dots dy_r.$$

Then F_r has moderate growth in τ. More precisely, there exist $A=A(r), B=B(r)$ and $C=C(l,r)>0$ such that: $\forall\tau\in\mathbb{R}\quad F_r(\tau)\le C\ (1+|\tau|)^{A|l|+B}$.

Remark: For $r=1$ a more precise version of lemma 1 can be found in [**MT**].

Proof of Lemma 1:

Set $\psi_r(l;\tau):=\int_{\mathbb{R}^r}\prod_{i=1}^r(1+|y_i|)^l e^{\frac{\pi}{2}\left(|\tau|-\sum_{i=1}^r|y_i|-|\tau-\sum_{i=1}^r y_i|\right)}\ dy_1\dots dy_r$. Since

$$\sum_{i=1}^k|\langle\boldsymbol{\mu}^i,\mathbf{y}\rangle|+\left|\tau-\sum_{i=1}^r y_i-\sum_{i=1}^k\langle\boldsymbol{\mu}^i,\mathbf{y}\rangle\right|\ge\left|\tau-\sum_{i=1}^r y_i\right|.$$

It follows that $F_r(\tau)\le\psi_r(l;\tau)$. So to prove the lemma it suffices to prove the asserted bound for $\psi_r(l;\tau)$. We do this by induction on r.

- $r=1$: It suffices to prove the inequality for $\psi_1^+(l;\tau)=\int_0^{+\infty}(1+y)^l e^{\frac{\pi}{2}(|\tau|-y-|\tau-y|)}\ dy$ and $\tau>1$.

For $\tau>1$ we have:

$$\begin{aligned}
\psi_1^+(l;\tau) &= \int_0^{+\infty}(1+y)^l e^{\frac{\pi}{2}(\tau-y-|\tau-y|)}\ dy\\
&= \int_0^{\tau}(1+y)^l e^{\frac{\pi}{2}(\tau-y-|\tau-y|)}\ dy+\int_\tau^{+\infty}(1+y)^l e^{\frac{\pi}{2}(\tau-y-|\tau-y|)}\ dy\\
&\ll_l \int_0^{\tau}(1+y)^l\ dy+e^{\pi\tau}\int_\tau^{+\infty}(1+y)^l e^{-\pi y}\ dy\\
&\ll_l (1+\tau)^{l+1}+\int_1^{+\infty}(t+\tau)^l e^{-\pi t}\ dt\\
&\ll_l (1+\tau)^{l+1}+(\tau+1)^l\int_1^{+\infty}(1+t)^l e^{-\pi t}\ dt\ll_l(1+\tau)^{l+1}.
\end{aligned}$$

• $r \geq 2$: Assume that the lemma is true for $r-1$. Thus there exist $A = A(r-1)$ and $B = B(r-1) > 0$ such that $\psi_{r-1}(l;\tau) \ll_l (1+|\tau|)^{A|l|+B}$ $(\tau \in \mathbb{R})$. It follows that we have uniformly in $\tau \in \mathbb{R}$:

$$\begin{aligned}
\psi_r(l;\tau) &:= \int_{\mathbb{R}^r} \prod_{i=1}^r (1+|y_i|)^l e^{\frac{\pi}{2}\left(|\tau| - \sum_{i=1}^r |y_i| - |\tau - \sum_{i=1}^r y_i|\right)}\, dy_1 \dots dy_r \\
&= \int_{\mathbb{R}} \psi_{r-1}(l;\tau - y_r)\,(1+|y_r|)^l e^{\frac{\pi}{2}(|\tau| - |\tau-y_r| - |y_r|)}\, dy_r \\
&\ll_{l,r} \int_{\mathbb{R}} (1+|\tau - y_r|)^{A|l|+B} (1+|y_r|)^l e^{\frac{\pi}{2}(|\tau| - |\tau-y_r| - |y_r|)}\, dy_r \\
&\qquad \text{(by the induction hypothesis)} \\
&\ll_{l,r} (1+|\tau|)^{A|l|+B} \int_{\mathbb{R}} (1+|y_r|)^{(A+1)|l|+B} e^{\frac{\pi}{2}(|\tau| - |y_r| - |\tau-y_r|)}\, dy_r \\
&\ll_{l,r} (1+|\tau|)^{A|l|+B} \psi_1\left((A+1)|l| + B;\tau\right).
\end{aligned}$$

We complete the proof by using the preceding estimate when $r = 1$. ◊

LEMMA 2. *Let $p \in \mathbb{R}$, $\mathbf{a} = (a_1, \dots, a_r) \in \mathbb{R}_+^{*n}$ and $\varepsilon \in (0, \inf_{i=1,\dots,r} a_i)$. Set $W(s;\mathbf{z}) = W(s; z_1, \dots, z_r) := \Gamma(s - p - z_1 - \dots - z_r)\,\Gamma(s)^{-1}\,\left(\prod_{i=1}^r \Gamma(a_i + z_i)\right)$. Then $(s;\mathbf{z}) \mapsto W(s;\mathbf{z})$ is holomorphic in the set*

$$\{(s,\mathbf{z}) = (\sigma + i\tau, \mathbf{x} + i\mathbf{y}) \in \mathbb{C} \times \mathbb{C}^r \mid \sigma > p - r\varepsilon \text{ and } |\Re(z_i)| < \varepsilon\ \forall i = 1, \dots, r\}$$

in which it satisfies the estimate:

$$\begin{aligned}
W(\sigma + i\tau; \mathbf{x} + i\mathbf{y}) &\ll_{\sigma,p,\mathbf{a},\varepsilon} (1+|\tau|)^{2|\sigma|+|p|+r\varepsilon+1} \prod_{i=1}^r (1+|y_i|)^{|\sigma|+|p|+a_i+(r+1)\varepsilon+1} \\
&\qquad \times e^{\frac{\pi}{2}\left(|\tau| - \sum_{i=1}^r |y_i| - |\tau - \sum_{i=1}^r y_i|\right)}.
\end{aligned}$$

Proof of lemma 2:

It is well known that the Euler function $z \mapsto \Gamma(z)$ is holomorphic and has no zeros in the half-plane $\{z \in \mathbb{C} \mid \Re(z) > 0\}$. Moreover for any x_1, x_2 verifying $x_2 > x_1 > 0$ we have uniformly in $x \in [x_1, x_2]$ and $y \in \mathbb{R}$:

(14) $$|\Gamma(x+iy)| = \sqrt{2\pi}(1+|y|)^{x-1/2} e^{-\pi|y|/2}\left(1 + O_{x_1,x_2}(|y|^{-1})\right) \text{ as } |y| \to \infty.$$

We deduce that $(s,\mathbf{z}) \mapsto W(s,\mathbf{z})$ is holomorphic in

$$\{(s,\mathbf{z}) = (\sigma + i\tau, \mathbf{x} + i\mathbf{y}) \in \mathbb{C} \times \mathbb{C}^r \mid \sigma > p - r\varepsilon \text{ and } |\Re(z_i)| < \varepsilon\ \forall i = 1, \dots, r\}$$

in which it satisfies the estimate:

$$\begin{aligned}
W(s;\mathbf{z}) &\ll_{\sigma,p,\mathbf{a},\varepsilon} (1+|\tau - y_1 - \dots - y_r|)^{\sigma - p - x_1 - \dots - x_r - \frac{1}{2}} (1+|\tau|)^{-\sigma + \frac{1}{2}} \\
&\qquad \times \left(\prod_{i=1}^r (1+|y_i|)^{a_i + x_i - \frac{1}{2}}\right) e^{\frac{\pi}{2}\left(|\tau| - \sum_{i=1}^r |y_i| - |\tau - \sum_{i=1}^r y_i|\right)} \\
&\ll_{\sigma,p,\mathbf{a},\varepsilon} (1+|\tau|)^{2|\sigma|+|p|+r\varepsilon+1} \prod_{i=1}^r (1+|y_i|)^{|\sigma|+|p|+a_i+(r+1)\varepsilon+1} \\
&\qquad \times e^{\frac{\pi}{2}\left(|\tau| - \sum_{i=1}^r |y_i| - |\tau - \sum_{i=1}^r y_i|\right)}.
\end{aligned}$$

This end the proof of lemma 2. ◊

4.1.2. *First crucial Lemma: Lemma* 3. Before stating the lemma, we first introduce some needed notations:
Let $\sigma_1 \in \mathbb{R}_+^*$, $q \in \mathbb{N}$, $r \in \mathbb{N}$ and $\boldsymbol{\phi} \in \mathbb{R}^r$.
Let I be a finite subset of $\mathbb{R}^r \setminus \{\mathbf{0}\}$ and $\mathbf{u} = (u(\boldsymbol{\alpha}))_{\boldsymbol{\alpha}\in I}$ be a vector of positive integers.
Set for all $\delta', \varepsilon' > 0$,

$$\mathcal{D}_r(\delta';\varepsilon') := \{(s,\mathbf{z}) \in \mathbb{C}\times\mathbb{C}^r \mid \sigma = \Re s > \sigma_1 - \delta' \text{ and } |\Re(z_i)| < \varepsilon' \ \forall i = 1,\dots,r\}.$$

Let $\varepsilon, \delta > 0$. Let $L(s;\mathbf{z})$ be a holomorphic function on $\mathcal{D}_r(2\delta;2\varepsilon)$.
Assume that there exist $A, B, w > 0$ and $\boldsymbol{\mu}^1,\dots,\boldsymbol{\mu}^p$ vectors of $\mathbb{R}^r$ such that we have uniformly in $(s;\mathbf{z}) = (\sigma+i\tau, \mathbf{x}+i\mathbf{y}) \in \mathcal{D}_r(2\delta;2\varepsilon)$:

$$\begin{aligned} L(s;\mathbf{z}) \quad &\ll_{\sigma,\mathbf{x}} \quad (1+|\tau|)^{A|\sigma|+B}\prod_{i=1}^r (1+|y_i|)^{A|\sigma|+B} \\ &\times e^{\frac{\pi}{2}\left(|w\tau| - \sum_{i=1}^r |y_i| - \sum_{i=1}^p |\langle \boldsymbol{\mu}^i, \mathbf{y}\rangle| - \left|w\tau - \sum_{i=1}^r y_i - \sum_{i=1}^p \langle \boldsymbol{\mu}^i, \mathbf{y}\rangle\right|\right)}. \end{aligned} \tag{15}$$

We denote by I_0 the set of real numbers which are the coordinates of at least one element of I, and by $\mathbb{Q}(I_0)$ the field generated by I_0 over $\mathbb{Q}$.
Set finally for all $\boldsymbol{\rho} = (\rho_1,\dots,\rho_r) \in]-\varepsilon,+\varepsilon[^n$ such that $\rho_1,\dots,\rho_r$ are $\mathbb{Q}(I_0)$-linearly independent,

$$T_r(s) := \frac{1}{(2\pi i)^r}\int_{\rho_1-i\infty}^{\rho_1+i\infty}\dots\int_{\rho_r-i\infty}^{\rho_r+i\infty} \frac{L(s;\mathbf{z})\,dz_1\dots dz_r}{(s-\sigma_1-\langle\boldsymbol{\phi},\mathbf{z}\rangle)^q \prod_{\boldsymbol{\alpha}\in I}\langle\boldsymbol{\alpha},\mathbf{z}\rangle^{u(\boldsymbol{\alpha})}}. \tag{16}$$

Lemma 1 imply that $s \mapsto T_r(s)$ converges absolutely in $\{\sigma > \sigma_1 + \sum_{i=1}^n |\phi_i\rho_i|\}$. The key lemma of this paper is the following:

LEMMA 3. *There exists $\eta > 0$ such that $s \mapsto T_r(s)$ has a meromorphic continuation with moderate growth to the half-plane $\{\Re(s) > \sigma_1 - \eta\}$ with at most a single pole at $s = \sigma_1$.*
If $s = \sigma_1$ is a pole of $T_r(s)$ then its order is at most

$$d_r := \sum_{\boldsymbol{\alpha}\in I} u(\boldsymbol{\alpha}) - rank(I) + q - \varepsilon_0(\boldsymbol{\phi};L),$$

Where the index $\varepsilon_0(\boldsymbol{\phi};L)$ is defined by:

(1) *$\varepsilon_0(\boldsymbol{\phi};L) = 1$ if $\boldsymbol{\phi} \in con^*(I)\setminus\{\mathbf{0}\}$ and if there exist two analytic functions K and W such that $\forall(s,\mathbf{z}) \in \mathcal{D}(2\delta;2\varepsilon)$ $L(s;\mathbf{z}) = K(s;\mathbf{z})W(s;\langle\boldsymbol{\alpha},\mathbf{z}\rangle_{\boldsymbol{\alpha}\in I})$ and $W(s;\mathbf{0}) \equiv 0$;*
(2) *$\varepsilon_0(\boldsymbol{\phi};L) = 0$ otherwise.*

Remark: Lemma 3 is used in this paper only with $\boldsymbol{\alpha} \in [0,\infty)^r$ and $\boldsymbol{\phi} \in (0,\infty)^n$. However, our proof of lemma 3 is by induction on r and the reduction to inductive hypothesis, after one application of residue calculation, produces a new $\boldsymbol{\alpha}'$ resp. $\boldsymbol{\phi}'$ which *need* not belong to $[0,\infty)^r$ resp. $(0,\infty)^n$.

Proof of lemma 3:
We proceed by induction on r.
Throughout the discussion, we use the following notations. Given $\mathbf{z} = (z_1,\dots,z_r)$ we set: $\mathbf{z}' := (z_1,\dots,z_{r-1})$ and $l(\mathbf{z}) := \frac{1}{z_r}\mathbf{z}' = \left(\frac{z_1}{z_r},\dots,\frac{z_{r-1}}{z_r}\right)$ if $z_r \neq 0$.

Step 1: Proof when $r = 1$:

Set $A = \prod_{\alpha\in I}\alpha^{u(\alpha)}$ and $c = \sum_{\alpha\in I} u(\alpha)$. We have:

$$2\pi i\, T_1(s) = \int_{\rho_1 - i\infty}^{\rho_1+i\infty} \frac{L(s;z)\, dz}{(s-\sigma_1-\phi z)^q \prod_{\alpha\in I}(\alpha z)^{u(\alpha)}} = \frac{1}{A}\int_{\rho_1 - i\infty}^{\rho_1+i\infty} \frac{L(s;z)\, dz}{(s-\sigma_1-\phi z)^q z^c}.$$

From our assumptions (see (15)) and lemma 1 it follows that $s\mapsto T_1(s)$ converges absolutely and defines a holomorphic function with moderate growth in $\{\sigma > \sigma_1 - \eta\}$ where $\eta = \inf(-\phi\rho_1, \delta)$.

• If $\phi\rho_1 < 0$ then $\eta > 0$. This proves the lemma in this case.

• if $\phi = 0$ then $T_1(s) = \frac{1}{(2\pi i)A(s-\sigma_1)^q}\int_{\rho_1 - i\infty}^{\rho_1+i\infty}\frac{L(s;z)\ dz}{z^c}$. It follows also from Lemma 1 and (15) that $s\mapsto T_1(s)$ has a meromorphic continuation with moderate growth to the half-plane $\{\sigma > \sigma_1 - \delta\}$ with at most one pole at $s = \sigma_1$ of order at most $q \le d_1$.

• We assume $\phi\rho_1 > 0$:

The residue theorem and lemma 1 imply that for $\sigma > \sigma_1 + \phi\rho_1$:

$T_1(s) = T_1'(s) + T_1''(s)$ where $T_1'(s) = \frac{1}{(2\pi i)A}\int_{-\rho_1 - i\infty}^{-\rho_1+i\infty}\frac{L(s;z)\ dz}{(s-\sigma_1-\phi z)^q z^c}$ and $T_1''(s) = \frac{1}{A} Res_{z=0}\frac{L(s;z)}{(s-\sigma_1-\phi z)^q z^c}$.

Lemma 1 and (15) imply that $s\mapsto T_1'(s)$ converges absolutely and defines a holomorphic function with moderate growth in the half-plane $\{\sigma > \sigma_1 - \eta\}$ where $\eta = \inf(\delta, \phi\rho_1)$.

If $c = 0$ then $T_1''(s) \equiv 0$. Thus $T_1(s) = T_1'(s)$ satisfies the conclusions of lemma 3. We assume now that $c \ge 1$. An easy computation shows that

$$T_1''(s) = \frac{1}{A}\sum_{k=0}^{c-1}\frac{\binom{-q}{k}(-\phi)^k}{(c-1-k)!}\ \frac{\partial_z{}^{(c-1-k)}L(s;0)}{(s-\sigma_1)^{q+k}}.$$

We deduce that $T_1(s)$ has a meromorphic continuation with moderate growth to $\{\sigma > \sigma_1 - \eta\}$ with at most one pole at $s = \sigma_1$ of order at most:

(1) $\mathrm{ord}_{s=\sigma_1} T_1(s) \le q + c - 1 = q + \sum_{\alpha\in I} u(\alpha) - rank(I) = d_1$ if $L(s,0) \neq 0$;
(2) $\mathrm{ord}_{s=\sigma_1} T_1(s) \le q + c - 2 = q + \sum_{\alpha\in I} u(\alpha) - rank(I) - 1 \le d_1$ if $L(s,0) = 0$;

This proves Lemma 3 if $r = 1$.

Step 2: Let $r \ge 2$. We assume that lemma 3 is true for any $r' \le r-1$. We will show that it also remains true for r:

We justify this assertion by induction on the integer

$$h = h(\boldsymbol{\phi}, \boldsymbol{\rho}) := \#\{i \in \{1,\ldots,r\} \mid \phi_i\rho_i \ge 0\} \in \{0,\ldots,r\}.$$

The idea is that use of residue theory in the z_r variable (moving from $\Re(z_r) = \rho_r$ to $\Re(z_r) = -\rho_r$) creates a contour integral over a vertical line to the left of the origin whereas one has started to the right. Thus $sgn(\phi_r\rho_r)$ changes in crossing $z_r = 0$. This reduces h, which can then be used as induction variable if the case $h = 0$ is true.

• **Proof of lemma 3 for $h = 0$:**

Since $h = 0$ then for each $i = 1,\ldots,r$, $\phi_i\rho_i < 0$. It follows from lemma 1 and (15) that $s\mapsto T_r(s)$ converges absolutely and defines a holomorphic function with moderate growth in the half-plane $\{\sigma > a - \eta\}$ where $\eta = \inf(-\langle\boldsymbol{\phi},\boldsymbol{\rho}\rangle, \delta) > 0$. Thus lemma 3 is also true in this case.

• Let $h \in \{1,\ldots,r\}$. We assume that lemma 3 is true for $h(\boldsymbol{\phi},\boldsymbol{\rho}) \le h-1$. We will prove that it remains true for $h(\boldsymbol{\phi},\boldsymbol{\rho}) = h$:

If $\boldsymbol{\phi} = \mathbf{0}$ then $T_r(s) = \frac{1}{(2\pi i)^r(s-\sigma_1)^q}\int_{\rho_1 - i\infty}^{\rho_1+i\infty}\cdots\int_{\rho_r - i\infty}^{\rho_r+i\infty}\frac{L(s;\mathbf{z})\ dz_1\ldots dz_r}{\prod_{\alpha\in I}\langle\boldsymbol{\alpha},\mathbf{z}\rangle^{u(\alpha)}}$. Since the ρ_i

are linearly independent over $\mathbb{Q}(I_0)$, lemma 1 and (15) imply that lemma 3 is true in this case, in the sense that there is at most one pole at $s = \sigma_1$ of order at most $q \leq d_r$.
If $\boldsymbol{\phi} \neq \mathbf{0}$ and $\phi_i\rho_i \leq 0$ for all $i = 1, \dots, r$, then there exists i_0 such that $\phi_{i_0}\rho_{i_0} < 0$. In this case, it is also easy to see that $s \mapsto T_r(s)$ is *holomorphic* with moderate growth in $\{\sigma > \sigma_1 - \eta\}$ where $\eta = \inf(\delta, -\phi_{i_0}\rho_{i_0}) > 0$. It follows that lemma 3 is also true in this case.
So to finish the proof of lemma 3 it suffices to consider the case where there exists $i \in \{1, \dots, r\}$ such that $\phi_i\rho_i > 0$.
Without loss of generality we can assume that $\phi_r\rho_r > 0$.

$$\text{(17)} \qquad \text{Set} \quad J := \left\{\boldsymbol{\alpha} \in I \mid \alpha_r \neq 0 \text{ and } \left|\frac{\alpha_1\rho_1 + \cdots + \alpha_{r-1}\rho_{r-1}}{\alpha_r}\right| < |\rho_r|\right\}.$$

Assume first that $J \neq \emptyset$.
Consider the equivalence relation $\mathcal{R}$ defined on J by:

$$\boldsymbol{\alpha}\mathcal{R}\boldsymbol{\gamma} \text{ iff } \frac{1}{\alpha_r}\boldsymbol{\alpha}' = \frac{1}{\gamma_r}\boldsymbol{\gamma}' \text{ iff } l(\boldsymbol{\alpha}) = l(\boldsymbol{\gamma}).$$

$$\text{(18)} \qquad \text{It's clear that } \boldsymbol{\alpha}\mathcal{R}\boldsymbol{\gamma} \text{ iff } \alpha_r\boldsymbol{\gamma} = \gamma_r\boldsymbol{\alpha} \text{ iff } \boldsymbol{\alpha}' = \alpha_r l(\boldsymbol{\gamma}).$$

Fix now $s = \sigma + i\tau \in \mathbb{C}$ and $\mathbf{z}' = (z_1, \dots, z_{r-1}) \in \mathbb{C}^{r-1}$ such that $\sigma > \sigma_1 + \sum_{i=1}^r |\phi_i\rho_i|$ and $\forall i = 1, \dots, r-1$, $\Re(z_i) = \rho_i$.
The function

$$\begin{aligned} z_r \mapsto I(z_r) \quad &:= \quad \frac{L(s;\mathbf{z})}{(s - \sigma_1 - \langle \boldsymbol{\phi}, \mathbf{z}\rangle)^q \prod_{\boldsymbol{\alpha}\in I} \langle \boldsymbol{\alpha}, \mathbf{z}\rangle^{u(\boldsymbol{\alpha})}} \\ &= \quad \frac{L(s;\mathbf{z}', z_r)}{(s - \sigma_1 - \langle \boldsymbol{\phi}', \mathbf{z}'\rangle - \phi_r z_r)^q \prod_{\boldsymbol{\alpha}\in I}(\langle \boldsymbol{\alpha}', \mathbf{z}'\rangle + \alpha_r z_r)^{u(\boldsymbol{\alpha})}} \end{aligned}$$

is meromorphic in $\{z_r \in \mathbb{C} \mid |\Re(z_r)| < |\rho_r|\}$ with at most poles at the points $z_r = -\frac{1}{\alpha_r}\langle \boldsymbol{\alpha}', \mathbf{z}'\rangle = -\langle l(\boldsymbol{\alpha}), \mathbf{z}'\rangle$ $(\boldsymbol{\alpha} \in J)$. Moreover since $\rho_1, \dots, \rho_r$ are $\mathbb{Q}(I_0)$-linearly independent, it follows that $\langle l(\boldsymbol{\alpha}), \mathbf{z}'\rangle = \langle l(\boldsymbol{\gamma}), \mathbf{z}'\rangle$ iff $l(\boldsymbol{\alpha}) = l(\boldsymbol{\gamma})$ iff $\boldsymbol{\alpha}\mathcal{R}\boldsymbol{\gamma}$.

Denote by $J_1, \dots, J_t$ the equivalence classes of $\mathcal{R}$ (they form a partition of J). Choose for each $k = 1, \dots, t$, an element $\boldsymbol{\alpha}^k \in J_k$ and set

$$\text{(19)} \qquad c_k := \sum_{\boldsymbol{\alpha}\in J_k} u(\boldsymbol{\alpha}).$$

The poles of $z_r \mapsto I(z_r)$ in $\{z_r \in \mathbb{C} \mid |\Re(z_r)| < |\rho_r|\}$ are the points

$$z_r = -\langle l(\boldsymbol{\alpha}^k), \mathbf{z}'\rangle \quad (k = 1, \dots, t)$$

and for any k, $ord_{z_r = -\langle l(\boldsymbol{\alpha}^k), \mathbf{z}'\rangle} I(z_r) = c_k$ (see (19)).

The residue theorem, (15) and lemma 1 imply then that there exist constants $A_1, \dots, A_t \in \mathbb{R}$ such that:

$$\text{(20)} \qquad \forall \sigma > \sigma_1 + \sum_{i=1}^r |\phi_i\rho_i|, \quad T_r(s) = T_r^0(s) + \sum_{k=1}^t A_k T_{r-1}^k(s)$$

where

$$T_r^0(s) := \frac{1}{(2\pi i)^r}\int_{\rho_1 - i\infty}^{\rho_1 + i\infty} \cdots \int_{\rho_{r-1} - i\infty}^{\rho_{r-1} + i\infty} \int_{-\rho_r - i\infty}^{-\rho_r + i\infty} \frac{L(s;\mathbf{z})\, dz_1 \dots dz_r}{(s - \sigma_1 - \langle \boldsymbol{\phi}, \mathbf{z}\rangle)^q \prod_{\boldsymbol{\alpha}\in I} \langle \boldsymbol{\alpha}, \mathbf{z}\rangle^{u(\boldsymbol{\alpha})}}$$

and for each $k = 1, \dots, t$:

$$T_{r-1}^k(s) := \int_{\rho_1 - i\infty}^{\rho_1+i\infty} \cdots \int_{\rho_{r-1}-i\infty}^{\rho_{r-1}+i\infty} \left(\frac{\partial}{\partial z_r}\right)^{c_k - 1} N(s, \mathbf{z}', z_r)|_{z_r = -\langle l(\boldsymbol{\alpha}^k), \mathbf{z}'\rangle}\, dz_1 \dots dz_{r-1},$$

where $N(s; \mathbf{z}', z_r) := \dfrac{L(s;\mathbf{z})}{(s - \sigma_1 - \langle \boldsymbol{\phi}, \mathbf{z}\rangle)^q \prod_{\boldsymbol{\alpha}\in I\setminus J_k} \langle \boldsymbol{\alpha}, \mathbf{z}\rangle^{u(\boldsymbol{\alpha})}}$.

If the set J defined in (17) verifies $J = \emptyset$, then it's clear from the previous that we have also $\forall \sigma > \sigma_1 + \sum_{i=1}^r |\phi_i \rho_i|, \quad T_r(s) = T_r^0(s)$.

Since $h(\boldsymbol{\phi}, (\rho_1, \dots, \rho_{r-1}, -\rho_r)) = h(\boldsymbol{\phi}, \boldsymbol{\rho}) - 1 = h - 1$, the induction hypothesis for $h-1$ implies that $s \mapsto T_r^0(s)$ satisfies the conclusions of lemma 3. So to conclude, it is enough to assume that $J \neq \emptyset$ and prove lemma 3 for each $s \mapsto T_{r-1}^k(s)$.

We then choose and fix any $k \in \{1, \dots, t\}$ for the rest of the discussion. An easy computation shows that:

$$T_{r-1}^k(s) = \sum_{u+v+\sum_{\boldsymbol{\alpha}\in I\setminus J_k} k_{\boldsymbol{\alpha}} = c_k - 1} w\left(u, v, (k_{\boldsymbol{\alpha}})\right) R_k\left(u, v, (k_{\boldsymbol{\alpha}}); s\right) \tag{21}$$

where u, v and the $k_{\boldsymbol{\alpha}}$ are in $\mathbb{N}_0$, each $w\left(u, v, (k_{\boldsymbol{\alpha}})\right) \in \mathbb{R}$ and

$$R_k\left(u, v, (k_{\boldsymbol{\alpha}}); s\right) := \frac{1}{(2\pi i)^{r-1}} \int_{\rho_1 - i\infty}^{\rho_1+i\infty} \cdots \int_{\rho_{r-1}-i\infty}^{\rho_{r-1}+i\infty} \frac{\frac{\partial^u L}{\partial {z_r}^u}(s; \mathbf{z}', -\langle l(\boldsymbol{\alpha}^k), \mathbf{z}'\rangle)\, dz_1 \dots dz_{r-1}}{\left(s - \sigma_1 - \langle \boldsymbol{\phi}' - \phi_r l(\boldsymbol{\alpha}^k), \mathbf{z}'\rangle\right)^{q+v} \prod_{\boldsymbol{\alpha}\in I\setminus J_k} \langle \boldsymbol{\alpha}' - \alpha_r l(\boldsymbol{\alpha}^k), \mathbf{z}'\rangle^{u(\boldsymbol{\alpha}) + k_{\boldsymbol{\alpha}}}}. \tag{22}$$

So to conclude it suffices to prove lemma 3 for each $R_k\left(u, v, (k_{\boldsymbol{\alpha}}); s\right)$.
We fix now $u, v \in \mathbb{N}_0$ and $(k_{\boldsymbol{\alpha}})_{\boldsymbol{\alpha}\in I\setminus J_k}$ such that

$$u + v + \sum_{\boldsymbol{\alpha}\in I\setminus J_k} k_{\boldsymbol{\alpha}} = c_k - 1. \tag{23}$$

It is clear that there exist $\delta, \varepsilon > 0$ such that $(s; \mathbf{z}') \mapsto \frac{\partial^u L}{\partial {z_r}^u}(s; \mathbf{z}', -\langle l(\boldsymbol{\alpha}^k), \mathbf{z}'\rangle)$ is holomorphic in $\mathcal{D}_{r-1}(2\delta, 2\varepsilon)$ and satisfies an estimate similar to (15) (the last assertion is justified by using Cauchy's integral formula).
The induction hypothesis on r implies that there exists $\eta > 0$ such that $s \mapsto R_k\left(u, v, (k_{\boldsymbol{\alpha}}); s\right)$ has meromorphic continuation with moderate growth to the half-plane $\{\sigma > \sigma_1 - \eta\}$ with at most one pole at $s = \sigma_1$ of order at most

$$\begin{aligned} ord_{s=\sigma_1} R_k\left(u, v, (k_{\boldsymbol{\alpha}}); s\right) \leq & \Big(\sum_{\boldsymbol{\alpha}\in I\setminus J_k} u(\boldsymbol{\alpha}) + k_{\boldsymbol{\alpha}}\Big) - rank(V) + (q+v) \\ & -\varepsilon_0\left(\boldsymbol{\phi}' - \phi_r l(\boldsymbol{\alpha}^k); \tilde{L}_u\right), \end{aligned} \tag{24}$$

where $V := \{\boldsymbol{\alpha}' - \alpha_r l(\boldsymbol{\alpha}^k) \mid \boldsymbol{\alpha} \in I \setminus J_k\}$ and $\tilde{L}_u(s; \mathbf{z}') := \frac{\partial^u L}{\partial {z_r}^u}\left(s; \mathbf{z}', -\langle l(\boldsymbol{\alpha}^k), \mathbf{z}'\rangle\right)$.

What now must be proved is that the upper bound in (24) is at most d_r. This requires a more careful analysis of the terms in the right side of (24) and especially the quantity $\varepsilon_0\left(\boldsymbol{\phi}' - \phi_r l(\boldsymbol{\alpha}^k); \tilde{L}_u\right)$.

Set $\tilde{V} := \left\{\boldsymbol{\alpha} - \frac{\alpha_r}{\alpha_r^k}\boldsymbol{\alpha}^k \mid \boldsymbol{\alpha} \in I \setminus J_k\right\} = \{\left(\boldsymbol{\alpha}' - \alpha_r l(\boldsymbol{\alpha}^k), 0\right) \mid \boldsymbol{\alpha} \in I \setminus J_k\} = V \times \{\mathbf{0}\}$.
It is clear that $rank(\tilde{V}) = rank(V)$. Moreover it follows from the definition of $\boldsymbol{\alpha}^k$

that $\alpha_r^k \neq 0$, and therefore $\boldsymbol{\alpha}^k \notin Vect_{\mathbb{R}}(\tilde{V}) := \sum_{\boldsymbol{\beta}\in\tilde{V}} \mathbb{R}\boldsymbol{\beta}$. We deduce that:

(25) $$rank(I) = rank(\tilde{V} \cup \{\boldsymbol{\alpha}^k\}) = rank(\tilde{V}) + 1 = rank(V) + 1.$$

So it follows from (24), (23) and (25) that

$$\begin{aligned} O_{\sigma_1} &:= ord_{s=\sigma_1} R_k\left(u, v, (k_{\boldsymbol{\alpha}}); s\right) \\ &\leq \Big(\sum_{\boldsymbol{\alpha}\in I\setminus J_k} u(\boldsymbol{\alpha}) + k_{\boldsymbol{\alpha}}\Big) - rank(V) + q + v - \varepsilon_0\left(\boldsymbol{\phi}' - \phi_r l(\boldsymbol{\alpha}^k); \tilde{L}_u\right) \\ &= c_k - 1 + \sum_{\boldsymbol{\alpha}\in I\setminus J_k} u(\boldsymbol{\alpha}) - rank(V) + q - u - \varepsilon_0\left(\boldsymbol{\phi}' - \phi_r l(\boldsymbol{\alpha}^k); \tilde{L}_u\right) \\ &\leq (c_k + \sum_{\boldsymbol{\alpha}\in I\setminus J_k} u(\boldsymbol{\alpha})) - (rank(V) + 1) + q - \varepsilon_0\left(\boldsymbol{\phi}' - \phi_r l(\boldsymbol{\alpha}^k); \tilde{L}_u\right) - u \\ &= \sum_{\boldsymbol{\alpha}\in I} u(\boldsymbol{\alpha}) - rank(I) + q - \varepsilon_0\left(\boldsymbol{\phi}' - \phi_r l(\boldsymbol{\alpha}^k); \tilde{L}_u\right) - u. \end{aligned}$$

Thus from the definition of d_r (see the statement of Lemma 3) we see that:

(26) $$O_{\sigma_1} \leq d_r + \varepsilon_0(\boldsymbol{\phi}; L) - \varepsilon_0\left(\boldsymbol{\phi}' - \phi_r l(\boldsymbol{\alpha}^k); \tilde{L}_u\right) - u.$$

We will now analyze carefully the quantity $\varepsilon_0\left(\boldsymbol{\phi}' - \phi_r l(\boldsymbol{\alpha}^k); \tilde{L}_u\right)$**:**
If $\varepsilon_0(\boldsymbol{\phi}; L) - \varepsilon_0\left(\boldsymbol{\phi}' - \phi_r l(\boldsymbol{\alpha}^k); \tilde{L}_u\right) \leq 0$ or $u \neq 0$, then $O_{\sigma_1} \leq d_r$. This finishes the proof of the lemma in all cases except the following possibility:

(27) $$u = 0,\ \varepsilon_0(\boldsymbol{\phi}; L) = 1 \text{ and } \varepsilon_0\left(\boldsymbol{\phi}' - \phi_r l(\boldsymbol{\alpha}^k); \tilde{L}_0\right) = 0.$$

We assume in the sequel that (27) holds.

Since $\varepsilon_0(\boldsymbol{\phi}, L) = 1$, it follows that $\boldsymbol{\phi} \in con^*(I) \setminus \{\mathbf{0}\}$ and that $\tilde{L}_0$ has the form:

(28) $$\tilde{L}_0(s, \mathbf{z}') = \tilde{K}(s; \mathbf{z}')\tilde{W}\left(s; \langle \boldsymbol{\mu}, \mathbf{z}'\rangle_{\boldsymbol{\mu}\in V}\right) \text{ with } \tilde{W}(s; \mathbf{0}) \equiv 0.$$

Thus, the fact that $\varepsilon_0\left(\boldsymbol{\phi}' - \phi_r l(\boldsymbol{\alpha}^k); \tilde{L}_0\right) = 0$ must now imply that:

(29)
$$\boldsymbol{\phi}' - \phi_r l(\boldsymbol{\alpha}^k) = \boldsymbol{\phi}' - \frac{\phi_r}{\alpha_r^k}(\boldsymbol{\alpha}^k)' = (\phi_1, \ldots, \phi_{r-1}) - \frac{\phi_r}{\alpha_r^k}(\alpha_1^k, \ldots, \alpha_{r-1}^k) \notin con^*(V) \setminus \{\mathbf{0}\}.$$

We next show:

Claim: (18) and (29) imply that $\boldsymbol{\phi} = \dfrac{\phi_r}{\alpha_r^k}\boldsymbol{\alpha}^k$.

Proof of the Claim: We first show that $\boldsymbol{\phi}' - \phi_r l(\boldsymbol{\alpha}^k) = 0$. To do so, (29) tells us that it suffices to show that $\boldsymbol{\phi}' - \phi_r l(\boldsymbol{\alpha}^k) \in con^*(V)$.
Since $\boldsymbol{\phi} \in con^*(I) \setminus \{\mathbf{0}\}$, there exist $\{\lambda_{\boldsymbol{\alpha}}\}_{\boldsymbol{\alpha}\in I} \subset \mathbb{R}_+^*$ such that $\boldsymbol{\phi} = \sum_{\boldsymbol{\alpha}\in I} \lambda_{\boldsymbol{\alpha}}\boldsymbol{\alpha}$. Thus, $\boldsymbol{\phi}' = \sum_{\boldsymbol{\alpha}\in I} \lambda_{\boldsymbol{\alpha}}\boldsymbol{\alpha}'$ and $\phi_r = \sum_{\boldsymbol{\alpha}\in I} \lambda_{\boldsymbol{\alpha}}\alpha_r$. A simple check now shows that (18) implies the following:

$$\begin{aligned} \boldsymbol{\phi}' - \phi_r l(\boldsymbol{\alpha}^k) &= \sum_{\boldsymbol{\alpha}\in I} \lambda_{\boldsymbol{\alpha}}\boldsymbol{\alpha}' - \sum_{\boldsymbol{\alpha}\in I} \lambda_{\boldsymbol{\alpha}}\alpha_r l(\boldsymbol{\alpha}^k) = \sum_{\boldsymbol{\alpha}\in I} \lambda_{\boldsymbol{\alpha}}\left(\boldsymbol{\alpha}' - \alpha_r l(\boldsymbol{\alpha}^k)\right) \\ &= \sum_{\boldsymbol{\alpha}\in I\setminus J_k} \lambda_{\boldsymbol{\alpha}}\left(\boldsymbol{\alpha}' - \alpha_r l(\boldsymbol{\alpha}^k)\right) \text{ (because } \boldsymbol{\alpha}' - \alpha_r l(\boldsymbol{\alpha}^k) = 0 \text{ if } \boldsymbol{\alpha} \in J_k). \end{aligned}$$

Thus, $\boldsymbol{\phi}' - \phi_r l(\boldsymbol{\alpha}^k) \in con^*(V)$, which, thanks to (29), now gives $\boldsymbol{\phi}' - \phi_r l(\boldsymbol{\alpha}^k) = \mathbf{0}$. Since $\boldsymbol{\phi} - \frac{\phi_r}{\alpha_r^k}\boldsymbol{\alpha}^k = (\boldsymbol{\phi}' - \phi_r l(\boldsymbol{\alpha}^k), 0)$, this finishes the proof of the claim. $\diamondsuit$
Moreover, since $\boldsymbol{\phi} \neq \mathbf{0}$, it follows from our claim that $\phi_r \neq 0$. Thus we must have

$$\mathbb{R}\boldsymbol{\alpha}^k = \mathbb{R}\boldsymbol{\phi} = Vect_{\mathbb{R}}(J_k).$$

As a result, we see that $\boldsymbol{\phi} \in con^*(I)$ implies

$$rank(I) = rank(I \setminus J_k), \text{ unless } I = J_k. \tag{30}$$

Assume that $I \neq J_k$:
Recalling that $u = 0$ is assumed, we first note that the identity $\boldsymbol{\phi}' - \phi_r l(\boldsymbol{\alpha}^k) = \mathbf{0}$, Lemma 1, and the expressions (22)-(23) imply:

$$ord_{s=\sigma_1} R_k\left(0, v, (k_{\boldsymbol{\alpha}}); s\right) \le q + v = q + c_k - 1 - \sum_{\boldsymbol{\alpha}\in I\setminus J_k} k_{\boldsymbol{\alpha}} \le q + \sum_{\boldsymbol{\alpha}\in J_k} u(\boldsymbol{\alpha}) - 1.$$

Combining this estimate with (30), we conclude:

$$\begin{aligned} ord_{s=\sigma_1} R_k\left(0, v, (k_{\boldsymbol{\alpha}}); s\right) &\le q + \sum_{\boldsymbol{\alpha}\in I} u(\boldsymbol{\alpha}) - rank(I) - \sum_{\boldsymbol{\alpha}\in I\setminus J_k} u(\boldsymbol{\alpha}) + rank(I) - 1 \\ &\le q + \sum_{\boldsymbol{\alpha}\in I} u(\boldsymbol{\alpha}) - rank(I) - 1 - \Big(\sum_{\boldsymbol{\alpha}\in I\setminus J_k} u(\boldsymbol{\alpha}) - rank(I \setminus J_k)\Big) \\ &\le q + \sum_{\boldsymbol{\alpha}\in I} u(\boldsymbol{\alpha}) - rank(I) - 1 - \left(\#\left(I \setminus J_k\right) - rank(I \setminus J_k)\right) \\ &\le q + \sum_{\boldsymbol{\alpha}\in I} u(\boldsymbol{\alpha}) - rank(I) - 1 \le d_r. \end{aligned}$$

Assume that $I = J_k$:
It is then clear that $V = \emptyset$ (see (24)). In this case, (28) implies that $\tilde{L}_0(s; \mathbf{z}') \equiv 0$. Recalling that $u = 0$ is assumed, (22) implies then that $R_k\left(0, v, (k_{\boldsymbol{\alpha}}); s\right) \equiv 0$. So, it is obvious that $ord_{s=\sigma_1} R_k\left(0, v, (k_{\boldsymbol{\alpha}}); s\right) \le d_r$.

We conclude that for any $u, v, (k_{\boldsymbol{\alpha}})$, $ord_{s=\sigma_1} R_k\left(u, v, (k_{\boldsymbol{\alpha}}); s\right) \le d_r$. This finishes the induction argument on h, therefore, also on r, and completes the proof of lemma 3. $\diamondsuit$

4.1.3. *Second crucial lemma: Lemma 4.*

LEMMA 4. *Let $\mathbf{a} = (a_1, \dots, a_r) \in \mathbb{R}_+^{*r}$ and $a = |\mathbf{a}| = a_1 + \cdots + a_r$. Let I be a finite nonempty subset of $\mathbb{R}_+^r \setminus \{\mathbf{0}\}$, $\mathbf{u} = (u(\boldsymbol{\beta}))_{\boldsymbol{\beta}\in I}$ a vector of positive integers, and $\mathbf{h} = (h_1, \dots, h_r) \in \mathbb{R}_+^{*r}$. Assume that: $\mathbf{1} \in con(I)$ and that $\langle \boldsymbol{\beta}, \mathbf{a}\rangle = 1\ \forall \boldsymbol{\beta} \in I$. Let $\boldsymbol{\rho} \in \mathbb{R}_+^{*r}$. For $\sigma = \Re(s) > a + |\boldsymbol{\rho}|$ set:*

$$\mathcal{R}(s) := \frac{1}{(2\pi i)^r} \int_{\rho_1 - i\infty}^{\rho_1 + i\infty} \cdots \int_{\rho_r - i\infty}^{\rho_r + i\infty} \frac{\Gamma(s - a - z_1 - \cdots - z_r)\, \prod_{i=1}^r \Gamma(a_i + z_i)\, d\mathbf{z}}{\Gamma(s) \prod_{k=1}^r h_k^{a_k + z_k}\, \prod_{\boldsymbol{\beta}\in I} \langle \boldsymbol{\beta}, \mathbf{z}\rangle^{u(\boldsymbol{\beta})}}. \tag{31}$$

Then there exists $\eta > 0$ such that $s \mapsto \mathcal{R}(s)$ has a meromorphic continuation to the half-plane $\{\sigma > a - \eta\}$ with exactly one pole at $s = a$ of order $\rho_0 := \sum_{\boldsymbol{\beta}\in I} u(\boldsymbol{\beta}) - rank\,(I) + 1$. Moreover we have $\mathcal{R}(s) \sim_{s\to a} \dfrac{A_0(I; \mathbf{u}; \mathbf{h})}{(s-a)^{\rho_0}}$, where $A_0(I; \mathbf{u}; \mathbf{h})$ is the volume constant (see §2.3.2) associated to I, $\mathbf{u}$ and $\mathbf{h}$.

Proof of Lemma 4:
We fix $\boldsymbol{\rho} \in \mathbb{R}_+^{*r}$. From lemma 2 and lemma 1 it follows easily that the integral $\mathcal{R}(s)$

converges for any sufficiently large σ.
Set $q := \sum_{\beta\in I} u(\beta)$ and define the vectors $\boldsymbol{\alpha}^1,\dots,\boldsymbol{\alpha}^q \in \mathbb{R}_+^r$ by:
$\{\boldsymbol{\alpha}^i \mid i=1,\dots,q\} = I$ and $\forall \boldsymbol{\beta}\in I\ \#\{i\in\{1,..,q\} \mid \boldsymbol{\alpha}^i = \boldsymbol{\beta}\} = u(\boldsymbol{\beta})$ (i.e. the family of vectors $(\boldsymbol{\alpha}^i)_{i=1,\dots,q}$ is obtained by repeating each vector $\boldsymbol{\beta}\in I$ $u(\boldsymbol{\beta})$ times).
We then define $\boldsymbol{\mu}^1,\dots,\boldsymbol{\mu}^r \in \mathbb{R}_+^q$ by setting: $\mu_i^k = \alpha_k^i$ $\forall i=1,\dots,q$ and $\forall k = 1,\dots,r$.
Define the generalized polynomial with positive coefficients:
$G(\mathbf{X}) = 1 + P_{(I;\mathbf{u};\mathbf{h})}(\mathbf{X}) := 1 + \sum_{k=1}^r h_k\ \mathbf{X}^{\boldsymbol{\mu}^k}$.
We note that G depends on all the variables $X_1,\dots,X_q$ since for any $i=1,\dots,q$ there exists $k\in\{1,\dots,r\}$ such that $\mu_i^k = \alpha_k^i \neq 0$. We will first prove:
Claim: For $\sigma \gg 1$,

$$\mathcal{R}(s) = \int_{[1,+\infty[^q} G^{-s}(\mathbf{x})\ d\mathbf{x}. \tag{32}$$

Proof of Claim: For $\sigma = \Re(s) \gg 1$ we have the following identities:

$$\begin{aligned}
\mathcal{R}(s) &:= \frac{1}{(2\pi i)^r}\int_{\rho_1-i\infty}^{\rho_1+i\infty}\cdots\int_{\rho_r-i\infty}^{\rho_r+i\infty} \frac{\Gamma(s-a-z_1-\cdots-z_r)\prod_{i=1}^r\Gamma(a_i+z_i)\ d\mathbf{z}}{\Gamma(s)\prod_{i=1}^r h_i^{a_i+z_i}\ \prod_{k=1}^q\langle\boldsymbol{\alpha}^k,\mathbf{z}\rangle}\\
&= \frac{1}{(2\pi i)^r}\int_{a_1+\rho_1-i\infty}^{a_1+\rho_1+i\infty}\cdots\int_{a_r+\rho_r-i\infty}^{a_r+\rho_r+i\infty} \frac{\Gamma(s-z_1-\cdots-z_r)\prod_{i=1}^r\Gamma(z_i)\ d\mathbf{z}}{\Gamma(s)\prod_{i=1}^r h_i^{z_i}\ \prod_{k=1}^q(\langle\boldsymbol{\alpha}^k,\mathbf{z}\rangle-1)}\\
&= \frac{1}{(2\pi i)^r}\int_{a_1+\rho_1-i\infty}^{a_1+\rho_1+i\infty}\cdots\int_{a_r+\rho_r-i\infty}^{a_r+\rho_r+i\infty} \Gamma(s-z_1-\cdots-z_r)\,\Gamma(s)^{-1}\prod_{i=1}^r\Gamma(z_i)\\
&\qquad\times\prod_{i=1}^r h_i^{-z_i}\left(\int_{[1,+\infty[^q}\prod_{k=1}^q x_k^{-\langle\boldsymbol{\alpha}^k,\mathbf{z}\rangle}\ dx_1\dots dx_q\right)dz_1\dots dz_r\\
&= \frac{1}{(2\pi i)^r}\int_{a_1+\rho_1-i\infty}^{a_1+\rho_1+i\infty}\cdots\int_{a_r+\rho_r-i\infty}^{a_r+\rho_r+i\infty} \Gamma(s-z_1-\cdots-z_r)\,\Gamma(s)^{-1}\prod_{i=1}^r\Gamma(z_i)
\end{aligned}$$

$$\times\prod_{i=1}^r h_i^{-z_i}\left(\int_{[1,+\infty[^q}\prod_{i=1}^r \mathbf{x}^{-z_i\boldsymbol{\mu}^i}\ dx_1\dots dx_q\right)dz_1\dots dz_r. \tag{33}$$

In the other hand, we have uniformly in $\mathbf{z}\in\mathbb{C}^r$ such that $\Re(z_i) = a_i+\rho_i$:

$$\forall\mathbf{x}\in[1,+\infty[^q\ |\prod_{i=1}^r\mathbf{x}^{-z_i\boldsymbol{\mu}^i}| = |\prod_{k=1}^q x_k^{-\langle\boldsymbol{\alpha}^k,\mathbf{z}\rangle}| = \prod_{k=1}^q x_k^{-\langle\boldsymbol{\alpha}^k,\mathbf{a}\rangle-\langle\boldsymbol{\alpha}^k,\boldsymbol{\rho}\rangle} = \prod_{k=1}^q x_k^{-1-\langle\boldsymbol{\alpha}^k,\boldsymbol{\rho}\rangle}.$$

We deduce that the integral $\int_{[1,+\infty[^q}\prod_{i=1}^r\mathbf{x}^{-z_i\boldsymbol{\mu}^i}\ dx_1\dots dx_q$ converges absolutely and uniformly in $\mathbf{z}\in\mathbb{C}^r$ such that $\Re(z_i) = a_i+\rho_i$. This, (33), lemma 2, and lemma 1 imply then that for $\sigma\gg 1$, we have

$$\begin{aligned}
\mathcal{R}(s) &= \int_{[1,+\infty[^q}\Big[\frac{1}{(2\pi i)^r}\int_{a_1+\rho_1-i\infty}^{a_1+\rho_1+i\infty}\cdots\int_{a_r+\rho_r-i\infty}^{a_r+\rho_r+i\infty}\Gamma(s-z_1-\cdots-z_r)\,\Gamma(s)^{-1}\\
&\qquad\times\prod_{i=1}^r\Gamma(z_i)\prod_{i=1}^r\left(h_i\ \mathbf{x}^{\boldsymbol{\mu}^i}\right)^{-z_i}dz_1\dots dz_r\Big]dx_1\dots dx_q.
\end{aligned}$$

Mellin's formula (13) implies then that for $\sigma\gg 1$, we have

$$\mathcal{R}(s) = \int_{[1,+\infty[^q}\left(1+\sum_{i=1}^r h_i\ \mathbf{x}^{\boldsymbol{\mu}^i}\right)^{-s}dx_1\dots dx_q = \int_{[1,+\infty[^n} G^{-s}(\mathbf{x})\ d\mathbf{x}.$$

This end the proof of the claim.
So to conclude it suffices to check that $s \mapsto Y(G;s) := \int_{[1,+\infty[^q} G^{-s}(\mathbf{x})\, d\mathbf{x}$ satisfies the assertions of Lemma 4.
Let $\mathcal{E}^\infty(G) = conv\left(supp(G) - \mathbb{R}_+^q\right)$ denote the Newton polyhedron at infinity of G. Denote by G_0 the smallest face that meets the diagonal. It follows from Sargos' result (see §2.3) that there exists $\eta > 0$ such that $Y(G;s)$ has a meromorphic continuation to the half-plane $\{\sigma > \sigma_0 - \eta\}$ (where $\sigma_0 = \sigma_0(G)$) with moderate growth and exactly one pole at $s = \sigma_0$ of order $\rho_0 := codim G_0$. Moreover σ_0 is characterized geometrically by: ${\sigma_0}^{-1}\mathbf{1} = \Delta \cap G_0 = \Delta \cap \mathcal{E}^\infty(G)$ and $Y(G;s) \sim_{s\to\sigma_0} \frac{A_0(G)}{(s-\sigma_0)^{\rho_0}}$ where $A_0(G)$ is the volume constant associated to the polynomial G. It is easy to see that in our case $A_0(G)$ is equal to the volume constant (see §2.3.2) $A_0(I;\mathbf{u};\mathbf{h})$ associated to I, $\mathbf{u}$ and $\mathbf{h}$.
By our hypothesis, we have $\mathbf{1} \in con(I) = con\left(\{\boldsymbol{\alpha}^k \mid k = 1,\ldots,q\}\right)$. Thus there exists $\mathbf{v} = (v_1,\ldots,v_q) \in \mathbb{R}_+^q \setminus \{\mathbf{0}\}$ such that $\mathbf{1} = \sum_{k=1}^q v_k \boldsymbol{\alpha}^k$. It follows that:

$$\forall i = 1,\ldots,r \quad \langle \mathbf{v}, \boldsymbol{\mu}^i\rangle = \sum_{k=1}^q v_k \mu_k^i = \sum_{k=1}^q v_k \alpha_i^k = 1.$$

Since $supp(G) = \{\boldsymbol{\mu}^i \mid i = 1,\ldots,r\} \cup \{\mathbf{0}\}$, we conclude that $L_{\mathbf{v}} := \{\mathbf{x} \in \mathbb{R}^q \mid \langle \mathbf{v},\mathbf{x}\rangle = 1\}$ is a support plane of $\mathcal{E}^\infty(G)$. Thus:
(34)
$F_{\mathbf{v}}^\infty := L_{\mathbf{v}} \cap \mathcal{E}^\infty(G) = conv\left(\{\boldsymbol{\mu}^i \mid i = 1,\ldots,r\}\right)$ is a face of the polyhedron $\mathcal{E}^\infty(G)$.

By our hypothesis we know that $\sum_{k=1}^r a_k \mu_i^k = \langle \mathbf{a}, \boldsymbol{\alpha}^i\rangle = 1\ \forall i = 1,\ldots,q$, which implies $\frac{1}{a}\mathbf{1} = \sum_{k=1}^r \frac{a_k}{a}\boldsymbol{\mu}^k \in conv\left(\{\boldsymbol{\mu}^i \mid i = 1,\ldots,r\}\right)$. Thus, $\frac{1}{a}\mathbf{1} \in F_{\mathbf{v}}^\infty \cap \Delta$, that is, the face $F_{\mathbf{v}}^\infty$ must meet the diagonal at $\frac{1}{a}\mathbf{1}$. It follows that $\sigma_0 = a$ and that $G_0 \subset F_{\mathbf{v}}^\infty$. Hence we deduce that:

$$\begin{aligned} ord_{s=a} Y(G;s) &= codim G_0 \geq codim F_{\mathbf{v}}^\infty = q - dim F_{\mathbf{v}}^\infty \\ &\geq q - rank\{\boldsymbol{\mu}^i \mid i = 1,..,r\} + 1 = q - rank\{\boldsymbol{\alpha}^i \mid i = 1,..,q\} + 1 \\ &\geq q - rank(I) + 1 = \sum_{\boldsymbol{\alpha}\in I} u(\boldsymbol{\alpha}) - rank(I) + 1. \end{aligned} \tag{35}$$

Using the relation $\Gamma(v+1) = v\Gamma(v)$ we also see that for $\sigma >> 1$,

$$\mathcal{R}(s) := \frac{1}{(2\pi i)^r}\int_{\rho_1 - i\infty}^{\rho_1+i\infty}\cdots\int_{\rho_r-i\infty}^{\rho_r+i\infty} \frac{L(s;\mathbf{z})\, d\mathbf{z}}{(s - a - \langle \mathbf{1},\mathbf{a}\rangle)\prod_{\boldsymbol{\alpha}\in I}\langle\boldsymbol{\alpha},\mathbf{z}\rangle^{u(\boldsymbol{\alpha})}},$$

where $L(s;\mathbf{z}) := \Gamma\left(s - (a-1) - z_1 - \cdots - z_r\right)\prod_{i=1}^r \Gamma(a_i + z_i)\,\Gamma(s)^{-1}\prod_{k=1}^r h_k^{-a_k - z_k}$. Lemma 2 imply that $L(s;\mathbf{z})$ satisfies the assumptions of lemma 3. In particular it imply that $L(s;\mathbf{z})$ satisfies the estimate (15). Therefore it follows from lemma 3 (with $\boldsymbol{\phi} = \mathbf{1}$, $q = 1$ and $\sigma_1 = a$) that $ord_{s=a}Y(G;s) = ord_{s=a}\mathcal{R}(s) \leq 1 + \sum_{\boldsymbol{\alpha}\in I} u(\boldsymbol{\alpha}) - rank(I)$. This and (35) imply that $ord_{s=a}Y(G;s) = \sum_{\boldsymbol{\alpha}\in I} u(\boldsymbol{\alpha}) - rank(I) + 1$. This completes the proof of lemma 4. $\diamondsuit$

4.2. Proof of the first part of theorem 3. Fix in the sequel of this proof a point $\mathbf{c} = (c_1,\ldots,c_n) \in \mathbb{R}_+^{*n}$ and a regularizing pair $\mathcal{T}_{\mathbf{c}} = (I_{\mathbf{c}};\mathbf{u})$ of $\mathcal{M}(f;.)$ at $\mathbf{c}$ as in Definition 2 in §3.2.1.

Since $\sum \frac{|f(m_1,\ldots,m_n)|}{(m_1\ldots m_n)^t} < +\infty$ for $t = 1 + \sup_i c_i$, we certainly have

$$f(m_1,\ldots,m_n) \ll (m_1\ldots m_n)^t \text{ uniformly in } \mathbf{m} \in \mathbb{N}^n. \tag{36}$$

Let $P := \sum_{k=1}^r b_k \mathbf{X}^{\gamma^k}$ be a generalized polynomial with positive coefficients, elliptic and homogeneous of degree $d > 0$. Since P is elliptic, we have $con^*(supp(P)) := \sum_{i=1}^r \mathbb{R}_+^* \gamma^k = \mathbb{R}_+^{*n}$. Therefore $\mathbf{c} \in con^*(supp(P))$ implies

$$\text{there exists } \mathbf{a} = (a_1,\ldots,a_r) \in \mathbb{R}_+^{*r}, \text{ such that } \mathbf{c} = \sum_{k=1}^r a_k \gamma^k. \tag{37}$$

It follows that we have uniformly in $\mathbf{x} \in [1,+\infty[^n$:

$$(x_1^{c_1}\ldots x_n^{c_n})^{1/|\mathbf{a}|} = \left(\mathbf{x}^{\sum_{k=1}^r (a_k/|\mathbf{a}|)\ \gamma^k}\right) \ll \sum_{k=1}^r \mathbf{x}^{\gamma^k} \ll P(\mathbf{x}) \ll (x_1\ldots x_n)^d. \tag{38}$$

It is clear that (36) and (38) imply that the series $Z(f;P;s) := \sum_{\mathbf{m}\in\mathbb{N}^n} \frac{f(m_1,\ldots,m_n)}{P(m_1,\ldots,m_n)^{s/d}}$ has an abscissa of convergence $\sigma_0 < +\infty$.
Moreover, (38) and Taylor's formula imply that $\forall M \in \mathbb{N}$ we have uniformly in $\mathbf{x} \in [1,+\infty[^n$ and $s \in \mathbb{C}$:

$$\begin{aligned} P(\mathbf{x})^{-s/d} &= (1+P(\mathbf{x})-1)^{-s/d} = (1+P(\mathbf{x}))^{-s/d}\left(1 - \frac{1}{1+P(\mathbf{x})}\right)^{-s/d} \\ &= \sum_{k=0}^M (-1)^k \binom{-s/d}{k} (1+P(\mathbf{x}))^{-(s+dk)/d} \\ &\quad + \mathcal{O}\left((1+|s|^{M+1})(1+P(\mathbf{x}))^{-(\Re(s)+dM+d)/d}\right). \end{aligned}$$

It follows that for $M \in \mathbb{N}$ and $\sigma > \sigma_0$:

$$\begin{aligned} Z(f;P;s) &= \sum_{k=0}^M (-1)^k \binom{-s/d}{k} Z(f;1+P;s+dk) \\ &\quad + \mathcal{O}\left((1+|s|^{M+1}) Z(|f|;1+P;\sigma+dM+d)\right). \end{aligned} \tag{39}$$

Thus, it suffices to prove the assertion of the theorem for $Z(f;1+P;s)$.

Let $\boldsymbol{\alpha}^1,\ldots,\boldsymbol{\alpha}^n$ be n elements of $\mathbb{R}_+^r \setminus \{\mathbf{0}\}$ defined by: $\alpha_k^i = \gamma_i^k$ for all $i = 1,\ldots,n$ and $k = 1,\ldots,r$. Since $\mathbf{c} = \sum_{k=1}^r a_k \boldsymbol{\gamma}^k$ we have

$$\forall i = 1,\ldots,n \quad \langle \boldsymbol{\alpha}^i, \mathbf{a} \rangle = c_i. \tag{40}$$

By using Mellin's formula (13) and (40) we obtain that for any $\boldsymbol{\rho} \in \mathbb{R}_+^{*n}$ and $\sigma > \sup(\sigma_0, d|\mathbf{a}| + d|\boldsymbol{\rho}|)$:

$$(2\pi i)^r\ Z(f; 1+P; s) = (2\pi i)^r \sum_{\mathbf{m}\in\mathbb{N}^n} \frac{f(m_1,\ldots,m_n)}{\left(1+\sum_{k=1}^r b_k \mathbf{m}^{\gamma^k}\right)^{s/d}}$$

$$= \sum_{\mathbf{m}\in\mathbb{N}^n} \int_{a_1+\rho_1-i\infty}^{a_1+\rho_1+i\infty} \cdots \int_{a_r+\rho_r-i\infty}^{a_r+\rho_r+i\infty} \frac{\Gamma(s/d - \sum_{i=1}^r z_i)}{\Gamma(s/d)} \prod_{i=1}^r \frac{\Gamma(z_i)}{b_i^{z_i}} \frac{f(m_1,\ldots,m_n)}{\left(\prod_{k=1}^r \mathbf{m}^{z_k\gamma^k}\right)} d\mathbf{z}$$

$$= \sum_{\mathbf{m}\in\mathbb{N}^n} \int_{a_1+\rho_1-i\infty}^{a_1+\rho_1+i\infty} \cdots \int_{a_r+\rho_r-i\infty}^{a_r+\rho_r+i\infty} \frac{\Gamma(s/d - \sum_{i=1}^r z_i)}{\Gamma(s/d)} \prod_{i=1}^r \frac{\Gamma(z_i)}{b_i^{z_i}} \frac{f(m_1,\ldots,m_n)}{\left(\prod_{i=1}^n m_i^{\langle \boldsymbol{\alpha}^i, \mathbf{z}\rangle}\right)} d\mathbf{z}$$

$$= \sum_{\mathbf{m}\in\mathbb{N}^n} \int_{\rho_1-i\infty}^{\rho_1+i\infty} \cdots \int_{\rho_r-i\infty}^{\rho_r+i\infty} \frac{\Gamma(s/d - |\mathbf{a}| - \sum_{i=1}^r z_i)}{\Gamma(s/d)} \prod_{i=1}^r \frac{\Gamma(a_i + z_i)}{b_i^{a_i+z_i}}$$

$$\times \frac{f(m_1,\ldots,m_n)}{\prod_{i=1}^n m_i^{c_i + \langle \boldsymbol{\alpha}^i, \mathbf{z}\rangle}}\ d\mathbf{z} \tag{41}$$

But for all $\boldsymbol{\beta} \in I_{\mathbf{c}}$ we have uniformly in $\mathbf{z} \in \mathbb{C}^r$ verifying $\Re(z_i) = \rho_i\ \forall i = 1, \ldots, r$:

$$\Re\left(\sum_{i=1}^n \beta_i(c_i + \langle \boldsymbol{\alpha}^i, \mathbf{z}\rangle)\right) = \langle \mathbf{c}, \boldsymbol{\beta}\rangle + \sum_{i=1}^n \beta_i \langle \boldsymbol{\alpha}^i, \boldsymbol{\rho}\rangle = 1 + \sum_{i=1}^n \beta_i \langle \boldsymbol{\alpha}^i, \boldsymbol{\rho}\rangle.$$

It follows then (see definition 2) that the series

$$\mathcal{M}\left(f; c_1 + \langle \boldsymbol{\alpha}^1, \mathbf{z}\rangle, \ldots, c_n + \langle \boldsymbol{\alpha}^n, \mathbf{z}\rangle\right) = \sum_{\mathbf{m}\in\mathbb{N}^n} \frac{f(m_1,\ldots,m_n)}{\prod_{i=1}^n m_i^{c_i + \langle \boldsymbol{\alpha}^i, \mathbf{z}\rangle}}$$

converges absolutely and uniformly in $\mathbf{z} \in \mathbb{C}^r$ verifying $\Re(z_i) = \rho_i\ \forall i = 1, \ldots, r$. This with (41), lemma 2 and lemma 1 imply then that for all $\boldsymbol{\rho} \in \mathbb{R}_+^{*n}$ and for all $\sigma > \sup(\sigma_0, d(|\mathbf{a}| + |\boldsymbol{\rho}|))$,

$$Z(f; 1+P; s) = \frac{1}{(2\pi i)^r} \int_{\rho_1-i\infty}^{\rho_1+i\infty} \cdots \int_{\rho_r-i\infty}^{\rho_r+i\infty} \Gamma(s/d - |\mathbf{a}| - z_1 - \cdots - z_r)\Gamma(s/d)^{-1}$$

$$\times \prod_{i=1}^r \frac{\Gamma(a_i + z_i)}{b_i^{a_i+z_i}}\ \mathcal{M}\left(f; c_1 + \langle \boldsymbol{\alpha}^1, \mathbf{z}\rangle, \ldots, c_n + \langle \boldsymbol{\alpha}^n, \mathbf{z}\rangle\right) d\mathbf{z} \tag{42}$$

We now use the hypothesis that $H(f; \mathcal{T}_{\mathbf{c}}; \mathbf{s}) = \left(\prod_{\boldsymbol{\beta}\in I_{\mathbf{c}}} \langle \boldsymbol{\beta}, \mathbf{s}\rangle^{u(\boldsymbol{\beta})}\right) \mathcal{M}(f; \mathbf{c} + s)$ (see (10) in §3.1) has a holomorphic continuation with moderate growth to $\{\mathbf{s} \in \mathbb{C}^n \mid \forall i\ \Re(s_i) > -\varepsilon\}$ for some positive ε. It is clear that this implies there exists $\varepsilon_1 > 0$ such that

$$\mathbf{z} \mapsto \mathcal{H}(\mathcal{T}_{\mathbf{c}}; \mathbf{z}) := H\left(f; \mathcal{T}_{\mathbf{c}}; \langle \boldsymbol{\alpha}^1, \mathbf{z}\rangle, \ldots, \langle \boldsymbol{\alpha}^n, \mathbf{z}\rangle\right) \tag{43}$$

has a holomorphic continuation with moderate growth to $\{\mathbf{z} \in \mathbb{C}^r \mid \forall i\ \Re(z_i) > -\varepsilon_1\}$.

We then rewrite the integrand factor involving $\mathcal{M}$ as follows. For all $\boldsymbol{\beta} \in I_{\mathbf{c}}$, set $\mu(\boldsymbol{\beta}) := \sum_{i=1}^n \beta_i \boldsymbol{\alpha}^i$, and also define:
(44)

$$I_{\mathbf{c}}^* = \{\mu(\boldsymbol{\beta}) \mid \boldsymbol{\beta} \in I_{\mathbf{c}}\}, u^*(\boldsymbol{\eta}) = \sum_{\{\boldsymbol{\beta}\in I_{\mathbf{c}};\ \mu(\boldsymbol{\beta})=\boldsymbol{\eta}\}} u(\boldsymbol{\beta})\ \forall \boldsymbol{\eta} \in I_{\mathbf{c}}^*, \text{ and } \mathbf{u}^* = (u^*(\boldsymbol{\eta}))_{\boldsymbol{\eta}\in I_{\mathbf{c}}^*}.$$

We conclude as follows:

$$(2\pi i)^r Z(f;1+P;s)$$
$$= \int_{\rho_1-i\infty}^{\rho_1+i\infty} \cdots \int_{\rho_r-i\infty}^{\rho_r+i\infty} \frac{\Gamma(s/d-|\mathbf{a}|-\sum_{i=1}^r z_i)}{\Gamma(s/d)} \prod_{i=1}^r \frac{\Gamma(a_i+z_i)}{b_i^{a_i+z_i}} \frac{\mathcal{H}(\mathcal{T}_{\mathbf{c}};\mathbf{z})\,d\mathbf{z}}{\prod_{\eta\in I_{\mathbf{c}}^*}(\langle\boldsymbol{\eta},\mathbf{z}\rangle)^{u^*(\boldsymbol{\eta})}}.$$

We deduce from this that for any $\boldsymbol{\rho}\in\mathbb{R}_+^{*r}$ and $\sigma>\sup(\sigma_0, d(|\mathbf{a}|+|\boldsymbol{\rho}|))$:
(45)

$$Z(f;1+P;s) = \frac{1}{(2\pi i)^r}\int_{\rho_1-i\infty}^{\rho_1+i\infty} \cdots \int_{\rho_r-i\infty}^{\rho_r+i\infty} \frac{U(s;\mathbf{z})\ \mathcal{H}(\mathcal{T}_{\mathbf{c}};\mathbf{z})\ d\mathbf{z}}{(s-d|\mathbf{a}|-\langle d\mathbf{1},\mathbf{z}\rangle)\prod_{\eta\in I_{\mathbf{c}}^*}(\boldsymbol{\eta},\mathbf{z}\rangle)^{u^*(\boldsymbol{\eta})}}$$

where

$$U(s;\mathbf{z}) := d\,(s/d-|\mathbf{a}|-\langle\mathbf{1},\mathbf{z}\rangle)\,\Gamma(s/d-|\mathbf{a}|-\langle\mathbf{1},\mathbf{z}\rangle)\Gamma(s/d)^{-1}\prod_{i=1}^r\Gamma(a_i+z_i)b_i^{-a_i-z_i}$$

$$(46) \qquad = d\,\Gamma(s/d-|\mathbf{a}|-\sum_{i=1}^r z_i+1)\ \Gamma(s/d)^{-1}\prod_{i=1}^r\Gamma(a_i+z_i)b_i^{-a_i-z_i}.$$

Remark: *The role played by the set I in lemmas 3 and 4 (see §4.1.2 and §4.1.3), is played here by the set $I_{\mathbf{c}}^*$ and the exponents $u(\boldsymbol{\alpha})$ become the $u^*(\boldsymbol{\eta})$ (see (44)).*

It is well known that the Euler function $z\mapsto\Gamma(z)$ is holomorphic and has no zeros in the half-plane $\{z\in\mathbb{C}\mid\Re(z)>0\}$. It follows that $(s,\mathbf{z})\mapsto U(s,z)$ is holomorphic in

$\mathcal{D}_r(2\delta_1,2\delta_1)=\{(s,\mathbf{z})=(\sigma+i\tau,\mathbf{x}+i\mathbf{y}\in\mathbb{C}\times\mathbb{C}^r\mid\sigma>d\,|\mathbf{a}|-2\delta_1 \text{ and } |\Re(z_i)|<2\delta_1\ \forall i\}$,

where $\delta_1:=\frac{1}{4}\inf(d\,|\mathbf{a}|,\frac{1+|\mathbf{a}|}{r+1/d},a_1,\ldots,a_r)>0$. Moreover lemma 2 implies that there exists $B_0=B_0(\mathbf{a},\mathbf{b},d,r)>0$ such that we have uniformly in $(s;\mathbf{z})\in\mathcal{D}(2\delta_1,2\delta_1)$ the estimate:
(47)

$$U(s;\mathbf{z})\ll_{\sigma,\mathbf{a},\mathbf{b},d,r}(1+|\tau|)^{2\frac{|\sigma|}{d}+B_0}\prod_{i=1}^r(1+|y_i|)^{\frac{|\sigma|}{d}+B_0}e^{\frac{\pi}{2}\left(|\frac{\tau}{d}|-\sum_{i=1}^r|y_i|-|\frac{\tau}{d}-\sum_{i=1}^r y_i|\right)}.$$

From (43) we also know that there exist $A_1,B_1>0$ such that $\mathcal{H}(\mathcal{T}_{\mathbf{c}};\mathbf{z})$ satisfies the following estimate on $\{\mathbf{z}\in\mathbb{C}^r\mid\forall i\ \Re(z_i)>-\varepsilon_1\}$:

$$(48)\qquad \mathcal{H}(\mathcal{T}_{\mathbf{c}};\mathbf{z})\ll_{\Re(\mathbf{z})}(1+|\Im(\mathbf{z})|)^{A_1|\Re(\mathbf{z})|+B_1}\ll_{\Re(\mathbf{z})}\prod_{i=1}^r(1+|\Im(z_i)|)^{A_1|\Re(\mathbf{z})|+B_1}.$$

Let $\delta:=\inf(\varepsilon_1/2,\delta_1)$. Putting together the two preceding estimates (47) and (48) we conclude that there exists $B>0$ such that :
$V(s,z):=U(s,z)\ \mathcal{H}(\mathcal{T}_{\mathbf{c}};\mathbf{z})$ is holomorphic in $\mathcal{D}_r(2\delta,2\delta)$ and satisfies in it:
(49)

$$V(s;\mathbf{z})\ll_\sigma(1+|\tau|)^{2\frac{|\sigma|}{d}+(|\mathbf{a}|+B}\prod_{i=1}^r(1+|y_i|)^{2\frac{|\sigma|}{d}+B}\times e^{\frac{\pi}{2}\left(|\frac{\tau}{d}|-\sum_{i=1}^r|y_i|-|\frac{\tau}{d}-\sum_{i=1}^r y_i|\right)}.$$

We can therefore apply Lemma 3 to (45) by setting

$$L(s,\mathbf{z})=V(s,\mathbf{z}),\quad \sigma_1=d\,|\mathbf{a}|,\quad \boldsymbol{\phi}=d\,\mathbf{1}\ \text{ and }\ q=1.$$

We conclude that there exists $\eta>0$ such that $s\mapsto Z(f;1+P;s)$ has a meromorphic continuation with moderate growth to the half-plane $\{\sigma>d|\mathbf{a}|-\eta\}$ with at most one pole at $s=d|\mathbf{a}|$ of order at most $\rho_0^*:=\sum_{\eta\in I_{\mathbf{c}}^*}u^*(\boldsymbol{\eta})-rank\,(I_{\mathbf{c}}^*)+1$.

Moreover it follows from the ellipticity of the polynomial P that $rank\left(\{\boldsymbol{\alpha}^1,\ldots,\boldsymbol{\alpha}^n\}\right) = rank\left(\{\boldsymbol{\gamma}^1,\ldots,\boldsymbol{\gamma}^r\}\right) = rank\left(supp(P)\right) = n$. This implies that $rank\left(I^*_{\mathbf{c}}\right) = rank(I_{\mathbf{c}})$. We conclude:

$$\rho_0^* := \sum_{\boldsymbol{\eta}\in I^*_{\mathbf{c}}} u^*(\boldsymbol{\eta}) - rank\left(I^*_{\mathbf{c}}\right) + 1 = \sum_{\boldsymbol{\beta}\in I_{\mathbf{c}}} u(\boldsymbol{\beta}) - rank\left(I_{\mathbf{c}}\right) + 1. \tag{50}$$

Since

$$|\mathbf{c}| = \langle \mathbf{1}, \mathbf{c}\rangle = \langle \mathbf{1}, \sum_{k=1}^r a_k \boldsymbol{\gamma}^k\rangle = \sum_{k=1}^r a_k|\boldsymbol{\gamma}^k| = \sum_{k=1}^r a_k d = d|\mathbf{a}|, \tag{51}$$

the proof of the first part of Theorem 3 now follows. ♢

4.3. Proof of the second part of Theorem 3. Notations used in §4.2 will also be used throughout this section.

We assume in addition that the point $\mathbf{c}\in\mathbb{R}_+^{*n}$ and the pair of regularization $\mathcal{T}_{\mathbf{c}} = (I_{\mathbf{c}};\mathbf{u})$ satisfy the two following assumptions:

(1) $\mathbf{1}\in con^*(I_{\mathbf{c}})$;
(2) there exists a function K (holomorphic in a tubular neighborhood of $\mathbf{0}$) such that: $H(f;\mathcal{T}_{\mathbf{c}};\mathbf{s}) = K\left(\langle\boldsymbol{\beta},\mathbf{s}\rangle_{\boldsymbol{\beta}\in I_{\mathbf{c}}}\right)$.

Recall also from (51) that $|\mathbf{c}| = d|\mathbf{a}|$.
From (39) we conclude that the proof of Theorem 3 will follow once we prove that $|\mathbf{c}|$ is a pole of $Z(f;1+P;s)$ of order at most ρ_0^* (see (50)) and

$$Z(f;1+P;s) = \frac{H(f;\mathcal{T}_{\mathbf{c}};\mathbf{0})d^{\rho_0^*}A_0(\mathcal{T}_{\mathbf{c}},P)}{(s-|\mathbf{c}|)^{\rho_0^*}} + O\left(\frac{1}{(s-|\mathbf{c}|)^{\rho_0^*-1}}\right) \text{ as } s\to|\mathbf{c}|.$$

We will prove this in the sequel. Our strategy is the following:

First we set $\tilde{H}(\mathbf{z}) := \mathcal{H}(\mathcal{T}_{\mathbf{c}};\mathbf{z}) - \mathcal{H}(\mathcal{T}_{\mathbf{c}};\mathbf{0})$, where $\mathcal{H}(\mathcal{T}_{\mathbf{c}};\mathbf{z})$ is defined by (43). Thus, $\mathcal{H}(\mathcal{T}_{\mathbf{c}},\mathbf{0}) = H(f;\mathcal{T}_{\mathbf{c}};\mathbf{0})$.
Let $U(s;\mathbf{z})$ be the function defined in (46). It follows that (45) can then be written in this way: $\forall\boldsymbol{\rho}\in\mathbb{R}_+^{*r}$ and $\sigma > \sup\left(\sigma_0, d(|\mathbf{a}|+|\boldsymbol{\rho}|)\right)$,

$$Z(f;1+P;s) = H(f;\mathcal{T}_{\mathbf{c}};\mathbf{0})\; Z_1(s) + Z_2(s), \tag{52}$$

where

$$\begin{aligned} Z_1(s) &= \frac{1}{(2\pi i)^r}\int_{\rho_1-i\infty}^{\rho_1+i\infty}\cdots\int_{\rho_r-i\infty}^{\rho_r+i\infty}\frac{U(s;\mathbf{z})\,d\mathbf{z}}{(s-d\,|\mathbf{a}|-\langle d\,\mathbf{1},\mathbf{z}\rangle)\prod_{\boldsymbol{\eta}\in I^*_{\mathbf{c}}}\left(\langle\boldsymbol{\eta},\mathbf{z}\rangle\right)^{u^*(\boldsymbol{\eta})}}\\ Z_2(s) &= \frac{1}{(2\pi i)^r}\int_{\rho_1-i\infty}^{\rho_1+i\infty}\cdots\int_{\rho_r-i\infty}^{\rho_r+i\infty}\frac{U(s;\mathbf{z})\tilde{H}(\mathbf{z})\,d\mathbf{z}}{(s-d\,|\mathbf{a}|-\langle d\,\mathbf{1},\mathbf{z}\rangle)\prod_{\boldsymbol{\eta}\in I^*_{\mathbf{c}}}\left(\langle\boldsymbol{\eta},\mathbf{z}\rangle\right)^{u^*(\boldsymbol{\eta})}}. \end{aligned}$$

In section §4.3.1, we will use Lemma 4 to prove that $|\mathbf{c}| = d|\mathbf{a}|$ is a pole of $Z_1(s)$ of order ρ_0^* and even to determine the top order term in the principal part of $Z_1(s)$ at $|\mathbf{c}|$.

The integral $Z_2(s)$ is more complicated and there is no hope to get for it a precise result like for $Z_1(s)$. *Moreover if we could only infer that $Z_2(s)$ has a a pole at $s=|\mathbf{c}|$ of order at most ρ_0^*, then, we would not yet be able to prove that $s=|\mathbf{c}|$ is a pole of $Z(f;1+P;s)$!* To get around this difficulty, we will use in §4.3.2 the crucial Lemma 3, which give a very precise estimate of the orfer of the possible pole $s=|\mathbf{c}|$ since it implies that $Z_2(s)$ has a pole at $s=|\mathbf{c}|$ of order at most $\rho_0^* - 1$. Combining this with the result on $Z_1(s)$ suffices to complete the proof of Theorem 3.

4.3.1. *The principal part of $Z_1(s)$ at its first pole.* It is easy to see that for $\sigma \gg 1$:

$$Z_1(s) = \frac{1}{(2\pi i)^r}\int_{\rho_1 - i\infty}^{\rho_1 + i\infty} \cdots \int_{\rho_r - i\infty}^{\rho_r + i\infty} \frac{\Gamma(s/d - |\mathbf{a}| - z_1 - \cdots - z_r)\prod_{k=1}^r \Gamma(a_k + z_k)\, d\mathbf{z}}{\Gamma(s/d)\prod_{k=1}^r b_k^{a_k + z_k}\prod_{\eta \in I_{\mathbf{c}}^*}(\langle \boldsymbol{\eta}, \mathbf{z}\rangle)^{u^*(\boldsymbol{\eta})}}.$$

Since $\mathbf{1} \in con^*(I_{\mathbf{c}})$, there exists a set $\{t_{\boldsymbol{\beta}}\}_{\boldsymbol{\beta} \in I_{\mathbf{c}}} \subset \mathbb{R}_+^*$ such that $\mathbf{1} = \sum_{\boldsymbol{\beta} \in I_{\mathbf{c}}} t_{\boldsymbol{\beta}} \boldsymbol{\beta}$ (i.e $\sum_{\boldsymbol{\beta} \in I_{\mathbf{c}}} t_{\boldsymbol{\beta}} \beta_i = 1\ \forall i = 1, \ldots, n$). Consequently we have:

$$\sum_{\boldsymbol{\beta} \in I_{\mathbf{c}}} t_{\boldsymbol{\beta}} \mu(\boldsymbol{\beta}) = \sum_{\boldsymbol{\beta} \in I_{\mathbf{c}}} t_{\boldsymbol{\beta}} \sum_{i=1}^n \beta_i \boldsymbol{\alpha}^i = \sum_{i=1}^n \Big(\sum_{\boldsymbol{\beta} \in I_{\mathbf{c}}} t_{\boldsymbol{\beta}} \beta_i\Big)\boldsymbol{\alpha}^i = \sum_{i=1}^n \boldsymbol{\alpha}^i = \sum_{k=1}^r \Big(\sum_{i=1}^n \gamma_i^k\Big)\mathbf{e}_k = d\mathbf{1}.$$

We conclude from this that

$$\mathbf{1} \in con^*\,(I_{\mathbf{c}}^*) \setminus \{\mathbf{0}\}. \tag{53}$$

By our hypotheses, we know also that $\langle \boldsymbol{\eta}, \mathbf{a}\rangle = \sum_{i=1}^n \beta_i \langle \boldsymbol{\alpha}^i, \mathbf{a}\rangle = \sum_{i=1}^n \beta_i c_i = \langle \boldsymbol{\beta}, \mathbf{c}\rangle = 1\ \forall \boldsymbol{\eta} = \mu(\boldsymbol{\beta}) \in I_{\mathbf{c}}^*$. Thus, it follows from lemma 4 that $s = d|\mathbf{a}| = |\mathbf{c}|$ is a pole of $Z_1(s)$ of order ρ_0^* (see (50)) and that

$$Z_1(s) \sim_{s \to |\mathbf{c}|} \frac{d^{\rho_0^*}\ A_0(I_{\mathbf{c}}^*; \mathbf{u}^*; \mathbf{b})}{(s - |\mathbf{c}|)^{\rho_0^*}} \tag{54}$$

where $A_0(I_{\mathbf{c}}^*; \mathbf{u}^*; \mathbf{b}) > 0$ is the *volume constant* associated to $I_{\mathbf{c}}^*, \mathbf{u}^*$ and $\mathbf{b}$ (see §2.3.2).

4.3.2. *A sharper estimate for $ord_{s=|\mathbf{c}|} Z_2(s)$ and end of the proof of Theorem 3.* The quantities $\sigma_1, \boldsymbol{\phi}$, and q, introduced during the proof of lemma 3 are assigned values here by setting: $\sigma_1 := |\mathbf{c}|$, $\boldsymbol{\phi} := d\mathbf{1} = (d, \ldots, d) \in \mathbb{R}^r$ and $q := 1$.
We also define $L(s; \mathbf{z}) := U(s; \mathbf{z})\ \tilde{H}(\mathbf{z})$ where $\tilde{H}(\mathbf{z}) := \mathcal{H}(\mathcal{T}_{\mathbf{c}}; \mathbf{z}) - \mathcal{H}(\mathcal{T}_{\mathbf{c}}; \mathbf{0})$ as above. As a result, we see that $Z_2(s) = T_r(s)$ (see (16) for the definition of $T_r(s)$).
It also follows that the role played by the set I in our key lemma 3 and 4, is played here by the set $I_{\mathbf{c}}^*$ and the exponents $u(\boldsymbol{\alpha})$ become the $u^*(\boldsymbol{\eta})$ (see (44)). Thus, the quantity d_r from lemma 3 is as follows (see (50)):

$$d_r = \sum_{\boldsymbol{\eta} \in I_{\mathbf{c}}^*} u^*(\boldsymbol{\eta}) - rank(I_{\mathbf{c}}^*) + 1 - \varepsilon_0(\boldsymbol{\phi}, L) = \rho_0^* - \varepsilon_0(\boldsymbol{\phi}, L).$$

The estimates (47) and (48) imply that there exists $\delta > 0$ such that $L(s; \mathbf{z})$ is holomorphic in $\mathcal{D}_r(2\delta, 2\delta)$, on which the estimate (15) is satisfied (with a suitable choice of $A, B > 0$). Lemma 3 implies then that there exists $\eta > 0$ such that $s \mapsto Z_2(s)$ has a meromorphic continuation to the half-plane $\{\sigma > \sigma_1 - \eta\}$ with only one possible pole at $s = \sigma_1 = |\mathbf{c}|$. Lemma 3 implies also the *crucial* estimate:

$$ord_{s=|\mathbf{c}|} Z_2(s) \le \rho_0^* - \varepsilon_0(\boldsymbol{\phi}, L). \tag{55}$$

Thus, we must evaluate $\varepsilon_0(\boldsymbol{\phi}, L)$. We do this as follows.

It follows from (53) that

$$\boldsymbol{\phi} := d\ \mathbf{1} \in con^*\,(I_{\mathbf{c}}^*) \setminus \{\mathbf{0}\}. \tag{56}$$

Furthermore, assumption 2 implies that there exists a function K (holomorphic in a tubular neighborhood of $\mathbf{0}$) such that $H(f; \mathcal{T}_{\mathbf{c}}, \mathbf{s}) = K\left(\langle \boldsymbol{\beta}, \mathbf{s}\rangle_{\boldsymbol{\beta} \in I_{\mathbf{c}}}\right)$. But for any

$\boldsymbol{\beta} \in I_{\mathbf{c}}$ and for any $\mathbf{z} \in \mathbb{C}^r$ we have,

$$\sum_{i=1}^{n} \beta_i \langle \boldsymbol{\alpha}^i, \mathbf{z} \rangle = \left\langle \sum_{i=1}^{n} \beta_i \boldsymbol{\alpha}^i, \mathbf{z} \right\rangle = \langle \mu(\boldsymbol{\beta}), \mathbf{z} \rangle.$$

It follows that: $\mathcal{H}(\mathcal{T}_{\mathbf{c}}; \mathbf{z}) := H\left(f; \mathcal{T}_{\mathbf{c}}; \langle \boldsymbol{\alpha}^1, \mathbf{z} \rangle, \ldots, \langle \boldsymbol{\alpha}^n, \mathbf{z} \rangle\right) = K\left(\langle \mu(\boldsymbol{\beta}), \mathbf{z} \rangle_{\boldsymbol{\beta} \in I_{\mathbf{c}}}\right)$. Consequently there exists a function $\tilde{K}$ (holomorphic in a tubular neighborhood of $\mathbf{0}$) such that: $\tilde{H}(\mathbf{z}) = \mathcal{H}(\mathcal{T}_{\mathbf{c}}; \mathbf{z}) - \mathcal{H}(\mathcal{T}_{\mathbf{c}}; \mathbf{0}) = \tilde{K}\left(\langle \boldsymbol{\eta}, \mathbf{z} \rangle_{\boldsymbol{\eta} \in I_{\mathbf{c}}^*}\right)$. Since in addition we have $\tilde{H}(\mathbf{0}) = 0$ and $\boldsymbol{\phi} = d\ \mathbf{1} \in con^*(I_{\mathbf{c}}^*) \setminus \{\mathbf{0}\}$, it follows from lemma 3 that $\varepsilon_0(\boldsymbol{\phi}, L) = 1$. Thus, we conclude from this and (55) that

$$ord_{s=|\mathbf{c}|} Z_2(s) \le \rho_0^* - 1.$$

Combining this with (54), (52), and (39) implies that

$$Z(f; P; s) = \frac{H(f; \mathcal{T}_{\mathbf{c}}; \mathbf{0}) d^{\rho_0^*} A_0(I_{\mathbf{c}}^*; \mathbf{u}^*; \mathbf{b})}{(s - |\mathbf{c}|)^{\rho_0^*}} + O\left(\frac{1}{(s - |\mathbf{c}|)^{\rho_0^* - 1}}\right) \quad \text{as } s \to |\mathbf{c}|.$$

This completes the proof of theorem 3, once one has also noted that $A_0(I_{\mathbf{c}}^*; \mathbf{u}^*; \mathbf{b}) = A_0(\mathcal{T}_{\mathbf{c}}, P)$ where $A_0(\mathcal{T}_{\mathbf{c}}, P)$ is the mixed volume constant associated to P and $\mathcal{T}_{\mathbf{c}}$ as in §2.3.3. ◇

5. Proofs of Proposition 2 and Theorems 1, 2

5.1. A Lemma from convex analysis and its proof. Using the definitions introduced in §2.2, I will first give a lemma from convex analysis.

LEMMA 5. *Let I be a nonempty subset of $\mathbb{R}_+^n \setminus \{\mathbf{0}\}$. Set $\mathcal{E}(I)$ to be its Newton polyhedron and denotes by $\mathcal{E}^o(I)$ its dual (see §2.2). Let F be a face of $\mathcal{E}(I)$ that is not contained in a coordinate hyperplane and $\mathbf{c} \in pol_0(F)$ a normalized polar vector of F. Then: F meets the diagonal if and only if $|\mathbf{c}| = \iota(I)$ where $\iota(I) = \min\{|\boldsymbol{\alpha}|;\ \boldsymbol{\alpha} \in \mathcal{E}^o(I) \cap \mathbb{R}_+^n\}$.*

Proof of Lemma 5: We first note that the definition of the normalized polar vector implies that $\mathbf{c} \in \mathcal{E}(I)^o \cap \mathbb{R}_+^n$.
• Assume first that the diagonal Δ meets the face F of the Newton Polyhedron $\mathcal{E}(I)$. Therefore, there exists $t_0 > 0$ such that $\Delta \cap F = \{t_0 \mathbf{1}\}$. Let $\boldsymbol{\alpha}^1, \ldots, \boldsymbol{\alpha}^r \in I \cap F$ and let J a subset (possibly empty) of $\{1, \ldots, n\}$ such that $F = convex\ hull\{\boldsymbol{\alpha}^1, \ldots, \boldsymbol{\alpha}^r\} + con\{\mathbf{e}_i \mid i \in J\}$. Thus there exist $\lambda_1, \ldots, \lambda_r \in \mathbb{R}_+$ verifying $\lambda_1 + \cdots + \lambda_r = 1$ and a finite family $(\mu_i)_{i \in J}$ of elements of $\mathbb{R}_+$ such that:

$$t_0 \mathbf{1} = \sum_{i=1}^{r} \lambda_i \boldsymbol{\alpha}^i + \sum_{j \in J} \mu_j \mathbf{e}_j. \tag{57}$$

But $\mathbf{c}$ is orthogonal to the vectors $\mathbf{e}_j$ $(j \in J)$ and $\langle \mathbf{c}, \boldsymbol{\alpha}^i \rangle = 1$ for all $i = 1, \ldots, r$. Thus it follows from the relation (57) that $t_0 |\mathbf{c}| = \langle \mathbf{c}, t_0 \mathbf{1} \rangle = \sum_{i=1}^{r} \lambda_i \langle \boldsymbol{\alpha}^i, \mathbf{c} \rangle + \sum_{j \in J} \mu_j \langle \mathbf{e}_j, \mathbf{c} \rangle = \sum_{i=1}^{r} \lambda_i = 1$. Consequently $|\mathbf{c}| = t_0^{-1}$.

Relation (57) implies also that for all $\mathbf{b} \in \mathcal{E}(I)^o \cap \mathbb{R}_+^n$:

$$|\mathbf{b}| = t_0^{-1} \langle \mathbf{b}, t_0 \mathbf{1}\rangle = t_0^{-1} \sum_{i=1}^r \lambda_i \langle \boldsymbol{\alpha}^i, \mathbf{b}\rangle + \sum_{j\in J} \mu_j b_j \geq t_0^{-1} \sum_{i=1}^r \lambda_i \langle \boldsymbol{\alpha}^i, \mathbf{b}\rangle$$
$$\geq |\mathbf{c}| \sum_{i=1}^r \lambda_i = |\mathbf{c}|.$$

This implies $|\mathbf{c}| = \iota(I)$.

• Conversely assume that $|\mathbf{c}| = \iota(I)$. We will show that $\Delta \cap F \neq \emptyset$.
Let G be a face of $\mathcal{E}(I)$ which meets the diagonal Δ. Since G is not included in the coordinate hyperplanes, it has a normalized polar vector $\mathbf{a} \in polo_0(G)$. Moreover there exist $\boldsymbol{\beta}^1, \dots, \boldsymbol{\beta}^k \in I \cap G$ and T a subset (possibly empty) of $\{1, \dots, n\}$ such that $G = convex\ hull\{\boldsymbol{\beta}^1, \dots, \boldsymbol{\beta}^k\} + con\{\mathbf{e}_i \mid i \in T\}$.
The proof of the first part shows then that $|\mathbf{a}| = \iota(I)$ and that there exist $\nu_1, \dots, \nu_k \in \mathbb{R}_+$ verifying $\nu_1 + \cdots + \nu_k = 1$ and a finite family $(\delta_j)_{j\in T}$ of elements of $\mathbb{R}_+$ such that if we set $t_0 := |\mathbf{a}|^{-1}$,then:

$$t_0 \mathbf{1} = \sum_{i=1}^k \nu_i \boldsymbol{\beta}^i + \sum_{j\in T} \delta_j \mathbf{e}_j \in G \cap \Delta. \tag{58}$$

Set $T' := \{j \in T \mid \delta_j \neq 0\}$. It follows from relation (58) that

$$t_0|\mathbf{c}| = \langle \mathbf{c}, t_0\mathbf{1}\rangle = \sum_{i=1}^k \nu_i \langle \boldsymbol{\beta}^i, \mathbf{c}\rangle + \sum_{j\in T'} \delta_j \langle \mathbf{e}_j, \mathbf{c}\rangle \geq \sum_{i=1}^k \nu_i \langle \boldsymbol{\beta}^i, \mathbf{c}\rangle \geq \sum_{i=1}^k \nu_i = 1.$$

But $t_0|\mathbf{c}| = |\mathbf{a}|^{-1}|\mathbf{c}| = \iota(I)^{-1}\iota(I) = 1$ so the intermediate inequalities must be equalities. This clearly forces $\langle \boldsymbol{\beta}^i, \mathbf{c}\rangle = 1\ \forall i = 1, \dots, k$ and $\langle \mathbf{c}, \mathbf{e}_j\rangle = 0\ \forall j \in T'$. Relation (58) implies then that:
$\langle t_0\mathbf{1}, \mathbf{c}\rangle = \sum_{i=1}^k \nu_i \langle \boldsymbol{\beta}^i, \mathbf{c}\rangle + \sum_{j\in J'} \delta_j \langle \mathbf{e}_j, \mathbf{c}\rangle = \sum_{i=1}^k \nu_i = 1$.
Since $t_0\mathbf{1} \in \mathcal{E}(I)$, it follows from the preceding discussion that $t_0\mathbf{1} \in F$ and therefore $\Delta \cap F \neq \emptyset$. This finishes the proof of Lemma 5. $\diamondsuit$

5.2. Proof of Proposition 2. $\forall \varepsilon \in \mathbb{R}$, set $U_\varepsilon := \{\mathbf{s} \in \mathbb{C}^n \mid \Re(s_i) > \varepsilon\ \forall i\}$.
• Let $\mathbf{c} \in polo_0(\mathcal{F}_0(f))$. The compactness of the face $\mathcal{F}_0(f)$ implies that $\mathbf{c} \in \mathbb{R}_+^{*n}$. Moreover it follows from the definition of $polo_0(\mathcal{F}_0(f))$ that $\forall \boldsymbol{\nu} \in S^*(g)$ $\langle \boldsymbol{\nu}, \mathbf{c}\rangle \geq 1$ with equality if and only if $\boldsymbol{\nu} \in I_f = \mathcal{F}_0(f) \cap S^*(g)$.
Set $\delta_0 := \frac{1}{2}\inf_{i=1,\dots,n} c_i > 0$. Fix also $N := \left[\frac{8}{\delta_0} + \sup_{\mathbf{x}\in\mathcal{F}_0(f)} |\mathbf{x}|\right] + 1 \in \mathbb{N}$. (Evidently, $N < \infty$ since $\mathcal{F}_0(f)$ is compact.)
It is easy to see that the following bound is uniform in p prime and $\mathbf{s} \in U_{-\delta_0} = \{\mathbf{s} \in \mathbb{C}^n \mid \forall i\ \Re(s_i) > -\delta_0\}$:

$$\sum_{|\boldsymbol{\nu}|\geq N+1} \frac{g(\boldsymbol{\nu})}{p^{\langle \mathbf{c}+\boldsymbol{\sigma}, \boldsymbol{\nu}\rangle}} \ll \sum_{|\boldsymbol{\nu}|\geq N+1} \frac{|\boldsymbol{\nu}|^M}{p^{\delta_0|\boldsymbol{\nu}|}} \ll \frac{1}{p^{\delta_0(N+1)/2}} \sum_{|\boldsymbol{\nu}|\geq N+1} \frac{|\boldsymbol{\nu}|^M}{2^{\delta_0|\boldsymbol{\nu}|/2}} \ll \frac{1}{p^{\delta_0(N+1)/2}} \ll \frac{1}{p^2}.$$

Thus, the following is uniform in p and $\mathbf{s}\in U_{-\delta_0}$:

$$\sum_{|\boldsymbol{\nu}|\geq 1}\frac{f(p^{\nu_1},\ldots,p^{\nu_n})}{p^{\langle \mathbf{c}+\mathbf{s},\boldsymbol{\nu}\rangle}}=\sum_{|\boldsymbol{\nu}|\geq 1}\frac{g(\boldsymbol{\nu})}{p^{\langle \mathbf{c}+\mathbf{s},\boldsymbol{\nu}\rangle}}=\sum_{1\leq|\boldsymbol{\nu}|\leq N}\frac{g(\boldsymbol{\nu})}{p^{\langle \mathbf{c}+\mathbf{s},\boldsymbol{\nu}\rangle}}+O\left(\frac{1}{p^2}\right)$$

$$(59)\qquad =\sum_{\boldsymbol{\nu}\in I_f}\frac{g(\boldsymbol{\nu})}{p^{1+\langle \mathbf{s},\boldsymbol{\nu}\rangle}}+\sum_{\substack{1\leq|\boldsymbol{\nu}|\leq N\\ \boldsymbol{\nu}\notin\mathcal{F}_0(f)}}\frac{g(\boldsymbol{\nu})}{p^{\langle \mathbf{c},\boldsymbol{\nu}\rangle+\langle \mathbf{s},\boldsymbol{\nu}\rangle}}+O\left(\frac{1}{p^2}\right)$$

Since I_f is a finite set and $\langle \mathbf{c},\boldsymbol{\nu}\rangle>1$ for all $\boldsymbol{\nu}\in S^*(g)\setminus\mathcal{F}_0(f)$, it follows from (59) that there exists $\delta_1\in]0,\delta_0[$ and $\varepsilon_1>0$ such that the following is uniform in p and $\mathbf{s}\in U_{-\delta_1}$:

$$(60)\qquad \sum_{|\boldsymbol{\nu}|\geq 1}\frac{f(p^{\nu_1},\ldots,p^{\nu_n})}{p^{\langle \mathbf{c}+\mathbf{s},\boldsymbol{\nu}\rangle}}=\sum_{|\boldsymbol{\nu}|\geq 1}\frac{g(\boldsymbol{\nu})}{p^{\langle \mathbf{c}+\mathbf{s},\boldsymbol{\nu}\rangle}}=\sum_{\boldsymbol{\nu}\in I_f}\frac{g(\boldsymbol{\nu})}{p^{1+\langle \mathbf{s},\boldsymbol{\nu}\rangle}}+O\left(\frac{1}{p^{1+\varepsilon_1}}\right)$$

The multiplicativity of f now implies that $\mathcal{M}(f;\mathbf{s})$ converges absolutely in $\Omega_{\mathbf{c}}:=\{\mathbf{s}\in\mathbb{C}^n\mid\forall i\ \Re(s_i)>c_i\}$ on which it can be written as follows:
(61)

$$\mathcal{M}(f;\mathbf{s}):=\sum_{\mathbf{m}\in\mathbb{N}^n}\frac{f(m_1,\ldots,m_n)}{m^{s_1}\ldots m_n^{s_n}}=\prod_p\Bigg(\sum_{\boldsymbol{\nu}\in\mathbb{N}_0^n}\frac{f(p^{\nu_1},\ldots,p^{\nu_n})}{p^{\langle\boldsymbol{\nu},\mathbf{s}\rangle}}\Bigg)=\prod_p\Bigg(\sum_{\boldsymbol{\nu}\in\mathbb{N}_0^n}\frac{g(\boldsymbol{\nu})}{p^{\langle\boldsymbol{\nu},\mathbf{s}\rangle}}\Bigg)$$

As in [**BEL**], we introduce the function $G(f;\mathbf{s})$ defined for all $\mathbf{s}\in U_0=\{\mathbf{s}\in\mathbb{C}^n\mid\forall i\ \Re(s_i)>0\}$ by:

$$G(f;\mathbf{s})\ :=\ \Bigg(\prod_{\boldsymbol{\nu}\in I_f}\zeta(1+\langle\boldsymbol{\nu},\mathbf{s}\rangle)^{-g(\boldsymbol{\nu})}\Bigg)\mathcal{M}(f;\mathbf{c}+\mathbf{s})$$

$$(62)\qquad =\ \prod_p\Bigg(\prod_{\boldsymbol{\nu}\in I_f}\left(1-\frac{1}{p^{1+\langle\boldsymbol{\nu},\mathbf{s}\rangle}}\right)^{g(\boldsymbol{\nu})}\Bigg)\Bigg(\sum_{\boldsymbol{\nu}\in\mathbb{N}_0^n}\frac{g(\boldsymbol{\nu})}{p^{\langle\boldsymbol{\nu},\mathbf{c}+\mathbf{s}\rangle}}\Bigg).$$

Combining (60) with (62) now implies that there exist $\delta_2\in]0,\delta_1[$ and $\varepsilon_2>0$ such that : $G(f;\mathbf{s})=\prod_p\left(1+O\left(\frac{1}{p^{1+\varepsilon_2}}\right)\right)$ uniformly in $\mathbf{s}\in U_{-\delta_2}$. It follows that the Euler product $\mathbf{s}\mapsto G(f;s)$ converges absolutely and defines a bounded holomorphic function on $U_{-\delta_2}$.

Moreover, since $G(f;\mathbf{0})=\prod_p\left(1-\frac{1}{p}\right)^{\sum_{\boldsymbol{\nu}\in I_f}g(\boldsymbol{\nu})}\cdot\Bigg(\sum_{\boldsymbol{\nu}\in\mathbb{N}_0^n}\frac{g(\boldsymbol{\nu})}{p^{\langle\boldsymbol{\nu},\mathbf{c}\rangle}}\Bigg)$ is a convergent infinite product whose general term is >0, we conclude that $G(f;\mathbf{0})>0$.

Moreover, for all $\mathbf{s}\in U_0=\{\mathbf{s}\in\mathbb{C}^n\mid\forall i\ \Re(s_i)>0\}$ we have:

$$H(f;\mathcal{T}_{\mathbf{c}};\mathbf{s})\ :=\ \Bigg(\prod_{\boldsymbol{\nu}\in I_f}\langle\boldsymbol{\nu},\mathbf{s}\rangle^{g(\boldsymbol{\nu})}\Bigg)\mathcal{M}(f;\mathbf{c}+\mathbf{s})$$

$$(63)\qquad =\ \Bigg(\prod_{\boldsymbol{\nu}\in I_f}(\langle\boldsymbol{\nu},\mathbf{s}\rangle\zeta(1+\langle\boldsymbol{\nu},\mathbf{s}\rangle))^{g(\boldsymbol{\nu})}\Bigg)G(f;\mathbf{s}).$$

Thus, by using the properties of the function $\mathbf{s}\mapsto G(f;\mathbf{s})$ established above and standard properties satisfied by the Riemann zeta function, it follows that there exists $\varepsilon_0>0$ such that $\mathbf{s}\mapsto H(f;\mathcal{T}_{\mathbf{c}};\mathbf{s})$ has a holomorphic continuation with moderate growth to $\{\mathbf{s}\in\mathbb{C}^n\mid\sigma_i>-\varepsilon_0\ \forall i\}$.
By (63) we conclude that $H(f;\mathcal{T}_{\mathbf{c}};\mathbf{0})=G(f;\mathbf{0})>0$.

Moreover, by definition, $\mathcal{F}_0(f)$ is the smallest face of $\mathcal{E}(f)$ which meets the diagonal $\Delta = \mathbb{R}_+\mathbf{1}$ and $\mathcal{F}_0(f) = conv\,(I_f)$. It follows that $\mathbf{1} \in con^*\,(I_f)$. This completes the proof of the first part of Proposition 2. $\diamondsuit$

• We assume now that $dim\,\mathcal{F}_0(f) = rank\,(S^*(g)) - 1$.

Set $r := rank(I_f)$ and fix in the sequel $\boldsymbol{\nu}^1, \dots, \boldsymbol{\nu}^r \in I_f$ such that $rank\{\boldsymbol{\nu}^1, \dots, \boldsymbol{\nu}^r\} = r$. Since we assume that $dim\,\mathcal{F}_0(f) = rank\,(S^*(g)) - 1$, it follows that

$$rank(I_f) = dim\,\mathcal{F}_0(f) + 1 = rank\,(S^*(g)).$$

Recalling the definition of $G(f;\mathbf{s})$ from (62), the proof of the proposition will follow once we prove the existence of a function $L(\mathbf{z})$ holomorphic on some open tubular neighborhood of $\mathbf{z} = \mathbf{0}$ in $\mathbb{C}^r$ such that $G(f;\mathbf{s}) = L(\langle \boldsymbol{\nu}^1, \mathbf{s}\rangle, \dots, \langle \boldsymbol{\nu}^r, \mathbf{s}\rangle)$ on some open tubular neighborhood of $\mathbf{s} = \mathbf{0}$ in $\mathbb{C}^n$.

We recall from the proof of the first part of Proposition 2 that there exists $\delta_2 > 0$ such that the Euler product $G(f;\mathbf{s})$ converges absolutely and defines a holomorphic function on $U_{-\delta_2} := \{\mathbf{s} \in \mathbb{C}^n \mid \Re(s_i) > -\delta_2\ \forall i = 1, \dots, n\}$.
We fix this δ_2 in the following discussion.

It follows from the equality $rank\{\boldsymbol{\nu}^1, \dots, \boldsymbol{\nu}^r\} = r$ that the linear function $\varphi : \mathbb{C}^n \to \mathbb{C}^r$, $\mathbf{s} \mapsto \varphi(\mathbf{s}) = (\langle \boldsymbol{\nu}^1, \mathbf{s}\rangle, \dots, \langle \boldsymbol{\nu}^r, \mathbf{s}\rangle)$ is onto. By a permutation of coordinates, if needed, we can then assume that the function $\psi : \mathbb{C}^r \to \mathbb{C}^r$, $\mathbf{s}' = (s_1, \dots, s_r) \mapsto \psi(\mathbf{s}') := \varphi(\mathbf{s}', \mathbf{0}) = (\sum_{i=1}^r \nu_i^1 s_i, \dots, \sum_{i=1}^r \nu_i^r s_i)$ is an isomorphism. In particular $\exists \boldsymbol{\beta}^1, \dots, \boldsymbol{\beta}^r \in \mathbb{Q}^r$ such that $\forall \mathbf{z} \in \mathbb{C}^r$, $\psi^{-1}(\mathbf{z}) = (\langle \boldsymbol{\beta}^1, \mathbf{z}\rangle, \dots, \langle \boldsymbol{\beta}^r, \mathbf{z}\rangle)$. Defining $\mathcal{R}(\varepsilon) := \{\mathbf{z} = (z_1, \dots, z_r) \in \mathbb{C}^r \mid |\Re(z_i)| < \varepsilon\ \forall i\}$, it follows that

$$\psi^{-1}\left(\mathcal{R}(\varepsilon_0)\right) \subset \mathcal{R}(\delta_2) \text{ for } \varepsilon_0 := \delta_2 \left(\max_{1 \le i \le r} |\boldsymbol{\beta}^i|\right)^{-1} > 0. \tag{64}$$

The fact that $rank(S^*(g)) = rank(I_f) = rank\{\boldsymbol{\nu}^1, \dots, \boldsymbol{\nu}^r\}$ then implies that for any $\boldsymbol{\nu} \in S^*(g)$ there exist $a_1(\boldsymbol{\nu}), \dots, a_r(\boldsymbol{\nu}) \in \mathbb{Q}$ such that $\boldsymbol{\nu} = a_1(\boldsymbol{\nu})\boldsymbol{\nu}^1 + \dots + a_r(\boldsymbol{\nu})\boldsymbol{\nu}^r$. It follows that for all $\mathbf{s} \in U_{-\delta_2}$, $G(f;s) = L\left(\langle \boldsymbol{\nu}^1, \mathbf{s}\rangle, \dots, \langle \boldsymbol{\nu}^r, \mathbf{s}\rangle\right)$ where
(65)

$$L(\mathbf{z}) = \prod_p \left(\prod_{\boldsymbol{\nu} \in I_f} \left(1 - \frac{1}{p^{1+\sum_{i=1}^r a_i(\boldsymbol{\nu}) z_i}}\right)^{g(\boldsymbol{\nu})} \right) \left(1 + \sum_{\boldsymbol{\nu} \in S^*(g)} \frac{g(\boldsymbol{\nu})}{p^{\langle \boldsymbol{\nu}, \mathbf{c}\rangle + \sum_{i=1}^r a_i(\boldsymbol{\nu}) z_i}}\right).$$

So to conclude, it suffices to prove that L converges and defines a holomorphic function in $\mathcal{R}(\varepsilon_0)$. But for all $\mathbf{s}' \in \psi^{-1}\left(\mathcal{R}(\varepsilon_0)\right)$, $\tilde{L}(\mathbf{s}') = L \circ \psi(\mathbf{s}') = L\left(\varphi(\mathbf{s}', \mathbf{0})\right) = G\left(f; (\mathbf{s}', \mathbf{0})\right)$, and by definition of ε_0 we have $\psi^{-1}\left(\mathcal{R}(\varepsilon_0)\right) \times \{\mathbf{0}\}^{n-r} \subset \mathcal{R}(\delta_2) \times \{\mathbf{0}\}^{n-r} \subset U_{-\delta_2}$. It follows that $\tilde{L} := L \circ \psi$ converges and defines a holomorphic function in $\psi^{-1}\left(\mathcal{R}(\varepsilon_0)\right)$. We deduce by composition that L converges and defines a holomorphic function in $\mathcal{R}(\varepsilon_0)$. This completes the proof of Proposition 2. $\diamondsuit$

5.3. Proofs of Theorems 1 and 2. Proof of Theorem 1:

By symmetry we have: $Z_{H_P}(U(\mathbf{A}); s) := \sum_{M \in U(\mathbf{A})} {H_P}^{-s}(M) = c(\mathbf{A}) \sum_{\mathbf{m} \in \mathbb{N}^n} \frac{f(\mathbf{m})}{P(\mathbf{m})^{s/d}}$,
where $c(\mathbf{A})$ is the constant defined in (8), and f is the function defined by:

1. (1) $f(m_1, \dots, m_n) = 1$ if $m_1^{a_{i,1}}..m_n^{a_{i,n}} = 1\ \forall i = 1, .., l$ and $\gcd(m_1, .., m_n) = 1$
2. (2) $f(m_1, \dots, m_n) = 0$ otherwise.

It is easy to see that f is a multiplicative function and that for any prime number p and any $\boldsymbol{\nu} \in \mathbb{N}_0^n$: $f(p^{\nu_1}, \dots, p^{\nu_n}) = g(\boldsymbol{\nu})$ where g is the characteristic function of

the set $T(\mathbf{A})$ defined in §3.1. So, it is obvious that f is a also uniform (see definition 3 in §3.2.3).
It now suffices to verify that the assumptions of Theorem 3 are satisfied.
By using the notations of §3.2.3, it is easy to check that

$$\mathcal{E}(f) = \mathcal{E}\left(T^*(\mathbf{A})\right) \text{ and } \mathcal{F}_0(f) = \mathcal{F}_0(\mathbf{A}).$$

Therefore, Proposition 2 implies that the first part of Theorem 1 follows from Theorem 3.

Let us now suppose that $dim\left(\mathcal{F}_0(\mathbf{A})\right) = dim X(\mathbf{A}) = n-1-l$. Since

$$dim\left(\mathcal{F}_0(\mathbf{A})\right)+1 = rank\left(\mathcal{F}_0(\mathbf{A}) \cap T^*(\mathbf{A})\right) \leq rank\left(T^*(\mathbf{A})\right) \leq n - rank(\mathbf{A}) = n-l,$$

it follows that $dim\left(\mathcal{F}_0(\mathbf{A})\right) = rank\left(T^*(\mathbf{A})\right) - 1 = rank\left(S^*(g)\right) - 1$. Consequently, Proposition 2 implies that the second part of Theorem 1 also follows from Theorem 3, once one has also noted that Lemma 5 implies $\iota(f) = \iota\left(\mathcal{E}(f)\right) = |\mathbf{c}|$ for any normalized polar vector $\mathbf{c}$ of $\mathcal{F}_0(f) = \mathcal{F}_0(\mathbf{A})$. $\diamondsuit$

Proof of Theorem 2:

Let $A_n(\mathbf{a})$ be the $1 \times n$ matrix $A_n(\mathbf{a}) := (a_1, \ldots, a_{n-1}, -q)$. It is then clear that $X_{n-1}(\mathbf{a}) = V\left(A_n(\mathbf{a})\right)$. Thus, Theorem 2 will follow directly from Corollary 2 once we show that all the hypotheses of the corollary are satisfied.
Set $c(\mathbf{a}) := \frac{1}{2} \# \left\{(\varepsilon_1, \ldots, \varepsilon_n) \in \{-1,+1\}^n \mid \varepsilon_1^{a_1} \ldots \varepsilon_{n-1}^{a_{n-1}} = \varepsilon_n^q\right\}$.
As above, by symmetry we have for all $t > 1$:

$$N_{H_P}(U_{n-1}(\mathbf{a}); t) = \#\{M \in U_{n-1}(\mathbf{a}) \mid H_P(M) \leq t\} = c(\mathbf{a}) \sum_{\{\mathbf{m} \in \mathbb{N}^{n-1}; \tilde{P}(\mathbf{m})^{1/d} \leq t\}} f(m_1, \ldots, m_{n-1})$$

where $\tilde{P}$ is defined by $\tilde{P}(X_1, \ldots, X_{n-1}) := P\left(X_1, \ldots, X_{n-1}, \prod_j X_j^{a_j/q}\right)$ and f is the function defined by: $f(m_1, .., m_{n-1}) = 1$ if $m_1^{a_1} \ldots m_{n-1}^{a_{n-1}}$ is the qth power of an integer and $\gcd(m_1, .., m_{n-1}) = 1$ and $f(m_1, \ldots, m_{n-1}) = 0$ otherwise.
It is easy to see that f is a multiplicative function and that for any prime number p and any $\boldsymbol{\nu} \in \mathbb{N}_0^{n-1}$: $f(p^{\nu_1}, \ldots, p^{\nu_{n-1}}) = g(\boldsymbol{\nu})$, where g is the characteristic function of the set $L_n(\mathbf{a}) \cup \{\mathbf{0}\}$ and $L_n(\mathbf{a}) := \left\{\boldsymbol{\nu} \in \mathbb{N}_0^{n-1} \setminus \{\mathbf{0}\};\ q | \langle \mathbf{a}, \boldsymbol{\nu} \rangle \text{ and } \nu_1 \ldots \nu_{n-1} = 0\right\}$.
So it is clear that f is also uniform.
By definition, we have that $\mathcal{E}(f) = \mathcal{E}(\mathbf{a}) = \mathcal{E}\left(L_n(\mathbf{a})\right)$. A simple check verifies that $rank\left(S^*(g)\right) = rank\left(L_n(\mathbf{a})\right) = n-1$. Moreover, the set $L_n(\mathbf{a})$ (and hence the polyhedron $\mathcal{E}(f)$) intersects all the coordinate axes $\mathbb{R}\mathbf{e}_i$ $(i = 1, \ldots, n-1)$. It follows that the face $\mathcal{F}_0(\mathbf{a}) = \mathcal{F}_0(f)$ is compact. Consequently the assumption $dim \mathcal{F}_0(f) = rank\left(S^*(g)\right) - 1$ is equivalent to the assumption that $\mathcal{F}_0(\mathbf{a})$ is a facet of $\mathcal{E}(\mathbf{a})$. As a result, Proposition 2 implies that Theorem 2 follows from Corollary 2, once one has also noted that Lemma 5 implies $\iota(f) = \iota\left(\mathcal{E}(f)\right) = |\mathbf{c}|$ for any normalized polar vector $\mathbf{c}$ of $\mathcal{F}_0(\mathbf{a})$. $\diamondsuit$

References

[BEL] **G. Bhowmik and D. Essouabri and B. Lichtin**. *Meromorphic Continuation of Multivariable Euler Products and Applications.* Forum Mathematicum, 19, no. 6 (2007). MR2367957 (2009m:11140)

[BT] **V.V. Batyrev and Yu. Tschinkel**. *Manin's conjecture for toric varieties.* Journal of Algebraic Geometry, p.15-53, (1998). MR1620682 (2000c:11107)

[Bre1] **R. de la Bretèche**. *Sur le nombre de points de hauteur bornée d'une certaine surface cubique singulière.* Astérisque, vol 251, no 2, p.51-57, (1998). MR1679839 (2000b:11074)

[Bre2] **R. de La Bretèche**. *Estimation de sommes multiples de fonctions arithmétiques.* Compositio Math., vol. 128, No. 3, 261-298 (2001). MR1858338 (2002j:11106)

[Bre3] **R. de la Bretèche**. *Compter des points d'une variété torique.* J. Number Theory, vol. 87, no 2, p. 315-331, (2001). MR1824152 (2002a:11067)

[Bro] **T.D. Browning**. *Quantitative arithmetic of projective varieties.* Progress in Mathematics, 277. Birkhauser Verlag, Basel, (2009). MR2559866 (2010i:11004)

[Ca1] **Pi. Cassou-Noguès**, *Abscisse de convergence de certaines séries de Dirichlet associées à un polynôme.* Progr. Math., 51, Birkhauser Boston, Boston, MA, (1984). MR791583 (87b:11114)

[Ca2] **Pi. Cassou-Noguès**, *Prolongement méromorphe des séries de Dirichlet associées à un polynôme à deux indéterminées.* J. Number Theory, vol. 23,no. 1, 1-54 (1986).

[Ch] **A. Chambert-Loir**, *Lectures on height zeta functions: At the confluence of algebraic geometry, algebraic number theory, and analysis.* arXiv:0812.0947v2 [math.NT] (2009). MR2647601 (2011g:11171)

[Co] **D. Cox**. *Recent developments in toric geometry.* Algebraic Geometry - Santa Cruz 1995, Proc. Sympos. Pure Math., 62, Part 2, Amer. Math. Soc., Providence, RI, 1997. MR1492541 (99d:14054)

[DNS] **J. Denef, J. Nicaise and P. Sargos**. *Oscillating integrals and Newton polyhedra.* J. Anal. Math. 95, 147-172 (2005). MR2145563 (2006a:58054)

[E1] **D. Essouabri**. *Singularités des séries de Dirichlet associées à des polynômes de plusieurs variables et application à la théorie analytique des nombres.* Annales de l'institut Fourier, vol. 47(2),429-484 (1997). MR1450422 (99d:11098)

[E2] **D. Essouabri**. *Prolongement analytique d'une classe de fonctions zêtas des hauteurs et applications.* Bulletin de la SMF, vol. 133, 297-329 (2005). MR2172269 (2006g:11134)

[F] **E. Fouvry** *Sur la hauteur des points d'une certaine surface cubique singulière.* Astérisque, vol 251, no 2, p.31-49, (1998). MR1679838 (2000b:11075)

[HBM] **D.R. Heath-Brown and B.Z. Moroz**. *The density of rational points on the cubic surface* $X_0^3 = X_1X_2X_3$. Math. Proc. Camb. Philos. Soc., vol. 125, no 3 (1999). MR1656797 (2000f:11080)

[K] **N. Kurokawa**. *On the meromorphy of Euler products I and II.* Proc. London Math. Soc., vol. 53, 1-47 and 209-236 (1986). MR842154 (88a:11084a)

[L] **B. Lichtin**. *Generalized Dirichlet series and b-functions.* Compositio Mathematica, vol. 65, No. 1, 81-120 (1988). MR930148 (89d:32030)

[Ma] **K. Mahler**. *Uber einer Satz von Mellin.* Math. Annalen, vol. 100, 384-395, (1928). MR1512491

[MT] **K. Matsumoto and Y. Tanigawa**. *The analytic continuation and the order estimate of multiple Dirichlet series*, J. Theorie des Nombres de Bordeaux, vol. 15, No.1 (2003). MR2019016 (2004i:11107)

[Me] **Hj. Mellin**. *Eine Formel fur den Logarithmus transzendenter Funktionen von endlichem Geschlecht.* Acta Math., vol. 25, 165-184 (1901).

[Mo] **B.Z. Moroz**. *Scalar product of L-functions with Grossencharacters: its meromorphic continuation and natural boundary.* J. reine. angew. Math., vol. 332 (1982). MR656857 (83j:12010)

[P1] **E. Peyre**. *Hauteurs et mesures de Tamagawa sur les variétés de Fano.* Duke Math.J., vol. 79, p. 101-218, (1995). MR1340296 (96h:11062)

[P2] **E. Peyre**. *Étude asymptotique des points de hauteur bornée.* Ecole d'été sur la géométrie des variétés toriques, Grenoble, (2002).

[R] **R. T. Rockafellar**. *Convex Analysis.* Princeton Univ. Press. Princeton, N.J., (1970). MR0274683 (43:445)

[Sal] **P. Salberger**. *Tamagawa measures on universal torsors and points of bounded height on Fano varieties.* Astérisque, vol. 251, no 2, p. 91-258, (1998). MR1679841 (2000d:11091)

[Sar1] **P. Sargos**. *Séries de Dirichlet associées à des polynômes de plusieurs variables.* Thèse d'Etat, Univ. Bordeaux 1 (1987). http://greenstone.refer.bf/collect/thef/index/assoc/HASHc0e2/84b2fdbf.dir/CS_00148.pdf

[Sar2] **P. Sargos**. *Sur le problème des diviseurs généralisés.* Publ. Math. Orsay 2 (1988). MR952871 (89i:11106)

[St] **B. Sturmfels**. *Equations defining toric varieties.* Algebraic Geometry - Santa Cruz 1995, Proc. Sympos. Pure Math., 62, Part 2, Amer. Math. Soc., Providence, RI, 1997. MR1492542 (99b:14058)

[Sw] **Sir. P. Swinnerton-Dyer**. *A canonical height on* $X_0^3 = X_1X_2X_3$. Proceedings du semestre Diophantine Geometry de Pise (2005).

PRES Université de Lyon, Université Jean-Monnet (Saint-Etienne), Faculté des Sciences, Département de Mathématiques, 23 rue du Docteur Paul Michelon, 42023 Saint-Etienne Cedex 2, France.

E-mail address: driss.essouabri@univ-st-etienne.fr

Contemporary Mathematics
Volume **566**, 2012
http://dx.doi.org/10.1090/conm/566/11217

Combinatorial cubic surfaces and reconstruction theorems

Yu. I. Manin

ABSTRACT. This note contains a solution to the following problem: reconstruct the definition field and the equation of a projective cubic surface, using only combinatorial information about the set of its rational points. This information is encoded in two relations: collinearity and coplanarity of certain subsets of points. We solve this problem, assuming mild "general position" properties. This study is motivated by an attempt to address the Mordell–Weil problem for cubic surfaces using essentially model theoretic methods. However, the language of model theory is not used explicitly.

Contents

0. Introduction and overview

0.1. Cubic hypersurfaces. Let K be a field, finite or infinite. In the finite case we assume cardinality of K to be sufficiently large, the exact lower boundary depending on various particular combinatorial constructions.

Let $P = \mathbf{P}^N_K$ be a projective space over K, with a projective coordinate system $(z_1 : z_2 : \cdots : z_{N+1})$. A *cubic hypersurface* $V \subset P$ defined over K is, by definition, the closed subscheme defined by an equation $c = 0$ where $c \in K[z_1 : z_2, \cdots : z_{N+1}]$ is a non–zero cubic form. There is a bijection between the set of such subschemes and the set $\mathbf{P}^M(K)$ of coefficients of c modulo K^*.

We will say that V is *generically reduced* if after extending K to an algebraic closure $\overline{K}$, c does not acquire a multiple factor.

In this paper, I will be interested in the following problem:

2010 *Mathematics Subject Classification.* Primary 14G05, 03C30.

0.1.1. Problem. *Assuming V generically reduced, reconstruct K and the subscheme $V \subset P = \mathbf{P}^N_K$ starting with the set of its K–points $V(K)$ endowed with some additional combinatorial structures of geometric origin.*

The basic combinatorial data that I will be using are subsets of smooth points of $V(K)$ lying upon various sections of V by projective subspaces of P defined over K. Thus, for the main case treated here, that of cubic surfaces ($N = 3$), I will deal combinatorially with the structure, consisting of

a) The subset of smooth (reduced, non–singular) points $S := V_{sm}(K)$.

b) A triple symmetric relation "collinearity": $\mathcal{L} \subset S^3 := S \times S \times S$.

c) A set $\mathcal{P}$ of subsets of S called "plane sections".

In the first approximation, one can imagine $\mathcal{L}$ (resp. $\mathcal{P}$) as simply subsets of collinear triples (resp. K–points of K–plane sections) of V. However, various limiting and degenerate cases must be treated with care as well.

For example, as a working definition of $\mathcal{L}$ we will adopt the following convention: $(p, q, r) \in S^3$ *belongs to* $\mathcal{L}$ *if either* $p + q + r$ *is the full intersection cycle of* V *with a* K*–line* $l \subset \mathbf{P}^N$ *(with correct multiplicities), or else if there exists a* K*–line* $l \subset V$ *such that* $p, q, r \in l$.

0.2. Geometric constraints. If an instance of the set–theoretic combinatorial structure such as $(S, \mathcal{L}, \mathcal{P})$ above, comes from a cubic surface V defined over a field K, we will call such a structure *geometric one.*

Geometric structures satisfy additional combinatorial constraints.

The reconstruction problem in this context consists of two parts:

(i) Find a list of constraints ensuring that each $(S, \mathcal{L}, \mathcal{P})$ satisfying these constraints is geometric.

(ii) Devise a combinatorial procedure that reconstructs K and $V \subset \mathbf{P}^3$ realizing $(S, \mathcal{L}, \mathcal{P})$ as a geometric one.

Besides, ideally we want the reconstruction procedure to be functorial: certain maps of combinatorial structures, in particular, their isomorphisms, must induce/be induced by morphisms of ground fields and K–linear maps of $\mathbf{P}^3$.

In the subsection 0.4, I will describe a classical archetype of reconstruction, – combinatorial characterization of projective planes. I will also explain the main motivation for trying to extend this technique to cubic surfaces: *the multidimensional weak Mordell-Weil problem.*

0.3. Reconstruction of K from curves and configurations of curves. One cannot hope to reconstruct the ground field K, if V is zero–dimensional or one–dimensional. Only starting with cubic surfaces ($N = 3$), this prospect becomes realistic.

In fact, if $N = 1$, we certainly cannot reconstruct K from any combinatorial information about one K–rational cycle of degree 3 on $\mathbf{P}^1_K$.

If $N = 2$, then for a smooth cubic curve V, the set $V(K)$ endowed with the collinearity relation is the same as $V(K)$ considered as a principal homogeneous space over the "Mordell–Weil" abelian group, unambiguously obtained from $(V(K), L)$ as soon as we arbitrarily choose the identity (or zero) point: cf. a recollection of classical facts in sec. 1 below. Generally, this group does not carry enough information to get hold of K, if K is finitely generated over $\mathbf{Q}$.

However, the situation becomes more promising, if we assume V geometrically irreducible and having just *one singular point* which is defined over K. More specifically, assume that this point is either an ordinary double point with two different branches/tangents defined over K each, or a cusp with triple tangent line, which is then automatically defined over K.

In the first case, we will say that V is a curve of *multiplicative type*, in the second, of *additive type*.

Then we can reconstruct, respectively, the multiplicative or the additive group of K, up to an isomorphism. In fact, these two groups are canonically identified with $V_{sm}(K)$ as soon as one smooth K–point is chosen, in the same way as the Mordell–Weil group is geometrically constructed from a smooth cubic curve with collinearity relation.

Finally for $N = 3$, now allowing V to be smooth, and under mild genericity restrictions, we can combine these two procedures and reconstruct both K and a considerable part of the whole geometric picture.

The idea, which is the main new contribution of this note, is this. Choose two points (p_m, p_a) in $V_{sm}(K)$, not lying on a line in V, whose tangent sections (C_m, C_a) are, respectively, of multiplicative and additive type. (To find such points, one might need to replace K by its finite extension first).

Now, one can intersect the tangent planes to p_m and p_a by elements of a K–rational pencil of planes, consisting of all planes containing p_m and p_a. This produces a birational identification of C_m and C_a.

The combinatorial information, used in this construction, can be extracted from the data $\mathcal{L}$ and $\mathcal{P}$. The resulting combinatorial object, carrying full information about both K^* and K^+, can be then processed into K, if a set of additional combinatorial constraints is satisfied.

Using four tangent plane sections in place of two, one can then unambiguously reconstruct the whole subscheme V.

For further information, cf. the main text.

0.4. Combinatorial projective planes and weak Mordell–Weil problem. My main motivation for this study was an analog of Mordell–Weil problem for cubic surfaces: cf. [**M3**], [**KaM**], [**Vi**].

Roughly speaking, the classical Mordell–Weil Theorem for elliptic curves can be stated as follows. Consider a smooth plane cubic curve C, i. e. a plane model of an elliptic curve, over a field K finitely generated over its prime subfield. Then the whole set $C(K)$ can be generated in the following way: start with a *finite* subset $U \subset C(K)$ and iteratively enlarge it, adding to already obtained points each point $p \circ q \in C(K)$ that is collinear with two points $p, q \in C(K)$ that were already constructed. If $p = q$, then the third collinear point, by definition, is obtained by drawing the tangent line to C at p.

In the case of a cubic surface V, say, not containing K–lines, there are two versions of this geometric process ("drawing secants and tangents"). We may allow to consecutively add only points collinear to $p, q \in V(K)$ when $p \neq q$. Alternatively, we may also allow to add all K–points of the plane section of V tangent to V at $p = q$.

I will call the respective two versions of finite generation conjecture *strong, resp. weak, Mordell–Weil problem* for cubic surfaces.

Computer experiments suggest that weak finite generation might hold at least for some cubic surfaces defined over $\mathbf{Q}$: see [**Vi**] for the latest data. The same experiments indicate however, that the "descent" procedure, by which Mordell–Weil is proved for cubic curves, will not work in two–dimensional case: a stable percentage of $\mathbf{Q}$–points of height $\leq H$ remains not expressible in the form $p \circ q$ with p, q of smaller height.

In view of this, I suggested in [**M3**], [**KaM**] to use a totally different approach to finite generation, based on the analogy with classical theory of abstract, or combinatorial, projective planes.

The respective finite generation statement can be stated as follows.

For any field K of finite type over its prime subfield, the whole set $\mathbf{P}^2(K)$ can be obtained by starting with a finite subset $U \subset \mathbf{P}^2(K)$ and consecutively adding to it lines through pairs of distinct points, already obtained, and intersection points of pairs of constructed lines.

The strategy of proof can be presented as a sequence of the following steps.

STEP 1. Define a *combinatorial projective plane* $(S, \mathcal{L})$ as an abstract set S whose elements are called *(combinatorial) points*, endowed with a set of subsets of points $\mathcal{L}$ called *(combinatorial) lines*, such that each two distinct points are contained in a single line, and each two distinct lines intersect at a single point.

STEP 2. Find combinatorial conditions upon $(S, \mathcal{L})$, that are satisfied for K–points of each geometric projective plane $\mathbf{P}^2(K)$, and that exactly characterize geometric planes, so that starting with $(S, \mathcal{L})$ satisfying these conditions, one can reconstruct from $(S, \mathcal{L})$ a field K and an isomorphism of $(S, \mathcal{L})$ with $(\mathbf{P}^2(K),\ \textit{projective}\ K - \textit{lines})$ unambiguously.

In fact, this reconstruction must be also functorial with respect to embeddings of projective planes $S \subset S'$ and the respective combinatorial lines.

These conditions are furnished by the beautiful *Pappus Theorem/Axiom* (at least, if cardinality of S is infinite or finite but large enough).

STEP 3. Given a geometric projective plane $(\mathbf{P}^2(K),\ \textit{projective}\ K - \textit{lines})$, start with four points in general position $U_0 \subset \mathbf{P}^2(K)$ and generate the minimal subset S of $\mathbf{P}^2(K)$ stable with respect to drawing lines through two points and taking intersection point of two lines.

This subset, with induced collinearity structure, is a combinatorial projective plane. It satisfies the Pappus Axiom, because it was satisfied for $\mathbf{P}^2(K)$. It is not difficult to deduce then that S is isomorphic to $\mathbf{P}^2(K_0)$, with $K_0 \subset K$ the prime subfield, and the embeddings $K_0 \to K$ and $S \to \mathbf{P}^2(K)$ are compatible with geometry.

STEP 4. Finally, one can iterate this procedure as follows. If K is finitely generated, there exists a finite sequence of subfields $K_0 \subset K_1 \subset \ldots \subset K_n = K$ such that each K_i is generated over K_{i-1} by one element, say θ_i. If we already know a finite generating set of points $U_{i-1} \subset \mathbf{P}^2(K_{i-1})$, define $U_i \subset \mathbf{P}^2(K_i)$ as $U_{i-1} \cup \{(\theta_i : 1 : 0)\}$. One easily sees that U_i generates $\mathbf{P}^2(K_i)$.

0.5. Results of this paper. As was explained in 0.3, results of this paper give partial versions for cubic surfaces of Steps 1 and 2 in the finite generation proof, sketched above. I can now reconstruct the ground field K and the total subscheme $V \subset \mathbf{P}^3_K$, under appropriate genericity assumptions, from the combinatorics of $V(K)$ geometric origin.

However, these results still fall short of a finite generation statement.

The reader must be aware that this approach is essentially model–theoretic, and it was inspired by the successes of [**HrZ**] and [**Z**].

My playground here is much more restricted, and I do not use explicitly the (meta)language of model theory, working in the framework of Bourbaki structures.

More precisely, constructions, explained in sec. 2 and 3, are oriented to the reconstruction of fields of finite type and cubic surfaces over them. According to [**HrZ**] and [**Z**], if one works over an algebraically closed ground field, one can reconstruct combinatorially (that is, in a model theoretic way) much of the classical algebraic geometry.

In sec. 4, I introduce the notion of a *large field*, tailor–made for cubic (hyper)surfaces, and show that large fields can be reconstructed even from (sets of rational points of) smooth plane cubic curves, endowed with collinearity relation and an additional structure consisting of *pencils of collinear points* on such a curve. Any field K having no non–trivial extensions of degree 2 and 3 is large, hence large fields lie between finitely generated and algebraically close ones.

1. Quasigroups and cubic curves

1.1. Definition. *Let S be a set and $\mathcal{L} \subset S \times S \times S$ be a subset of triples with the following properties:*

(i) $\mathcal{L}$ is invariant with respect to permutations of factors S.

(ii) Each pair $p, q \in S$ uniquely determines $r \in S$ such that $(p, q, r) \in \mathcal{L}$.

Then $(S, \mathcal{L})$ is called a symmetric quasigroup.

This structure in fact defines a binary composition law

$$\circ : S \times S \to S : \ p \circ q = r \Longleftrightarrow (p, q, r) \in \mathcal{L}. \tag{1.1}$$

Properties of $\mathcal{L}$ stated in the Definition 1.1 can be equivalently rewritten in terms of $\circ$: for all $p, q \in S$

$$p \circ q = q \circ p, \ p \circ (p \circ q) = q. \tag{1.2}$$

The structure $(S, \circ)$, satisfying (1.2), will also be called a symmetric quasigroup. The importance of $\mathcal{L}$ for us is that, together with its versions, it naturally comes from geometry.

In terms of $(S, \circ)$, we can define the following groups. For each $p \in S$, the map $t_p : q \mapsto p \circ q$ is an involutive permutation of S: $t_p^2 = \mathrm{id}_S$.

Denote by $\Gamma = \Gamma(S, \mathcal{L})$ the group generated by all $t_p, p \in S$. Let $\Gamma^0 \subset \Gamma$ be its subgroup, consisting of products of an even number of involutions t_p.

1.2. Theorem–Definition. *A symmetric quasigroup $(S, \circ)$ is called abelian, if it satisfies any (and thus all) of the following equivalent conditions:*

(i) There exists a structure of abelian group on S, $(p, q) \mapsto pq$, and an element $u \in S$ such that for all $p, q \in S$ we have $p \circ q = up^{-1}q^{-1}$.

(ii) The group Γ^0 *is abelian.*

(iii) For all $p, q, r \in S$, $(t_p t_q t_r)^2 = 1$.

(iv) For any element $u \in S$, *the composition law* $pq := u \circ (p \circ q)$ *turns* S *into an abelian group.*

(v) The same as (iv) for some fixed element $u \in S$.

Under these conditions, S *is a principal homogeneous space over* Γ^0.

For a proof, cf. [**M1**], Ch. I, sec. 1,2, especially Theorem 2.1.

1.3. Example: plane cubic curves. Let K be a field, $C \subset \mathbf{P}^2_K$ an absolutely irreducible cubic curve defined over K. Denote by $S = C_{sm}(K) \subset C(K)$ the set of non–singular K–points of C. Define *the collinearity relation* $\mathcal{L}$ by the following condition:

$$(1.3) \quad (p, q, r) \in \mathcal{L} \iff p+q+r \text{ is the intersection cycle of } C \text{ with a } K-\text{line}.$$

Then $(S, \mathcal{L})$ is an abelian symmetric quasigroup. This is a classical result.

More precisely, we have the following alternatives. C might be non–singular over an algebraic closure of K. Then C is the plane model of an abstract elliptic curve defined over K, the group Γ^0 can be identified with K–points of its Picard group. We call the latter also the Mordell–Weil group of C over K.

Singular curves will be more interesting for us, because they carry more information about the ground field K. Each geometrically irreducible singular cubic curve has exactly one singular geometric point, say p, and it is rational over K. More precisely, we will distinguish three cases.

(I) C is of *multiplicative type.* This means that p is a double point two tangents to which at p are rational over K.

(II) C is of *additive type.* This means that p is a cusp: a point with triple tangent.

(III) C is of *twisted type.* This means that p is a double point p two tangents to which at p are rational and conjugate over a quadratic extension of K.

The structure of quasigroups related to singular cubic curves is clarified by the following elementary and well known statement.

1.3.1. Lemma. *(i) If* C *is of multiplicative type,* Γ^0 *is isomorphic to* K^*.

(ii) If C *is of additive type,* Γ^0 *is isomorphic to* K^+.

(iii) If C *is of twisted type,* Γ^0 *is isomorphic to the group of* K*–points of a form of* G_m *or* G_a *that splits over the respective quadratic extension of* K. *The first case occurs when* $\mathrm{char} K \neq 2$, *the second one when* $\mathrm{char} K = 2$.

Proof. *(Sketch.)* In all cases, the group law $pq := u \circ (p \circ q)$, for an arbitrary fixed $u \in S$ determines the structure of an algebraic group over K upon the curve C_0 which can be defined as the normalization of C with preimage(s) of p deleted. An one–dimensional geometrically connected algebraic group becomes isomorphic to G_m or G_a over any field of definition of its points "at infinity".

In the next section, we will recall more precise information about the respective isomorphisms in the non–twisted cases.

2. Reconstruction of the ground field and a cubic surface from combinatorics of tangent sections

2.1. The key construction. Let K be a field of cardinality ≥ 4. Then the set $H := \mathbf{P}^1(K)$ consists of ≥ 5 points.

Consider a family of five pairwise distinct points in H for which we choose the following suggestive notation:

$$0_a, \infty_a, 0_m, 1_m, \infty_m \in \mathbf{P}^1(K). \tag{2.1}$$

In view of its origin, the set $H \setminus \{\infty_a\}$ has a special structure of abelian group A (written additively, with zero 0_a). In fact, the choice of any affine coordinate x_a on $\mathbf{P}^1_K$ with zero at 0_a and pole at ∞_a defines this structure: it sends $p \in H \setminus \{\infty_a\}$ to the value of x_a at p, and addition is addition in K^+. The structure does not depend on x_a, but x_a determines the isomorphism of G_a with K^+, and this isomorphism *does* depend on x_a: the set of all x_a's is the principal homogeneous space over K^*.

Similarly, the set $H \setminus \{0_m, \infty_m\}$ has a special structure of abelian group M, with identity 1_m. A choice of affine coordinate x_m on $\mathbf{P}^1$, with divisor supported by$(0_m, \infty_m)$ and taking value $1 \in K$ at 1_m, defines this structure. Again, it does not depend on x_m, but x_m determines its isomorphism with K^*, and this isomorphism does depend on x_m. There are, however, only two choices: x_m and x_m^{-1}. They differ by renaming $0_m \leftrightarrow \infty_m$.

Having said this, consider now *an abstract set* H with a subfamily of five elements denoted as in (2.1). Moreover, assume in addition that we are given composition laws $+$ on $H \setminus \{\infty_a\}$ and $\cdot$ on $H \setminus \{0_m, \infty_m\}$ turning these sets into two abelian groups, A (written additively, with zero 0_a) and M (written multiplicatively, with identity 1_m). Define the inversion map $i : M \to M$ using this multiplication law: $i(p) = p^{-1}$.

We will encode this extended version of (2.1), with additional data recorded in the notation M, A, as a bijection

$$\mu : \ M \cup \{0_m, \infty_m\} \to A \cup \{\infty_a\} \tag{2.2}$$

It is convenient to extend the multiplication and inversion, resp. addition and sign reversal, to commutative partial composition laws on two sets (2.2) by the usual rules: for $p \in M$, $q \in A$, we set

$$p \cdot 0_m := 0_m, \ p \cdot \infty_m := \infty_m, \ i(0_m) := \infty_m, \ i(\infty_m) := 0_m, \tag{2.3}$$

$$q \pm \infty_a := \infty_a. \tag{2.4}$$

The following two lemmas are our main tool in this section.

2.2. Lemma. *If (2.2) comes from a projective line as above, then the map*

$$\nu : \ M \cup \{0_m, \infty_m\} \to A \cup \{\infty_a\},$$

$$\nu(p) := \mu\{\mu^{-1}[\mu(p) - \mu(0_m)] \cdot i \circ \mu^{-1}[\mu(p) - \mu(\infty_m)]\} \tag{2.5}$$

is a well defined bijection.

Moreover,

$$\nu(0_m) = 0_a := 0, \ \nu(\infty_m) = \infty_a := \infty. \tag{2.6}$$

Finally, identifying $M \cup \{0_m, \infty_m\}$ and $A \cup \{\infty_a\}$ with the help of ν and combining addition and multiplication, now (partially) defined on H, we get upon $H \setminus \{\infty\}$

a structure of the commutative field, with zero 0 *and identity* $1 := \nu(1_m)$. *This field is isomorphic to the initial field* K .

Proof. In the situation (2.1), if A is identified with K^+ using an affine coordinate x_a, and M is identified with K^* using another affine coordinate x_m as above, these coordinates are connected by the evident fractional linear transformation, bijective on $\mathbf{P}^1(K)$:

$$x_a = c \cdot (x_m - x_m(0_m)) \cdot (x_m - x_m(\infty_m))^{-1}, \ c \in K^*.$$

The definition (2.5) is just a fancy way to render this relation, taking into account that now we have to add and to multiply in two different locations, passing back and forth via μ and μ^{-1}. Instead of multiplying by c, we normalize multiplication so that $\nu(1_\infty)$ becomes identity.

This observation makes all the statements evident.

The same arguments read in reverse direction establish the following result:

2.3. Lemma on Reconstruction. *Conversely, let* M *and* A *be two abstract abelian groups, extended by "improper elements" to the sets with partial composition laws* $M \cup \{0_m, \infty_m\}$ *and* $A \cup \{\infty_a\}$, *as in (2.3), (2.4). Assume that we are given a bijection* μ *as in (2.2), mapping* $1, 0_m$, *and* ∞_m *to* A. *Assume moreover that:*

(i) The respective mapping ν *defined by (2.5) is a well defined bijection.*

(ii) The set A *endowed with its own addition, and multiplication transported by* ν *from* M, *is a commutative field* K.

Then we get a natural identification $H = \mathbf{P}^1(K)$. *This construction is inverse to the one described in sec. 2.1.*

2.4. Combinatorial projective lines and functoriality. Let us call an instance of the data (2.2)–(2.4), satisfying the constraints of Lemma 2.2, *a combinatorial projective line* (this name will be better justified in the remainder of this section). Let us call triples $(K, \mathbf{P}^1(K), j)$ where j is a subfamily of five points in $\mathbf{P}^1(K)$ as in (2.1), *geometric projective lines.*

The constructions we sketched above are obviously functorial with respect to various natural maps such as:

a) *On the geometric side:* Morphisms of fields, naturally extended to projective lines with marked points. Fractional linear transformations of $\mathbf{P}^1(K)$, naturally acting upon j and identical on K.

b) *On the combinatorial side:* Embeddings of groups $M \to M', A \to A'$, compatible with (μ, μ') and on improper points. Automorphisms of (M, A), supplied with compatibly changed μ and improper points.

These statements can be made precise and stated as equivalence of categories. We omit details here.

Now we turn to the description of a bare–bones geometric situation, that can be obtained (in many ways) from a cubic surface, directly producing combinatorial projective lines.

2.5. (C_m, C_a)**–configurations**. Consider a family of subschemes in $\mathbf{P}^3_K$, that we will call *a configuration:*

$$Conf := (p_m, p_a; C_m, C_a; P_m, P_a) \tag{2.7}$$

It consists of the following data:

(i) Two distinct K–points $p_m, p_a \in \mathbf{P}^3(K)$.

(ii) Two distinct K–planes $P_m, P_a \subset \mathbf{P}^3$ *such that* $p_m \in P_m, p_m \notin P_a$ *and* $p_a \in P_a, p_a \notin P_m$.

(iii) Two geometrically irreducible cubic K–curves $C_m \subset P_m$, $C_a \subset P_a$.

We impose on these data the following constraints:

(A) $p_m \in C_m(K)$ is a double point, and C_m if of multiplicative type, in the sense of 1.3.

(B) $p_a \in C_a(K)$ is a cusp, and C_a is of additive type.

(C) Let $l := P_m \cap P_a$. Denote by $0_m, \infty_m \in l$ the intersection points with l of two tangents to C_m at x_m (in the chosen order). Denote by $0_a \in l$ the intersection point with l of the tangent to C_a at x_a. These three points are pairwise distinct.

Let $M := C_{m,sm}(K)$, $A := C_{a,sm}(K)$ be the respective sets of smooth points, with their group structure, induced by collinearity relation and a choice of 1_m, resp. 0_a, as in sec. 1.

Define the bijection $\alpha : \widetilde{C}_m(K) \to l(K)$, where $\widetilde{C}_m$ is the normalization of C_m, by mapping each smooth point $q \in C(K)$ to the intersection point with l of the line, passing through p_m and q. The two tangent lines at p_m define the images of two points of $\tilde{C}_m$ lying over p_m.

Similarly, define the bijection $\beta : C_a(K) \to l(K)$, by mapping each smooth point $q \in C(K)$ to the intersection point with l of the line, passing through p_a and q. The point on l where the triple tangent at cusp intersects it, is denoted ∞_a.

Finally, put

$$\mu := \beta^{-1} \circ \alpha : \quad M \cup \{0_m, \infty_m\} \to A \cup \{\infty_a\} \tag{2.8}$$

Thus $l(K)$ acquires both structures: of a combinatorial line and of a geometric line.

2.6. (C_m, C_a)–configurations from cubic surfaces. Let V be a smooth cubic surface defined over K. At each non–singular point $p \in V(K)$, there exists a well defined tangent plane to V defined over K. The intersection of this plane with V, for p outside of a proper Zariski closed subset, is a geometrically irreducible curve C, having p as its single singular point.

Again, generically it is of twisted multiplicative type, if char$K \neq 2$, and of twisted additive type, when p lies on a curve in V.

Therefore, under these *genericity conditions*, replacing K by its finite extension if need be, and renaming this new field K, we can find two tangent plane sections of V that form a (C_m, C_a)–configuration in the ambient projective space.

2.6.1. Example. Consider the diagonal cubic surface $\sum_{i=1}^4 a_i z_i^3 = 0$ over a field K of characteristic $\neq 3$. Then the discriminant of the quadratic equation defining directions of two tangents of the tangent section at $(z_1 : z_2 : z_3 : z_4)$, up to a factor in K^{*2}, is

$$D := \prod_{i=1}^4 a_i z_i.$$

Hence the set of points of (twisted) additive type consists of four elliptic curves

$$E_i:\ z_i = \sum_{j\neq i} a_j z_j^3 = 0, \quad i = 1,\dots,4.$$

The remaining points (outside 27 lines) are of (twisted) multiplicative type. Those for which $D \in K^{*2}$ are of purely multiplicative type.

2.7. Reconstruction of the configuration itself. Returning to the map (2.8), we see that K can be reconstructed from the (C_m, C_a) configuration, using *only* the collinearity relation on the set

$$\widetilde{C}_m(K) \cup C_a(K) \cup l(K). \tag{2.9}$$

Moreover, we get the canonical structure of a projective line over K on l, together with the family of five K–points on it.

To reconstruct the whole configuration, as a K–scheme up to an isomorphism, from the same data, it remains to give in addition two 0–cycles on l: its intersection with C_m and C_a respectively. Again, passing to a finite extension of K, if need be, we may and will assume that all intersection points in $C_m \cap l$, $C_a \cap l$ are defined over K. This again means that these cycles belong to the respective collinearity relation on $C_m(K) \cup C_a(K) \cup l(K)$.

To show that knowing these cycles, we can reconstruct C_m and C_a in their respective projective planes, let us look at the equations of these curves.

In P_m, choose projective coordinates $(z_1 : z_2 : z_3)$ over K in such a way that l is given by the equation $z_3 = 0$, p_m is $(0:0:1)$, equations of two tangents at p_m are $z_1 = 0$, $z_2 = 0$, and the points $0_m, \infty_m$ are respectively $(0:1:0)$ and $(1:0:0)$ Then the equation of C_m must be of the form

$$z_1 z_2 z_3 + c(z_1, z_2) = 0$$

where c is a cubic form. To give the intersection $C_m \cap l$ is the same as to give the linear factors of c. Since z_i are defined up to multiplication by constants from K^*, this defines (C_m, P_m) up to isomorphism.

Similar arguments work for C_a; its equation in coordinates $(z_1' : z_2' : z_3')$ on P_a such that l is defined by $z_3' = 0$, will now be

$$z_1'^2 z_3 + c'(z_1', z_2') = 0.$$

We may normalize z_2' by the condition that $0_a = (1:0:0)$, and then reconstruct linear factors of c' from the respective intersection cycle $C_a \cap l$.

2.8. Reconstruction of V from a tangent tetrahedral configuration. Let now V be a cubic surface over K. Assume that $V(K)$ contains four points $p_i, i = 1,\dots,4$, such that tangent planes P_i at them are pairwise distinct. Moreover, assume that tangent sections C_i are either of multiplicative, or of additive type, and each of these two types is represented by some C_i. One can certainly find such p_i defined over a finite extension of K.

We will call such a family of subschemes (p_i, C_i, P_i) *a tetrahedral configuration*, even when we do not assumed a priori that it comes from a V. If it comes from a V, we will say that it is *a tangent tetrahedral configuration.*

Without restricting generality, we may choose in the ambient $\mathbf{P}^3_K$ a coordinate system $(z_1 : \dots : z_4)$ in such a way that $z_i = 0$ is an equation of P_i.

If the configuration is tangent to V, let $F(z_1, \dots, z_4) = 0$ be the equation of V. Here F is a cubic form with coefficients in K determined by V up to a scalar factor. For each $i \in \{1, \dots, 4\}$, write F in the form

$$F = \sum_{a=0}^{3} z_i^a f_{3-a}^{(i)}(z_j \,|\, j \neq i), \tag{2.10}$$

where $f_b^{(i)}$ is a form of degree b in remaining variables.

Clearly, $f_3^{(i)} = 0$ is an equation of C_i in the plane P_i. Hence K and this equation can be reconstructed, up to a common factor, from a part of the tetrahedral configuration consisting of P_i, another plane P_j with tangent section of different type, and the induced relation of collinearity on them.

Consider the graph $G = G(V; p_1, \dots, p_4)$ with four vertices labeled $(1, \dots, 4)$, in which i and $j \neq i$ are connected by an edge, if there is a cubic monomial in $(z_k \,|\, k \neq i, j)$, that enters with nonzero coefficients in both $f_3^{(i)}$ and $f_3^{(j)}$. We want this graph to be connected. This will hold, for example, if in F all four coefficients at z_i^3 do not vanish. It is clear from this remark that connectedness of G is an open condition holding on a Zariski dense subset of all tangent configurations.

2.8.1. Proposition. *If the tetrahedral configuration is tangent to V, with connected graph G, then this V is unique.*

Proof. Let $g^{(i)}$ be a cubic form in $z_k, k \neq i$, such that $z_i = 0, g^{(i)} = 0$ are equations of C_i. We may change $g^{(i)}$ multiplying them by non–vanishing constants $c_i \in K$. If our configuration is tangent to V, given by (2.10), we may find c_i in such a way that $c_i g^{(i)} = f_3^{(i)}$. The obtained family of forms $\{c_i g^{(i)}\}$ is *compatible* in the following sense: if a cubic monomial in only two variables has non–zero coefficients in two $g^{(i)}$'s, then these coefficients coincide. In fact, they are equal to the coefficient of the respective monomial in F.

Conversely, if such a compatible system exists, and moreover, the graph G is connected, then (c_i) is unique up to a common factor. From such $c_i g^{(i)}$ one can reconstruct a cubic form of four variables, which will be necessarily proportional to F: coefficient at any cubic monomial m in $(z_1, \dots, z_4)$ in it will be equal to the coefficient of this monomial in any of $c_i g^{(i)}$, for which z_i does not divide m.

2.9. Summary. This section was dedicated to several key constructions that show how and under what conditions a cubic surface V considered as a scheme, together with a ground field K, can be reconstructed from its set of K–points, endowed with some combinatorial data.

The main part of the data was the collinearity relation on $V_{sm}(K)$, and this relation, when it came from geometry, satisfied some strong conditions stated in Lemma 2.2.

However, this Lemma and the data used in 2.8 made appeal also to information about points on the lines of intersections of tangent planes: cf. specifically constructions of maps α and β before formula (2.8).

We want to get rid of this extra datum and work *only* with points of $V_{sm}(K)$.

This must be compensated by taking in account, besides the collinearity relation, an additional *coplanarity relation* on $V(K)$, essentially given by the sets of K–points of (many) non–tangent plane sections.

The next section is dedicated principally to a description of the relevant abstract combinatorial framework. The geometric situations are used mainly to motivate or illustrate combinatorial definitions and axioms.

3. Combinatorial and geometric cubic surfaces

3.1. Definition. *A combinatorial cubic surface is an abstract set S endowed with two structures:*

(i) A symmetric ternary relation "collinearity": $\mathcal{L} \subset S^3$. We will say that triples $(p,q,r) \in \mathcal{L}$ are collinear.

(ii) A set $\mathcal{P}$ of subsets $C \subset S$ called plane sections.

These relations must satisfy the axioms made explicit below in the subsections 3.2 and 3.3. Until all the axioms are stated and imposed, we may call a structure $(S, \mathcal{L}, \mathcal{P})$ a *cubic pre–surface.*

3.2. Collinearity Axioms. *(i) For any $(p,q) \in S^2$, there exists an $r \in S$ such that $(p,q,r) \in L$.*

Call the triple (p,q,r) strictly collinear, if r is unique with this property, and p,q,r are pairwise distinct.

(ii) The subset $\mathcal{L}_s \subset \mathcal{L}$ of strictly collinear triples is a symmetric ternary relation.

(iii) Assume that $p \neq q$ and that there are two distinct $r_1, r_2 \in S$ with $(p,q,r_1) \in \mathcal{L}$ and $(p,q,r_2) \in \mathcal{L}$. Denote by $l = l(p,q)$ the set of all such r's. Then $l^3 \subset \mathcal{L}$, that is any triple (r_1, r_2, r_3) of points in l is collinear.

Such sets l are called lines in S.

3.2.1. Example: combinatorial cubic surfaces of geometric origin. Let K be a field, and V a cubic surface in $\mathbf{P}^3$ over K. Denote by $S = V_{sm}(K)$ the set of nonsingular K–points of V.

We endow S with the following relations:

(a) $(p,q,r) \in \mathcal{L}$ iff either $p+q+r$ is the complete intersection cycle of V with a line in $\mathbf{P}^3$ defined over K (K–line), or else if p,q,r lie on a K–line $\mathbf{P}^1_K$, entirely contained in V.

(b) Let $\mathbf{P} \subset \mathbf{P}^3$ be a K–plane. Assume that it either contains at least two distinct points of S, or is tangent to a K–point p, or else contains the tangent line to one of the branches of the tangent section of multiplicative type. Then $C := \mathbf{P}(K) \cap S$ is an element of $\mathcal{P}$. All elements of $\mathcal{P}$ are obtained in this way.

3.3. Plane sections. We now return to the general combinatorial situation. Let $(S, \mathcal{L}, \mathcal{P})$ be a cubic pre–surface.

For any $p \in S$, put

$$C_p = C_p(S) := \{\, q \,|\, (p,p,q) \in \mathcal{L}\} \cup \{p\}. \tag{3.1}$$

3.3.1. Tangent Plane Sections Axiom. *For each $p \in S$, we have $C_p \in \mathcal{P}$. Such plane sections are called tangent ones.*

The next geometric property of plane sections of geometric cubics can now be rephrased combinatorially as follows.

3.3.2. Composition Axiom. *(i) Let $C \in \mathcal{P}$ be a non–tangent plane section containing no lines in S. Then the collinearity relation $\mathcal{L}$ induces on such C a structure of Abelian symmetric quasigroup (cf. Theorem–Definition 1.2).*

(ii) Let $C_p = C_p(S)$ be a tangent plane section containing no lines. Then $\mathcal{L}$ induces on $C_p^0 := C_p \setminus \{p\}$ a structure of Abelian symmetric quasigroup.

Choosing a zero/identity point in C, resp. $C_p \setminus \{p\}$, we get in this way a structure of abelian group on each of these sets.

3.3.3. Pencils of Plane Sections Axiom. *Let $\lambda := (p, q, r) \in \mathcal{L}$. Assume that at least two of the points p, q, r are distinct. Denote by $\Pi_\lambda \subset \mathcal{P}$ the set*

$$\Pi_\lambda := \{C \in \mathcal{P} \,|\, p, q, r \in P\}. \tag{3.2}$$

and call such Π_λ's pencils of plane sections.

Then we have:

(i) If (p, q, r) do not lie on a line in S, then

$$S \setminus \{p, q, r\} = \coprod_{C \in \Pi_\lambda} (C \setminus \{p, q, r\}) \tag{3.3}$$

(disjoint union).

(ii) If (p, q, r) lie on a line l, then

$$S \setminus l = \coprod_{C \in \Pi_\lambda} (C \setminus l) \tag{3.4}$$

(disjoint union).

3.4. Combinatorial plane sections C_p of multiplicative/additive types. First of all we must postulate $(p, p, p) \in \mathcal{L}$, since in the geometric case $(p, p, p) \notin \mathcal{L}$ can happen only in a twisted case.

There are two different approaches to the tentative distinction between multiplicative and additive types. In one, we may try to prefigure the future realization of C_m and C_a as essentially the multiplicative (resp. additive) groups of a field K to be constructed.

Then, restricting ourselves for simplicity by fields of characteristic zero, we see that $C_p \setminus \{p\}$ which is of additive type after a choice of 0_a must become a vector space over $\mathbf{Q}$ (be uniquely divisible), whereas the respective group of multiplicative type is never uniquely divisible.

However, these restrictions are too weak.

Instead, we will define *pairs* of combinatorial tangent plane sections modeled on (C_m, C_a)–configurations of sec. 2. After this is done, we will be able to "objectively", independently of another member of the pair, distinguish between C_m and C_a using e.g. the divisibility criterion.

3.5. Combinatorial (C_m, C_a)–configurations. We can now give a combinatorial version of those (C_m, C_a)–configurations, that in the geometric case consist of two tangent plane sections of a cubic surface, one of additive, another of multiplicative type.

The main point is to see, how to use combinatorial plane sections in place of "external" lines $l = P_m \cap P_a$. This is possible, because the set of points of this line will now be replaced by bijective to it set of plane sections, belonging to a pencil, defined in terms of (C_m, C_a), and geometrically consisting just of all sections containing p_m and p_a.

Let $(S, \mathcal{L}, \mathcal{P})$ be a combinatorial pre–surface, satisfying Axioms 3.2, 3.3.1, 3.3.2, 3.3.3.

Start with two distinct points of S, not lying on a line in S, and respective tangent sections of S

$$(p_m, p_a; C_{p_m}, C_{p_a}) \tag{3.5}$$

Let $r \in S$ be the unique third point such that $(p_m, p_a, r) \in \mathcal{L}$, $\lambda := \{p_m, p_a, r\}$. Put $C^0_{p_m} := C_{p_m} \setminus \{p_m\}$, $C^0_{p_a} := C_{p_a} \setminus \{p_a\}$.

Denote by Π_λ the respective pencil of plane sections. Consider the following binary relation:

$$R \subset C_{p_a} \times C_{p_m} : \ (p, q) \in R \Longleftrightarrow \exists P \in \Pi_\lambda,\ p, q \in P. \tag{3.6}$$

3.5.1. Definition. $(p_m, p_a; C_{p_m}, C_{p_a})$ *is called a* (C_m, C_a)*–configuration, if the following conditions are satisfied.*

(i) R is a graph of some function

$$\lambda : \ C_{p_a} \to C_{p_m} \tag{3.7}$$

This function must be a bijection outside of two distinct points $0_m, \infty_m \in C^0_{p_a}$ which are mapped to p_m. Besides, we must have $\lambda(p_a) \in C^0_{p_m}$.

Assuming (i), put

$$A := C^0_{p_a},\ M := C^0_{p_m}$$

Introduce on these sets the structures of abelian groups using the the Composition Axiom 3.3.2 and some choices of zero and identity

$$0_a \in C^0_{p_a},\ 1_m \in C^0_{p_m}.$$

Define the map

$$\mu : \ M \cup \{0_m, \infty_m\} \to A \cup \{\infty_a\} \tag{3.8}$$

which is λ^{-1} on M and identical on $0_m, \infty_m$. Then

(ii) Conditions of Lemma 2.2 must be satisfied for this μ.

(iii) $C_{p_m} \cap C_{p_a}$ consists of three pairwise distinct points.

Thus, if $(p_m, p_a; C_{p_m}, C_{p_a})$ is a (C_m, C_a)–configuration, then we can combinatorially reconstruct the ground field and the isomorphic geometric configuration.

However, passing to tetrahedral configurations, we have to impose additional combinatorial compatibility conditions, that in the geometric case were automatic.

They are of two types:

(a) If two planes P_i, P_j of the tetrahedron carry plane sections of the same type (both additive or both multiplicative), we must write combinatorially maps, establishing their isomorphism, and postulate this fact in the combinatorial setup.

This can be done similarly to the case of (C_m, C_a)–configurations.

(b) If a schematic tangent plane section C_i can be reconstructed from two different pairs of tetrahedral plane sections C_i, C_j and (C_i, C_k), the results must be naturally isomorphic.

It is clear, how to do it in principle, and the respective constraints must be stated explicitly.

I abstain from elaborating all details here for the following reason.

If, as a main application of this technique, one tries to imitate the approach to weak Mordell–Weil problem following the scheme of 0.4, then the necessary combinatorial constraints will probably hold automatically for finitely generated combinatorial subsurfaces of an initial geometric surface.

The real problem is: how to recognize that a given (say, finitely generated) combinatorial subsurface of a geometric surface is actually the whole geometric surface.

I do not know any answer to this problem.

It is well known, however, that such proper combinatorial subsurfaces do exist.

For example, when $K = \mathbf{R}$ and $V(\mathbf{R})$ is not connected, one of the components can be a combinatorial cubic surface in its own right. More generally, some unions of classes of the universal equivalence relation ([**M1**]) are closed with respect to collinearity and coplanarity relations: this can be extracted from the results of [**Sw–D1**].

4. Cubic curves and combinatorial cubic curves over large fields

4.1. Large fields and smooth cubic curves. Consider a smooth cubic curve $C \subset \mathbf{P}^2_K$ defined over K. Put $S := C(K)$ and endow S with the collinearity relation $\mathcal{L} \subset S^3$ defined by (1.3). Let $\mathcal{L}^0 := L/\mathbf{S}_3$, the set of orbits of $\mathcal{L}$ with respect to the permutations. We may and will represent the image in $\mathcal{L}^0$ of $(p,q,r) \in \mathcal{L}$ as the 0–cycle $p+q+r$.

Now assume that K has no non–trivial extensions of degree 2 and 3. Then all intersection points of any K–line with C lie in $C(K)$. Therefore, we have the canonical bijection

$$(4.1)\qquad \xi : \{K - lines\ in\ \mathbf{P}^2(K)\} \to \mathcal{L}^0 : \quad l \mapsto\ intersection\ cycle\ l \cap C.$$

In this approach, K–points of $\mathbf{P}^2(K)\}$ have to be characterized in terms of pencils of all lines passing through a given point. Therefore, it is more natural to work with the dual projective plane from the start.

Let $\widehat{\mathbf{P}}^2$ be the projective plane dual to the plane in which C lies. Combinatorially, K–points $\widehat{l}$ (resp. lines $\widehat{p}$) of $\widehat{\mathbf{P}}^2$ are K–lines l (resp. points p) of $\mathbf{P}^2_K$, with inverted incidence relation: $\widehat{l} \in \widehat{p}$ iff $p \in l$. Thus, (4.1) turns into a bijection

$$(4.2)\qquad \widehat{\xi} : \{K - points\ in\ \widehat{\mathbf{P}}^2(K)\} \to \mathcal{L}^0 : \quad \widehat{l} \mapsto\ intersection\ cycle\ l \cap C.$$

Thus, $\widehat{\xi}$ sends lines in $\widehat{\mathbf{P}}^2$ to certain subsets in $\mathcal{L}^0$ that we also may call *pencils.*

This geometric situation motivates the following definition.

Let $(S, \mathcal{L})$ be an abelian symmetric quasigroup. Put $\mathcal{L}^0 = \mathcal{L}/\mathbf{S}_3$.

Assume that $\mathcal{L}^0$ is endowed with a set of its subsets $\mathcal{P}^0$, called *pencils,* which turns it into a combinatorial projective plane, with pencils as lines. This means that besides the trivial incidence conditions, Pappus Axiom is also valid. Hence we can reconstruct from $(\mathcal{L}^0, \mathcal{P}^0)$ a field K such that $\mathcal{L}^0 = \mathbf{P}^2(K)$, $\mathcal{P}^0 =$ the set of K–lines in $\mathbf{P}^2(K)$.

The following Definition, inspired by geometry, encodes the interaction between the structures $(S, \mathcal{L})$ and $(\mathcal{L}^0, \mathcal{P}^0)$.

4.2. Definition. *The structure $(S, \mathcal{L}, \mathcal{P}^0)$ is called a combinatorial cubic curve over a large field, if the following conditions are satisfied.*

(i) For each fixed $p \in S$, the set of cycles $p + q + r \in \mathcal{L}^0$, $q, r \in S$, is a pencil Π_p.

(ii) If a pencil Π is not of the type Π_p, then any two distinct elements in Π do not intersect (as unordered triples of S–points).

(iii) For each pencil Π_q (resp. each pencil not of type Π_q) and any $p \in S$, (resp. any $p \neq q$) there exists a unique cycle in $\mathcal{L}^0$ contained in Π and containing p.

Obviously, each geometric smooth cubic curve over a large field defines the respective combinatorial object.

4.3. Question. Are there such combinatorial curves not coming from a geometric one? In particular, are fields K coming from such combinatorial objects necessarily "large" in the sense of algebraic definition above (closed under taking square and cubic roots)?

Similar constructions can be done and question asked for cubic surfaces. Notice that over a large field, any point on a smooth surface, not lying on a line, is of either multiplicative, or additive type.

APPENDIX. Mordell–Weil and height: numerical evidence

Let V be a geometrically irreducible cubic curve or a cubic surface over a field K, with the standard geometric collinearity relation (1.3) for curves, (3.2.1a) for surfaces, and the binary composition law (1.1) for curves. For surfaces, we will state the following fancy definition.

A.1. Definition. Let $S \subset V(K)$, and $X_1, \dots, X_N, \dots$ free commuting but nonassociative variables,

$$w = (\dots(X_{i_1} \circ X_{i_2}) \circ (X_{i_3} \circ \dots (\dots \circ X_{i_k}) \dots)$$

a finite word in these variables,

$$ev : \{X_1, \dots, X_N, \dots\} \to S$$

an evaluation map.

a) A point $p \in V(K)$ is called *the strong value of w at (S, ev)* if during the inductive calculation of

$$p = ev(W) := (\dots(ev(X_{i_1}) \circ ev(X_{i_2})) \circ (ev(X_{i_3})) \circ \dots (\dots \circ ev(X_{i_k}) \dots)$$

we never land in a situation where the result of composition is not uniquely defined, that is $x \circ x$ with singular x for a curve, or $x \circ y$ where $y = x$ or the line through x, y lies in V for a surface.

b) A point $p \in S(K)$ is called *a weak value of w at (S, ev)* if during the inductive calculation of

$$p := ev_{weak}(W) = (\dots(ev(X_{i_1}) \circ ev(X_{i_2})) \circ (ev(X_{i_3})) \circ \dots (\dots \circ ev(X_{i_k}) \dots)$$

whenever we land in a situation where $\circ$ is not defined, we are allowed to choose as a value of $y \circ z$ (resp. $y \circ y$) any point of the line $\overline{yz}$ (resp. any point of intersection of a tangent line to V at y with V.)

Thus, weak evaluation produces a whole set of answers.

A.2. Definition. *(i) A subset $S \subset V(K)$* strongly generates *$V(K)$, if $V(K)$ coincides with the set of all strong S–values of all words w as above.*

(i) A subset $S \subset V(K)$ weakly generates *$V(K)$, if $V(K)$ coincides with the set of all weak S–values of all words w.*

Now we can state two versions of Mordell–Weil problem for cubic surfaces.

Strong Mordell–Weil problem for V: *Is there a finite S that strongly generates $V(K)$?*

Weak Mordell–Weil problem for V: *Is there a finite S that weakly generates $V(K)$?*

For curves, one often calls the weak Mordell–Weil theorem the statement that $C(K)/2C(K)$ is finite (referring to the group structure $p + q = e \circ (p \circ q)$).

A.3. Proving strong Mordell–Weil for smooth cubic curves over number fields. The classical strategy of proof includes two ingredients.

(a) Introduce an arithmetic *height function* $h : \mathbf{P}^2(K) \to \mathbf{R}$. E.g. for

$$K = \mathbf{Q},\ p = (x_0 : x_1 : x_2) \in \mathbf{P}^2(\mathbf{Q}),\ x_i \in \mathbf{Z},\ \text{g.c.d.}(x_i) = 1$$

put

$$h(p) := \max_i |x_i|.$$

(b) Prove *the descent property*: $\exists H_0$ such that if $h(p) > H_0,\ p \in C(K)$, then $p = q \circ r$ for some $q, r \in C(K)$ with $h(q), h(r) < h(p)$.

The same strategy works for general finitely generated fields. For larger fields, the strong Mordell–Weil generally fails, but the weak one might survive.

Let K be algebraically closed, or $\mathbf{R}$, or a finite extension of $\mathbf{Q}_p$. Let C be a smooth cubic curve, V a smooth cubic surface over K.

Then:

– $V(K)$ is weakly finitely generated, but not strongly.

– $C(K)$, if non–empty, is not finitely generated.

A.4. Point count on cubic curves. It is well known that as $H \to \infty$,

$$\operatorname{card}\{p \in C(K) \,|\, h(p) \le H\} = const \cdot (\log H)^{r/2}(1 + o(1)),$$

$$r := \operatorname{rk} C(K) = \operatorname{rk} \operatorname{Pic} C. \tag{A.1}$$

A.5. Point count on cubic surfaces. Here we have only a conjecture and some partial approximations to it:

Conjecture: as $H \to \infty$,

$$\operatorname{card}\{p \in V_0(K) \,|\, h(p) \le H\} = const \cdot H(\log H)^{r-1}(1 + o(1)),$$

$$V_0 := V \setminus \{all\ lines\},\ r := \operatorname{rk} \operatorname{Pic} V \tag{A.2}$$

A proof of (A.1) can be obtained by a slight strengthening of the technique used in the finite generation proof. Namely, one shows that $\log h(p)$ is "almost a quadratic form" on $C(K)$. In fact, it differs from a positive defined quadratic form by $O(1)$, so that (A.1) follows from the count of lattice point in an ellipsoid.

The descent property used for Mordell–Weil ensures that this quadratic form is positive definite.

How could one attack the conjecture (A.2)? For the circle method, there are too few variables. Moreover, connections with Mordell–Weil for cubic surfaces are totally missing.

Nevertheless, the inequality

$$\mathrm{card}\,\{p \in V_0(K) \,|\, h(p) \leq H\} > const \cdot H(\log H)^{r-1}$$

is proved in [**SlSw–D**] for cubic surfaces over $\mathbf{Q}$ with two rational skew lines. There are also results for singular surfaces: cf. [**Br**], [**BrD1**], [**BrD2**].

A.6. Some numerical data. Here I will survey some numerical evidence computed by Bogdan Vioreanu, cf. [**Vi**].

In the following tables, the following notation is used.

Input/table head: code $[\mathbf{a_1}, \mathbf{a_2}, \mathbf{a_3}, \mathbf{a_4}]$ of the surface

$$V : \sum_{i=1}^{4} a_i x_i^3 = 0.$$

Outputs:

(i) **GEN**: Conjectural list of weak generators

$$p := (x_1 : x_2 : x_3 : x_4) \in V(\mathbf{Q}).$$

(ii) **Nr**: The length of the list $List_{good}$ of all points x, ordered by the increasing height $h(p) := \sum_i |x_i|$, such that any point of the height $\leq$ maximal height in $List_{good}$, is weakly generated by **GEN**.

(iii) $\mathbf{H}_{bad}$: the height of the first point that was NOT shown to be generated by **GEN**.

(iv) **L**: the maximal length of a non–associative word with generators in $(\mathbf{GEN}, \circ)$ one of whose weak values produced an entry in $List_{good}$.

Example: For $V = [\mathbf{1}, \mathbf{2}, \mathbf{3}, \mathbf{4}]$, we have:

$\mathbf{GEN} = \{p^0 := (1 : -1 : -1 : 1)\}$

$\mathbf{Nr} = 8521$: the first 8521 points in the list of points of increasing height are weakly generated by the single point p^0.

$\mathbf{L} = 13$: the maximal length of a non–associative word in $(p^0, \circ)$ representing some point of the $List_{good}$ was 13.

$\mathbf{H}_{bad} = 24677$: the first point that was not found to be generated by p^0 was of height 24677.

SELECTED DATA

$[\mathbf{1},\mathbf{2},\mathbf{3},\mathbf{5}]$, $\operatorname{rk}\operatorname{Pic}=1$

GEN	Nr	L	$\mathbf{H}_{bad}$
(0:1:1:-1) (1:1:-1:0) (2:-2:1:1)	15222	12	23243

$[\mathbf{1},\mathbf{1},\mathbf{5},\mathbf{25}]$, $\operatorname{rk}\operatorname{Pic}=2$

GEN	Nr	L	$\mathbf{H}_{bad}$
(1:-1:0:0) (1:4:-2:-1)	32419	9	30072

$[\mathbf{1},\mathbf{1},\mathbf{7},\mathbf{7}]$, $\operatorname{rk}\operatorname{Pic}=2$

GEN	Nr	L	$\mathbf{H}_{bad}$
(0:0:1:-1) (1:-1:0:0) (1:-1:-1:1) (1:-1:1:-1)	16063	7	2578

A.7. Discussion of other numerical data. Bogdan Vioreanu studied all in all 16 diagonal cubic surfaces V; he compiled lists of all points up to height 10^5, for some of them up to height $3\cdot 10^5$.

The conjectural asymptotics (A.2) seems to be confirmed.

There is a *good conjectural expression* for the constant in (A.2) (for appropriately normalized height, not the naive one we used). It goes back to works of E. Peyre. Its theoretical structure very much reminds the Birch–Swinnerton–Dyer constant for elliptic curves. For theory and numerical evidence, see [**PeT1**], [**PeT2**], [**Sw–D2**], [**Ch-L**].

The (weak) finite generation looks confirmed for most of the considered surfaces, but some stubbornly resist, most notably $[\mathbf{17},\mathbf{18},\mathbf{19},\mathbf{20}]$, $[\mathbf{4},\mathbf{5},\mathbf{6},\mathbf{7}]$, $[\mathbf{9},\mathbf{10},\mathbf{11},\mathbf{12}]$.

If one is willing to believe in weak finite generation (as I am), the reason for failure might be the following observable fact:

When one manages to represent a "bad" point p of large height as a non–associative word in the generators $(\mathbf{GEN},\circ)$, the height of intermediate results (represented by subwords) tends to be much higher than $h(p)$, and hence outside of the compiled list of points.

Finally, the relative density of points p for which "one–step descent" works, that is,

$$p=q\circ r,\quad h(q),h(r)<h(p),$$

seems to tend to a certain value $0<d(V)<1$.

Question: *Can one guess a theoretical expression for $d(V)$?*

Notice that on each smooth cubic curve C, the "one–step descent" works for all points of sufficiently large height, so that $d(C) = 1$.

References

[Br] T. D. Browning. *Quantitative arithmetic of projective varieties.* Progress in Math. 277, Birkhäuser, 2009. MR2559866 (2010i:11004)

[BrD1] T. D. Browning, U. Derenthal. *Manin's conjecture for a quartic del Pezzo surface with A_4 singularity.* Ann. Inst. Fourier (Grenoble) 59 (2009), no. 3, 1231–1265. MR2543667 (2010g:11105)

[BrD2] T. D. Browning, U. Derenthal. *Manin's conjecture for a cubic surface with D_5 singularity.* Int. Math. Res. Not. IMRN 2009, no. 14, 2620–2647. MR2520769 (2011a:14041)

[Ch-L] A. Chambert-Loir. *Lectures on height zeta functions: At the confluence of algebraic geometry, algebraic number theory, and analysis.* arXiv:0812.0947 MR2647601 (2011g:11171)

[HrZ] E. Hrushovski, B. Zilber. *Zariski geometries.* Journ. AMS, 9:1 (1996), 1–56. MR1311822 (96c:03077)

[K1] D. S. Kanevski. *Structure of groups, related to cubic surfaces*, Mat. Sb. 103:2, (1977), 292–308 (in Russian); English. transl. in Mat. USSR Sbornik, Vol. 32:2 (1977), 252–264. MR0466144 (57:6025)

[K2] D. S. Kanevsky, *On cubic planes and groups connected with cubic surfaces.* J. Algebra 80:2 (1983), 559–565. MR691814 (85e:14014)

[KaM] D. S. Kanevsky, Yu. I. Manin. *Composition of points and Mordell–Weil problem for cubic surfaces.* In: Rational Points on Algebraic Varieties (ed. by E. Peyre, Yu. Tschinkel), Progress in Mathematics, vol. 199, Birkhäuser, Basel, 2001, 199–219. Preprint math.AG/0011198 MR1875175 (2002m:14018)

[M1] Yu. I. Manin. *Cubic Forms: Algebra, Geometry, Arithmetic.* North Holland, 1974 and 1986. MR833513 (87d:11037)

[M2] Yu. I. Manin. *On some groups related to cubic surfaces.* In: Algebraic Geometry. Tata Press, Bombay, 1968, 255–263. MR0257083 (41:1737)

[M3] Yu. I. Manin. *Mordell–Weil problem for cubic surfaces.* In: Advances in the Mathematical Sciences—CRM's 25 Years (L. Vinet, ed.) CRM Proc. and Lecture Notes, vol. 11, Amer. Math. Soc., Providence, RI, 1997, pp. 313–318. MR1479681 (99a:14029)

[PeT1] E. Peyre, Yu. Tschinkel. *Tamagawa numbers of diagonal cubic surfaces, numerical evidence.* In: Rational Points on Algebraic Varieties, Progr. Math., 199. Birkhäuser, Basel, 2001, 275–305. arXiv:9809054 MR1875177 (2003a:11076)

[PeT2] E. Peyre, Yu. Tschinkel. *Tamagawa numbers of diagonal cubic surfaces of higher rank.* arXiv:0009092 MR1875177 (2003a:11076)

[Pr] S. J. Pride. *Involutary presentations, with applications to Coxeter groups, NEC-Groups, and groups of Kanevsky.* J. of Algebra 120 (1989), 200–223. MR977867 (90g:20047)

[SlSw–D] J.Slater, H. P. F. Swinnerton–Dyer. *Counting points on cubic surfaces I.* In: Astérisque 251 (1998), 1–12. MR1679836 (2000d:11087)

[Sw–D1] H. P. F. Swinnerton–Dyer. *Universal equivalence for cubic surfaces over finite and local fields.* Symp. Math., Bologna 24 (1981), 111–143. MR619244 (82k:14019)

[Sw–D2] H. P. F. Swinnerton–Dyer. *Counting points on cubic surfaces II.* In: Geometric Methods in Algebra and Number Theory. Progr. Math., 235, Birkhäuser, Boston, 2005, pp. 303–309. MR2166089 (2006e:11088)

[Sw–D3] H. P. F. Swinnerton–Dyer. *Universal equivalence for cubic surfaces over finite and local fields.* Symp. Math., Bologna 24 (1981), 111–143. MR619244 (82k:14019)

[Vi] B. G. Vioreanu. *Mordell–Weil problem for cubic surfaces, numerical evidence.* arXiv:0802.0742 MR2498063 (2010i:11092)

[Z] B. I. Zilber. *Algebraic geometry via model theory.* Contemp. Math., vol. 131, Part 3 (1992), 523–537. MR1175905 (93j:03020)

Max–Planck–Institut für Mathematik, Bonn, Germany, and

Northwestern University, Evanston, USA

Contemporary Mathematics
Volume **566**, 2012
http://dx.doi.org/10.1090/conm/566/11218

Height zeta functions of equivariant compactifications of semi-direct products of algebraic groups

Sho Tanimoto and Yuri Tschinkel

ABSTRACT. We apply the theory of height zeta functions to study the asymptotic distribution of rational points of bounded height on projective equivariant compactifications of semi-direct products.

Introduction

Let X be a smooth projective variety over a number field F and L a very ample line bundle on X. An adelic metrization $\mathcal{L} = (L, \|\cdot\|)$ on L induces a height function

$$\mathsf{H}_{\mathcal{L}} \colon X(F) \to \mathbb{R}_{>0},$$

let

$$\mathsf{N}(X^\circ, \mathcal{L}, \mathsf{B}) := \#\{x \in X^\circ(F) \,|\, \mathsf{H}_{\mathcal{L}}(x) \leq \mathsf{B}\}, \quad X^\circ \subset X,$$

be the associated counting function for a subvariety X°. Manin's program, initiated in [**21**] and significantly developed over the last 20 years, relates the asymptotic of the counting function $\mathsf{N}(X^\circ, \mathcal{L}, \mathsf{B})$, as $\mathsf{B} \to \infty$, for a suitable Zariski open $X^\circ \subset X$, to global geometric invariants of the underlying variety X. By general principles of diophantine geometry, such a connection can be expected for varieties with sufficiently positive anticanonical line bundle $-K_X$, e.g., for Fano varieties. Manin's conjecture asserts that

$$\mathsf{N}(X^\circ, -\mathcal{K}_X, \mathsf{B}) = c \cdot \mathsf{B} \log(\mathsf{B})^{r-1}, \tag{0.1}$$

where r is the rank of the Picard group $\mathrm{Pic}(X)$ of X, at least over a finite extension of the ground field. The constant c admits a conceptual interpretation, its main ingredient is a Tamagawa-type number introduced by Peyre [**25**].

For recent surveys highlighting different aspects of this program, see, e.g., [**37**], [**9**], [**7**], [**8**].

Several approaches to this problem have evolved:

- passage to (universal) torsors combined with lattice point counts;
- variants of the circle method;
- ergodic theory and mixing;
- height zeta functions and spectral theory on adelic groups.

2000 *Mathematics Subject Classification.* Primary 11G35.

The universal torsor approach has been particularly successful in the treatment of del Pezzo surfaces, especially the singular ones. This method works best over $\mathbb{Q}$; applying it to surfaces over more general number fields often presents insurmountable difficulties, see, e.g., [**14**]. Here we will explain the basic principles of the method of height zeta functions of equivariant compactifications of linear algebraic groups and apply it to semi-direct products; this method is insensitive to the ground field. The spectral expansion of the height zeta function involves 1-dimensional as well as infinite-dimensional representations, see Section 3 for details on the spectral theory. We show that the main term appearing in the spectral analysis, namely, the term corresponding to 1-dimensional representations, matches precisely the predictions of Manin's conjecture, i.e., has the form (0.1). The analogous result for the universal torsor approach can be found in [**26**] and for the circle method applied to universal torsors in [**27**].

Furthermore, using the tools developed in Section 3, we provide new examples of rational surfaces satisfying Manin's conjecture.

Acknowledgments. We are grateful to the referee and to A. Chambert-Loir for useful suggestions which helped us improve the exposition. The second author was partially supported by NSF grants DMS-0739380 and 0901777.

1. Geometry

In this section, we collect some general geometric facts concerning equivariant compactifications of solvable linear algebraic groups. Here we work over an algebraically closed field of characteristic 0.

Let G be a connected linear algebraic group. In dimension 1, the only examples are the additive group $\mathbb{G}_a$ and the multiplicative group $\mathbb{G}_m$. Let

$$\mathfrak{X}^*(G) := \mathrm{Hom}(G, \mathbb{G}_m)$$

be the group of algebraic characters of G. For any connected linear algebraic group G, this is a torsion-free $\mathbb{Z}$-module of finite rank (see [**36**, Lemma 4]).

Let X be a projective equivariant compactification of G. If X is normal, then it follows from Hartogs' theorem that the boundary

$$D := X \setminus G,$$

is a Weil divisor. Moreover, after applying equivariant resolution of singularities, if necessary, we may assume that X is smooth and that the boundary

$$D = \cup_\iota D_\iota,$$

is a divisor with normal crossings. Here D_ι are irreducible components of D. Let $\mathrm{Pic}^G(X)$ be the group of equivalence classes of G-linearized line bundles on X.

Generally, we will identify divisors, associated line bundles, and their classes in $\mathrm{Pic}(X)$, resp. $\mathrm{Pic}^G(X)$.

PROPOSITION 1.1. *Let X be a smooth and proper equivariant compactification of a connected solvable linear algebraic group G. Then,*

(1) *we have an exact sequence*

$$0 \to \mathfrak{X}^*(G) \to \mathrm{Pic}^G(X) \to \mathrm{Pic}(X) \to 0,$$

(2) $\mathrm{Pic}^G(X) = \oplus_{\iota \in \mathcal{I}} \mathbb{Z} D_\iota$, *and*

(3) *the closed cone of pseudo-effective divisors of X is spanned by the boundary components:*

$$\Lambda_{\mathrm{eff}}(X) = \sum_{\iota \in \mathcal{I}} \mathbb{R}_{\geq 0} D_\iota.$$

PROOF. The first claim follows from the proof of [**24**, Proposition 1.5]. The crucial point is to show that the Picard group of G is trivial. As an algebraic variety, a connected solvable group is a product of an algebraic torus and an affine space. The second assertion holds since every finite-dimensional representation of a solvable group has a fixed vector. For the last statement, see [**22**, Theorem 2.5]. □

PROPOSITION 1.2. *Let X be a smooth and proper equivariant compactification for the left action of a connected linear algebraic group. Then the right invariant top degree differential form ω on $X^\circ := G \subset X$ satisfies*

$$-\mathrm{div}(\omega) = \sum_{\iota \in \mathcal{I}} d_\iota D_\iota,$$

where $d_\iota > 0$. The same result holds for the right action and the left invariant form.

PROOF. This fact was proved in [**22**, Theorem 2.7] or [**11**, Lemma 2.4]. Suppose that X has the left action. Let $\mathfrak{g}$ be the Lie algebra of G. For any $\partial \in \mathfrak{g}$, the global vector field ∂^X on X is defined by

$$\partial^X(f)(x) = \partial_g f(g \cdot x)|_{g=1},$$

where $f \in \mathcal{O}_X(U)$ and U is a Zariski open subset of X. Note that this is a right invariant vector field on $X^\circ = G$. Let $\partial_1, \cdots, \partial_n$ be a basis for $\mathfrak{g}$. Consider a global section of $\det \mathcal{T}_X$,

$$\delta := \partial_1^X \wedge \cdots \wedge \partial_n^X,$$

which is the dual of ω on X°. The proof of [**11**, Lemma 2.4] implies that δ vanishes along the boundary. Thus our assertion follows. □

PROPOSITION 1.3. *Let X be a smooth and proper equivariant compactification of a connected linear algebraic group. Let $f : X \to Y$ be a birational morphism to a normal projective variety Y. Then Y is an equivariant compactification of G such that the contraction map f is a G-morphism.*

PROOF. This fact was proved in [**22**, Corollary 2.4]. Choose an embedding $Y \hookrightarrow \mathbb{P}^N$, and let L be the pullback of $\mathcal{O}(1)$ on X. Since Y is normal, Zariski's main theorem implies that the image of the complete linear series $|L|$ is isomorphic to Y. According to [**24**, Corollary 1.6], after replacing L by a multiple of L, if necessary, we may assume that L carries G-linearizations. Fix one G-linearization of L. This defines the action of G on $\mathrm{H}^0(X, L)$ and on $\mathbb{P}(\mathrm{H}^0(X, L)^*)$. Now note that the morphism

$$\Phi_{|L|} : X \to \mathbb{P}(\mathrm{H}^0(X, L)^*),$$

is a G-morphism with respect to this action. Thus our assertion follows. □

The simplest solvable groups are $\mathbb{G}_a$ and $\mathbb{G}_m$, as well as their products. New examples arise as semi-direct products. For example, let

$$\begin{array}{rcl} \varphi_d : \mathbb{G}_m & \to & \mathbb{G}_m = \mathrm{GL}_1, \\ a & \mapsto & a^d \end{array}$$

and put

$$G_d := \mathbb{G}_a \rtimes_{\varphi_d} \mathbb{G}_m,$$

where the group law is given by

$$(x, a) \cdot (y, b) = (x + \varphi_d(a)y, ab).$$

It is easy to see that $G_d \simeq G_{-d}$.

One of the central themes in birational geometry is the problem of classification of algebraic varieties. The classification of G-varieties, i.e., varieties with G-actions, is already a formidable task. The theory of toric varieties, i.e., equivariant compactifications of $G = \mathbb{G}_m^n$, is very rich, and provides a testing ground for many conjectures in algebraic and arithmetic geometry. See [**22**] for first steps towards a classification of equivariant compactifications of $G = \mathbb{G}_a^n$, as well as [**33**], [**2**], [**1**] for further results in this direction.

Much less is known concerning equivariant compactifications of other solvable groups; indeed, classifying equivariant compactifications of G_d is already an interesting open question. We now collect several results illustrating specific phenomena connected with noncommutativity of G_d and with the necessity to distinguish actions on the left, on the right, or on both sides. These play a role in the analysis of height zeta functions in following sections. First of all, we have

LEMMA 1.4. *Let X be a biequivariant compactification of a semi-direct product $G \rtimes H$ of linear algebraic groups. Then X is a one-sided (left- or right-) equivariant compactification of $G \times H$.*

PROOF. Fix one section $s : H \to G \rtimes H$. Define a left action by

$$(g, h) \cdot x = g \cdot x \cdot s(h)^{-1},$$

for any $g \in G$, $h \in H$, and $x \in X$. □

In particular, there is no need to invoke noncommutative harmonic analysis in the treatment of height zeta functions of biequivariant compactifications of general solvable groups since such groups are semi-direct products of tori with unipotent groups and the lemma reduces the problem to a one-sided action of the *direct* product. Height zeta functions of direct products of additive groups and tori can be treated by combining the methods of [**4**] and [**5**] with [**11**], see Theorem 2.1. However, Manin's conjectures are still open for one-sided actions of unipotent groups, even for the Heisenberg group.

The next observation is that the projective plane $\mathbb{P}^2$ is an equivariant compactification of G_d, for any d. Indeed, the embedding

$$(x, a) \mapsto (x : a : 1) \in \mathbb{P}^2$$

defines a left-sided equivariant compactification, with boundary a union of two lines. The left action is given by

$$(x, a) \cdot (x_0 : x_1 : x_2) \mapsto (a^d x_0 + x x_2 : a x_1 : x_2).$$

In contrast, we have

PROPOSITION 1.5. *If $d \neq 1, 0,$ or -1, then $\mathbb{P}^2$ is not a biequivariant compactification of G_d.*

PROOF. Assume otherwise. Proposition 1.1 implies that the boundary must consist of two irreducible components. Let D_1 and D_2 be the two irreducible boundary components. Since $\mathcal{O}(K_{\mathbb{P}^2}) \cong \mathcal{O}(-3)$, it follows from Proposition 1.2 that either both components D_1 and D_2 are lines or one of them is a line and the other a conic. Let ω be a right invariant top degree differential form. Then $\omega/\varphi_d(a)$ is a left invariant differential form. If one of D_1 and D_2 is a conic, then the divisor of ω takes the form

$$-\mathrm{div}(\omega) = -\mathrm{div}(\omega/\varphi_d(a)) = D_1 + D_2,$$

but this is a contradiction. If D_1 and D_2 are lines, then without loss of generality, we can assume that

$$-\mathrm{div}(\omega) = 2D_1 + D_2 \quad \text{and} \quad -\mathrm{div}(\omega/\varphi_d(a)) = D_1 + 2D_2.$$

However, $\mathrm{div}(a)$ is a multiple of $D_1 - D_2$, which is also a contradiction. □

Combining this result with Proposition 1.3, we conclude that a del Pezzo surface is not a biequivariant compactification of G_d, for $d \neq 1, 0, \text{or}, -1$. Another sample result in this direction is:

PROPOSITION 1.6. *Let S be the singular quartic del Pezzo surface of type $A_3 + A_1$ defined by*

$$x_0^2 + x_0x_3 + x_2x_4 = x_1x_3 - x_2^2 = 0$$

Then S is a one-sided equivariant compactification of G_1, but not a biequivariant compactification of G_d if $d \neq 0$.

PROOF. For the first assertion, see [**20**, Section 5]. Assume that S is a biequivariant compactification of G_d. Let $\pi : \widetilde{S} \to S$ be its minimal desingularization. Then $\widetilde{S}$ is also a biequivariant compactification of G_d because the action of G_d must fix the singular locus of S. See [**20**, Lemma 4]. It has three (-1)-curves L_1, L_2, and L_3, which are the strict transforms of

$$\{x_0 = x_1 = x_2 = 0\}, \quad \{x_0 + x_3 = x_1 = x_2 = 0\}, \quad \text{and} \quad \{x_0 = x_2 = x_3 = 0\},$$

respectively, and has four (-2)-curves R_1, R_2, R_3, and R_4. The nonzero intersection numbers are given by:

$$L_1.R_1 = L_2.R_1 = R_1.R_2 = R_2.R_3 = R_3.L_3 = L_3.R_4 = 1.$$

Since the cone of curves is generated by the components of the boundary, these negative curves must be in the boundary because each generates an extremal ray. Since the Picard group of $\widetilde{S}$ has rank six, it follows from Proposition 1.1 that the number of boundary components is seven. Thus, the boundary is equal to the union of these negative curves.

Let $f : \widetilde{S} \to \mathbb{P}^2$ be the birational morphism which contracts L_1, L_2, L_3, R_2, and R_3. According to Proposition 1.3, this induces a biequivariant compactification on $\mathbb{P}^2$. The birational map $f \circ \pi^{-1} : S \dashrightarrow \mathbb{P}^2$ is given by

$$S \ni (x_0 : x_1 : x_2 : x_3 : x_4) \mapsto (x_2 : x_0 : x_3) \in \mathbb{P}^2.$$

The images of R_1 and R_4 are $\{y_0 = 0\}$ and $\{y_2 = 0\}$ and we denote them by D_0 and D_2, respectively. The images of L_1 and L_2 are $(0 : 0 : 1)$ and $(0 : 1 : -1)$, respectively; so that the induced group action on $\mathbb{P}^2$ must fix $(0 : 0 : 1)$, $(0 : 1 : -1)$, and $D_0 \cap D_2 = (0 : 1 : 0)$. Thus, the group action must fix the line D_0, and this fact implies that all left and right invariant vector fields vanish along D_0. It follows that

$$-\mathrm{div}(\omega) = -\mathrm{div}(\omega/\varphi_d(a)) = 2D_0 + D_2,$$

which contradicts $d \neq 0$. □

EXAMPLE 1.7. Let $l \geq d \geq 0$. The Hirzebruch surface $\mathbb{F}_l = \mathbb{P}_{\mathbb{P}^1}((\mathcal{O} \oplus \mathcal{O}(l))^*)$ is a biequivariant compactification of G_d. Indeed, we may take the embedding

$$\begin{array}{ccc} G_d & \hookrightarrow & \mathbb{F}_l \\ (x, a) & \mapsto & ((a:1), [1 \oplus x\sigma_1^l]), \end{array}$$

where σ_1 is a section of the line bundle $\mathcal{O}(1)$ on $\mathbb{P}^1$ such that

$$\mathrm{div}(\sigma_1) = (1:0).$$

Let $\pi : \mathbb{F}_l \to \mathbb{P}^1$ be the $\mathbb{P}^1$-fibration. The right action is given by

$$((x_0 : x_1), [y_0 \oplus y_1\sigma_1^l]) \mapsto ((ax_0 : x_1), [y_0 \oplus (y_1 + (x_0/x_1)^d xy_0)\sigma_1^l]),$$

on $\pi^{-1}(U_0 = \mathbb{P}^1 \setminus \{(1:0)\})$ and

$$((x_0 : x_1), [y_0 \oplus y_1\sigma_0^l]) \mapsto ((ax_0 : x_1), [a^l y_0 \oplus (y_1 + (x_1/x_0)^{l-d} xy_0)\sigma_0^l]),$$

on $U_1 = \pi^{-1}(\mathbb{P}^1 \setminus \{(0:1)\})$. Similarly, one defines the left action. The boundary consists of three components: two fibers $f_0 = \pi^{-1}((0:1))$, $f_1 = \pi^{-1}((0:1))$ and the special section D characterized by $D^2 = -l$.

EXAMPLE 1.8. Consider the right actions in Examples 1.7. When $l > d > 0$, these actions fix the fiber f_0 and act multiplicatively, i.e., with two fixed points, on the fiber f_1. Let X be the blowup of two points (or more) on f_0 and of one fixed point P on $f_1 \setminus D$. Then X is an equivariant compactification of G_d which is neither a toric variety nor a $\mathbb{G}_a^2$-variety. Indeed, there are no equivariant compactifications of $\mathbb{G}_m^2$ on $\mathbb{F}_l$ fixing f_0, so X cannot be toric. Also, if X were a $\mathbb{G}_a^2$-variety, we would obtain an induced $\mathbb{G}_a^2$-action on $\mathbb{F}_l$ fixing f_0 and P. However, the boundary consists of two irreducible components and must contain f_0, D, and P because D is a negative curve. This is a contradiction.

For $l = 2$ and $d = 1$, blowing up two points on f_0 we obtain a quintic del Pezzo surface with an A_2 singularity. Manin's conjecture for this surface is proved in [**19**].

In Section 5, we prove Manin's conjecture for X with $l \geq 3$.

2. Height zeta functions

Let F be a number field, $\mathfrak{o}_F$ its ring of integers, and Val_F the set of equivalence classes of valuations of F. For $v \in \mathrm{Val}_F$ let F_v be the completions of F with respect to v, for nonarchimedean v, let $\mathfrak{o}_v$ be the corresponding ring of integers and $\mathfrak{m}_v$ the maximal ideal. Let $\mathbb{A} = \mathbb{A}_F$ be the adele ring of F.

Let X be a smooth and projective right-sided equivariant compactification of a split connected solvable linear algebraic group G over F, i.e., the toric part T of G is isomorphic to $\mathbb{G}_m^n$. Moreover, we assume that the boundary $D = \cup_{\iota \in \mathcal{I}} D_\iota$ consists of geometrically irreducible components meeting transversely. We are interested in the asymptotic distribution of rational points of bounded height on $X^\circ = G \subset X$, with respect to adelically metrized ample line bundles $\mathcal{L} = (L, (\|\cdot\|_{\mathbb{A}}))$ on X. We now recall the method of height zeta functions; see [**37**, Section 6] for more details and examples.

Step 1. Define an adelic height pairing

$$\mathsf{H}\colon \mathrm{Pic}^G(X)_{\mathbb{C}} \times G(\mathbb{A}_F) \to \mathbb{C},$$

whose restriction to

$$\mathsf{H}\colon \mathrm{Pic}^G(X) \times G(F) \to \mathbb{R}_{\geq 0},$$

descends to a *height system* on $\mathrm{Pic}(X)$ (see [**26**, Definition 2.5.2]). This means that the restriction of H to an $L \in \mathrm{Pic}^G(X)$ defines a Weil height corresponding to some adelic metrization of $L \in \mathrm{Pic}^G(X)$, and that it does not depend on the choice of a G-linearization on L. Such a pairing appeared in [**3**] in the context of toric varieties, the extension to general solvable groups is straightforward.

Concretely, by Proposition 1.1, we know that $\mathrm{Pic}^G(X)$ is generated by boundary components D_ι, for $\iota \in \mathcal{I}$. The v-adic analytic manifold $X(F_v)$ admits a "partition of unity", i.e., a decomposition into charts $X_{I,v}$, labeled by $I \subseteq \mathcal{I}$, such that in each chart the local height function takes the form

$$\mathsf{H}_v(\mathbf{s}, x_v) = \phi(x_v) \cdot \prod_{\iota \in I} |x_{\iota,v}|_v^{s_\iota},$$

where for each $\iota \in I$, x_ι is the local coordinate of D_ι in this chart,

$$\mathbf{s} = \sum_{\iota \in \mathcal{I}} s_\iota D_\iota,$$

and $\log(\phi)$ is a bounded function, equal to 1 for almost all v (see [**13**, Section 2] for more details). Note that, locally, the height function

$$\mathsf{H}_{\iota,v}(x_v) := |x_{\iota,v}|_v$$

is simply the v-adic distance to the boundary component D_ι. To visualize $X_{I,v}$ (for almost all v) consider the partition induced by

$$X(F_v) = X(\mathfrak{o}_v) \xrightarrow{\rho} \sqcup_{I \subset \mathcal{I}} X_I^\circ(\mathbb{F}_q),$$

where

$$X_I := \cup_{\iota \in I} D_I, \quad X_I^\circ := X_I \setminus \cup_{I' \supsetneq I} X_{I'},$$

is the stratification of the boundary and ρ is the reduction map; by convention $X_\emptyset = G$. Then $X_{I,v}$ is the preimage of $X_I^\circ(\mathbb{F}_q)$ in $X(F_v)$, and in particular, $X_{\emptyset,v} = G(\mathfrak{o}_v)$, for almost all v.

Since the action of G lifts to integral models of G, X, and L, the nonarchimedean local height pairings are invariant with respect to a compact subgroup $\mathbf{K}_v \subset G(F_v)$, which is $G(\mathfrak{o}_v)$, for almost all v.

Step 2. The height zeta function

$$\mathsf{Z}(\mathbf{s}, g) := \sum_{\gamma \in G(F)} \mathsf{H}(\mathbf{s}, \gamma g)^{-1},$$

converges absolutely to a holomorphic function, for $\Re(\mathbf{s})$ sufficiently large, and defines a continuous function in $\mathsf{L}^1(G(F)\backslash G(\mathbb{A}_F)) \cap \mathsf{L}^2(G(F)\backslash G(\mathbb{A}_F))$. Formally, we have the spectral expansion

$$\mathsf{Z}(\mathbf{s}, g) = \sum_{\pi} \mathsf{Z}_\pi(\mathbf{s}, g), \tag{2.1}$$

where the "sum" is over irreducible unitary representations occurring in the right regular representation of $G(\mathbb{A}_F)$ in $\mathsf{L}^2(G(F)\backslash G(\mathbb{A}_F))$. The invariance of the global height pairing under the action of a compact subgroup $\mathbf{K} \subset G(\mathbb{A}_F)$, on the side of the action, insures that Z_π are in $\mathsf{L}^2(G(F)\backslash G(\mathbb{A}_F))^{\mathbf{K}}$.

Step 3. Ideally, we would like to obtain a meromorphic continuation of Z to a tube domain

$$\mathsf{T}_\Omega = \Omega + i\,\mathrm{Pic}(X)_\mathbb{R} \subset \mathrm{Pic}(X)_\mathbb{C},$$

where $\Omega \subset \mathrm{Pic}(X)_\mathbb{R}$ is an open neighborhood of the anticanonical class $-K_X$. It is expected that Z is holomorphic for

$$\Re(\mathbf{s}) \in -K_X + \Lambda^\circ_{\mathrm{eff}}(X)$$

and that the polar set of the shifted height zeta function $\mathsf{Z}(\mathbf{s} - K_X, g)$ is the same as that of

$$\mathcal{X}_{\Lambda_{\mathrm{eff}}(X)}(\mathbf{s}) := \int_{\Lambda^*_{\mathrm{eff}}(X)} e^{-\langle \mathbf{s},\mathbf{y}\rangle} \mathrm{d}y, \tag{2.2}$$

the Laplace transform of the set-theoretic characteristic function of the dual cone $\Lambda_{\mathrm{eff}}(X)^* \subset \mathrm{Pic}(X)^*_\mathbb{R}$. Here the Lebesgue measure $\mathrm{d}y$ is normalized by the dual lattice $\mathrm{Pic}(X)^* \subset \mathrm{Pic}(X)^*_\mathbb{R}$. In particular, for

$$\kappa = -K_X = \sum_\iota \kappa_\iota D_\iota,$$

the restriction of the height zeta function $\mathsf{Z}(\mathbf{s}, \mathrm{id})$ to the one-parameter zeta function $\mathsf{Z}(s\kappa, \mathrm{id})$ should be holomorphic for $\Re(s) > 1$, admit a meromorphic continuation to $\Re(s) > 1 - \epsilon$, for some $\epsilon > 0$, with a unique pole at $s = 1$, of order $r = \mathrm{rk}\,\mathrm{Pic}(X)$. Furthermore, it is desirable to have some growth estimates in vertical strips. In this case, a Tauberian theorem implies Manin's conjecture (0.1) for the counting function; the quality of the error term depends on the growth rate in vertical strips. Finally, the leading constant at the pole of $\mathsf{Z}(s\kappa, \mathrm{id})$ is essentially the Tamagawa-type number defined by Peyre. We will refer to this by saying that the height zeta function Z satisfies Manin's conjecture; a precise definition of this class of functions can be found in [**10**, Section 3.1].

This strategy has worked well and lead to a proof of Manin's conjecture for the following varieties:

- toric varieties [**3**], [**4**], [**5**];
- equivariant compactifications of additive groups $\mathbb{G}_a^n$ [**11**];
- equivariant compactifications of unipotent groups [**32**], [**31**];
- wonderful compactifications of semi-simple groups of adjoint type [**30**].

Moreover, applications of Langlands' theory of Eisenstein series allowed to prove Manin's conjecture for flag varieties [**21**], their twisted products [**34**], and horospherical varieties [**35**], [**10**].

The analysis of the spectral expansion (2.1) is easier when every automorphic representation π is 1-dimensional, i.e., when G is abelian: $G = \mathbb{G}_a^n$ or $G = T$, an algebraic torus. In these cases, (2.1) is simply the Fourier expansion of the height zeta function and we have, at least formally,

$$\mathsf{Z}(\mathbf{s}, \mathrm{id}) = \int \widehat{\mathsf{H}}(\mathbf{s}, \chi) \mathrm{d}\chi, \tag{2.3}$$

where

$$\hat{\mathsf{H}}(\mathbf{s},\chi)=\int_{G(\mathbb{A}_F)}\mathsf{H}(\mathbf{s},g)^{-1}\bar{\chi}(g)\mathrm{d}g, \tag{2.4}$$

is the Fourier transform of the height function, χ is a character of $G(F)\backslash G(\mathbb{A}_F)$, and $\mathrm{d}\chi$ an appropriate measure on the space of automorphic characters. For $G=\mathbb{G}_a^n$, the space of automorphic characters is $G(F)$ itself, for G an algebraic torus it is (noncanonically) $\mathfrak{X}(G)^*_{\mathbb{R}}\times\mathcal{U}_G$, where $\mathcal{U}_G$ is a discrete group.

The v-adic integration technique developed by Igusa, Denef, Denef and Loeser (see, e.g., [**23**], [**16**], [**17**], and [**18**]) allows to compute local Fourier transforms of height functions, in particular, for the trivial character $\chi=1$ and almost all v we obtain

$$\widehat{\mathsf{H}}_v(\mathbf{s},1)=\int_{G(F_v)}\mathsf{H}(\mathbf{s},g)^{-1}\mathrm{d}g=\tau_v(G)^{-1}\left(\sum_{I\subset\mathcal{I}}\frac{\#X_I^\circ(\mathbb{F}_q)}{q^{\dim(X)}}\prod_{\iota\in I}\frac{q-1}{q^{s_\iota-\kappa_\iota+1}-1}\right),$$

where X_I are strata of the stratification described in Step 1 and $\tau_v(G)$ is the local Tamagawa number of G,

$$\tau_v(G)=\frac{\#G(\mathbb{F}_q)}{q^{\dim(G)}}.$$

Such height integrals are geometric versions of Igusa's integrals; a comprehensive theory in the analytic and adelic setting can be found in [**13**].

The computation of Fourier transforms at nontrivial characters requires a finer partition of $X(F_v)$ which takes into account possible zeroes of the *phase* of the character in $G(F_v)$; see [**11**, Section 10] for the the additive case and [**3**, Section 2] for the toric case. The result is that in the neighborhood of

$$\kappa=\sum_\iota\kappa_\iota D_\iota\in\mathrm{Pic}^G(X),$$

the Fourier transform is regularized as follows

$$\widehat{\mathsf{H}}(\mathbf{s},\chi)=\begin{cases}\prod_{v\notin S(\chi)}\prod_{\iota\in\mathcal{I}(\chi)}\zeta_{F,v}(s_\iota-\kappa_\iota+1)\prod_{v\in S(\chi)}\phi_v(\mathbf{s},\chi) & G=\mathbb{G}_a^n,\\ \prod_{v\notin S(\chi)}\prod_{\iota\in\mathcal{I}}\mathsf{L}_{F,v}(s_\iota-\kappa_\iota+1+im(\chi),\chi_u)\prod_{v\in S(\chi)}\phi_v(\mathbf{s},\chi) & G=T,\end{cases}$$

where

- $\mathcal{I}(\chi)\subsetneq\mathcal{I}$;
- $S(\chi)$ is a finite set of places, which, in general, depends on χ;
- $\zeta_{F,v}$ is a local factor of the Dedekind zeta function of F and $\mathsf{L}_{F,v}$ a local factor of a Hecke L-function;
- $m(\chi)$ is the "coordinate" of the automorphic character χ of $G=T$ under the embedding $\mathfrak{X}(G)^*_{\mathbb{R}}\hookrightarrow\mathrm{Pic}^G(X)_{\mathbb{R}}$ in the exact sequence (1) in Proposition 1.1 and χ_u is the "discrete" component of χ;
- and $\phi_v(\mathbf{s},\chi)$ is a function which is holomorphic and bounded.

In particular, each $\widehat{\mathsf{H}}(\mathbf{s},\chi)$ admits a meromorphic continuation as desired and we can control the poles of each term. Moreover, at archimedean places we may use integration by parts with respect to vector fields in the universal enveloping algebra of the corresponding real of complex group to derive bounds in terms of the "phase" of the occurring oscillatory integrals, i.e., in terms of "coordinates" of χ.

So far, we have not used the fact that X is an equivariant compactification of G. Only at this stage do we see that the **K**-invariance of the height is an important, in fact, crucial, property that allows to establish uniform convergence of the right side of the expansion (2.1); it insures that

$$\widehat{\mathsf{H}}(\mathbf{s},\chi)=0,$$

for all χ which are *nontrivial* on $\mathbf{K}$. For $G=\mathbb{G}_a^n$ this means that the trivial representation is *isolated* and that the integral on the right side of Equation (2.3) is in fact a *sum* over a *lattice* of integral points in $G(F)$. Note that Manin's conjecture *fails* for nonequivariant compactifications of the affine space, there are counterexamples already in dimension three [**6**]. The analytic method described above fails precisely because we cannot insure the convergence on the Fourier expansion.

A similar effect occurs in the noncommutative setting; one-sided actions *do not* guarantee bi-$\mathbf{K}$-invariance of the height, in contrast with the abelian case. Analytically, this translates into subtle convergence issues of the spectral expansion, in particular, for infinite-dimensional representations.

THEOREM 2.1. *Let G be an extension of an algebraic torus T by a unipotent group N such that $[G,G]=N$ over a number field F. Let X be an equivariant compactification of G over F and*

$$\mathsf{Z}(\mathbf{s},g)=\sum_{\gamma\in G(F)}\mathsf{H}(\mathbf{s},\gamma g)^{-1},$$

the height zeta function with respect to an adelic height pairing as in Step 1. Let

$$\mathsf{Z}_0(\mathbf{s},g)=\int \mathsf{Z}_\chi(\mathbf{s},g)\,\mathrm{d}\chi,$$

be the integral over all 1-dimensional automorphic representations of $G(\mathbb{A}_F)$ occurring in the spectral expansion (2.1). Then Z_0 satisfies Manin's conjecture.

PROOF. Let

$$1\to N\to G\to T\to 1$$

be the defining extension. One-dimensional automorphic representations of $G(\mathbb{A}_F)$ are precisely those which are trivial on $N(\mathbb{A}_F)$, i.e., these are automorphic characters of T. The $\mathbf{K}$-invariance of the height (on one side) insures that only unramified characters, i.e., $\mathbf{K}_T$-invariant characters contribute to the spectral expansion of Z_0.

Let $M=\mathfrak{X}^*(G)$ be the group of algebraic characters. We have

$$\begin{aligned}\mathsf{Z}_0(\mathbf{s},\mathrm{id})&=\int_{M_\mathbb{R}\times\mathcal{U}_T}\int_{G(F)\backslash G(\mathbb{A}_F)}\mathsf{Z}(\mathbf{s},g)\bar{\chi}(g)\,\mathrm{d}g\mathrm{d}\chi\\&=\int_{M_\mathbb{R}\times\mathcal{U}_T}\int_{G(\mathbb{A}_F)}\mathsf{H}(\mathbf{s},g)^{-1}\bar{\chi}(g)\,\mathrm{d}g\mathrm{d}\chi\\&=\int_{M_\mathbb{R}}\mathsf{F}(\mathbf{s}+im(\chi))\,\mathrm{d}m,\end{aligned}$$

where

$$\mathsf{F}(\mathbf{s}):=\sum_{\chi\in\mathcal{U}_T}\widehat{\mathsf{H}}(\mathbf{s},\chi_u).$$

Computations of local Fourier transforms explained above show that F can be regularized as follows:

$$\mathsf{F}(\mathbf{s})=\prod_{\iota\in\mathcal{I}}\zeta_F(s_\iota-\kappa_\iota+1)\cdot\mathsf{F}_\infty(\mathbf{s}),$$

where F_∞ is holomorphic for $\Re(s_\iota) - \kappa_\iota > -\epsilon$, for some $\epsilon > 0$, with growth control in vertical strips. Now we have placed ourselves into the situation considered in [**10**, Section 3]: Theorem 3.1.14 establishes analytic properties of integrals

$$\int_{M_{\mathbb{R}}} \frac{1}{\prod_{\iota\in\mathcal{I}}(s_\iota - \kappa_\iota + im_\iota)} \cdot \mathsf{F}_\infty(\mathbf{s} + im)\, dm,$$

where the image of $\iota\colon M_{\mathbb{R}} \hookrightarrow \mathbb{R}^{\#\mathcal{I}}$ intersects the simplicial cone $\mathbb{R}^{\#\mathcal{I}}_{\geq 0}$ only in the origin. The main result is that the analytic properties of such integrals match those of the $\mathcal{X}$-function (2.2) of the *image* cone under the projection

$$\begin{array}{ccccccccc} 0 & \longrightarrow & M_{\mathbb{R}} & \xrightarrow{\iota} & \mathbb{R}^{\#\mathcal{I}} & \xrightarrow{\pi} & \mathbb{R}^{\#\mathcal{I}-\dim(M)} & \longrightarrow & 0 \\ & & \| & & \| & & \| & & \\ 0 & \longrightarrow & \mathfrak{X}^*(G)_{\mathbb{R}} & \longrightarrow & \mathrm{Pic}^G(X)_{\mathbb{R}} & \longrightarrow & \mathrm{Pic}(X)_{\mathbb{R}} & \longrightarrow & 0; \end{array}$$

according to Proposition 1.1, the image of the simplicial cone $\mathbb{R}^{\#\mathcal{I}}_{\geq 0}$ under π is precisely $\Lambda_{\mathrm{eff}}(X) \subset \mathrm{Pic}(X)_{\mathbb{R}}$. □

3. Harmonic analysis

In this section we study the local and adelic representation theory of

$$G := \mathbb{G}_a \rtimes_\varphi \mathbb{G}_m,$$

an extension of $T := \mathbb{G}_m$ by $N := \mathbb{G}_a$ via a homomorphism $\varphi : \mathbb{G}_m \to \mathrm{GL}_1$. The group law given by

$$(x, a) \cdot (y, b) = (x + \varphi(a)y, ab).$$

We fix the standard Haar measures

$$\mathrm{d}x = \prod_v \mathrm{d}x_v \quad \text{and} \quad \mathrm{d}a^\times = \prod_v \mathrm{d}a_v^\times,$$

on $N(\mathbb{A}_F)$ and $T(\mathbb{A}_F)$. Note that $G(\mathbb{A}_F)$ is not unimodular, unless φ is trivial. The product measure $\mathrm{d}g := \mathrm{d}x\mathrm{d}a^\times$ is a right invariant measure on $G(\mathbb{A}_F)$ and $\mathrm{d}g/\varphi(a)$ is a left invariant measure.

Let ϱ be the right regular unitary representation of $G(\mathbb{A}_F)$ on the Hilbert space:

$$\mathcal{H} := \mathsf{L}^2(G(F)\backslash G(\mathbb{A}_F), \mathrm{d}g).$$

We now discuss the decomposition of $\mathcal{H}$ into irreducible representations. Let

$$\psi = \prod_v \psi_v : \mathbb{A}_F \to \mathbb{S}^1,$$

be the standard automorphic character and ψ_n the character defined by

$$x \to \psi(nx),$$

for $n \in F^\times$. Let

$$W := \ker(\varphi : F^\times \to F^\times),$$

and

$$\pi_n := \mathrm{Ind}_{N(\mathbb{A}_F)\times W}^{G(\mathbb{A}_F)}(\psi_n),$$

for $n \in F^\times$. More precisely, the underlying Hilbert space of π_n is $\mathsf{L}^2(W\backslash T(\mathbb{A}_F))$, and that the group action is given by

$$(x, a) \cdot f(b) = \psi_n(\varphi(b)x) f(ab),$$

where f is a square-integrable function on $T(\mathbb{A}_F)$. The following proposition we learned from J. Shalika [**29**].

PROPOSITION 3.1. *Irreducible automorphic representations, i.e., irreducible unitary representations occurring in* $\mathcal{H} = \mathsf{L}^2(G(F)\backslash G(\mathbb{A}_F))$, *are parametrized as follows:*

$$\mathcal{H} = \mathsf{L}^2(T(F)\backslash T(\mathbb{A}_F)) \oplus \widehat{\bigoplus}_{n\in(F^\times/\varphi(F^\times))} \pi_n,$$

REMARK 3.2. Up to unitary equivalence, the representation π_n does not depend on the choice of a representative $n \in F^\times/\varphi(F^\times)$.

PROOF. Define

$$\mathcal{H}_0 := \{\phi \in \mathcal{H}\,|\,\phi((x,1)g) = \phi(g)\},$$

and let $\mathcal{H}_1$ be the orthogonal complement of $\mathcal{H}_0$. It is straightforward to prove that

$$\mathcal{H}_0 \cong \mathsf{L}^2(T(F)\backslash T(\mathbb{A}_F)).$$

Lemma 3.4 concludes our assertion. □

LEMMA 3.3. *For any* $\phi \in \mathsf{L}^1(G(F)\backslash G(\mathbb{A}_F)) \cap \mathcal{H}$, *the projection of* ϕ *onto* $\mathcal{H}_0$ *is given by*

$$\phi_0(g) := \int_{N(F)\backslash N(\mathbb{A}_F)} \phi((x,1)g)dx \text{ a.e..}$$

PROOF. It is easy to check that $\phi_0 \in \mathcal{H}_0$. Also, for any $\phi' \in \mathcal{H}_0$, we have

$$\int_{G(F)\backslash G(\mathbb{A}_F)} (\phi - \phi_0)\phi' \mathrm{d}g = 0.$$

□

LEMMA 3.4. *We have*

$$\mathcal{H}_1 \cong \widehat{\bigoplus}_{n\in F^\times/\varphi(F^\times)} \pi_n.$$

PROOF. For $\phi \in \mathsf{C}_c^\infty(G(F)\backslash G(\mathbb{A}_F)) \cap \mathcal{H}_1$, define

$$f_{n,\phi}(a) := \int_{N(F)\backslash N(\mathbb{A}_F)} \phi(x,a)\bar{\psi}_n(x)\,\mathrm{d}x.$$

Then,

$$\begin{aligned}
\| \phi \|_{\mathsf{L}^2}^2 &= \int_{T(F)\backslash T(\mathbb{A}_F)} \int_{N(F)\backslash N(\mathbb{A}_F)} |\phi(x,a))|^2 \,\mathrm{d}x\mathrm{d}a^\times \\
&= \int_{T(F)\backslash T(\mathbb{A}_F)} \sum_{\alpha\in F} \left| \int_{N(F)\backslash N(\mathbb{A}_F)} \phi(x,a)\bar{\psi}(\alpha x)\mathrm{d}x \right|^2 \mathrm{d}a^\times \\
&= \int_{T(F)\backslash T(\mathbb{A}_F)} \sum_{\alpha\in F^\times} \left| \int_{N(F)\backslash N(\mathbb{A}_F)} \phi(x,a)\bar{\psi}(\alpha x)\mathrm{d}x \right|^2 \mathrm{d}a^\times \\
&= \int_{T(F)\backslash T(\mathbb{A}_F)} \sum_{\alpha\in F^\times} \sum_{n\in F^\times/\varphi(F^\times)} \frac{1}{\#W} \left| \int_{N(F)\backslash N(\mathbb{A}_F)} \phi(x,a)\bar{\psi}(n\varphi(\alpha)x)\mathrm{d}x \right|^2 \mathrm{d}a^\times \\
&= \int_{T(F)\backslash T(\mathbb{A}_F)} \sum_{\alpha\in F^\times} \sum_{n\in F^\times/\varphi(F^\times)} \frac{1}{\#W} \left| \int_{N(F)\backslash N(\mathbb{A}_F)} \phi(x,\alpha a)\bar{\psi}(nx)\mathrm{d}x \right|^2 \mathrm{d}a^\times \\
&= \sum_{n\in F^\times/\varphi(F^\times)} \frac{1}{\#W} \int_{T(\mathbb{A}_F)} \left| \int_{N(F)\backslash N(\mathbb{A}_F)} \phi(x,a)\bar{\psi}_n(x)dx \right|^2 \mathrm{d}a^\times \\
&= \sum_{n\in F^\times/\varphi(F^\times)} \| f_{n,\phi} \|_{\mathsf{L}^2}^2 .
\end{aligned}$$

The second equality is the Plancherel theorem for $N(F)\backslash N(\mathbb{A}_F)$. Third equality follows from Lemma 3.3. The fifth equality follows from the left $G(F)$-invariance of ϕ. Thus, we obtain an unitary operator:

$$I : \mathcal{H}_1 \to \widehat{\bigoplus}_{n\in F^\times/\varphi(F^\times)} \pi_n.$$

Compatibility with the group action is straightforward, so I is actually a morphism of unitary representations. We construct the inverse map of I explicitly. For $f \in \mathsf{C}_c^\infty(W\backslash T(\mathbb{A}_F))$, define

$$\phi_{n,f}(x,a) := \frac{1}{\#W} \sum_{\alpha\in F^\times} \psi_n(\varphi(\alpha)x)f(\alpha a).$$

The orthogonality of characters implies that

$$\begin{aligned}
&\textstyle\int_{N(F)\backslash N(\mathbb{A}_F)} \phi_{n,f}(x,a)\cdot\overline{\phi_{n,f}(x,a)}\,\mathrm{d}x \\
&= (\#W)^2 \textstyle\int_{N(F)\backslash N(\mathbb{A}_F)} (\sum_{\alpha\in F^\times} \psi_n(\varphi(\alpha)x)f(\alpha a))\cdot(\sum_{\alpha\in F^\times} \bar{\psi}_n(\varphi(\alpha)x)\overline{f}(\alpha a))\,\mathrm{d}x \\
&= \tfrac{1}{\#W} \textstyle\sum_{\alpha\in F^\times} |f(\alpha a)|^2.
\end{aligned}$$

Substituting, we obtain

$$\begin{aligned}
\| \phi_{n,f} \|^2 &= \int_{T(F)\backslash T(\mathbb{A}_F)} \int_{N(F)\backslash N(\mathbb{A}_F)} |\phi_{n,f}(x,a)|^2 \,\mathrm{d}x\mathrm{d}a^\times \\
&= \frac{1}{\#W} \int_{T(F)\backslash T(\mathbb{A}_F)} \sum_{\alpha\in F^\times} |f(\alpha a)|^2 \mathrm{d}a^\times = \| f \|_n^2 .
\end{aligned}$$

Lemma 3.3 implies that $\phi_f \in \mathcal{H}_1$ and we obtain a morphism

$$\Theta : \widehat{\bigoplus}_{n\in F^\times/\varphi(F^\times)} \pi_n \to \mathcal{H}_1.$$

Now we only need to check that $\Theta I = id$ and $I\Theta = id$. The first follows from the Poisson formula: For any $\phi \in \mathsf{C}_c^\infty(G(F)\backslash G(\mathbb{A}_F)) \cap \mathcal{H}_1$,

$$\begin{aligned}
\Theta I\phi &= \sum_{n\in F^\times/\varphi(F^\times)} \frac{1}{\#W} \sum_{\alpha\in F^\times} \psi_n(\varphi(\alpha)x) \int_{N(F)\backslash N(\mathbb{A}_F)} \phi(y,\alpha a)\bar{\psi}_n(y)\,\mathrm{d}y \\
&= \sum_{n\in F^\times/\varphi(F^\times)} \frac{1}{\#W} \sum_{\alpha\in F^\times} \int_{N(F)\backslash N(\mathbb{A}_F)} \phi(\varphi(\alpha)y,\alpha a)\bar{\psi}_n(\varphi(\alpha)(y-x))\,\mathrm{d}y \\
&= \sum_{n\in F^\times/\varphi(F^\times)} \frac{1}{\#W} \sum_{\alpha\in F^\times} \int_{N(F)\backslash N(\mathbb{A}_F)} \phi((y+x,a))\bar{\psi}_n(\varphi(\alpha)y)\,\mathrm{d}y \\
&= \sum_{\alpha\in F} \int_{N(F)\backslash N(\mathbb{A}_F)} \phi((y,1)(x,a))\bar{\psi}(\alpha y)\,\mathrm{d}y = \phi(x,a),
\end{aligned}$$

where we apply Poisson formula for the last equality. The other identity, $I\Theta = id$ is checked by a similar computation. □

To simplify notation, we now restrict to $F = \mathbb{Q}$. For our applications in Sections 4 and 5, we need to know an explicit orthonormal basis for the unique infinite-dimensional representation $\pi = \mathsf{L}^2(\mathbb{A}_\mathbb{Q}^\times)$ of $G = G_1$. For any $n \geq 1$, define compact subgroups of $G(\mathbb{Z}_p)$

$$G(p^n\mathbb{Z}_p) := \{(x,a) \,|\, x \in p^n\mathbb{Z}_p,\ a \in 1 + p^n\mathbb{Z}_p\}.$$

Let $v_p : \mathbb{Q}_p \to \mathbb{Z}$ be the discrete valuation on $\mathbb{Q}_p$.

LEMMA 3.5. *Let* $\mathbf{K}_p = G(p^n\mathbb{Z}_p)$.

- *When* $n = 0$, *an orthonormal basis for* $\mathsf{L}^2(\mathbb{Q}_p^\times)^{\mathbf{K}_p}$ *is given by*
$$\{\mathbf{1}_{p^j\mathbb{Z}_p^\times} \,|\, j \geq 0\}.$$
- *When* $n \geq 1$, *an orthonormal basis for* $\mathsf{L}^2(\mathbb{Q}_p^\times)^{\mathbf{K}_p}$ *is given by*
$$\{\lambda_p(\cdot/p^j)\mathbf{1}_{p^j\mathbb{Z}_p^\times} \,|\, j \geq -n,\ \lambda_p \in \mathsf{M}_p\},$$
where M_p *is the set of characters on* $\mathbb{Z}_p^\times/(1+p^n\mathbb{Z}_p)$.

Moreover, let $\mathbf{K}_{\mathrm{fin}} = \prod_p \mathbf{K}_p$, *where the local compact subgroups are given by* $\mathbf{K}_p = G(p^{n_p}\mathbb{Z}_p)$, *with* $n_p = 0$ *for almost all* p. *Let* S *be the set of primes with* $n_p \neq 0$ *and* $N = \prod_p p^{n_p}$. *Then an orthonormal basis for* $\mathsf{L}^2(\mathbb{A}_{\mathbb{Q},\mathrm{fin}}^\times)^{\mathbf{K}_{\mathrm{fin}}}$ *is given by the functions*

$$(a_p)_p \mapsto \prod_{p\in S} \lambda_p(a_p \cdot p^{-v_p(a_p)})\mathbf{1}_{\frac{m}{N}\mathbb{Z}_p^\times}(a_p) \cdot \prod_{p\notin S} \mathbf{1}_{m\mathbb{Z}_p^\times}(a_p), \qquad m \in \mathbb{N},\ \lambda_p \in \mathsf{M}_p.$$

PROOF. For the first assertion, let $f \in \mathsf{L}^2(\mathbb{Q}_p^\times)^{\mathbf{K}_p}$ where $\mathbf{K}_p = G(\mathbb{Z}_p)$. Since it is $\mathbf{K}_p$-invariant, we have

$$f(b_p \cdot a_p) = f(a_p),$$

for any $b \in \mathbb{Z}_p^\times$. Hence f takes the form of

$$f = \sum_{j=-\infty}^{\infty} c_j \mathbf{1}_{p^j\mathbb{Z}_p^\times}$$

where $c_j = f(p^j)$ and $\sum_{j=-\infty}^{\infty} |c_j|^2 < +\infty$. On the other hand, we have

$$\psi_p(a_p \cdot x_p) f(a_p) = f(a_p)$$

for any $x_p \in \mathbb{Z}_p$. This implies that $f(p^j) = 0$ for any $j < 0$. Thus the first assertion follows. The second assertion is treated similarly. The last assertion follows from the first and the second assertions. □

We denote these vectors by $\mathbf{v}_{m,\lambda}$ where $m \in \mathbb{N}$ and $\lambda \in \mathsf{M} := \prod_{p \in S} \mathsf{M}_p$. Note that M is a finite set. Also we define

$$\begin{aligned}\theta_{m,\lambda,t}(g) &:= \Theta(\mathbf{v}_{m,\lambda} \otimes |\cdot|_\infty^{it})(g) \\ &= \sum_{\alpha \in \mathbb{Q}^\times} \psi(\alpha x) \mathbf{v}_{m,\lambda}(\alpha a_{\mathrm{fin}}) \, |\alpha a_\infty|_\infty^{it}.\end{aligned}$$

The following proposition is a combination of Lemma 3.5 and the standard Fourier analysis on the real line:

PROPOSITION 3.6. *Let $f \in \mathcal{H}_1^{\mathbf{K}}$. Suppose that*

(1) *$I(f)$ is integrable, i.e.,*

$$I(f) \in \mathsf{L}^2(\mathbb{A}^\times)^{\mathbf{K}} \cap \mathsf{L}^1(\mathbb{A}^\times),$$

(2) *the Fourier transform of f is also integrable i.e.*

$$\int_{-\infty}^{+\infty} |(f, \theta_{m,\lambda,t})| \, \mathrm{d}t < +\infty,$$

for any $m \in \mathbb{N}$ and $\lambda \in \mathbf{M}$.

Then we have

$$f(g) = \sum_{\lambda \in \mathbf{M}} \sum_{m=1}^{\infty} \frac{1}{4\pi} \int_{-\infty}^{+\infty} (f, \theta_{m,\lambda,t}) \theta_{m,\lambda,t}(g) \, \mathrm{d}t \quad \text{a.e.},$$

where

$$(f, \theta_{m,\lambda,t}) = \int_{G(\mathbb{Q}) \backslash G(\mathbb{A}_\mathbb{Q})} f(g) \bar{\theta}_{m,\lambda,t}(g) \, \mathrm{d}g.$$

PROOF. For simplicity, we assume that $n_p = 0$ for all primes p. Let $I(f) = h \in \mathsf{L}^2(\mathbb{A}^\times)^{\mathbf{K}} \cap \mathsf{L}^1(\mathbb{A}^\times)$. It follows from the proof of Lemma 3.4 that

$$f(g) = \Theta(h)(g) = \sum_{\alpha \in \mathbb{Q}^\times} \psi(\alpha x) h(\alpha a).$$

Note that this infinite sum exists in both L^1 and L^2 sense. It is easy to check that

$$\int_{\mathbb{A}^\times} h(a) \mathbf{v}_m(a_{\mathrm{fin}}) |a_\infty|_\infty^{-it} \, \mathrm{d}a^\times = (f, \theta_{m,t}).$$

Write

$$h = \sum_m \mathbf{v}_m \otimes h_m,$$

where $h_m \in \mathsf{L}^2(\mathbb{R}_{>0}, \mathrm{d}a_\infty^\times)$. The first and the second assumptions imply that h_m and the Fourier transform of $\widehat{h}_m$ both are integrable. Hence the inverse formula of Fourier transformation on the real line implies that

$$h(a) = \sum_m \frac{1}{4\pi} \int_{-\infty}^{+\infty} (f, \theta_{m,t}) \mathbf{v}_m(a_{\mathrm{fin}}) |a_\infty|_\infty^{it} \, \mathrm{d}t \quad \text{a.e.}.$$

Apply Θ to both sides, and our assertion follows. □

We recall some results regarding Igusa integrals with rapidly oscillating phase, studied in [**12**]:

PROPOSITION 3.7. *Let p be a finite place of $\mathbb{Q}$ and $d, e \in \mathbb{Z}$. Let*

$$\Phi : \mathbb{Q}_p^2 \times \mathbb{C}^2 \to \mathbb{C},$$

be a function such that for each $(x, y) \in \mathbb{Q}_p^2$, $\Phi((x, y), \mathbf{s})$ is holomorphic in $\mathbf{s} = (s_1, s_2) \in \mathbb{C}^2$. Assume that the function $(x, y) \mapsto \Phi(x, y, \mathbf{s})$ belongs to a bounded subset of the space of smooth compactly supported functions when $\Re(\mathbf{s})$ belongs to a fixed compact subset of $\mathbb{R}^2$. Let Λ be the interior of a closed convex cone generated by

$$(1, 0), (0, 1), (d, e).$$

Then, for any $\alpha \in \mathbb{Q}_p^\times$,

$$\eta_\alpha(\mathbf{s}) = \int_{\mathbb{Q}_p^2} |x|_p^{s_1} |y|_p^{s_2} \psi_p(\alpha x^d y^e)\Phi(x, y, \mathbf{s}) \mathrm{d}x_p^\times \mathrm{d}y_p^\times,$$

is holomorphic on T_Λ. The same argument holds for the infinite place when Φ is a smooth function with compact supports.

PROOF. For the infinite place, use integration by parts and apply the convexity principle. For finite places, assume that d, e are both negative. Let $\delta(x, y) = 1$ if $|x|_p = |y|_p = 1$ and 0 else. Then we have

$$\begin{aligned}\eta_\alpha(\mathbf{s}) &= \sum_{n,m\in\mathbb{Z}} \int_{\mathbb{Q}_p^2} |x|_p^{s_1} |y|_p^{s_2} \psi_p(\alpha x^d y^e)\Phi(x, y, \mathbf{s})\delta(p^{-n}x, p^{-m}y)\, \mathrm{d}x_p^\times \mathrm{d}y_p^\times \\ &= \sum_{n,m\in\mathbb{Z}} p^{-(ns_1+ms_2)} \cdot \eta_{\alpha,n,m}(\mathbf{s}),\end{aligned}$$

where

$$\eta_{\alpha,n,m}(\mathbf{s}) = \int_{|x|_p=|y|_p=1} \psi_p(\alpha p^{nd+me} x^d y^e)\Phi(p^n x, p^m y, \mathbf{s})\, \mathrm{d}x_p^\times \mathrm{d}y_p^\times.$$

Fix a compact subset of $\mathbb{C}^2$ and assume that $\Re(\mathbf{s})$ is in that compact set. The assumptions in our proposition mean that the support of $\Phi(\cdot, \mathbf{s})$ is contained in a fixed compact set in $\mathbb{Q}_p^2$, so there exists an integer N_0 such that $\eta_{\alpha,n,m}(\mathbf{s}) = 0$ if $n < N_0$ or $m < N_0$. Moreover, our assumptions imply that there exists a positive real number δ such that $\Phi(\cdot, \mathbf{s})$ is constant on any ball of radius δ in $\mathbb{Q}_p^2$. This implies that if $1/p^n < \delta$, then for any $u \in \mathbb{Z}_p^\times$,

$$\begin{aligned}\eta_{\alpha,n,m}(\mathbf{s}) &= \int \psi_p(\alpha p^{nd+me} x^d y^e u^d)\Phi(p^n xu, p^m y, \mathbf{s})\, \mathrm{d}x_p^\times \mathrm{d}y_p^\times \\ &= \int \psi_p(\alpha p^{nd+me} x^d y^e u^d)\Phi(p^n x, p^m y, \mathbf{s})\, \mathrm{d}x_p^\times \mathrm{d}y_p^\times \\ &= \int\int_{\mathbb{Z}_p^\times} \psi_p(\alpha p^{nd+me} x^d y^e u^d)\, \mathrm{d}u^\times \Phi(p^n x, p^m y, \mathbf{s})\, \mathrm{d}x_p^\times \mathrm{d}y_p^\times,\end{aligned}$$

and the last integral is zero if n is sufficiently large because of [**12**, Lemma 2.3.5]. Thus we conclude that there exists an integer N_1 such that $\eta_{\alpha,n,m}(\mathbf{s}) = 0$ if $n > N_1$

or $m > N_1$. Hence we obtained that

$$\eta_\alpha(\mathbf{s}) = \sum_{N_0 \leq n,m \leq N_1} p^{-(ns_1+ms_2)} \cdot \eta_{\alpha,n,m}(\mathbf{s}),$$

and this is holomorphic everywhere.

The case of $d < 0$ and $e = 0$ is treated similarly.

Next assume that $d < 0$ and $e > 0$. Then again we have a constant c such that $\eta_{\alpha,n,m}(\mathbf{s}) = 0$ if $1/p^n < \delta$ and $n|d| - me > c$. We may assume that c is sufficiently large so that the first condition is unnecessary. Then we have

$$\begin{aligned}|\eta_\alpha(\mathbf{s})| &\leq \sum_{N_0 \leq n} \sum_{m} p^{-\frac{n}{e}(e\Re(s_1)+|d|\Re(s_2))} \cdot p^{\frac{(n|d|-me)}{e}\Re(s_2)} \cdot |\eta_{\alpha,n,m}(\mathbf{s})| \\ &\leq \sum_{N_0 \leq n} p^{-\frac{n}{e}(e\Re(s_1)+|d|\Re(s_2))} \cdot \frac{p^{\frac{c}{e}\Re(s_2)}}{1 - p^{-\frac{\Re(s_2)}{e}}}\end{aligned}$$

Thus $\eta_\alpha(\mathbf{s})$ is holomorphic on T_Λ. □

From the proof of Proposition 3.7, we can claim more for finite places:

PROPOSITION 3.8. *Let $\epsilon > 0$ be any small positive real number. Fix a compact subset K of Λ, and assume that $\Re(\mathbf{s})$ is in K. Define:*

$$\kappa(K) := \begin{cases} \max\left\{0, -\frac{\Re(s_1)}{|d|}, -\frac{\Re(s_2)}{|e|}\right\} & \text{if } d < 0 \text{ and } e < 0, \\ \max\left\{0, -\frac{\Re(s_1)}{|d|}\right\} & \text{if } d < 0 \text{ and } e \geq 0,. \end{cases}$$

Then we have

$$|\eta_\alpha(\mathbf{s})| \ll 1/|\alpha|_p^{\kappa(K)+\epsilon}$$

as $|\alpha|_p \to 0$.

PROOF. Let $|\alpha|_p = p^{-k}$, and assume that both d, e are negative. By changing variables, if necessary, we may assume that N_0 in the proof of Proposition 3.7 is zero. If k is sufficiently large, then one can prove that there exists a constant c such that $\eta_{\alpha,n,m}(\mathbf{s}) = 0$ if $n|d| + m|e| > k + c$. Also it is easy to see that

$$|p^{-(ns_1+ms_2)}| \leq p^{(n|d|+m|e|)\kappa(K)}.$$

Hence we can conclude that

$$|\eta_\alpha(\mathbf{s})| \ll k^2 1/|\alpha|_p^{\kappa(K)} \ll 1/|\alpha|_p^{\kappa(K)+\epsilon}.$$

The case of $d < 0$ and $e = 0$ is treated similarly.

Assume that $d < 0$ and $e > 0$. Then we have a constant c such that $\eta_{\alpha,n,m}(\mathbf{s}) = 0$ if $n|d| - me > k + c$. Thus we can conclude that

$$\begin{aligned}|\eta_\alpha(\mathbf{s})| &\leq \sum_{m \geq 0} \sum_{n \geq 0} p^{-(n\Re(s_1)+m\Re(s_2))} |\eta_{\alpha,n,m}(\mathbf{s})| \\ &\ll \sum_{m \geq 0} p^{-m\Re(s_2)} (me + k) p^{(me+k)\kappa(K,s_2)} \\ &\ll k 1/|\alpha|_p^{\kappa(K,s_2)} \sum_{m \geq 0} (m+1) p^{-m(\Re(s_2) - e\kappa(K,s_2))} \\ &\ll 1/|\alpha|_p^{\kappa(K,s_2)+\epsilon}.\end{aligned}$$

where

$$\kappa(K,s_2)=\max\left\{0,-\frac{\Re(s_1)}{|d|}:(\Re(s_1),\Re(s_2))\in K\right\}.$$

Thus we can conclude that

$$|\eta_\alpha(\mathbf{s})|\ll 1/|\alpha|_p^{\kappa(K,s_2)+\epsilon}\ll 1/|\alpha|_p^{\kappa(K)+\epsilon}.$$

□

4. The projective plane

In this section, we implement the program described in Section 2 for the simplest equivariant compactifications of $G=G_1=\mathbb{G}_a\rtimes\mathbb{G}_m$, namely, the projective plane $\mathbb{P}^2$, for a *one-sided*, right, action of G given by

$$G\ni(x,a)\mapsto[x_0:x_1:x_2]=(a:a^{-1}x:1)\in\mathbb{P}^2.$$

The boundary consists of two lines, D_0 and D_2 given by the vanishing of x_0 and x_2. We will use the following identities:

$$\begin{aligned}\mathrm{div}(a)&=D_0-D_2,\\ \mathrm{div}(x)&=D_0+D_1-2D_2,\\ \mathrm{div}(\omega)&=-3D_2,\end{aligned}$$

where D_1 is given by the vanishing of x_1 and ω is the right invariant top degree form. The height functions are given by

$$\begin{aligned}\mathsf{H}_{D_0,p}(a,x)&=\frac{\max\{|a|_p,|a^{-1}x|_p,1\}}{|a|_p}, & \mathsf{H}_{D_2,p}(a,x)&=\max\{|a|_p,|a^{-1}x|_p,1\},\\ \mathsf{H}_{D_0,\infty}(a,x)&=\frac{\sqrt{|a|^2+|a^{-1}x|^2+1}}{|a|}, & \mathsf{H}_{D_2,p}(a,x)&=\sqrt{|a|^2+|a^{-1}x|^2+1},\\ \mathsf{H}_{D_0}&=\prod_p\mathsf{H}_{D_0,p}\times\mathsf{H}_{D_0,\infty}, & \mathsf{H}_{D_2}&=\prod_p\mathsf{H}_{D_2,p}\times\mathsf{H}_{D_2,\infty},\end{aligned}$$

and the height pairing by

$$\mathsf{H}(\mathbf{s},g)=\mathsf{H}_{D_0}^{s_0}(g)\mathsf{H}_{D_2}^{s_2}(g),$$

for $\mathbf{s}=s_0D_0+s_2D_2$ and $g\in G(\mathbb{A})$. The height zeta function takes the form

$$\mathsf{Z}(\mathbf{s},g)=\sum_{\gamma\in G(\mathbb{Q})}\mathsf{H}(\mathbf{s},\gamma g)^{-1}.$$

The proof of Northcott's theorem shows that the number of points of height $\le\mathsf{B}$ grows at most polynomially in B, Consequently, the Dirichlet series $\mathsf{Z}(\mathbf{s},g)$ converges absolutely and normally to a holomorphic function, for $\Re(\mathbf{s})$ is sufficiently large, which is continuous in $g\in G(\mathbb{A})$. Moreover, if $\Re(\mathbf{s})$ is sufficiently large, then

$$\mathsf{Z}(\mathbf{s},g)\in\mathsf{L}^2(G(\mathbb{Q})\backslash G(\mathbb{A}))\cap\mathsf{L}^1(G(\mathbb{Q})\backslash G(\mathbb{A})),$$

(see Lemma 5.2 for a proof). According to Proposition 3.1, we have the following decomposition:

$$\mathsf{L}^2(G(\mathbb{Q})\backslash G(\mathbb{A}))=\mathsf{L}^2(\mathbb{G}_m(\mathbb{Q})\backslash\mathbb{G}_m(\mathbb{A}))\oplus\pi,$$

and we can write

$$\mathsf{Z}(\mathbf{s},g)=\mathsf{Z}_0(\mathbf{s},g)+\mathsf{Z}_1(\mathbf{s},g).$$

The analysis of $\mathsf{Z}_0(\mathbf{s}, \mathrm{id})$ is a special case of our considerations in Section 2, in particular Theorem 2.1 (for further details, see [**4**] and [**13**]). The conclusion here is that there exist a $\delta > 0$ and a function h which is holomorphic on the tube domain $\mathsf{T}_{>3-\delta}$ such that

$$\mathsf{Z}_0(\mathbf{s}, \mathrm{id}) = \frac{\mathsf{h}(s_0 + s_2)}{(s_0 + s_2 - 3)}.$$

The analysis of $\mathsf{Z}_1(\mathbf{s}, \mathrm{id})$, i.e., of the contribution from the unique infinite-dimensional representation occurring in $\mathsf{L}^2(G(\mathbb{Q})\backslash G(\mathbb{A}))$, is the main part of this section. Define

$$\mathbf{K} = \prod_p \mathbf{K}_p \cdot \mathbf{K}_\infty = \prod_p G(\mathbb{Z}_p) \cdot \{(0, \pm 1)\}.$$

Since the height functions are $\mathbf{K}$-invariant,

$$\mathsf{Z}_1(\mathbf{s}, g) \in \pi^{\mathbf{K}} \simeq \mathsf{L}^2(\mathbb{A}^\times)^{\mathbf{K}}.$$

Lemma 3.5 provides a choice of an orthonormal basis for $\mathsf{L}^2(\mathbb{A}_{\mathrm{fin}}^\times)$. Combining with the Fourier expansion at the archimedean place, we obtain the following spectral expansion of Z_1:

LEMMA 4.1. *Assume that* $\Re(\mathbf{s})$ *is sufficiently large. Then*

$$\mathsf{Z}_1(\mathbf{s}, g) = \sum_{m \geq 1} \frac{1}{4\pi} \int_{-\infty}^{\infty} (\mathsf{Z}(\mathbf{s}, g), \theta_{m,t}(g)) \theta_{m,t}(g)\, dt,$$

where $\theta_{m,t}(g) = \Theta(\mathbf{v}_m \otimes |\cdot|^{it})(g)$.

PROOF. We use Proposition 3.6. To check the validity of assumptions of that proposition, in particular, the integrability in t, we invoke Lemma 5.3. □

It is easy to see that

$$\begin{aligned}
(\mathsf{Z}(\mathbf{s}, g), \theta_{m,t}(g)) &= \int_{G(\mathbb{Q})\backslash G(\mathbb{A})} \mathsf{Z}(\mathbf{s}, g) \bar{\theta}_{m,t}(g)\, dg \\
&= \int_{G(\mathbb{A})} \mathsf{H}(\mathbf{s}, g)^{-1} \bar{\theta}_{m,t}(g)\, dg \\
&= \sum_{\alpha \in \mathbb{Q}^\times} \int_{G(\mathbb{A})} \mathsf{H}(\mathbf{s}, g)^{-1} \bar{\psi}(\alpha x) \mathbf{v}_m(\alpha a_{\mathrm{fin}}) |\alpha a_\infty|^{-it}\, dg \\
&= \sum_{\alpha \in \mathbb{Q}^\times} \prod_p \mathsf{H}'_p(\mathbf{s}, m, \alpha) \cdot \mathsf{H}'_\infty(\mathbf{s}, t, \alpha),
\end{aligned}$$

where

$$\mathsf{H}'_p(\mathbf{s}, m, \alpha) = \int_{G(\mathbb{Q}_p)} \mathsf{H}_p(\mathbf{s}, g_p)^{-1} \bar{\psi}_p(\alpha x_p) \mathbf{1}_{m\mathbb{Z}_p^\times}(\alpha a_p)\, dg_p,$$

$$\mathsf{H}'_\infty(\mathbf{s}, t, \alpha) = \int_{G(\mathbb{R})} \mathsf{H}_\infty(\mathbf{s}, g_\infty)^{-1} \bar{\psi}_\infty(\alpha x_\infty) |\alpha a_\infty|^{-it}\, dg_\infty.$$

Note that $\theta_{m,t}(\mathrm{id}) = 2|m|^{it}$. Hence we can conclude that

$$\mathsf{Z}_1(\mathbf{s}, \mathrm{id}) = \sum_{\alpha\in\mathbb{Q}^\times} \sum_{m=1}^{\infty} \frac{1}{2\pi} \int_{-\infty}^{\infty} \prod_p \mathsf{H}'_p(\mathbf{s}, m, \alpha) \cdot \mathsf{H}'_\infty(\mathbf{s}, t, \alpha)|m|^{it}\, \mathrm{d}t$$

$$= \sum_{\alpha\in\mathbb{Q}^\times} \sum_{m=1}^{\infty} \frac{1}{2\pi} \int_{-\infty}^{\infty} \prod_p \int_{G(\mathbb{Q}_p)} \mathsf{H}_p(\mathbf{s}, g_p)^{-1} \bar{\psi}_p(\alpha x_p) \mathbf{1}_{m\mathbb{Z}_p^\times}(\alpha a_p)|\alpha a|_p^{-it} \mathrm{d}g_p \cdot \mathsf{H}'_\infty(\mathbf{s}, t, \alpha)\, \mathrm{d}t$$

$$= \sum_{\alpha\in\mathbb{Q}^\times} \frac{1}{2\pi} \int_{-\infty}^{\infty} \prod_p \widehat{\mathsf{H}}_p(\mathbf{s}, \alpha, t) \cdot \widehat{\mathsf{H}}_\infty(\mathbf{s}, \alpha, t)\, \mathrm{d}t,$$

where

$$\widehat{\mathsf{H}}_p(\mathbf{s}, \alpha, t) = \int_{G(\mathbb{Q}_p)} \mathsf{H}_p(\mathbf{s}, g_p)^{-1} \bar{\psi}_p(\alpha x_p) \mathbf{1}_{\mathbb{Z}_p}(\alpha a_p)|a_p|_p^{-it}\, \mathrm{d}g_p,$$

$$\widehat{\mathsf{H}}_\infty(\mathbf{s}, \alpha, t) = \int_{G(\mathbb{R})} \mathsf{H}_\infty(\mathbf{s}, g_\infty)^{-1} \bar{\psi}_\infty(\alpha x_\infty)|a_\infty|^{-it}\, \mathrm{d}g_\infty.$$

Note that the summation over m absorbed into the Euler product, see Proposition 5.4. It is clear that

$$\widehat{\mathsf{H}}_p(\mathbf{s}, \alpha, t) = \widehat{\mathsf{H}}_p((s_0 - it, s_2 + it), \alpha, 0),$$

so we only need to study $\widehat{\mathsf{H}}_p(\mathbf{s}, \alpha) = \widehat{\mathsf{H}}_p(\mathbf{s}, \alpha, 0)$. To do this, we introduce some notation. We have the canonical integral model of $\mathbb{P}^2$ over $\mathrm{Spec}(\mathbb{Z})$, and for any prime p, we have the reduction map modulo p:

$$\rho : G(\mathbb{Q}_p) \subset \mathbb{P}^2(\mathbb{Q}_p) = \mathbb{P}^2(\mathbb{Z}_p) \to \mathbb{P}^2(\mathbb{F}_p)$$

This is a continuous map from $G(\mathbb{Q}_p)$ to $\mathbb{P}^2(\mathbb{F}_p)$. Consider the following open sets:

$$\begin{aligned} U_\emptyset &= \rho^{-1}(\mathbb{P}^2 \setminus (D_0 \cup D_2)) = \{|a|_p = 1, |x|_p \leq 1\} \\ U_{D_0} &= \rho^{-1}(D_0 \setminus (D_0 \cap D_2)) = \{|a|_p < 1, |a^{-1}x|_p \leq 1\} \\ U_{D_2} &= \rho^{-1}(D_2 \setminus (D_0 \cap D_2)) = \{|a|_p > 1, |a^{-2}x|_p \leq 1\} \\ U_{D_0,D_2} &= \rho^{-1}(D_0 \cap D_2) = \{|a^{-1}x|_p > 1, |a^{-2}x|_p > 1\}. \end{aligned}$$

The height functions have a partial left invariance, i.e., they are invariant under the left action of the compact subgroup $\{(0, b) \mid b \in \mathbb{Z}_p^\times\}$. This implies that

$$\widehat{\mathsf{H}}_p(\mathbf{s}, \alpha) = \int_{G(\mathbb{Q}_p)} \mathsf{H}_p(\mathbf{s}, g)^{-1} \int_{\mathbb{Z}_p^\times} \bar{\psi}_p(\alpha b x) \mathrm{d}b^\times \mathbf{1}_{\mathbb{Z}_p}(\alpha a) \mathrm{d}g.$$

We record the following useful lemma (see, e.g., [**11**, Lemma 10.3] for a similar integral with respect to the *additive* measure):

LEMMA 4.2.

$$\int_{\mathbb{Z}_p^\times} \bar{\psi}_p(bx)\, \mathrm{d}b^\times = \begin{cases} 1 & \text{if } |x|_p \leq 1, \\ -\frac{1}{p-1} & \text{if } |x|_p = p, \\ 0 & \text{otherwise.} \end{cases}$$

LEMMA 4.3. *Assume that* $|\alpha|_p = 1$. *Then*

$$\widehat{\mathsf{H}}_p(\mathbf{s},\alpha) = \frac{\zeta_p(s_0+1)\zeta_p(2s_0+s_2)}{\zeta_p(s_0+s_2)}.$$

PROOF. We apply Lemma 4.2 and obtain

$$\widehat{\mathsf{H}}_p(\mathbf{s},\alpha) = \int_{U_\emptyset} + \int_{U_{D_0}} + \int_{U_{D_0,D_2}}$$

$$= 1 + \frac{p^{-(s_0+1)}}{1-p^{-(s_0+1)}} + \frac{p^{-(2s_0+s_2)} - p^{-(s_0+s_2)}}{(1-p^{-(s_0+1)})(1-p^{-(2s_0+s_2)})}$$

$$= \frac{\zeta_p(s_0+1)\zeta_p(2s_0+s_2)}{\zeta_p(s_0+s_2)}.$$

□

LEMMA 4.4. *Assume that* $|\alpha|_p > 1$. *Let* $|\alpha|_p = p^k$. *Then*

$$\widehat{\mathsf{H}}_p(\mathbf{s},\alpha) = p^{-k(s_0+1)}\widehat{\mathsf{H}}_p(\mathbf{s},1).$$

PROOF. Using Lemma 4.3 we obtain that

$$\begin{aligned}\widehat{\mathsf{H}}_p(\mathbf{s},\alpha) &= \int_{U_{D_0}} + \int_{U_{D_0,D_2}} \\ &= \frac{p^{-k(s_0+1)}}{1-p^{-(s_0+1)}} + p^{-k(s_0+1)}\frac{p^{-(2s_0+s_2)} - p^{-(s_0+s_2)}}{(1-p^{-(s_0+1)})(1-p^{-(2s_0+s_2)})} \\ &= p^{-k(s_0+1)}\widehat{\mathsf{H}}_p(\mathbf{s},1).\end{aligned}$$

□

LEMMA 4.5. *Assume that* $|\alpha|_p < 1$. *Let* $|\alpha|_p = p^{-k}$. *Then* $\widehat{\mathsf{H}}_p(\mathbf{s},\alpha)$ *is holomorphic on the tube domain* $\mathsf{T}_\Lambda = \Lambda + i\mathbb{R}^2$ *over the cone*

$$\Lambda = \{s_0 > -1, s_0+s_2 > 0, 2s_0+s_2 > 0\}.$$

Moreover, for any compact subset of Λ, *there exists a constant* $C > 0$ *such that*

$$|\widehat{\mathsf{H}}_p(\mathbf{s},\alpha)| \le Ck\max\{1, p^{-\frac{k}{2}\Re(s_2-2)}\}$$

for any $\mathbf{s}$ *with real part in this compact set.*

PROOF. It is easy to see that

$$\widehat{\mathsf{H}}_p(\mathbf{s},\alpha) = \int_{U_\emptyset} + \int_{U_{D_0}} + \int_{U_{D_2}} + \int_{U_{D_0,D_2}}$$

$$= 1 + \frac{p^{-(s_0+1)}}{1-p^{-(s_0+1)}} + \int_{U_{D_2}} + \int_{U_{D_0,D_2}}.$$

On U_{D_2}, we choose $(1 : x_1 : x_2)$ as coordinates, then we have

$$\int_{U_{D_2}} = \int_{|x_1|_p \leq 1, |x_2|_p < 1} |x_2|_p^{s_2-2} \int_{\mathbb{Z}_p^\times} \bar{\psi}_p(\alpha b \frac{x_1}{x_2^2}) \mathrm{d}b^\times \mathbf{1}_{\mathbb{Z}_p}(\alpha x_2^{-1}) \mathrm{d}x_1 \, \mathrm{d}x_2^\times$$

$$= \int_{p^{-\frac{k}{2}} \leq |x_2|_p < 1} |x_2|_p^{s_2-2} \mathrm{d}x_2^\times$$

Hence this integral is holomorphic everywhere, and we have

$$\left| \int_{U_{D_2}} \right| < k \max\{1, p^{-\frac{k}{2}\Re(s_2-2)}\}.$$

On U_{D_0,D_2}, we choose $(x_0 : 1 : x_2)$ as coordinates and obtain

$$\frac{1}{1-p^{-1}} \int_{U_{D_0,D_2}} = \int_{|x_0|_p, |x_2|_p < 1} |x_0|_p^{s_0+1} |x_2|_p^{s_2-2} \int_{\mathbb{Z}_p^\times} \bar{\psi}_p(\alpha b \frac{x_0}{x_2^2}) \mathrm{d}b^\times \mathrm{d}x_0^\times \, \mathrm{d}x_2^\times$$

$$= -\frac{p^{-1}}{1-p^{-1}} p^{(k+1)(s_0+1)} \frac{p^{-(1-[-\frac{k}{2}])(2s_0+s_2)}}{1-p^{-(2s_0+s_2)}} + \int |x_0|_p^{s_0+1} |x_2|_p^{s_2-2} \mathrm{d}x_0^\times \, \mathrm{d}x_2^\times,$$

where the last integral is over

$$\{|x_0|_p < 1, |x_2|_p < 1, |\alpha x_0|_p \leq |x_2|_p^2\},$$

and is equal to

$$\sum_{\substack{j,l \geq 1 \\ 2j-l \leq k}} p^{-l(s_0+1)-j(s_2-2)}$$

$$= \sum_{j \leq [\frac{k}{2}]} p^{-j(s_2-2)} \frac{p^{-(s_0+1)}}{1-p^{-(s_0+1)}} + \sum_{j > [\frac{k}{2}]} p^{-j(s_2-2)} \frac{p^{-(2j-k)(s_0+1)}}{1-p^{-(s_{D_0}+1)}}$$

$$= \sum_{j \leq [\frac{k}{2}]} p^{-j(s_2-2)} \frac{p^{-(s_0+1)}}{1-p^{-(s_0+1)}} + \frac{p^{-([\frac{k}{2}]+1)(2s_0+s_2)}}{1-p^{-(2s_0+s_2)}} \frac{p^{k(s_0+1)}}{1-p^{-(s_0+1)}}.$$

From this we can see that $\int_{U_{D_0,D_2}}$ is holomorphic on T_Λ and that for any compact subset of Λ, we can find a constant $C > 0$ such that

$$\left| \int_{U_{D_0,D_2}} \right| < Ck \max\{1, p^{-\frac{k}{2}\Re(s_2-2)}\},$$

for $\Re(\mathbf{s})$ in this compact. $\square$

Next, we study the local integral at the real place. Again,

$$\widehat{\mathsf{H}}_\infty(\mathbf{s}, \alpha, t) = \widehat{\mathsf{H}}_\infty((s_0 - it, s_2 + it), \alpha, 0),$$

and we start with

$$\widehat{\mathsf{H}}_\infty(\mathbf{s}, \alpha) = \widehat{\mathsf{H}}_\infty(\mathbf{s}, \alpha, 0).$$

LEMMA 4.6. *The function*

$$\mathbf{s} \mapsto \widehat{\mathsf{H}}_\infty(\mathbf{s}, \alpha),$$

is holomorphic on the tube domain $\mathsf{T}_{\Lambda'}$ *over*

$$\Lambda' = \{s_0 > -1, s_2 > 0, 2s_0 + s_2 > 0\}.$$

Moreover, for any $r \in \mathbb{N}$ *and any compact subset of*

$$\Lambda'_r = \{s_0 > -1 + r, s_2 > 0\},$$

there exists a constant $C > 0$ *such that*

$$|\widehat{\mathsf{H}}_\infty(\mathbf{s}, \alpha)| < \frac{C}{|\alpha|_\infty^r},$$

for any $\mathbf{s}$ *in the tube domain over this compact.*

PROOF. Let $U_\emptyset = X(\mathbb{R}) \setminus (D_0 \cup D_2)$, U_{D_i} be a small tubular neighborhood of D_i minus $D_0 \cap D_2$, and U_{D_0,D_2} be a small neighborhood of $D_0 \cap D_2$. Then $\{U_\emptyset, U_{D_0}, U_{D_2}, U_{D_0,D_2}\}$ is an open covering of $X(\mathbb{R})$, and consider the partition of unity for this covering; $\theta_\emptyset, \theta_{D_0}, \theta_{D_2}, \theta_{D_0,D_2}$. Then we have

$$\widehat{\mathsf{H}}_\infty(\mathbf{s}, \alpha) = \int_{U_\emptyset} \mathsf{H}_\infty^{-1} \bar{\psi}_\infty(\alpha x_\infty) \theta_\emptyset \mathrm{d}g_\infty + \int_{U_{D_0}} + \int_{U_{D_2}} + \int_{U_{D_0,D_2}}.$$

On U_{D_0,D_2}, we choose $(x_0 : 1 : x_2)$ as analytic coordinates and obtain

$$\int_{U_{D_0,D_2}} = \int_{\mathbb{R}^2} |x_0|^{s_0+1} |x_2|^{s_2-2} \bar{\psi}(\alpha \frac{x_0}{x_2^2}) \phi(\mathbf{s}, x_0, x_2) \mathrm{d}x_0^\times dx_2^\times,$$

where ϕ is a smooth bounded function with compact support. Such oscillatory integrals have been studied in [**12**], in our case the integral is holomorphic if $\Re(s_0) > -1$ and $\Re(2s_0 + s_2) > 0$. Assume that $\Re(\mathbf{s})$ is sufficiently large. Integration by parts implies that

$$\int_{U_{D_0,D_2}} = \frac{1}{\alpha^r} \int_{\mathbb{R}^2} |x_0|^{s_0+1-r} |x_2|^{s_2-2+2r} \bar{\psi}(\alpha \frac{x_0}{x_2^2}) \phi'(\mathbf{s}, x_0, x_2) dx_0^\times \mathrm{d}x_2^\times,$$

and this integral is holomorphic if $\Re(s_0) > -1 + r$ and $\Re(s_2) > 2 - 2r$. Thus, our second assertion follows. The other integrals are studied similarly. □

LEMMA 4.7. *For any compact set* $K \subset \Lambda'_2$, *there exists a constant* $C > 0$ *such that*

$$|\widehat{\mathsf{H}}_\infty(\mathbf{s}, \alpha, t)| < \frac{C}{|\alpha|^2(1 + t^2)},$$

for any $\mathbf{s} \in \mathsf{T}_K$.

PROOF. Consider a left invariant differential operator $\partial_a = a\partial/\partial a$. Integrating by parts we obtain that

$$\widehat{\mathsf{H}}_\infty(\mathbf{s}, \alpha, t) = -\frac{1}{t^2} \int_{G(\mathbb{R})} \partial_a^2 \mathsf{H}_\infty(\mathbf{s}, g_\infty)^{-1} \bar{\psi}_\infty(\alpha x_\infty) |a_\infty|^{-it} \mathrm{d}g_\infty,$$

According to [**11**],

$$\partial_a^2 \mathsf{H}_\infty(\mathbf{s}, g_\infty)^{-1} = \mathsf{H}_\infty(\mathbf{s}, g_\infty)^{-1} \times \text{(a bounded smooth function)},$$

so we can apply the discussion of the previous proposition. □

LEMMA 4.8. *The Euler product*

$$\prod_p \widehat{\mathsf{H}}_p(\mathbf{s},\alpha,t)\cdot\widehat{\mathsf{H}}_\infty(\mathbf{s},\alpha,t),$$

is holomorphic on the tube domain T_Ω *over*

$$\Omega=\{s_0>0,s_2>0,2s_0+s_2>1\}.$$

Moreover, let $\alpha=\frac{\beta}{\gamma}$, *where* $\gcd(\beta,\gamma)=1$. *Then for any* $\epsilon>0$ *and any compact set*

$$K\subset\Omega'=\{s_0>1,s_2>0,2s_0+s_2>1\},$$

there exists a constant $C>0$ *such that*

$$|\prod_p \widehat{\mathsf{H}}_p(\mathbf{s},\alpha,t)\cdot\widehat{\mathsf{H}}_\infty(\mathbf{s},\alpha,t)|<C\cdot\frac{\max\{1,\sqrt{|\beta|}^{-\Re(s_2-2)}\}}{|\beta|^{2-\epsilon}|\gamma|^{\Re(s_0-1)}},$$

for all $\mathbf{s}\in\mathsf{T}_K$.

THEOREM 4.9. *There exists* $\delta>0$ *such that* $\mathsf{Z}_1(s_0+s_2,\mathrm{id})$ *is holomorphic on* $\mathsf{T}_{>3-\delta}$.

PROOF. Let $\delta>0$ be a sufficiently small real number, and define

$$\Lambda=\{s_0>2+\delta,s_2>1-2\delta\}.$$

It follows from the previous proposition that for any $\epsilon>0$ and any compact set $K\subset\Lambda$, there exists a constant $C>0$ such that

$$|\prod_p \widehat{\mathsf{H}}_p(\mathbf{s},\alpha,t)\cdot\widehat{\mathsf{H}}_\infty(\mathbf{s},\alpha,t)|<\frac{C}{(1+t^2)|\beta|^{\frac{3}{2}-\epsilon-\delta}|\gamma|^{1+\delta}}.$$

From this inequality, we can conclude that the integral

$$\int_{-\infty}^{\infty}\prod_p \widehat{\mathsf{H}}_p(\mathbf{s},\alpha,t)\cdot\widehat{\mathsf{H}}_\infty(\mathbf{s},\alpha,t)dt,$$

converges uniformly and absolutely to a holomorphic function on T_K. Furthermore, we have

$$\left|\int_{-\infty}^{\infty}\prod_p \widehat{\mathsf{H}}_p(\mathbf{s},\alpha,t)\cdot\widehat{\mathsf{H}}_\infty(\mathbf{s},\alpha,t)dt\right|<\frac{C'}{|b|^{\frac{3}{2}-\epsilon-\delta}|c|^{1+\delta}}.$$

For sufficiently small $\epsilon>0$ and $\delta>0$, the sum

$$\sum_{\alpha\in\mathbb{Q}^\times}\frac{1}{2\pi}\int_{-\infty}^{\infty}\prod_p \widehat{\mathsf{H}}_p(\mathbf{s},\alpha,t)\cdot\widehat{\mathsf{H}}_\infty(\mathbf{s},\alpha,t)dt,$$

converges absolutely and uniformly to a function in s_0+s_2. This concludes the proof of our theorem. □

5. Geometrization

In this section we geometrize the method described in Section 4. Our main theorem is:

THEOREM 5.1. *Let X be a smooth projective equivariant compactification of $G = G_1$ over $\mathbb{Q}$, under the right action. Assume that the boundary divisor has strict normal crossings. Let $a, x \in \mathbb{Q}(X)$ be rational functions, where (x, a) are the standard coordinates on $G \subset X$. Let E be the Zariski closure of $\{x = 0\} \subset G$. Assume that:*

- *the union of the boundary and E is a divisor with strict normal crossings,*
- $\mathrm{div}(a)$ *is a reduced divisor, and*
- *for any pole D_ι of a, one has*

$$-\mathrm{ord}_{D_\iota}(x) > 1.$$

Then Manin's conjecture holds for X.

The remainder of this section is devoted to a proof of this fact. Blowing up the zero-dimensional subscheme

$$\mathrm{Supp}(\mathrm{div}_0(a)) \cap \mathrm{Supp}(\mathrm{div}_\infty(a)),$$

if necessary, we may assume that

$$\mathrm{Supp}(\mathrm{div}_0(a)) \cap \mathrm{Supp}(\mathrm{div}_\infty(a)) = \emptyset.$$

Here div_0 and div_∞ stand for the divisor of zeroes, respectively poles, of the rational function a on X. The local height functions are invariant under the right action of some compact subgroup $\mathbf{K}_p \subset G(\mathbb{Z}_p)$. Moreover, we can assume that $\mathbf{K}_p = G(p^{n_p}\mathbb{Z}_p)$, for some $n_p \in \mathbb{Z}_{\geq 0}$. Let S be the set of bad places for X; *a priori*, this set depends on a choice of an integral model for X and for the action of G. Specifically, we insist that for $p \notin S$, the reduction of X at p is smooth, the reduction of the boundary is a union of smooth geometrically irreducible divisors with normal crossings, and the action of G lifts to the integral models. In particular, we insist that $n_p = 0$, for all $p \notin S$. The proof works with S being any, sufficiently large, finite set. For simplicity, we assume that the height function at the infinite place is invariant under the action of $\mathbf{K}_\infty = \{(0, \pm 1)\}$.

LEMMA 5.2. *We have*

$$\mathsf{Z}(\mathbf{s}, g) \in \mathsf{L}^2(G(\mathbb{Q})\backslash G(\mathbb{A}))^{\mathbf{K}} \cap \mathsf{L}^1(G(\mathbb{Q})\backslash G(\mathbb{A})).$$

PROOF. First it is easy to see that

$$\begin{aligned}\int_{G(\mathbb{Q})\backslash G(\mathbb{A})} |\mathsf{Z}(\mathbf{s}, g)| \, dg &\leq \int_{G(\mathbb{Q})\backslash G(\mathbb{A})} \sum_{\gamma \in G(\mathbb{Q})} |\mathsf{H}(\mathbf{s}, \gamma g)|^{-1} \, dg \\ &= \int_{G(\mathbb{A})} \mathsf{H}(\Re(\mathbf{s}), g)^{-1} \, dg,\end{aligned}$$

and the last integral is bounded when $\Re(\mathbf{s})$ is sufficiently large. (See [**13**, Proposition 4.3.4].) Hence it follows that $\mathsf{Z}(\mathbf{s}, g)$ is integrable. To conclude that $\mathsf{Z}(\mathbf{s}, g)$ is square-integrable, we prove that $\mathsf{Z}(\mathbf{s}, g) \in \mathsf{L}^\infty$ for $\Re(\mathbf{s})$ sufficiently large. Let u, v be sufficiently large positive real numbers. Assume that $\Re(\mathbf{s})$ is in a fixed compact subset of $\mathrm{Pic}^G(X) \otimes \mathbb{R}$ and sufficiently large. Then we have

$$\mathsf{H}(\Re(\mathbf{s}), g)^{-1} \ll \mathsf{H}_1(a)^{-u} \cdot \mathsf{H}_2(x)^{-v}$$

where

$$\mathsf{H}_1(a) = \prod_p \max\{|a_p|_p, |a_p|_p^{-1}\} \cdot \sqrt{|a_\infty|_\infty^2 + |a_\infty|_\infty^{-2}}$$

$$\mathsf{H}_2(x) = \prod_p \max\{1, |x_p|_p\} \cdot \sqrt{1 + |x_\infty|_\infty^2}.$$

Since $\mathsf{Z}(\mathbf{s}, g)$ is $G(\mathbb{Q})$-periodic, we may assume that $|a_p|_p = 1$ where $g_p = (x_p, a_p)$. Then we obtain that

$$\sum_{\gamma \in G(\mathbb{Q})} \mathsf{H}(\Re(\mathbf{s}), \gamma g)^{-1} \ll \sum_{\alpha \in \mathbb{Q}^\times} \sum_{\beta \in \mathbb{Q}} \mathsf{H}_1(\alpha a)^{-u} \cdot \mathsf{H}_2(\alpha x + \beta)^{-v}$$
$$\leq \sum_{\alpha \in \mathbb{Q}^\times} \mathsf{H}_{1,\mathrm{fin}}(\alpha)^{-u} \mathsf{Z}_2(\alpha x),$$

where

$$\mathsf{Z}_2(x) = \sum_{\beta \in \mathbb{Q}} \mathsf{H}_2(x + \beta)^{-v}.$$

It is known that Z_2 is a bounded function for sufficiently large v, (see [**11**]) so we can conclude that $\mathsf{Z}(\mathbf{s}, g)$ is also a bounded function because

$$\sum_{\alpha \in \mathbb{Q}^\times} \mathsf{H}_{1,\mathrm{fin}}(\alpha)^{-u} < +\infty,$$

for sufficiently large u. □

By Proposition 3.1, the height zeta function decomposes as

$$\mathsf{Z}(\mathbf{s}, \mathrm{id}) = \mathsf{Z}_0(\mathbf{s}, \mathrm{id}) + \mathsf{Z}_1(\mathbf{s}, \mathrm{id}).$$

Analytic properties of $\mathsf{Z}_0(\mathbf{s}, \mathrm{id})$ were established in Section 2. It remains to show that $\mathsf{Z}_1(\mathbf{s}, \mathrm{id})$ is holomorphic on a tube domain over an open neighborhood of the shifted effective cone $-K_X + \Lambda_{\mathrm{eff}}(X)$. To conclude this, we use the spectral decomposition of Z_1:

LEMMA 5.3. *We have*

$$\mathsf{Z}_1(\mathbf{s}, \mathrm{id}) = \sum_{\lambda \in \mathbf{M}} \sum_{m=1}^{\infty} \frac{1}{4\pi} \int_{-\infty}^{+\infty} (\mathsf{Z}(\mathbf{s}, g), \theta_{m,\lambda,t}) \theta_{m,\lambda,t}(\mathrm{id})\, \mathrm{d}t.$$

PROOF. To apply Proposition 3.6, we need to check that Z_1 satisfies the assumptions of Proposition 3.6. The proof of Lemma 3.4 implies that

$$I(\mathsf{Z}_1) = \int_{N(\mathbb{Q}) \backslash N(\mathbb{A})} \mathsf{Z}(\mathbf{s}, g) \psi(x)\, \mathrm{d}x.$$

Thus we have

$$\begin{aligned}\int_{T(\mathbb{A})} |I(\mathsf{Z}_1)|\,\mathrm{d}a^\times &= \int_{T(\mathbb{A})} \left| \int_{N(\mathbb{Q})\backslash N(\mathbb{A})} \mathsf{Z}(\mathbf{s},g)\psi(x)\,\mathrm{d}x \right| \mathrm{d}a^\times \\ &\leq \sum_{\alpha\in\mathbb{Q}^\times} \int_{T(\mathbb{A})} \left| \int_{N(\mathbb{A})} \mathsf{H}(\mathbf{s},g)^{-1}\psi(\alpha x)\,\mathrm{d}x \right| \mathrm{d}a^\times \\ &= \sum_{\alpha\in\mathbb{Q}^\times} \prod_p \int_{T(\mathbb{Q}_p)} \left| \int_{N(\mathbb{Q}_p)} \mathsf{H}_p(\mathbf{s},g_p)^{-1}\psi_p(\alpha x_p)\,\mathrm{d}x_p \right| \mathrm{d}a_p^\times \\ &\qquad \times \int_{T(\mathbb{R})} \left| \int_{N(\mathbb{R})} \mathsf{H}_\infty(\mathbf{s},g_\infty)^{-1}\psi_\infty(\alpha x)\,\mathrm{d}x_\infty \right| \mathrm{d}a_\infty^\times .\end{aligned}$$

Assume that $p \notin S$. Since the height function is right $\mathbf{K}_p$-invariant, we obtain that for any $y_p \in \mathbb{Z}_p$,

$$\begin{aligned}\int_{N(\mathbb{Q}_p)} \mathsf{H}_p(\mathbf{s},g_p)^{-1}\psi_p(\alpha x_p)\,\mathrm{d}x_p &= \int_{N(\mathbb{Q}_p)} \mathsf{H}_p(\mathbf{s},(x_p+a_py_p,a_p))^{-1}\psi_p(\alpha x_p)\,\mathrm{d}x_p \\ &= \int_{N(\mathbb{Q}_p)} \mathsf{H}_p(\mathbf{s},g_p)^{-1}\psi_p(\alpha x_p) \int_{\mathbb{Z}_p} \bar{\psi}_p(\alpha a_p y_p)\,\mathrm{d}y_p\,\mathrm{d}x_p \\ &= 0 \qquad \text{if } |\alpha a_p|_p > 1.\end{aligned}$$

Hence we can conclude that

$$\int_{T(\mathbb{Q}_p)} \left| \int_{N(\mathbb{Q}_p)} \mathsf{H}_p(\mathbf{s},g_p)^{-1}\psi_p(\alpha x_p)\,\mathrm{d}x_p \right| \mathrm{d}a_p^\times \leq \int_{G(\mathbb{Q}_p)} \mathsf{H}_p(\Re(\mathbf{s}),g_p)^{-1}\mathbf{1}_{\mathbb{Z}_p}(\alpha a_p)\,\mathrm{d}g_p.$$

Similarly, for $p \in S$, we can conclude that

$$\int_{T(\mathbb{Q}_p)} \left| \int_{N(\mathbb{Q}_p)} \mathsf{H}_p(\mathbf{s},g_p)^{-1}\psi_p(\alpha x_p)\,\mathrm{d}x_p \right| \mathrm{d}a_p^\times \leq \int_{G(\mathbb{Q}_p)} \mathsf{H}_p(\Re(\mathbf{s}),g_p)^{-1}\mathbf{1}_{\frac{1}{N}\mathbb{Z}_p}(\alpha a_p)\,\mathrm{d}g_p.$$

Then the convergence of the following sum

$$\sum_{\alpha\in\mathbb{Q}^\times} \prod_p \int_{G(\mathbb{Q}_p)} \mathsf{H}_p^{-1}\mathbf{1}_{\frac{1}{N}\mathbb{Z}_p}(\alpha a_p)\,\mathrm{d}g_p \cdot \int_{T(\mathbb{R})} \left| \int_{N(\mathbb{R})} \mathsf{H}_\infty^{-1}\psi_\infty(\alpha x)\,\mathrm{d}x_\infty \right| \mathrm{d}a_\infty^\times,$$

can be verified from the detailed study of the local integrals which we will conduct later. See proofs of Lemmas 5.6, 5.9, and 5.10.

Next we need to check that

$$\int_{-\infty}^{+\infty} |(\mathsf{Z}(\mathbf{s},g),\theta_{m,\lambda,t})|\,\mathrm{d}t < +\infty.$$

It is easy to see that

$$\begin{aligned}(\mathsf{Z}(\mathbf{s},g),\theta_{m,\lambda,t}) &= \int_{G(\mathbb{Q})\backslash G(\mathbb{A}_\mathbb{Q})} \mathsf{Z}(\mathbf{s},g)\bar{\theta}_{m,\lambda,t}\,\mathrm{d}g \\ &= \int_{G(\mathbb{A}_\mathbb{Q})} \mathsf{H}(\mathbf{s},g)^{-1}\bar{\theta}_{m,\lambda,t}\,\mathrm{d}g \\ &= \sum_{\alpha\in\mathbb{Q}^\times}\int_{G(\mathbb{A}_\mathbb{Q})} \mathsf{H}(\mathbf{s},g)^{-1}\bar{\psi}(\alpha x)\bar{\mathbf{v}}_{m,\lambda}(\alpha a_{\mathrm{fin}})|\alpha a_\infty|_\infty^{-it}\,\mathrm{d}g \\ &= \sum_{\alpha\in\mathbb{Q}^\times}\prod_p \mathsf{H}'_p(\mathbf{s},m,\lambda,\alpha)\cdot\mathsf{H}'_\infty(\mathbf{s},t,\alpha),\end{aligned}$$

where $\mathsf{H}'_p(\mathbf{s},m,\lambda,\alpha)$ is given by

$$\begin{aligned} &= \int_{G(\mathbb{Q}_p)} \mathsf{H}_p(\mathbf{s},g_p)^{-1}\bar{\psi}_p(\alpha x_p)\mathbf{1}_{m\mathbb{Z}_p^\times}(\alpha a_p)\,\mathrm{d}g_p, && p\notin S \\ &= \int_{G(\mathbb{Q}_p)} \mathsf{H}_p(\mathbf{s},g_p)^{-1}\bar{\psi}_p(\alpha x_p)\bar{\lambda}_p(\alpha a_p/p^{v_p(\alpha a_p)})\mathbf{1}_{\frac{m}{N}\mathbb{Z}_p^\times}(\alpha a_p)\,\mathrm{d}g_p, && p\in S\end{aligned}$$

and

$$\mathsf{H}'_\infty(\mathbf{s},t,\alpha) = \int_{G(\mathbb{R})} \mathsf{H}_\infty(\mathbf{s},g_\infty)^{-1}\bar{\psi}_\infty(\alpha x_\infty)|\alpha a_\infty|_\infty^{-it}\,\mathrm{d}g_\infty.$$

The integrability follows from the proof of Lemma 5.9. Thus we can apply Proposition 3.6, and the identity in our statement follows from the continuity of $\mathsf{Z}(\mathbf{s},g)$. □

We obtained that

$$\begin{aligned}\mathsf{Z}_1(\mathbf{s},\mathrm{id}) &= \sum_{\lambda\in\mathbf{M}}\sum_{m=1}^{\infty}\frac{1}{4\pi}\int_{-\infty}^{+\infty}(\mathsf{Z}(\mathbf{s},g),\theta_{m,\lambda,t})\theta_{m,\lambda,t}(\mathrm{id})\,\mathrm{d}t \\ &= \sum_{\lambda\in\mathbf{M},\,\lambda(-1)=1}\sum_{m=1}^{\infty}\frac{1}{2\pi}\int_{-\infty}^{+\infty}(\mathsf{Z}(\mathbf{s},g),\theta_{m,\lambda,t})\prod_{p\in S}\lambda_p\left(\frac{m}{N}\cdot p^{-v_p(m/N)}\right)\left|\frac{m}{N}\right|_\infty^{it}\,\mathrm{d}t.\end{aligned}$$

We will use the following notation:

$$\begin{aligned}\lambda_S(\alpha a_p) &:= \prod_{q\in S}\bar{\lambda}_q(p^{v_p(\alpha a_p)}), && p\notin S \\ \lambda_{S,p}(\alpha a_p) &:= \bar{\lambda}_p\left(\frac{\alpha a_p}{p^{v_p(\alpha a_p)}}\right)\prod_{q\in S\backslash p}\bar{\lambda}_q(p^{v_p(\alpha a_p)}), && p\in S.\end{aligned}$$

PROPOSITION 5.4. *If $\Re(s)$ is sufficiently large, then*

$$\mathsf{Z}_1(\mathbf{s},id) = \sum_{\lambda\in\mathbf{M},\,\lambda(-1)=1}\sum_{\alpha\in\mathbb{Q}^\times}\frac{1}{2\pi}\int_{-\infty}^{+\infty}\prod_p\widehat{\mathsf{H}}_p(\mathbf{s},\lambda,t,\alpha)\cdot\widehat{\mathsf{H}}_\infty(\mathbf{s},t,\alpha)\,\mathrm{d}t,$$

where $\widehat{\mathsf{H}}_p(\mathbf{s},\lambda,t,\alpha)$ is given by

$$\begin{aligned}&\int_{G(\mathbb{Q}_p)}\mathsf{H}_p(\mathbf{s},g_p)^{-1}\bar{\psi}_p(\alpha x_p)\lambda_S(\alpha a_p)\mathbf{1}_{\mathbb{Z}_p}(\alpha a_p)|a_p|_p^{-it}\,\mathrm{d}g_p, && p\notin S \\ &\int_{G(\mathbb{Q}_p)}\mathsf{H}_p(\mathbf{s},g_p)^{-1}\bar{\psi}_p(\alpha x_p)\lambda_{S,p}(\alpha a_p)\mathbf{1}_{\frac{1}{N}\mathbb{Z}_p}(\alpha a_p)|a_p|_p^{-it}\,\mathrm{d}g_p, && p\in S\end{aligned}$$

and

$$\widehat{\mathsf{H}}_\infty(\mathbf{s},t,\alpha) = \int_{G(\mathbb{R})} \mathsf{H}_\infty(\mathbf{s},g_\infty)^{-1}\bar{\psi}_\infty(\alpha x_\infty)|a_\infty|_\infty^{-it}\,\mathrm{d}g_\infty$$

PROOF. For simplicity, we assume that $S=\emptyset$. We have seen that

$$\mathsf{Z}_1(\mathbf{s},\mathrm{id}) = \sum_{m=1}^{\infty}\sum_{\alpha\in\mathbb{Q}^\times}\frac{1}{2\pi}\int_{-\infty}^{+\infty}\prod_p \mathsf{H}'_p(\mathbf{s},m,\alpha)\cdot\mathsf{H}'_\infty(\mathbf{s},t,\alpha)|m|_\infty^{it}\,\mathrm{d}t.$$

On the other hand, it is easy to see that

$$\widehat{\mathsf{H}}_p(\mathbf{s},t,\alpha) = \sum_{j=0}^{\infty}\int_{G(\mathbb{Q}_p)}\mathsf{H}_p(\mathbf{s},g_p)^{-1}\bar{\psi}_p(\alpha x_p)\mathbf{1}_{p^j\mathbb{Z}_p^\times}(\alpha a_p)\left|\frac{p^j}{\alpha}\right|_p^{-it}\mathrm{d}g_p.$$

Hence we have the formal identity:

$$\prod_p\widehat{\mathsf{H}}_p(\mathbf{s},t,\alpha)\cdot\widehat{\mathsf{H}}_\infty(\mathbf{s},t,\alpha) = \sum_{m=1}^{\infty}\prod_p\mathsf{H}'_p(\mathbf{s},m,\alpha)\cdot\mathsf{H}'_\infty(\mathbf{s},t,\alpha)|m|_\infty^{it},$$

and our assertion follows from this. To justify the above identity, we need to address convergence issues; this will be discussed below (see the proof of Lemma 5.6). □

Thus we need to study the local integrals in Proposition 5.4. We introduce some notation:

$$\mathcal{I}_1 = \{\iota\in\mathcal{I}\,|\,D_\iota\subset\mathrm{Supp}(\mathrm{div}_0(a))\}$$
$$\mathcal{I}_2 = \{\iota\in\mathcal{I}\,|\,D_\iota\subset\mathrm{Supp}(\mathrm{div}_\infty(a))\}$$
$$\mathcal{I}_3 = \{\iota\in\mathcal{I}\,|\,D_\iota\not\subset\mathrm{Supp}(\mathrm{div}(a))\}.$$

Note that $\mathcal{I}=\mathcal{I}_1\sqcup\mathcal{I}_2\sqcup\mathcal{I}_3$ and $\mathcal{I}_1\neq\emptyset$. Also $D_\iota\subset\mathrm{Supp}(\mathrm{div}_\infty(x))$ for any $\iota\in\mathcal{I}_3$ because

$$D=\cup_{\iota\in\mathcal{I}}D_\iota=\mathrm{Supp}(\mathrm{div}(a))\cup\mathrm{Supp}(\mathrm{div}_\infty(x)).$$

Let

$$-\mathrm{div}(\omega)=\sum_{\iota\in\mathcal{I}}d_\iota D_\iota,$$

where $\omega=\mathrm{d}x\mathrm{d}a/a$ is the top degree right invariant form on G. Note that ω defines a measure $|\omega|$ on an analytic manifold $G(\mathbb{Q}_v)$, and for any finite place p,

$$|\omega|=\left(1-\frac{1}{p}\right)\mathrm{d}g_p,$$

where $\mathrm{d}g_p$ is the standard Haar measure defined in Section 3.

LEMMA 5.5. *Consider an open convex cone Ω in $\mathrm{Pic}^G(X)_\mathbb{R}$, defined by the following relations:*

$$\begin{cases} s_\iota-d_\iota+1>0 & \text{if } \iota\in\mathcal{I}_1\\ s_\iota-d_\iota+1+e_\iota>0 & \text{if } \iota\in\mathcal{I}_2\\ s_\iota-d_\iota+1>0 & \text{if } \iota\in\mathcal{I}_3\end{cases}$$

where $e_\iota=|\mathrm{ord}_{D_\iota}(x)|$. Then $\widehat{\mathsf{H}}_p(\mathbf{s},\lambda,t,\alpha)$ and $\widehat{\mathsf{H}}_\infty(\mathbf{s},t,\alpha)$ are holomorphic on T_Ω.

PROOF. First we prove our assertion for $\widehat{\mathsf{H}}_\infty$. We can assume that

$$\widehat{\mathsf{H}}_v(\mathbf{s}, t) = \widehat{\mathsf{H}}_v(\mathbf{s} - it\mathbf{m}(a), 0),$$

where $\mathbf{m}(a) \in \mathfrak{X}^*(G) \subset \mathrm{Pic}^G(X)$ is the character associated to the rational function a (by choosing an appropriate height function). It suffices to discuss the case when $t = 0$. Choose a finite covering $\{U_\eta\}$ of $X(\mathbb{R})$ by open subsets and local coordinates y_η, z_η on U_η such that the union of the boundary divisor D and E is locally defined by $y_\eta = 0$ or $y_\eta \cdot z_\eta = 0$. Choose a partition of unity $\{\theta_\eta\}$; the local integral takes the form

$$\widehat{\mathsf{H}}_\infty(\mathbf{s}, \alpha) = \sum_\eta \int_{G(\mathbb{R})} \mathsf{H}_\infty(\mathbf{s}, g_\infty)^{-1} \bar{\psi}_\infty(\alpha x_\infty)\theta_\eta \, \mathrm{d}g_\infty.$$

Each integral is a oscillatory integral in the variables y_η, z_η. For example, assume that U_η meets $D_\iota, D_{\iota'}$, where $\iota, \iota' \in \mathcal{I}_2$. Then

$$\int_{G(\mathbb{R})} \mathsf{H}_\infty(\mathbf{s}, g_\infty)^{-1} \bar{\psi}_\infty(\alpha x_\infty)\theta_\eta \, \mathrm{d}g_\infty$$

$$= \int_{\mathbb{R}^2} |y_\eta|^{s_\iota - d_\iota} |z_\eta|^{s_{\iota'} - d_{\iota'}} \bar{\psi}_\infty\left(\frac{\alpha f}{y_\eta^{e_\iota} z_\eta^{e_{\iota'}}}\right) \phi(\mathbf{s}, y_\eta, z_\eta) \, \mathrm{d}y_\eta \, \mathrm{d}z_\eta,$$

where ϕ is a smooth function with compact support and f is a nonvanishing analytic function. Shrinking U_η and changing variables, if necessary, we may assume that f is a constant. Proposition 3.7 implies that this integral is holomorphic everywhere. The other integrals can be studied similarly.

Next we consider finite places. Let p be a prime of good reduction. Since

$$\mathrm{Supp}(\mathrm{div}_0(a)) \cap \mathrm{Supp}(\mathrm{div}_\infty(a)) = \emptyset,$$

the smooth function $\mathbf{1}_{\mathbb{Z}_p}(\alpha a_p)$ extends to a smooth function h on $X(\mathbb{Q}_p)$. Let

$$U = \{h = 1\}.$$

Then

$$\widehat{\mathsf{H}}_p(\mathbf{s}, \lambda, \alpha) = \int_U \mathsf{H}_p(\mathbf{s}, g_p)^{-1} \bar{\psi}_p(\alpha x_p) \lambda_S(\alpha a_p) \mathrm{d}g_p.$$

Now the proof of [**13**, Lemma 4.4.1] implies that this is holomorphic on T_Ω because $U \cap (\cup_{\iota \in \mathcal{I}_2} D_\iota(\mathbb{Q}_p)) = \emptyset$. Places of bad reduction are treated similarly. □

LEMMA 5.6. *Let $|\alpha|_p = p^k > 1$. Then, for any compact set in Ω and for any $\delta > 0$, there exists a constant $C > 0$ such that*

$$|\widehat{\mathsf{H}}_p(\mathbf{s}, \lambda, t, \alpha)| < C|\alpha|_p^{-\min_{\iota \in \mathcal{I}_1}\{\Re(s_\iota) - d_\iota + 1 - \delta\}},$$

for $\Re(\mathbf{s})$ in that compact set.

PROOF. First assume that p is a good reduction place. Let $\rho : \mathcal{X}(\mathbb{Z}_p) \to \mathcal{X}(\mathbb{F}_p)$ be the reduction map modulo p where $\mathcal{X}$ is a smooth integral model of X over $\mathrm{Spec}(\mathbb{Z}_p)$. Note that

$$\rho(\{|a|_p < 1\}) \subset \cup_{\iota \in \mathcal{I}_1} \mathcal{D}_\iota(\mathbb{F}_p),$$

where $\mathcal{D}_\iota$ is the Zariski closure of D_ι in $\mathcal{X}$. Thus $\widehat{\mathsf{H}}_p(\mathbf{s}, \lambda, \alpha)$ is given by

$$\widehat{\mathsf{H}}_p(\mathbf{s}, \lambda, \alpha) = \sum_{\tilde{x} \in \cup_{\iota \in \mathcal{I}_1} \mathcal{D}_\iota(\mathbb{F}_p)} \int_{\rho^{-1}(\tilde{x})} \mathsf{H}_p(\mathbf{s}, g_p)^{-1} \bar{\psi}_p(\alpha x_p) \lambda_S(\alpha a_p) \mathbf{1}_{\mathbb{Z}_p}(\alpha a_p) \mathrm{d}g_p.$$

Let $\tilde{x} \in \mathcal{D}_\iota(\mathbb{F}_p)$ for some $\iota \in \mathcal{I}_1$, but $\tilde{x} \notin \mathcal{D}_{\iota'}(\mathbb{F}_p)$ for any $\iota' \in \mathcal{I} \setminus \{\iota\}$. Since p is a good reduction place, we can find analytic coordinates y, z such that

$$\begin{aligned}\left|\int_{\rho^{-1}(\tilde{x})}\right| &\leq \int_{\rho^{-1}(\tilde{x})} \mathsf{H}_p(\Re(\mathbf{s}), g_p)^{-1}\mathbf{1}_{\mathbb{Z}_p}(\alpha a_p)\mathrm{d}g_p \\ &= \left(1-\frac{1}{p}\right)^{-1}\int_{\rho^{-1}(\tilde{x})} \mathsf{H}_p(\Re(\mathbf{s})-\mathbf{d}, g_p)^{-1}\mathbf{1}_{\mathbb{Z}_p}(\alpha a_p)\mathrm{d}\tau_{X,p} \\ &= \left(1-\frac{1}{p}\right)^{-1}\int_{\mathfrak{m}_p^2} |y|_p^{\Re(s_\iota)-d_\iota}\mathbf{1}_{\mathbb{Z}_p}(\alpha y)\mathrm{d}y_p\mathrm{d}z_p \\ &= \frac{1}{p}\cdot\frac{p^{-k(\Re(s_\iota)-d_\iota+1)}}{1-p^{-(\Re(s_\iota)-d_\iota+1)}},\end{aligned}$$

where $\mathrm{d}\tau_{X,p}$ is the local Tamagawa measure (see [**13**, Section 2] for the definition). For the construction of such local analytic coordinates, see [**38**], [**15**], or [**28**]. If $\tilde{x} \in \mathcal{D}_\iota(\mathbb{F}_p) \cap \mathcal{D}_{\iota'}(\mathbb{F}_p)$ for $\iota \in \mathcal{I}_1$, $\iota' \in \mathcal{I}_3$, then we can find local analytic coordinates y, z such that

$$\begin{aligned}\left|\int_{\rho^{-1}(\tilde{x})}\right| &\leq \left(1-\frac{1}{p}\right)\int_{\mathfrak{m}_p^2} |y|_p^{\Re(s_\iota)-d_\iota+1}|z|_p^{\Re(s_{\iota'})-d_{\iota'}+1}\mathbf{1}_{\mathbb{Z}_p}(\alpha y)\mathrm{d}y_p^\times\mathrm{d}z_p^\times \\ &= \left(1-\frac{1}{p}\right)\frac{p^{-k(\Re(s_\iota)-d_\iota+1)}}{1-p^{-(\Re(s_\iota)-d_\iota+1)}}\frac{p^{-(\Re(s_{\iota'})-d_{\iota'}+1)}}{1-p^{-(\Re(s_{\iota'})-d_{\iota'}+1)}}.\end{aligned}$$

If $\tilde{x} \in \mathcal{D}_\iota(\mathbb{F}_p) \cap \mathcal{D}_{\iota'}(\mathbb{F}_p)$ for $\iota, \iota' \in \mathcal{I}_1$, $\iota \neq \iota'$, then we can find analytic coordinates x, y such that

$$\begin{aligned}\left|\int_{\rho^{-1}(\tilde{x})}\right| &\leq \left(1-\frac{1}{p}\right)\int_{\mathfrak{m}_p^2} |y|_p^{\Re(s_\iota)-d_\iota+1}|z|_p^{\Re(s_{\iota'})-d_{\iota'}+1}\mathbf{1}_{\mathbb{Z}_p}(\alpha yz)\,\mathrm{d}y_p^\times\mathrm{d}z_p^\times \\ &\leq \left(1-\frac{1}{p}\right)\int_{\mathfrak{m}_p^2} |yz|_p^{\min\{\Re(s_\iota)-d_\iota+1,\,\Re(s_{\iota'})-d_{\iota'}+1\}}\mathbf{1}_{\mathbb{Z}_p}(\alpha yz)\,\mathrm{d}y_p^\times\mathrm{d}z_p^\times \\ &= \left(1-\frac{1}{p}\right)\left((k-1)\frac{p^{-kr}}{1-p^{-r}}+\frac{p^{-(k+1)r}}{(1-p^{-r})^2}\right),\end{aligned}$$

where

$$r = \min\{\Re(s_\iota)-d_\iota+1,\ \Re(s_{\iota'})-d_{\iota'}+1\}.$$

It follows from these inequalities and Lemma 9.4 in [**11**] that there exists a constant $C > 0$, independent of p, satisfying the inequality in the statement.

Next assume that p is a bad reduction place. Choose an open covering $\{U_\eta\}$ of $\cup_{\iota\in\mathcal{I}_1} D_\iota(\mathbb{Q}_p)$ such that

$$(\cup_\eta U_\eta) \cap (\cup_{\iota\in\mathcal{I}_2} D_\iota(\mathbb{Q}_p)) = \emptyset,$$

and each U_η has analytic coordinates y_η, z_η. Moreover, we can assume that the boundary divisor is defined by $y_\eta = 0$ or $y_\eta \cdot z_\eta = 0$ on U_η. Let V be the complement of $\cup_{\iota\in\mathcal{I}_1} D_\iota(\mathbb{Q}_p)$, and consider the partition of unity for $\{U_\eta, V\}$ which we denote by $\{\theta_\eta, \theta_V\}$. If k is sufficiently large, then

$$\{\mathbf{1}_{\frac{1}{N}\mathbb{Z}_p}(\alpha a) = 1\} \cap \mathrm{Supp}(\theta_V) = \emptyset.$$

Hence if k is sufficiently large, then

$$|\widehat{\mathsf{H}}_p(\mathbf{s}, \lambda, \alpha)| \leq \sum_\eta \int_{U_\eta} \mathsf{H}_p(\Re(\mathbf{s}), g_p)^{-1} \mathbf{1}_{\frac{1}{N}\mathbb{Z}_p}(\alpha a_p) \cdot \theta_\eta \, \mathrm{d}g_p.$$

When U_η meets only one component $D_\iota(\mathbb{Q}_p)$ for $\iota \in \mathcal{I}_1$, then

$$\int_{U_\eta} \leq \int_{\mathbb{Q}_p^2} |y_\eta|_p^{\Re(s_\iota)-d_\iota} \mathbf{1}_{c\mathbb{Z}_p}(\alpha y_\eta)\phi(\mathbf{s}, y_\eta, z_\eta) \, \mathrm{d}y_{\eta,p}\mathrm{d}z_{\eta,p} \ll p^{-k(\Re(s_\iota)-d_\iota+1)},$$

as $k \to \infty$, where c is some rational number and ϕ is a smooth function with compact support. Other integrals are treated similarly. $\square$

We record the following useful lemma (see, e.g., [**12**, Lemma 2.3.1]):

LEMMA 5.7. *Let d be a positive integer and $a \in \mathbb{Q}_p$. If $|a|_p > p$ and $p \nmid d$, then*

$$\int_{\mathbb{Z}_p^\times} \bar{\psi}_p(ax^d) \, \mathrm{d}x_p^\times = 0.$$

Moreover, if $|a|_p = p$ and $d = 2$, then

$$\int_{\mathbb{Z}_p^\times} \bar{\psi}_p(ax^d)\mathrm{d}x_p^\times = \begin{cases} \frac{\sqrt{p}-1}{p-1} \text{ or } \frac{i\sqrt{p}-1}{p-1} & \text{if } pa \text{ is a quadratic residue,} \\ \frac{-\sqrt{p}-1}{p-1} \text{ or } \frac{-i\sqrt{p}-1}{p-1} & \text{if } pa \text{ is a quadratic non-residue.} \end{cases}$$

LEMMA 5.8. *Let $|\alpha|_p = p^{-k} < 1$. Consider an open convex cone Ω_ϵ in $\mathrm{Pic}(X)_\mathbb{R}$, defined by the following relations:*

$$\begin{cases} s_\iota - d_\iota + 1 > 0 & \text{if } \iota \in \mathcal{I}_1 \\ s_\iota - d_\iota + 2 + \epsilon > 0 & \text{if } \iota \in \mathcal{I}_2 \\ s_\iota - d_\iota + 1 > 0 & \text{if } \iota \in \mathcal{I}_3 \end{cases}$$

where $0 < \epsilon < 1/3$. Then, for any compact set in Ω_ϵ, there exists a constant $C > 0$ such that

$$|\widehat{\mathsf{H}}_p(\mathbf{s}, \lambda, t, \alpha)| < C|\alpha|_p^{-\frac{2}{3}(1+2\epsilon)},$$

for $\Re(\mathbf{s})$ in that compact set.

PROOF. First assume that p is a good reduction place and that $p \nmid e_\iota$, for any $\iota \in \mathcal{I}_2$. We have

$$\widehat{\mathsf{H}}_p(\mathbf{s}, \lambda, \alpha) = \sum_{\tilde{x} \in \mathcal{X}(\mathbb{F}_p)} \int_{\rho^{-1}(\tilde{x})} \mathsf{H}_p(\mathbf{s}, g_p)^{-1} \bar{\psi}_p(\alpha x_p)\lambda_S(\alpha a_p)\mathbf{1}_{\mathbb{Z}_p}(\alpha a_p) \, \mathrm{d}g_p.$$

A formula of J. Denef (see [**15**, Theorem 3.1] or [**13**, Proposition 4.1.7]) and Lemma 9.4 in [**11**] give us a uniform bound:

$$|\sum_{\tilde{x} \notin \cup_{\iota \in \mathcal{I}_2} \mathcal{D}_\iota(\mathbb{F}_p)}| \leq \sum_{\tilde{x} \notin \cup_{\iota \in \mathcal{I}_2} \mathcal{D}_\iota(\mathbb{F}_p)} \int_{\rho^{-1}(\tilde{x})} \mathsf{H}_p(\Re(\mathbf{s}), g_p)^{-1} \, \mathrm{d}g_p.$$

Hence we need to study

$$\sum_{\tilde{x} \in \cup_{\iota \in \mathcal{I}_2} \mathcal{D}_\iota(\mathbb{F}_p)} \int_{\rho^{-1}(\tilde{x})} \mathsf{H}_p(\mathbf{s}, g_p)^{-1} \bar{\psi}_p(\alpha x_p)\lambda_S(\alpha a_p)\mathbf{1}_{\mathbb{Z}_p}(\alpha a_p) \, \mathrm{d}g_p.$$

Let $\tilde{x} \in \mathcal{D}_\iota(\mathbb{F}_p)$ for some $\iota \in \mathcal{I}_2$, but $\tilde{x} \notin \mathcal{D}_{\iota'}(\mathbb{F}_p) \cup \mathcal{E}(\mathbb{F}_p)$ for any $\iota' \in \mathcal{I} \setminus \{\iota\}$, where $\mathcal{E}$ is the Zariski closure of E in $\mathcal{X}$. Then we can find local analytic coordinates y, z such that

$$\int_{\rho^{-1}(\tilde{x})} = \left(1 - \frac{1}{p}\right)^{-1} \int_{\mathfrak{m}_p^2} |y|_p^{s_\iota - d_\iota} \bar{\psi}_p(\alpha f / y^{e_\iota}) \lambda_S(\alpha y^{-1}) \mathbf{1}_{\mathbb{Z}_p}(\alpha y^{-1}) \, \mathrm{d}y_p \mathrm{d}z_p,$$

where $f \in \mathbb{Z}_p[[y, z]]$ such that $f(0) \in \mathbb{Z}_p^\times$. Since p does not divide e_ι, there exists $g \in \mathbb{Z}_p[[y, z]]$ such that $f = f(0) g^{e_\iota}$. After a change of variables, we can assume that $f = u \in \mathbb{Z}_p^\times$. Lemma 5.7 implies that

$$\begin{aligned}
\int_{\rho^{-1}(\tilde{x})} &= \frac{1}{p} \int_{\mathfrak{m}_p} |y|_p^{s_\iota - d_\iota + 1} \lambda_S(\alpha y^{-1}) \int_{\mathbb{Z}_p^\times} \bar{\psi}_p(\alpha u b^{e_\iota} / y^{e_\iota}) \mathrm{d}b_p^\times \mathbf{1}_{\mathbb{Z}_p}(\alpha y^{-1}) \mathrm{d}y_p^\times \\
&= \frac{1}{p} \int_{p^{-(k+1)} \le |y^{e_\iota}|_p} |y|_p^{s_\iota - d_\iota + 1} \lambda_S(\alpha y^{-1}) \int_{\mathbb{Z}_p^\times} \bar{\psi}_p(\alpha u b^{e_\iota} / y^{e_\iota}) \, \mathrm{d}b_p^\times \mathrm{d}y_p^\times
\end{aligned}$$

Thus it follows from the second assertion of Lemma 5.7 that

$$\begin{aligned}
\left| \int_{\rho^{-1}(\tilde{x})} \right| &\le \frac{1}{p} \int_{p^{-(k+1)} \le |y^{e_\iota}|} |y|_p^{\Re(s_\iota) - d_\iota + 1} \left| \int_{\mathbb{Z}_p^\times} \bar{\psi}_p(\alpha u b^{e_\iota} / y^{e_\iota}) \mathrm{d}b_p^\times \right| \mathrm{d}y_p^\times \\
&\le \frac{1}{p} k p^{\frac{k}{e_\iota}(1+\epsilon)} + \frac{1}{p} p^{\frac{k+1}{e_\iota}(1+\epsilon)} \times \begin{cases} 1 & \text{if } e_\iota > 2 \\ \frac{1}{\sqrt{p} - 1} & \text{if } e_\iota = 2 \end{cases} \\
&\ll \frac{1}{p} k p^{\frac{2}{3} k (1+\epsilon)}.
\end{aligned}$$

If $\tilde{x} \in \mathcal{D}_\iota(\mathbb{F}_p) \cap \mathcal{E}(\mathbb{F}_p)$, for some $\iota \in \mathcal{I}_2$, then we have

$$\begin{aligned}
\int_{\rho^{-1}(\tilde{x})} &= \left(1 - \frac{1}{p}\right)^{-1} \int_{\mathfrak{m}_p^2} |y|_p^{s_\iota - d_\iota} \bar{\psi}_p(\alpha z / y^{e_\iota}) \lambda_S(\alpha y^{-1}) \mathbf{1}_{\mathbb{Z}_p}(\alpha y^{-1}) \, \mathrm{d}y_p \mathrm{d}z_p \\
&= \int_{\mathfrak{m}_p} |y|_p^{s_\iota - d_\iota + 1} \lambda_S(\alpha y^{-1}) \mathbf{1}_{\mathbb{Z}_p}(\alpha y^{-1}) \int_{\mathfrak{m}_p} \bar{\psi}_p(\alpha z / y^{e_\iota}) \mathrm{d}z_p \mathrm{d}y_p^\times \\
&= \frac{1}{p} \int_{p^{-(k+1)} \le |y|_p^{e_\iota} < 1} |y|_p^{s_\iota - d_\iota + 1} \lambda_S(\alpha y^{-1}) \, \mathrm{d}y_p^\times.
\end{aligned}$$

Hence we obtain that

$$\left| \int_{\rho^{-1}(\tilde{x})} \right| \le \frac{1}{p} \int_{p^{-(k+1)} \le |y|_p^{e_\iota} < 1} |y|_p^{\Re(s_\iota) - d_\iota + 1} \, \mathrm{d}y_p^\times \le k p^{\frac{k}{e_\iota}(1+\epsilon)} < k p^{\frac{2}{3} k (1+\epsilon)}.$$

If $\tilde{x} \in \mathcal{D}_\iota(\mathbb{F}_p) \cap \mathcal{D}_{\iota'}(\mathbb{F}_p)$ for some $\iota \in \mathcal{I}_2$ and $\iota' \in \mathcal{I}_3$, then it follows from Lemma 5.7

$$\begin{aligned}
\int_{\rho^{-1}(\tilde{x})} &= \left(1 - \frac{1}{p}\right)^{-1} \int_{\mathfrak{m}_p^2} |y|_p^{s_\iota - d_\iota} |z|_p^{s_{\iota'} - d_{\iota'}} \bar{\psi}_p\left(\frac{\alpha u}{y^{e_\iota} z^{e_{\iota'}}}\right) \lambda_S(\alpha y^{-1}) \mathbf{1}_{\mathbb{Z}_p}(\alpha y^{-1}) \, \mathrm{d}y_p \mathrm{d}z_p \\
&= \left(1 - \frac{1}{p}\right)^{-1} \int |y|_p^{s_\iota - d_\iota} |z|_p^{s_{\iota'} - d_{\iota'}} \lambda_S(\alpha y^{-1}) \int_{\mathbb{Z}_p^\times} \bar{\psi}_p\left(\frac{\alpha u b^{e_\iota}}{y^{e_\iota} z^{e_{\iota'}}}\right) \mathrm{d}b_p^\times \, \mathrm{d}y_p \mathrm{d}z_p,
\end{aligned}$$

where the last integral is over the domain

$$\{(y, z) \in \mathfrak{m}_p^2 : p^{-(k+1)} \le |y^{e_\iota} z^{e_{\iota'}}|_p\}.$$

We conclude that

$$\begin{aligned}\left|\int_{\rho^{-1}(\tilde{x})}\right| &\leq \left(1-\frac{1}{p}\right)^{-1}\int_{p^{-(k+1)}\leq|y^{e_\iota}z^{e_{\iota'}}|_p}|y|_p^{\Re(s_\iota)-d_\iota}|z|_p^{\Re(s_{\iota'})-d_{\iota'}}\,\mathrm{d}y_p\mathrm{d}z_p\\ &\leq \left(1-\frac{1}{p}\right)^{-1}\int_{p^{-k}\leq|y^{e_\iota}|_p<1}|y|_p^{\Re(s_\iota)-d_\iota}\,\mathrm{d}y_p\int_{\mathfrak{m}_p}|z|_p^{\Re(s_{\iota'})-d_{\iota'}}\,\mathrm{d}z_p\\ &\leq kp^{\frac{k}{e_\iota}(1+\epsilon)}\frac{p^{-(\Re(s_{\iota'})-d_{\iota'}+1)}}{1-p^{-(\Re(s_{\iota'})-d_{\iota'}+1)}}.\end{aligned}$$

If $\tilde{x}\in\mathcal{D}_\iota(\mathbb{F}_p)\cap\mathcal{D}_{\iota'}(\mathbb{F}_p)$ for some $\iota,\iota'\in\mathcal{I}_2$, then the local integral on $\rho^{-1}(\tilde{x})$ is:

$$\begin{aligned}&\left(1-\frac{1}{p}\right)^{-1}\int_{\mathfrak{m}_p^2}|y|_p^{s_\iota-d_\iota}|z|_p^{s_{\iota'}-d_{\iota'}}\bar{\psi}_p\left(\frac{\alpha u}{y^{e_\iota}z^{e_{\iota'}}}\right)\lambda_S(\alpha y^{-1}z^{-1})\mathbf{1}_{\mathbb{Z}_p}(\alpha y^{-1}z^{-1})\,\mathrm{d}y_p\mathrm{d}z_p\\ &\quad=\left(1-\frac{1}{p}\right)\int_{\mathfrak{m}_p^2}|y|_p^{s_\iota-d_\iota}|z|_p^{s_{\iota'}-d_{\iota'}}\lambda_S(\alpha y^{-1}z^{-1})\int_{\mathbb{Z}_p^\times}\bar{\psi}_p\left(\frac{\alpha ub^{e_\iota}}{y^{e_\iota}z^{e_{\iota'}}}\right)\mathrm{d}b_p^\times\mathrm{d}y_p^\times\mathrm{d}z_p^\times.\end{aligned}$$

We can assume that $e_\iota\leq e_{\iota'}$. Then we can conclude that

$$\begin{aligned}\left|\int_{\rho^{-1}(\tilde{x})}\right| &\leq \int_{p^{-k}\leq|y^{e_\iota}z^{e_{\iota'}}|_p}|y^{e_\iota}z^{e_{\iota'}}|_p^{-\frac{1}{e_\iota}(1+\epsilon)}\,\mathrm{d}y_p^\times\mathrm{d}z_p^\times\\ &\quad+\int_{p^{-(k+1)}=|y^{e_\iota}z^{e_{\iota'}}|_p}|y^{e_\iota}z^{e_{\iota'}}|_p^{-\frac{1}{e_\iota}(1+\epsilon)}\left|\int_{\mathbb{Z}_p^\times}\bar{\psi}_p\left(\frac{\alpha ub^{e_\iota}}{y^{e_\iota}z^{e_{\iota'}}}\right)\mathrm{d}b^\times\right|\mathrm{d}y_p^\times\mathrm{d}z_p^\times\\ &\leq k^2p^{\frac{k}{e_\iota}(1+\epsilon)}+kp^{\frac{k+1}{e_\iota}(1+\epsilon)}\times\begin{cases}1 & \text{if } e_\iota>2\\ \frac{1}{\sqrt{p}-1} & \text{if } e_\iota=2\end{cases}\\ &\ll k^2p^{\frac{2}{3}k(1+\epsilon)}.\end{aligned}$$

Thus our assertion follows from these estimates and Lemma 9.4 in [**11**].

Next assume that p is a place of bad reduction or that p divides e_ι, for some $\iota\in\mathcal{I}_2$. Fix a compact subset of Ω_ϵ and assume that $\Re(\mathbf{s})$ is in that compact set. Choose a finite open covering $\{U_\eta\}$ of $\cup_{\iota\in\mathcal{I}_2}D_\iota(\mathbb{Q}_p)$ with analytic coordinates y_η, z_η such that the union of the boundary $D(\mathbb{Q}_p)$ and $E(\mathbb{Q}_p)$ is defined by $y_\eta=0$ or $y_\eta\cdot z_\eta=0$. Let V be the complement of $\cup_{\iota\in\mathcal{I}_2}D_\iota(\mathbb{Q}_p)$, and consider a partition of unity $\{\theta_\eta,\theta_V\}$ for $\{U_\eta,V\}$. Then it is clear that

$$\int_V \mathsf{H}_p(\mathbf{s},g_p)^{-1}\bar{\psi}_p(\alpha x_p)\lambda_{S,p}(\alpha a_p)\mathbf{1}_{\frac{1}{N}\mathbb{Z}_p}(\alpha a_p)\theta_V\mathrm{d}g_p,$$

is bounded, so we need to study

$$\int_{U_\eta} \mathsf{H}_p(\mathbf{s},g_p)^{-1}\bar{\psi}_p(\alpha x_p)\lambda_{S,p}(\alpha a_p)\mathbf{1}_{\frac{1}{N}\mathbb{Z}_p}(\alpha a_p)\theta_{U_\eta}\mathrm{d}g_p.$$

Assume that U_η meets only one $D_\iota(\mathbb{Q}_p)$ for some $\iota\in\mathcal{I}_2$. Then, the above integral looks like

$$\int_{U_\eta}=\int_{\mathbb{Q}_p^2}|y_\eta|_p^{s_\iota-d_\iota}\bar{\psi}_p(\alpha f/y_\eta^{e_\iota}))\lambda_{S,p}(\alpha g/y_\eta)\mathbf{1}_{\frac{1}{N}\mathbb{Z}_p}(\alpha g/y_\eta)\Phi(\mathbf{s},y_\eta,z_\eta)\mathrm{d}y_{\eta,p}\mathrm{d}z_{\eta,p},$$

where f and g are nonvanishing analytic functions, and Φ is a smooth function with compact support. By shrinking U_η and changing variables, if necessary, we can assume that f and g are constant. The proof of Proposition 3.8 implies our assertion for this integral. Other integrals are treated similarly. □

LEMMA 5.9. *For any compact set in an open convex cone Ω', defined by*

$$\begin{cases} s_\iota - d_\iota - 1 > 0 & \text{if } \iota \in \mathcal{I}_1 \\ s_\iota - d_\iota + 3 > 0 & \text{if } \iota \in \mathcal{I}_2 \\ s_\iota - d_\iota + 1 > 0 & \text{if } \iota \in \mathcal{I}_3 \end{cases}$$

there exists a constant $C > 0$ such that

$$|\widehat{\mathsf{H}}_\infty(\mathbf{s}, t, \alpha)| < \frac{C}{|\alpha|^2(1+t^2)},$$

for $\Re(\mathbf{s})$ in that compact set.

PROOF. Consider the left invariant differential operators $\partial_a = a\partial/\partial a$ and $\partial_x = a\partial/\partial x$. Assume that $\Re(\mathbf{s}) \gg 0$. Integrating by parts, we have

$$\begin{aligned} \widehat{\mathsf{H}}_\infty(\mathbf{s}, t, \alpha) &= -\frac{1}{t^2}\int_{G(\mathbb{R})} \partial_a^2 \mathsf{H}_\infty(\mathbf{s}, g_\infty)^{-1}\bar{\psi}_\infty(\alpha x_\infty)|a_\infty|_\infty^{-it}\, dg_\infty \\ &= \frac{1}{(2\pi)^2|\alpha|^2t^2}\int_{G(\mathbb{R})} \frac{\partial^2}{\partial x^2}(\partial_a^2\mathsf{H}_\infty(\mathbf{s}, g_\infty)^{-1})\bar{\psi}_\infty(\alpha x_\infty)|a_\infty|_\infty^{-it}\, dg_\infty. \end{aligned}$$

According to Proposition 2.2. in [**11**],

$$\begin{aligned} \frac{\partial^2}{\partial x^2}(\partial_a^2\mathsf{H}_\infty(\mathbf{s}, g_\infty)^{-1}) &= |a|^{-2}\partial_x^2\partial_a^2\mathsf{H}_\infty(\mathbf{s}, g_\infty)^{-1} \\ &= \mathsf{H}_\infty(\mathbf{s} - 2\mathbf{m}(a), g_\infty)^{-1} \times (\text{a bounded smooth function}). \end{aligned}$$

Moreover, Lemma 4.4.1. of [**13**] tells us that

$$\int_{G(\mathbb{R})} \mathsf{H}_\infty(\mathbf{s} - 2\mathbf{m}(a), g_\infty)^{-1} dg_\infty,$$

is holomorphic on $\mathsf{T}_{\Omega'}$. Thus we can conclude our lemma. □

LEMMA 5.10. *The Euler product*

$$\prod_p \widehat{\mathsf{H}}_p(\mathbf{s}, \lambda, t, \alpha) \cdot \widehat{\mathsf{H}}_\infty(\mathbf{s}, t, \alpha)$$

is holomorphic on $\mathsf{T}_{\Omega'}$.

PROOF. First we prove that the Euler product is holomorphic on $\mathsf{T}_{\Omega'}$. To conclude this, we only need to discuss:

$$\prod_{p \notin S \cup S_3,\, |\alpha|_p = 1,} \widehat{\mathsf{H}}_p(\mathbf{s}, \lambda, t, \alpha),$$

where $S_3 = \{p : p \mid e_\iota \text{ for some } \iota \in \mathcal{I}_3\}$. Let p be a prime such that $p \notin S \cup S_3$ and $|\alpha|_p = 1$. Fix a compact subset of Ω', and assume that $\Re(s)$ is sitting in that compact set. From the definition of Ω', there exists $\epsilon > 0$ such that

$$\begin{cases} s_\iota - d_\iota + 1 > 2 + \epsilon & \text{for any } \iota \in \mathcal{I}_1 \\ s_\iota - d_\iota + 1 > \epsilon & \text{for any } \iota \in \mathcal{I}_3. \end{cases}$$

Since we have

$$\{|a|_p \leq 1\} = X(\mathbb{Q}_p) \setminus \rho^{-1}(\cup_{\iota \in \mathcal{I}_2}\mathcal{D}_\iota(\mathbb{F}_p)),$$

we can conclude that

$$\widehat{\mathsf{H}}_p(\mathbf{s},\lambda,\alpha)=\sum_{\tilde{x}\notin\cup_{\iota\in\mathcal{I}_2}\mathcal{D}_\iota(\mathbb{F}_p)}\int_{\rho^{-1}(\tilde{x})}\mathsf{H}_p(\mathbf{s},g_p)^{-1}\bar{\psi}_p(\alpha x_p)\lambda_S(a_p)\mathrm{d}g_p.$$

It is easy to see that

$$\sum_{\tilde{x}\notin\cup_{\iota\in\mathcal{I}}\mathcal{D}_\iota(\mathbb{F}_p)}\int_{\rho^{-1}(\tilde{x})}=\int_{G(\mathbb{Z}_p)}1\,\mathrm{d}g_p=1.$$

Also it follows from a formula of J. Denef (see [**15**, Theorem 3.1] or [**13**, Proposition 4.1.7]) and Lemma 9.4 in [**11**] that there exists an uniform bound $C>0$ such that for any $\tilde{x}\in\cup_{\iota\in\mathcal{I}_1}\mathcal{D}_\iota(\mathbb{F}_p)$,

$$\left|\int_{\rho^{-1}(\tilde{x})}\right|<\int_{\rho^{-1}(\tilde{x})}\mathsf{H}_p(\Re(\mathbf{s}),g_p)^{-1}\mathrm{d}g_p<\frac{C}{p^{2+\epsilon}}.$$

Hence we need to obtain uniform bounds of $\int_{\rho^{-1}(\tilde{x})}$ for

$$\tilde{x}\in\cup_{\iota\in\mathcal{I}_3}\mathcal{D}_\iota(\mathbb{F}_p)\setminus\cup_{\iota\in\mathcal{I}_1\cup\mathcal{I}_2}\mathcal{D}_\iota(\mathbb{F}_p).$$

Let $\tilde{x}\in\mathcal{D}_\iota(\mathbb{F}_p)$ for some $\iota\in\mathcal{I}_3$, but $\tilde{x}\notin\cup_{\iota\in\mathcal{I}_1\cup\mathcal{I}_2}\mathcal{D}_\iota(\mathbb{F}_p)\cup\mathcal{E}(\mathbb{F}_p)$. Then it follows from Lemmas 4.2 and 5.7 that

$$\begin{aligned}\int_{\rho^{-1}(\tilde{x})}&=\left(1-\frac{1}{p}\right)^{-1}\int_{\mathfrak{m}_p^2}|y|_p^{s_\iota-d_\iota}\bar{\psi}_p(u/y^{e_\iota})\,\mathrm{d}y_p\mathrm{d}z_p\\&=\frac{1}{p-1}\int_{\mathfrak{m}_p}|y|_p^{s_\iota-d_\iota}\int_{\mathbb{Z}_p^\times}\bar{\psi}_p(ub^{e_\iota}/y^{e_\iota})\,\mathrm{d}b_p^\times\mathrm{d}y_p\\&=\begin{cases}0&\text{if } e_\iota>1\\-\frac{p^{-(s_\iota-d_\iota+2)}}{p-1}&\text{if } e_\iota=1.\end{cases}\end{aligned}$$

If $\tilde{x}\in\mathcal{D}_\iota(\mathbb{F}_p)\cap\mathcal{E}(\mathbb{F}_p)$ for some $\iota\in\mathcal{I}_3$, then we have

$$\begin{aligned}\int_{\rho^{-1}(\tilde{x})}&=\left(1-\frac{1}{p}\right)^{-1}\int_{\mathfrak{m}_p^2}|y|_p^{s_\iota-d_\iota}\bar{\psi}_p(z/y^{e_\iota})\mathrm{d}y_p\mathrm{d}z_p\\&=\left(1-\frac{1}{p}\right)^{-1}\int_{m_p}|y|_p^{s_\iota-d_\iota}\int_{\mathfrak{m}_p}\bar{\psi}_p(z/y^{e_\iota})\,\mathrm{d}z_p\mathrm{d}y_p\\&=\begin{cases}0&\text{if } e_\iota>1\\p^{-(s_\iota-d_\iota+2)}&\text{if } e_\iota=1.\end{cases}\end{aligned}$$

If $\tilde{x}\in\mathcal{D}_\iota(\mathbb{F}_p)\cap\mathcal{D}_{\iota'}(\mathbb{F}_p)$ for some $\iota,\iota'\in\mathcal{I}_3$, then it follows from Lemma 5.7 that

$$\begin{aligned}\int_{\rho^{-1}(\tilde{x})}&=\left(1-\frac{1}{p}\right)^{-1}\int_{\mathfrak{m}_p^2}|y|_p^{s_\iota-d_\iota}|z|_p^{s_{\iota'}-d_{\iota'}}\bar{\psi}_p\left(\frac{u}{y^{e_\iota}z^{e_{\iota'}}}\right)\mathrm{d}y_p\mathrm{d}z_p\\&=\left(1-\frac{1}{p}\right)^{-1}\int_{\mathfrak{m}_p^2}|y|_p^{s_\iota-d_\iota}|z|_p^{s_{\iota'}-d_{\iota'}}\int_{\mathbb{Z}_p^\times}\bar{\psi}_p\left(\frac{ub^{e_\iota}}{y^{e_\iota}z^{e_{\iota'}}}\right)\mathrm{d}b_p^\times\,\mathrm{d}y_p\mathrm{d}z_p\\&=0.\end{aligned}$$

Thus we can conclude from these estimates and Lemma 9.4 in [**11**] that there exists an uniform bound $C' > 0$ such that

$$\left|\widehat{\mathsf{H}}_p(\mathbf{s},\lambda,t,\alpha) - 1\right| < \frac{C'}{p^{1+\epsilon}}$$

Our assertion follows from this. □

LEMMA 5.11. *Let Ω'_ϵ be an open convex cone, defined by*

$$\begin{cases} s_\iota - d_\iota - 2 - \epsilon > 0 & \text{if } \iota \in \mathcal{I}_1 \\ s_\iota - d_\iota + 2 + 2\epsilon > 0 & \text{if } \iota \in \mathcal{I}_2 \\ s_\iota - d_\iota + 1 > 0 & \text{if } \iota \in \mathcal{I}_3 \end{cases}$$

where $\epsilon > 0$ is sufficiently small. Fix a compact subset of Ω'_ϵ and $\epsilon \gg \delta > 0$. Then there exists a constant $C > 0$ such that

$$|\prod_p \widehat{\mathsf{H}}_p(\mathbf{s},\lambda,t,\alpha)\cdot\widehat{\mathsf{H}}_\infty(\mathbf{s},\alpha,t)| < \frac{C}{(1+t^2)|\beta|^{\frac{4}{3}-\frac{8}{3}\epsilon-\delta}|\gamma|^{1+\epsilon-\delta}},$$

for $\Re(s)$ in that compact set, where $\alpha = \frac{\beta}{\gamma}$ with $\gcd(\beta,\gamma) = 1$.

PROOF. This lemma follows from Lemmas 5.6, 5.8, and 5.9, and from the proof of Lemma 5.10. □

THEOREM 5.12. *The zeta function $\mathsf{Z}_1(\mathbf{s},\mathrm{id})$ is holomorphic on the tube domain over an open neighborhood of the shifted effective cone $-K_X + \Lambda_{\mathrm{eff}}(X)$.*

PROOF. Let $1 \gg \epsilon \gg \delta > 0$. Lemma 5.11 implies that

$$\mathsf{Z}_1(\mathbf{s},\mathrm{id}) = \sum_{\lambda\in\mathbf{M},\,\lambda(-1)=1}\ \sum_{\alpha\in\mathbb{Q}^\times} \frac{1}{2\pi}\int_{-\infty}^{+\infty} \prod_p \widehat{\mathsf{H}}_p(\mathbf{s},\lambda,t,\alpha)\cdot\widehat{\mathsf{H}}_\infty(\mathbf{s},t,\alpha)\,dt,$$

is absolutely and uniformly convergent on Ω'_ϵ, so $\mathsf{Z}_1(\mathbf{s},\mathrm{id})$ is holomorphic on $\mathsf{T}_{\Omega'_\epsilon}$. Now note that the image of Ω'_ϵ by $\mathrm{Pic}^G(X) \to \mathrm{Pic}(X)$ contains an open neighborhood of $-K_X + \Lambda_{\mathrm{eff}}(X)$. This concludes the proof of our theorem. □

References

1. I. V. Arzhantsev, *Flag varieties as equivariant compactifications of $\mathbb{G}_a^n$*, Proc. Amer. Math. Soc. **139** (2011), no. 3, 783–786. MR2745631
2. I. V. Arzhantsev and E. V. Sharoyko, *Hassett-Tschinkel correspondence: modality and projective hypersurfaces*, 2009, arXiv 0912.1474.
3. V. Batyrev and Y. Tschinkel, *Rational points of bounded height on compactifications of anisotropic tori*, Internat. Math. Res. Notices (1995), no. 12, 591–635. MR1369408 (97a:14021)
4. ———, *Manin's conjecture for toric varieties*, J. Algebraic Geom. **7** (1998), no. 1, 15–53. MR1620682 (2000c:11107)
5. V. Batyrev and Yu. Tschinkel, *Height zeta functions of toric varieties*, J. Math. Sci. **82** (1996), no. 1, 3220–3239, Algebraic geometry, 5. MR1423638 (98b:11067)
6. Victor V. Batyrev and Yuri Tschinkel, *Rational points on some Fano cubic bundles*, C. R. Acad. Sci. Paris Sér. I Math. **323** (1996), no. 1, 41–46. MR1401626 (97j:14023)
7. T. D. Browning, *An overview of Manin's conjecture for del Pezzo surfaces*, Analytic number theory, Clay Math. Proc., vol. 7, Amer. Math. Soc., Providence, RI, 2007, pp. 39–55. MR2362193 (2008j:14041)
8. ———, *Quantitative arithmetic of projective varieties*, Progress in Mathematics, vol. 277, Birkhäuser Verlag, Basel, 2009. MR2559866 (2010i:11004)

9. A. Chambert-Loir, *Lectures on height zeta functions: At the confluence of algebraic geometry, algebraic number theory, and analysis*, Algebraic and analytic aspects of zeta functions and L-functions, MSJ Mem., vol. 21, Math. Soc. Japan, Tokyo, 2010, pp. 17–49. MR2647601
10. A. Chambert-Loir and Y. Tschinkel, *Fonctions zêta des hauteurs des espaces fibrés*, Rational points on algebraic varieties, Progr. Math., vol. 199, Birkhäuser, Basel, 2001, pp. 71–115. MR1875171 (2003a:11079)
11. ______, *On the distribution of points of bounded height on equivariant compactifications of vector groups*, Invent. Math. **148** (2002), no. 2, 421–452. MR1906155 (2003d:11094)
12. ______, *Integral points of bounded height on partial equivariant compactifications of vector groups*, 2009, `arXiv:0912.4751`.
13. ______, *Igusa integrals and volume asymptotics in analytic and adelic geometry*, Confluentes Mathematici **2, no. 3** (2010), 351–429. MR2740045
14. Régis de la Bretèche and Étienne Fouvry, *L'éclaté du plan projectif en quatre points dont deux conjugués*, J. Reine Angew. Math. **576** (2004), 63–122. MR2099200 (2005f:11131)
15. J. Denef, *On the degree of Igusa's local zeta function*, Amer. J. Math. **109** (1987), no. 6, 991–1008. MR919001 (89d:11108)
16. J. Denef and F. Loeser, *Motivic Igusa zeta functions*, J. Algebraic Geom. **7** (1998), no. 3, 505–537. MR1618144 (99j:14021)
17. ______, *Germs of arcs on singular algebraic varieties and motivic integration*, Invent. Math. **135** (1999), no. 1, 201–232. MR1664700 (99k:14002)
18. ______, *Definable sets, motives and p-adic integrals*, J. Amer. Math. Soc. **14** (2001), no. 2, 429–469 (electronic). MR1815218 (2002k:14033)
19. U. Derenthal, *Manin's conjecture for a quintic del Pezzo surface with A_2 singularity*, 2007, `arXiv 0710.1583`. MR2318651 (2008i:14053)
20. U. Derenthal and D. Loughran, *Singular del Pezzo surfaces that are equivariant compactifications*, J. of Math. Sci. **171** (2010), no. 6, 714–724. MR2753646 (2012b:14068)
21. J. Franke, Y. I. Manin, and Y. Tschinkel, *Rational points of bounded height on Fano varieties*, Invent. Math. **95** (1989), no. 2, 421–435. MR974910 (89m:11060)
22. B. Hassett and Y. Tschinkel, *Geometry of equivariant compactifications of $\mathbf{G}_a^n$*, Internat. Math. Res. Notices (1999), no. 22, 1211–1230. MR1731473 (2000j:14073)
23. J.-I. Igusa, *An introduction to the theory of local zeta functions*, AMS/IP Studies in Advanced Mathematics, vol. 14, American Mathematical Society, Providence, RI, 2000. MR1743467 (2001j:11112)
24. D. Mumford, J. Fogarty, and F. Kirwan, *Geometric invariant theory*, third ed., Ergebnisse der Mathematik und ihrer Grenzgebiete (2) [Results in Mathematics and Related Areas (2)], vol. 34, Springer-Verlag, Berlin, 1994. MR1304906 (95m:14012)
25. E. Peyre, *Hauteurs et mesures de Tamagawa sur les variétés de Fano*, Duke Math. J. **79** (1995), no. 1, 101–218. MR1340296 (96h:11062)
26. ______, *Terme principal de la fonction zêta des hauteurs et torseurs universels*, Astérisque (1998), no. 251, 259–298, Nombre et répartition de points de hauteur bornée (Paris, 1996). MR1679842 (2000f:11081)
27. ______, *Torseurs universels et méthode du cercle*, Rational points on algebraic varieties, Progr. Math., vol. 199, Birkhäuser, Basel, 2001, pp. 221–274. MR1875176 (2003d:11092)
28. P. Salberger, *Tamagawa measures on universal torsors and points of bounded height on Fano varieties*, Astérisque (1998), no. 251, 91–258, Nombre et répartition de points de hauteur bornée (Paris, 1996). MR1679841 (2000d:11091)
29. J. Shalika, *Informal notes*.
30. J. Shalika, R. Takloo-Bighash, and Y. Tschinkel, *Rational points on compactifications of semi-simple groups*, J. Amer. Math. Soc. **20** (2007), no. 4, 1135–1186 (electronic). MR2328719 (2008g:14034)
31. J. Shalika and Y. Tschinkel, *Height zeta functions of equivariant compactifications of unipotent groups*, preprint.
32. ______, *Height zeta functions of equivariant compactifications of the Heisenberg group*, Contributions to automorphic forms, geometry, and number theory, Johns Hopkins Univ. Press, Baltimore, MD, 2004, pp. 743–771. MR2058627 (2005c:11110)
33. E. V. Sharoĭko, *The Hassett-Tschinkel correspondence and automorphisms of a quadric*, Mat. Sb. **200** (2009), no. 11, 145–160. MR2590000 (2011a:14095)

34. M. Strauch, *Arithmetic stratifications and partial Eisenstein series*, Rational points on algebraic varieties, Progr. Math., vol. 199, Birkhäuser, Basel, 2001, pp. 335–355. MR1875180 (2002m:11057)
35. M. Strauch and Y. Tschinkel, *Height zeta functions of toric bundles over flag varieties*, Selecta Math. (N.S.) **5** (1999), no. 3, 325–396. MR1723811 (2001h:14028)
36. H. Sumihiro, *Equivariant completion*, J. Math. Kyoto Univ. **14** (1974), 1–28. MR0337963 (49:2732)
37. Y. Tschinkel, *Algebraic varieties with many rational points*, Arithmetic geometry, Clay Math. Proc., vol. 8, Amer. Math. Soc., Providence, RI, 2009, pp. 243–334. MR2498064 (2010j:14039)
38. A. Weil, *Adeles and algebraic groups*, Progress in Mathematics, vol. 23, Birkhäuser Boston, Mass., 1982, With appendices by M. Demazure and Takashi Ono. MR670072 (83m:10032)

Courant Institute, NYU, 251 Mercer Str., New York, NY 10012, USA
E-mail address: `tanimoto@cims.nyu.edu`

Courant Institute, NYU, 251 Mercer Str., New York, NY 10012, USA
E-mail address: `tschinkel@cims.nyu.edu`

Part III: Motivic zeta functions, Poincaré series, complex monodromy and knots

Contemporary Mathematics
Volume **566**, 2012
http://dx.doi.org/10.1090/conm/566/11220

Singularity invariants related to Milnor fibers: survey

Nero Budur

ABSTRACT. This brief survey of some singularity invariants related to Milnor fibers should serve as a quick guide to references. We attempt to place things into a wide geometric context while leaving technicalities aside. We focus on relations among different invariants and on the practical aspect of computing them.

CONTENTS

Trivia: in how many different ways can the log canonical threshold of a polynomial be computed ? At least 6 ways in general, plus 4 more ways with some luck.

Singularity theory is a subject deeply connected with many other fields of mathematics. We give a brief survey of some singularity invariants related to Milnor fibers that should serve as a quick guide to references. This is by no means an exhaustive survey and many topics are left out. What we offer in this survey is an attempt to place things into a wide geometric context while leaving technicalities aside. We focus on relations among different invariants and on the practical aspect of computing them. Along the way we recall some questions to serve as food for thought.

To achieve the goals we set with this survey, we pay a price. This is not a historical survey, in the sense that general references are mentioned when available, rather than pinpointing the important contributions made along the way to the current shape of a certain result. We stress that this is not a comprehensive survey and the choices of reflect bias.

In the first part we are concerned with theoretical aspects: definitions and relations. In the second part we focus on the practical aspect of computing singularity invariants and we review certain classes of singularities.

2010 *Mathematics Subject Classification.* Primary 14-02, 14B05, 14J17, 32S05, 32S35, 32S40, 32S45, 32S55, 58K10.
This work was partially supported by NSF and NSA.

I would like to thank A. Dimca, G.-M. Greuel, K. Sugiyama, K. Takeuchi, and W. Veys for their help, comments, and suggestions. Also I would like to thank Université de Nice for their hospitality during the writing of this article.

1. Theoretical aspects

1.1. Topology.

Milnor fiber and monodromy. Let f be a hypersurface singularity germ at the origin in $\mathbb{C}^n$.

Let

$$M_t := f^{-1}(t) \cap B_\epsilon,$$

where B_ϵ is a ball of radius ϵ around the origin. Small values of ϵ and even smaller values of $|t|$ do not change the diffeomorphism class of M_t, the *Milnor fiber of f at 0* [**89, 80**].

Fix a Milnor fiber M_t and let

$$M_{f,0} := M_t.$$

The cohomology groups $H^i(M_{f,0}, \mathbb{C})$ admit an action T called *monodromy* generated by going once around a loop starting at t around 0. The eigenvalues of the monodromy action T are roots of unity, [**89, 80**].

The *monodromy zeta function of f at* 0 is

$$Z_0^{mon}(s) := \prod_{j\in\mathbb{Z}} \det(1 - sT, H^j(M_{f,0}, \mathbb{C}))^{(-1)^j}.$$

The *m-th Lefschetz number of f at* 0 is

$$\Lambda(T^m) := \sum_{j\in\mathbb{Z}} (-1)^j \mathrm{Trace}\,(T^m, H^j(M_{f,0}, \mathbb{C})).$$

These numbers recover the monodromy zeta function: if $\Lambda(T^m) = \sum_{i|m} s_i$ for $m \geq 1$, then $Z_0^{mon}(s) = \prod_{i\geq 1}(1 - s^i)^{s_i/i}$, [**39**].

When f has an isolated singularity,

$$\dim_{\mathbb{C}} H^j(M_{f,0}, \mathbb{C}) = \begin{cases} 0 & \text{for } j \neq 0, n-1, \\ 1 & \text{for } j = 0, \\ \dim_{\mathbb{C}} \mathbb{C}[[x_1, \ldots, x_n]]/\left(\frac{\partial f}{\partial x_1}, \ldots, \frac{\partial f}{\partial x_n}\right) & \text{for } j = n-1. \end{cases}$$

The last value for $j = n - 1$ is denoted $\mu(f)$ and called the *Milnor number of f*, [**89, 80**].

The most recent and complete textbook on the basics, necessary to understand many of the advanced topics here is [**60**].

Constructible sheaves. Let X be a nonsingular complex variety and Z a closed subscheme.

The Milnor fiber and the monodromy can be generalized to this setting. Let $D_c^b(X)$ be the derived category of bounded complexes of sheaves of $\mathbb{C}$-vector spaces with constructible cohomology in the analytic topology of X, [**39**].

If Z is a hypersurface given by a regular function f, one has Deligne's *nearby cycles functor*. This is the composition of derived functors

$$\psi_f := i^* p_* p^* : D^b_c(X) \to D^b_c(Z_{red}),$$

where i is the inclusion of Z_{red} in X, $\tilde{\mathbb{C}}^*$ is the universal cover of $\mathbb{C}^*$, and $p : X \times_{\mathbb{C}} \tilde{\mathbb{C}}^* \to X$ is the natural projection. If i_x is the inclusion of a point x in Z_{red} and $M_{f,x}$ is the Milnor fiber of f at x, then

$$H^i(i_x^* \psi_f \mathbb{C}_X) = H^i(M_{f,x}, \mathbb{C})$$

and there is an induced action recovering the monodromy, [**39**].

When Z is closed subscheme one has *Verdier's specialization functor* Sp_Z. This is defined by using ψ_t, where $t : \mathcal{X} \to \mathbb{C}$ is the deformation to the normal cone of Z in X. This functor recovers the nearby cycles functor ψ_f in the case when $Z = \{f = 0\}$, [**127**]. Another functor, also recovering the nearby cycles functor, seemingly depending on equations $f = (f_1, \dots, f_r)$ for Z, is *Sabbah's specialization functor* ${}^A\psi_f$. This is defined by replacing in the definition of the nearby cycles functor $\mathbb{C}$ and $\mathbb{C}^*$ with $\mathbb{C}^r$ and $(\mathbb{C}^*)^r$, respectively, [**107**]. It would be interesting to understand the differences between these two functors.

1.2. Analysis.

Asymptotic expansions. Let f be a hypersurface germ with an isolated singularity in $\mathbb{C}^n$.

Let σ be a top relative holomorphic form on the Milnor fibration $M \to S$, where S is a small disc and $M = \cup_{t \in S} M_t$. Let δ_t be a flat family of cycles in $H_{n-1}(M_t, \mathbb{C})$ for $t \neq 0$. Then

$$\lim_{t \to 0} \int_{\delta_t} \sigma = \sum_{\alpha \in \mathbb{Q}, k \in \mathbb{N}} a(\sigma, \delta, \alpha, k) \cdot t^{\alpha} (\log t)^k$$

where a are constants. The infimum of rational numbers α that can appear in such expansion for some σ and δ is Arnold's *complex oscillation index.* This is an analytic invariant, [**3, 79**].

L^2-multipliers. Let f be a collection of polynomials $f_1, \dots, f_r$ in $\mathbb{C}[x_1, \dots, x_n]$, and let c be a positive real number.

The *multiplier ideal of f with coefficient* c of Nadel is the ideal sheaf $\mathcal{J}(f^c)$ consisting locally of holomorphic functions g such that $|g|^2/(\sum_i |f_i|^2)^c$ is locally integrable. This is a coherent ideal sheaf. The intuition behind this analytic invariant is: the smaller the multiplier ideals are, the worse the singularities of the zero locus of f are, [**81**].

The smallest c such that $\mathcal{J}(f^c) \neq \mathcal{O}_X$, i.e. $1 \notin \mathcal{J}(f^c)$, is called the *log canonical threshold* of f and is denoted $lct\,(f)$.

Log canonical thresholds are a special set of numbers: for a fixed n, the set $\{lct\,(f) \mid f \in \mathbb{C}[x_1, \dots, x_n]\}$ satisfies the ascending chain condition, [**30**].

When f is only one polynomial with an isolated singularity, the log canonical threshold coincides with 1+ Arnold's complex oscillation index [**79**].

The definition of the multiplier ideal generalizes and patches up to define, globally on a nonsingular variety X with a subscheme Z, a *multiplier ideal sheaf* $\mathcal{J}(X, c \cdot Z)$ in $\mathcal{O}_X$. In fact, the multiplier ideal $\mathcal{J}(X, c \cdot Z)$ depends only on the integral closure of the ideal of Z in X. One has similarly a *log canonical threshold* for Z in X, denoted $lct\,(X, Z)$, [**81**].

1.3. Geometry.

Resolution of singularities. Let X be a nonsingular complex variety and Z a closed subscheme.

Let $\mu : Y \to X$ be a *log resolution* of (X, Z). This means that Y is nonsingular, μ is birational and proper, and the inverse image of Z together with the support of the determinant of the Jacobian of μ is a simple normal crossings divisor. This exists by Hironaka. Denote by $K_{Y/X} = \sum_{i\in S} k_i E_i$ the divisor given by the determinant of the Jacobian of μ. Denote by $E = \sum_{i\in S} a_i E_i$ the divisor in Y given by Z. Here E_i are irreducible divisors. For $I \subset S$, let $E_I^o := \cap_{i\in I} E_i - \cup_{i\notin S} E_i$.

Let $c \in \mathbb{R}_{>0}$. Then Nadel's multiplier ideal equals

$$\mathcal{J}(X, c\cdot Z) = \mu_* \mathcal{O}_Y(K_{Y/X} - \lfloor c\cdot E\rfloor).$$

Here $\lfloor\cdot\rfloor$ takes the round-down of the coefficients of the irreducible components of a divisor, [**81**].

In particular, the log canonical threshold is given by

$$lct\,(X, Z) = \min_i \left\{ \frac{k_i + 1}{a_i} \right\}. \tag{1.1}$$

When $Z = \{f = 0\}$ is a hypersurface and $x \in Z$ is a point, the monodromy zeta function at x and the Lefschetz numbers can be computed from the log resolution by *A'Campo formula*:

$$\Lambda(T^m) = \sum_{a_i | m} a_i \cdot \chi(E_i^o \cap \mu^{-1}(x)),$$

where χ is the topological Euler characteristic, [**39**].

One can imitate the construction via log resolutions to define *multiplier ideals for any linear combination of subschemes*, or equivalently, *of ideals*:

$$\mathcal{J}(X, c_1\cdot Z_1 + \ldots + c_r\cdot Z_r) = \mathcal{J}(X, I_{Z_1}^{c_1}\cdot\ldots\cdot I_{Z_r}^{c_r}).$$

If $X = \mathbb{C}^n$ and the ambient dimension n is < 3, every integrally closed ideal is a multiplier ideal [**84**]. This is not so if $n \geq 3$, [**82**].

The *jumping numbers of Z in X* are those numbers c such that

$$\mathcal{J}(c\cdot Z) \neq \mathcal{J}((c-\epsilon)\cdot Z)$$

for all $\epsilon > 0$. The log canonical threshold $lct\,(X, Z)$ is the smallest jumping number. The list of jumping numbers is another numerical analytic invariant of the singularities of Z in X. The list contains finitely many numbers in any compact interval, all rational numbers, and is periodic. If $lct\,(f) = c_1 < c_2 < \ldots$ denotes the list of jumping numbers then

$$c_{i+1} \leq c_1 + c_i.$$

The standard reference for jumping numbers and multiplier ideals is [**81**].

For a point x in Z, the *inner jumping multiplicity of c at x* is the vector space dimension

$$m_{c,x} := \dim_{\mathbb{C}} \mathcal{J}(X, (c-\epsilon)\cdot Z)/\mathcal{J}(X, (c-\epsilon)\cdot Z + \delta\cdot\{x\}),$$

where $0 < \epsilon \ll \delta \ll 1$. This multiplicity measures the contribution of the singular point x to the jumping number c, [**18**].

Another interesting singularity invariant is the Denef-Loeser *topological zeta function*. This is the rational function of complex variable s defined as

$$Z_Z^{top}(s) := \sum_{I\subset S} \chi(E_I^o)\cdot \prod_{i\in I} \frac{1}{a_i s + k_i + 1}.$$

This is independent of the choice of log resolution, [**36**]. In spite of the name, $Z_f^{top}(s)$ is not a topological invariant, [**5**].

When $Z = \{f = 0\}$ is a hypersurface, the *Monodromy Conjecture* states that if c is pole of the topological zeta function, then $e^{2\pi i c}$ is an eigenvalue of the Milnor monodromy of f at some point in $f^{-1}(0)$, [**36**]. A similar conjecture, using Verdier's specialization functor Sp_Z, can be made when Z is not a hypersurface, [**126**].

Mixed Hodge structures. Let X be a nonsingular complex variety and Z a closed subscheme.

The topological package consisting of the Milnor fibers, monodromy, nearby cycles functor, specialization functor can be enhanced to take into account natural mixed Hodge structures, [**108**].

Consider the case when Z is a hypersurface given by one polynomial $f \in \mathbb{C}[x_1, \ldots, x_n]$ with the origin included in the singular locus. The *Hodge spectrum of f at* 0 of Steenbrink is

$$Sp(f, 0) = \sum_{c>0} n_{c,0}(f)\cdot t^c,$$

where the *spectrum multiplicities*

$$n_{c,0}(f) := \sum_{i\in\mathbb{Z}} (-1)^{n-1-i} \dim_{\mathbb{C}} \mathrm{Gr}_F^{\lfloor n-c\rfloor} \widetilde{H}^i(M_{f,0}, \mathbb{C})_{e^{-2\pi i c}}$$

record the generalized Euler characteristic on the $\lfloor n-c\rfloor$-graded piece of the Hodge filtration on the $\exp(-2\pi i c)$-monodromy eigenspace on the reduced cohomology of the Milnor fiber. These invariants can be refined by considering the weight filtration as well, [**80**].

In the case of isolated hypersurface singularities, the spectrum recovers the Milnor number

$$\mu(f) = \sum_c n_{c,0}(f)$$

and, by M. Saito, the *geometric genus* of the singularity

$$\sum_{0<c\leq 1} n_{c,0}(f) = p_g(f) := \begin{cases} \dim_{\mathbb{C}}(R^{n-2}p_*\mathcal{O}_{\tilde{Z}})_0 & \text{if } n \geq 3, \\ \dim_{\mathbb{C}}(p_*\mathcal{O}_{\tilde{Z}}/\mathcal{O}_Z)_0 & \text{if } n = 2, \end{cases}$$

where $p : \tilde{Z} \to Z$ is a log resolution of Z, [**80**]. The spectrum also satisfies a symmetry $n_{c,0}(f) = n_{n-c,0}(f)$, and a semicontinuity property. It is thus useful in the classification of such singularities, [**80**]. The smallest spectral number c equals the log canonical threshold $lct(f)$. Let $c_1 \leq \ldots \leq c_{\mu(f)}$ denote the list of spectral numbers counted with the spectrum multiplicities. An open question is *Hertling's Conjecture* stating that

$$\frac{1}{\mu(f)}\sum_{i=1}^{\mu(f)}\left(c_i-\frac{n}{2}\right)^2 \leq \frac{c_{\mu(f)}-c_1}{12}.$$

This has been solved for quasi-homogeneous singularities [**69**], where equality holds. This is due to a duality with the spectrum of the Milnor fiber at infity, for which in general a similar conjecture is made but with reversed sign, [**38**]. Other solved cases are: irreducible plane curves [**110**] and Newton nondegenerate polynomials of two variables [**13**]. Another open question is *Durfee's Conjecture* of [**43**] that for $n=3$,

$$6p_g(f) \leq \mu(f).$$

This was shown to be true in the following cases: quasi-homogeneous [**131**], weakly elliptic, $f = g(x,y)+z^N$ [**8, 97**], double point [**123**], triple point [**7**], absolutely isolated [**88**].

The jumping numbers are also related to Milnor fibers and monodromy. If the singularity is isolated, the spectrum recovers all the jumping numbers in $(0,1)$. In general, when the singularities are not necessarily isolated, we have more precisely that the spectrum multiplicities for $c \in (0,1]$ are computed in terms of the inner jumping multiplicities of jumping numbers: $m_{c,x}(f) = n_{c,x}(f)$, [**18**].

A sufficient condition for symmetry of the spectrum of a homogeneous polynomial in the non-isolated case is given in [**41**]-Prop. 4.1.

The Hodge spectrum has a generalization to any subscheme Z in a nonsingular variety X, using Verdier's specialization functor and M. Saito's mixed Hodge modules. There is a relation between the multiplier ideals and the specialization functor, [**40**].

Jets and arcs. Let X be a nonsingular complex variety of dimension n and Z a closed subscheme.

The *scheme of m-jets* and the *arc space* of Z are

$$Z_m := \mathrm{Hom}(\mathrm{Spec}\ \mathbb{C}[t]/(t^{m+1}), Z) \quad \text{respectively}$$
$$Z_\infty := \mathrm{Hom}(\mathrm{Spec}\ \mathbb{C}[[t]], Z).$$

Jets compute log canonical thresholds by Mustaţă's formula [**91**]:

$$lct(X,Z) = \min_m \left\{\frac{\mathrm{codim}\,(Z_m, X_m)}{m+1}\right\}.$$

If $Z = \{f=0\}$ is a hypersurface, one has the *Denef-Loeser motivic zeta function*:

$$Z_f^{mot}(s) := \sum_{m\geq 1}[X_{m,1}][\mathbb{A}^1]^{-mn}s^m,$$

where $[.]$ denotes the class of a variety in an appropriate Grothendieck ring, and $X_{m,1}$ consists of the m-jets ϕ of X such that $f(\phi)=t^m$. The motivic zeta function is a rational function. The monodromy zeta function, the Hodge spectrum, and the topological zeta function can be recovered from the motivic zeta function. Thus these singularity invariants can be computed from jets. In fact, one has the *motivic Milnor fiber*

$$\mathcal{S}_f := -\lim_{s\to\infty} Z_f^{mot}(s),$$

which is a common generalization of the monodromy zeta function and of the Hodge spectrum, [**36**].

The motivic zeta function can be defined also when Z is a closed subscheme with equations $f = (f_1, \dots, f_r)$. The *Monodromy Conjecture* can be stated for the motivic zeta function and implies the previous version, [**98**]. The analog of the motivic Milnor fiber $\mathcal{S}_f$ is related in this case with Sabbah's specialization functor ${}^A\psi_f$: it recovers the generalization via ${}^A\psi_f$ of the monodromy zeta function, [**62**].

The biggest pole of $Z_f^{mot}(s)$ gives the negative of the log canonical threshold, [**66**]-p.18.

The motivic zeta function is a *motivic integral.* Without explaining what this is, a motivic integral enjoys a change of variables formula. In practice this means that a motivic integral can be computed from a log resolution. From this very advanced point of view, one can see more naturally A'Campo's formula for the monodromy zeta function and the connection between jumping numbers and the Hodge spectrum, [**36**].

The change of variables formula can be streamlined and the motivic integration eliminated. The *m-th contact locus of Z in X* is the subset of X_∞ consisting of arcs of order m along Z. The contact loci can also be expressed in terms of log resolutions and exceptional divisors in log resolutions give rise to components of contact loci. In many cases this also gives a back-and-forth pass between arc-theoretic invariants and invariants defined by log resolutions, such as Mustaţă's result on the log canonical threshold. Multiplier ideals and jumping numbers can also be interpreted arc-theoretically, [**44**].

If Z is a normal local complete intersection variety, then Z_m is equidimensional (respectively irreducible, normal) for every m if and only if Z has log canonical (canonical, terminal) singularities, [**46**].

1.4. Algebra.

Riemann-Hilbert correspondence. Let X be nonsingular complex variety of dimension n.

The *sheaf of algebraic differential operators* $\mathcal{D}_X$ is locally given in affine coordinates by the Weil algebra $\mathbb{C}[x_1, \dots, x_n, \partial/\partial x_1, \dots, \partial/\partial x_n]$. An important class of (left) $\mathcal{D}_X$-modules consists of those *regular and holonomic.* Let $D^b_{rh}(\mathcal{D}_X)$ be the bounded derived category of complexes of $\mathcal{D}_X$-modules with regular holonomic cohomology, [**11**].

One of the main reasons why the theory of $\mathcal{D}$-modules has become important recently is because of its suitability for computer calculations.

The topological package, consisting of the bounded derived category of constructible sheaves $D^b_c(X)$ and the natural functors attached to it, has an algebraic counterpart. There is a well-defined functor

$$DR : D^b_{rh}(\mathcal{D}_X) \to D^b_c(X)$$

which is an equivalence of categories commuting with the usual functors, [**11**].

The $\mathcal{D}$-module theoretic counterpart of the nearby cycles functor ψ_f, hence of the Milnor monodromy of f, is achieved by the *V-filtration along f* of Malgrange-Kashiwara. For $c \in (0,1)$,

$$\psi_{f,\lambda}\mathbb{C}_X[-1] = DR(\mathrm{Gr}_V^c \widetilde{\mathcal{O}_X}),$$

where $\lambda = e^{-2\pi i c}$, $\psi_f = \oplus_{\lambda'}\psi_{f,\lambda'}$ is the functor decomposition corresponding to the eigenspace decomposition of the semisimple part of the Milnor monodromy, and $\widetilde{\mathcal{O}_X} = \mathcal{O}_X[\partial_t]$ is the $\mathcal{D}$-module push-forward of $\mathcal{O}_X$ under the graph embedding of f, [**19**].

In algebraic geometry, integral (co)homology groups are endowed with additional structure: mixed Hodge structures, [**104**]. The modern point of view is M. Saito's theory of *mixed Hodge modules*. The derived category of mixed Hodge modules $D^b(MHM(X))$ has natural forgetful functors to $D^b_{rh}(\mathcal{D}_X)$ and $D^b_c(X)$ and recovers Deligne's mixed Hodge structures on the usual (co)homology groups. When Z is a closed subvariety of X, the Verdier specialization functor Sp_Z also exists in the framework of mixed Hodge modules, [**108**].

Let Z be a closed subscheme of X, and let $\widetilde{\mathcal{O}_X}$ be the $\mathcal{D}$-module push-forward of $\mathcal{O}_X$ under the graph embedding of a set of local generators of the ideal of Z in X. The smallest nontrivial piece of the Hodge filtration of the V-filtration on $\widetilde{\mathcal{O}_X}$ gives the multiplier ideals:

$$\mathcal{J}(X,(c-\epsilon)\cdot Z) = F^{n-1}V^c\widetilde{\mathcal{O}_X},$$

with $0<\epsilon\ll 1$. This is another point of view on the relation between multiplier ideals, mixed Hodge structures, and Milnor monodromy, [**20**].

b-functions. Let X be a nonsingular complex variety of dimension n and Z a closed subscheme.

Suppose first that Z is given by an ideal $f=(f_1,\ldots,f_r)$ with $f_i\in\mathbb{C}[x_1,\ldots,x_n]$. Let g be another polynomial in n variables. The *generalized b-function of f twisted by g*, also called the *generalized Bernstein-Sato polynomial of f twisted by g* and denoted $b_{f,g}(s)$, is the nonzero monic polynomial of minimal degree among those $b\in\mathbb{C}[s]$ such that

$$b(s_1+\ldots+s_r)g\prod_{i=1}^r f_i^{s_i} = \sum_{k=1}^r P_k(gf_k\prod_{i=1}^r f_i^{s_i}),$$

for some algebraic operators $P_k \in \mathbb{C}[x_1,\ldots,x_n,\partial/\partial x_1,\ldots,\partial/\partial x_n][s_{ij}]_{1\le i,j\le r}$, where s_{ij} are defined as follows. First, let the operator t_i act by leaving s_j alone if $i\neq j$, and replacing s_i with s_i+1. For example: $t_j\prod_{i=1}^r f_i^{s_i} = f_j\prod_{i=1}^r f_i^{s_i}$. Then $s_{ij} := s_it_i^{-1}t_j$. The generalized b-function is independent of the choice of local generators $f_1,\ldots,f_r$ for the ideal of Z, [**20**].

The *b-function of the ideal f* is $b_f(s) := b_{f,1}(s)$. When Z is a hypersurface, $b_f(s)$ is the usual b-function of Bernstein and Sato, satisfying the relation

$$b_f(s)f^s = Pf^{s+1}$$

for some operator $P_k\in\mathbb{C}[x_1,\ldots,x_n,\partial/\partial x_1,\ldots,\partial/\partial x_n][s]$.

For a scheme Z, the *b-function of* Z is the polynomial $b_Z(s)$ obtained by replacing s with $s - \mathrm{codim}\,(Z,X)$ in the lowest common multiple of the polynomials $b_f(s)$ obtained by varying local charts of a closed embedding of Z into a nonsingular X. This polynomial depends only on Z. The roots of $b_{f,g}(s)$ and $b_f(s)$ are negative rational numbers, [**20**].

For simplicity, say X is affine from now and the ideal of Z is f. The b-function recovers the monodromy eigenvalues, as observed originally by Malgrange and Kashiwara. If Z is a hypersurface, the set consisting of $e^{2\pi i c}$, where c are roots of $b_f(s)$, is the set of eigenvalues of the Milnor monodromy at points along Z. In higher codimension, a similar statement holds for eigenvalues related with the specialization functor Sp_Z, [**20**].

The *Strong Monodromy Conjecture* states that if c is a pole of the topological zeta function $Z_f^{top}(s)$ of the ideal f, then $b_f(c) = 0$. It can be stated for the motivic zeta function as well. It implies the Monodromy Conjecture.

The biggest root of the b-function $b_f(s)$ of the ideal of Z is the negative of the log canonical threshold of (X, Z), [**79, 20**].

If $lct\,(X,Z) \leq c < lct\,(X,Z) + 1$ and c is a jumping number of Z in X, then $b_f(-c) = 0$, [**45, 20**].

Generalized b-functions recover multiplier ideals, [**20**]:

$$\mathcal{J}(X, c \cdot Z) =_{loc} \{g \in \mathcal{O}_X \mid c < \alpha \text{ if } b_{f,g}(-\alpha) = 0\}.$$

Generalized b-functions are related to the V-filtration along f. More precisely, $b_{f,g}(s)$ is the minimal polynomial of the action of

$$s = -(\partial_1 t_1 + \ldots \partial_r t_r) \text{ on } V^0\mathcal{D}_Y(g \otimes 1)/V^1\mathcal{D}_Y(g \otimes 1).$$

Here $Y = X \times \mathbb{C}^r$, the coordinate functions on $\mathbb{C}^r$ are $t_1, \ldots, t_r$, the operator ∂_j is $\partial/\partial t_j$, $\widetilde{\mathcal{O}_X} = \mathcal{O}_X[\partial_1, \ldots, \partial_r]$ is viewed as a $\mathcal{D}_Y$-module via the graph embedding of f, $g \otimes 1 \in \widetilde{\mathcal{O}_X}$, and $V^i\mathcal{D}_Y$ consists of operators P in $\mathcal{D}_Y$ such that $P(t_1, \ldots, t_r)^i \subset (t_1, \ldots, t_r)^{i+j}$ for all $j \in \mathbb{Z}$, [**20**].

The b-function of a polynomial can sometimes be calculated via *microlocal calculus*. This method has been successful for computation of relative invariants of irreducible regular prehomogeneous vector spaces, see below, [**76**].

1.5. Arithmetic.

K-log canonical thresholds. Let $f \in K[x_1, \ldots, x_n]$ be a polynomial with coefficients in a complete field K of characteristic zero.

By Hironaka, f admits a K-analytic log resolution of the zero locus of f in K^n. In general this might be too small to be a log resolution over an algebraic closure of K, in the sense that one only sees K-rational points of such a resolution. We can define vanishing orders a_i of f and k_i of $dx_1 \wedge \ldots \wedge dx_n$ along the hypersurfaces in this K-analytic log resolution. Then one defines the *K-log canonical threshold of* f, which we denote $lct_K(f)$, by the formula (1.1). We have that $lct\,(f) = lct_{\mathbb{C}}(f)$ for any embedding $K \subset \mathbb{C}$, but in general $lct\,(f) \leq lct_K(f)$, due to vanishing $E_i(K) = \emptyset$ of the K-rational points of divisors in a log resolution over $\mathbb{C}$. This can be generalized to the case of an ideal f of polynomials with coefficients in K, [**129**].

If $K = \mathbb{R}$ one can also define the *real jumping numbers* of f. It can happen that a real jumping number is not an usual jumping number. However, any real jumping number smaller than $lct_{\mathbb{R}}(f) + 1$ is a root of $b_f(-s)$, [**112**].

p-adic local zeta functions. Let p be a prime number and K a finite extension field of $\mathbb{Q}_p$ with a fixed embedding into $\mathbb{C}$. Let f be an ideal of polynomials in $K[x_1, \ldots, x_n]$.

If f is a single polynomial with coefficients in $\mathbb{Q}_p$, the *Igusa p-adic local zeta function of f* is defined for a character χ on the units of $\mathbb{Z}_p$ as

$$Z^p_{f,\chi}(s) := \int_{\mathbb{Z}_p^n} |f(x)|^s \chi(ac(f(x)))dx \ ,$$

where $|t| = p^{-ord_p t}$, $ac(t) = |t|t$, and dx is the Haar measure normalized such that the measure of $p^{m_1}\mathbb{Z}_p \times \ldots \times p^{m_n}\mathbb{Z}_p$ is $p^{-(m_1+\ldots+m_n)}$. Let $Z^p_f(s) = Z^p_{f,1}(s)$, [**32**].

The poles of $Z^p_{f,\chi}(s)$ determine by [**73**] the asymptotic expansion as $|t| \to 0$ of the numbers

$$N_m(t) := \{x \in (\mathbb{Z}/p^m\mathbb{Z})^n \mid f(x) \equiv t \bmod p^m\} \quad (m \gg 0).$$

The definition of $Z^p_f(s)$ can be made more generally for an ideal f of polynomials with coefficients in a finite extension K of $\mathbb{Q}_p$. The p-adic local zeta functions are rational, [**71**].

If $K = \mathbb{Q}_p$, the K-log canonical threshold is determined by the numbers $N_m := N_m(0)$:

$$\lim_{m\to\infty} (N_m)^{1/m} = p^{\,n-lct_K(f)}.$$

A similar statement holds for a finite extension K of $\mathbb{Q}_p$, see [**129**].

If c is the pole of $Z^p_f(s)$ with the biggest real part, then $lct_K(f) = -Re(c)$, [**129**].

The motivic zeta function $Z^{mot}_f(s)$ of Denef-Loeser determines the p-adic local zeta function $Z^p_f(s)$, [**36**].

If f is a single polynomial, Igusa's original *Monodromy Conjecture* states that if c is a pole of $Z^p_f(s)$ then $e^{2\pi i Re(c)}$ is an eigenvalue of the Milnor monodromy of $f_{\mathbb{C}}$ at some point of $f_{\mathbb{C}}^{-1}(0)$, where $f_{\mathbb{C}}$ is f viewed as a polynomial with complex coefficients. If f is an ideal defining a subscheme Z, the Monodromy Conjecture is stated via Verdier's specialization functor Sp_Z. The *Strong Monodromy Conjecture* states that if f is an ideal and c is a pole of $Z^p_f(s)$, then $b_f(Re(c)) = 0$, [**74**].

Test ideals. Let p be a prime number and f be an ideal of polynomials in $\mathbb{F}_p[x_1, \ldots, x_n]$.

The Hara-Yoshida *test ideal of f with coefficient* c, where c is positive real number, is

$$\tau(f^c) := \left(f^{\lceil cp^e \rceil}\right)^{[1/p^e]}, \quad e \gg 0,$$

where for an ideal I, the ideal $I^{[1/p^e]}$ is defined as follows. This is the unique smallest ideal J such that $I \subset \{u^{p^e} \mid u \in J\}$, [**10**].

The *F-jumping numbers* of f are the positive real numbers c such that

$$\tau(f^c) \neq \tau(f^{c-\epsilon})$$

for all $\epsilon > 0$. The Takagi-Watanabe *F-pure threshold* of f is the smallest F-jumping number and is denoted $fpt\,(f)$, [**10**].

Test ideals, F-jumping numbers, and the F-threshold are positive characteristic analogs of multiplier ideals, jumping numbers, and respectively, the log canonical threshold, [**68**]. More precisely, let now f be an ideal of polynomials in $\mathbb{Q}[x_1, \dots, x_n]$. For large prime numbers p, let $f_p \subset \mathbb{F}_p[x_1, \dots, x_n]$ denote the reduction modulo p of f. Fix $c > 0$. Then for $p \gg 0$,

$$\tau(f_p^c) = \mathcal{J}(f^c)_p$$

and

$$\lim_{p\to\infty} fpt\,(f_p) = lct\,(f).$$

The *Hara-Watanabe Conjecture* [**67**] states that there are infinitely many prime numbers p such that for all $c > 0$,

$$\tau(f_p^c) = \mathcal{J}(f^c)_p.$$

The list of F-jumping numbers enjoys similar properties as the list of jumping numbers: rationality, discreteness, and periodicity, [**10**]. However, in any ambient dimension, every ideal is a test ideal, in contrast with the speciality of the multiplier ideals, [**96**].

There are results connecting test ideals with b-functions. If $f \in \mathbb{Q}[x_1, \dots, x_n]$ is a single polynomial and c is an F-jumping number of the reduction f_p for some $p \gg 0$, then $\lceil cp^e \rceil - 1$ is a root of $b_f(s)$ modulo p, [**95, 94**].

1.6. Remarks and questions.

Answer to the trivia question. In how many ways can the log canonical threshold of a polynomial be computed? We summarize some of the things we have talked about so far. The lct can be computed, theoretically, via: the L^2 condition, the orders of vanishing on a log resolution, the growth of the codimension of jet schemes, the poles of the motivic zeta function, the b-function, and the test ideals. If a log resolution over $\mathbb{C}$ is practically the same as a K-analytic log resolution over a p-adic field K containing all the coefficients of the polynomial, i.e. if $lct_K(f) = lct\,(f)$, then there are two more ways: via the poles of p-adic local zeta functions and via the asymptotics of the number of solutions modulo p^m. If the singularity is isolated it can also be done via Arnold's complex oscillation index and via the Hodge spectrum. So, 6+2+2 ways.

What topics were left out. May topics are left out from this survey: singularities of varieties inside singular ambient spaces, the Milnor fiber at infinity, the characteristics classes point of view on singularities, other invariants such as polar and Le numbers, the theory of Brieskorn lattices, local systems, archimedean local zeta functions, deformations, equisingularity, etc.

Questions. We have already mentioned the Monodromy Conjecture and its Strong version, the Hertling Conjecture, the Durfee Conjecture, and the Hara-Watanabe Conjecture.

It is not known how to relate b-functions with jets and arcs. In principle, this would help with the Strong Monodromy Conjecture.

We know little about the most natural singularity invariant, the *multiplicity*. *Zariski conjecture* states that if two reduced hypersurface singularity germs in $\mathbb{C}^n$ are embedded-topologically equivalent then their multiplicities are the same. Even the isolated singularity case is not known, [**50**]. It known to be true for semi-quasihomogeneous singularities: [**59**], and slightly weaker, [**103**].

We can raise the same question for log canonical thresholds. Can one find an example of two reduced hypersurface singularity germs in $\mathbb{C}^n$ that are embedded-topologically equivalent but have different log canonical thresholds?

Is the biggest pole of the topological zeta function $Z_f^{top}(s)$ of a polynomial f equal to $-lct\,(f)$? This true for 2 variables, [**128**].

For a polynomial f with coefficients in a complete field K of characteristic zero, define K-jumping numbers and prove the ones $< lct_K(f) + 1$ are roots of $b_f(-s)$, as in the cases when K is $\mathbb{C}$ or $\mathbb{R}$.

Can microlocal calculus, which provides a method for computation of the b-function of a polynomial, be made to work for b-functions of ideals ?

Let $f = (f_1, \ldots, f_r)$ be a collection of polynomials. What are the differences between Verdier's and Sabbah's specialization functor for f? Does the motivic object $\mathcal{S}_f$, the higher-codimensional analog of the motivic Milnor fiber of a hypersurface, recover the generalized Hodge spectrum of f?

Can the geometric genus p_g of a normal isolated singularity can be recovered from the generalized Hodge spectrum, in analogy with the isolated hypersurface case? This would be relevant to the original, more general form of Durfee's Conjecture, which was stated for isolated complete intersection singularities.

2. Practical aspects

2.1. General rules. We mention some rules that apply for calculation of singularity invariants or help approximate singularity invariants. Whenever a geometric construction is available, one can look for the formula describing the change in a singularity invariant. We have already talked about log resolutions and jet schemes.

An *additive Thom-Sebastiani rule* describes a singularity invariant for $f(x) + g(y)$ in terms of the invariants for f and g, when $f(x)$ and $g(y)$ are polynomials in two disjoint sets of variables. This rule is available: for the motivic Milnor fiber, and hence for the monodromy zeta function and the Hodge spectrum, [**36**]; for the poles of the p-adic zeta functions, [**34**]; and for the b-function when both polynomials have isolated singularities and g is also quasihomogeneous, [**132**].

An *additive Thom-Sebastiani rule for ideals* describes a singularity invariant for a sum of two ideals in two disjoint sets of variables. Equivalently, this rule describes a singularity invariant of a product of schemes. This rule is the easiest one to obtain. It is available for example for motivic zeta functions [**36**], multiplier ideals, jumping numbers [**81**], b-functions [**20**], and test ideals [**120**].

A *multiplicative Thom-Sebastiani rule* describes a singularity invariant for $f(x)\cdot g(y)$ in terms of the invariants for f and g, when $f(x)$ and $g(y)$ are polynomials in two disjoint sets of variables. This rule is available for the Milnor monodromy of homogeneous polynomials [**41**]-Thm. 1.4, and, in the even greater generality when f and g are ideals, for multiplier ideals and jumping numbers [**81**].

A more general idea is to describe singularity invariants of $F(f_1,\ldots,f_r)$, where F is a nice polynomial and $f_1,\ldots,f_r$ are polynomials in distinct sets of variables. For results in this direction for the motivic Milnor fiber see [**63, 64, 65**].

It is hard to say what a *summation rule* should be in general. This is available for multiplier ideals [**92**] and test ideals [**120**]:

$$\mathcal{J}((f+g)^c)=\sum_{\lambda+\mu=c}\mathcal{J}(f^\lambda\cdot g^\mu),$$

and similarly for test ideals, where f,g are ideals of polynomials in the same set of variables. One can ask if a similar rule exists for the Verdier specialization functor or motivic zeta functions.

A *restriction rule* says that an invariant of a hyperplane section of a singularity germ is the same or worse, reflecting more complicated singularities, than the one of the original singularity. For example, log canonical and F-pure thresholds get smaller upon restriction. Also multiplier ideals [**81**] and test ideals [**68**] get smaller upon restriction. These invariants also satisfy a *generic restriction rule* saying that they remain the same upon restriction to a general hyperplane section. This is related to the *semicontinuity rule* stating that singularities get worse at special points in a family. The Hodge spectrum of an isolated hypersurface singularity satisfies a semicontinuity property, [**80**].

For more geometric transformation rules for multiplier ideals see [**81**], for test ideals see [**12, 115**], for nearby cycles functors, motivic Milnor fibers see [**39, 64, 65**], for jet schemes see [**47**], for log canonical thresholds see [**29**].

2.2. Ambient dimension two. For a germ of a reduced and irreducible curve f in $(\mathbb{C}^2,0)$ one has a set of *Puiseux pairs* $(k_1,n_1;\ldots;k_g,n_g)$ defined via a parametrization of the curve

$$\begin{aligned}y=&\sum_{1\le i\le\lfloor\frac{k_1}{n_1}\rfloor}c_{0,i}x^i+\sum_{0\le i\le\lfloor\frac{k_2}{n_2}\rfloor}c_{1,i}x^{(k_1+i)/n_1}+\\&+\sum_{0\le i\le\lfloor\frac{k_3}{n_3}\rfloor}c_{2,i}x^{k_1/n_1+(k_2+i)/n_1n_2}+\ldots\\&\ldots+\sum_{0\le i}c_{g,i}x^{k_1/n_1+k_2/n_1n_2+\ldots+(k_g+i)/n_1\ldots n_g},\end{aligned}$$

where $c_{j,i}\in\mathbb{C}$, $c_{j,0}\neq0$ for $j\neq0$, $k_j,n_j\in\mathbb{Z}_+$, $(k_j,n_j)=1$, $n_j>1$, and $k_1>n_1$. The Puiseux pairs determine the embedded topological type. For every plane curve there is a *minimal log resolution*, [**27**].

The Hodge spectrum can be written in terms of the Puiseux pairs for irreducible curves [**110**], and in terms of the graph and the vanishing orders of the minimal log resolution for any curves, [**122**].

Example. If f is a irreducible curve germ, the numbers $c < 1$ appearing in the Hodge spectrum of f, counted with their spectrum multiplicity $n_{c,0}(f)$, are

$$\left\{ \frac{1}{n_{s+1}\dots n_g} \cdot \left(\frac{i}{n_s} + \frac{j}{w_s}\right) + \frac{r}{n_{s+1}\dots n_g} \right\}$$

where: $w_1 = k_1$, $w_i = w_{i-1}n_{i-1}n_i + k_i$ for $i > 1$, $0 < i < n_s$, $0 < j < w_s$, $0 \leq r < n_{s+1}\dots n_g$, and $1 \leq s \leq g$ such that $i/n_s + j/w_s < 1$.

Jumping numbers of reduced curves can be written in terms of the graph and the vanishing orders of the minimal log resolution by [**18**], which reduced the problem to the Hodge spectrum. The above example also gives the jumping numbers $c < 1$ of an irreducible germ together with their inner jumping multiplicities $m_{c,0}(f)$.

For any plane curve germ there exists a local system of coordinates such that the log canonical threshold is $1/t$ where (t, t) is the intersection of the boundary of the Newton polytope (see 2.3) with the diagonal line, [**4, 2**].

Jumping numbers for a complete ideal of finite colength in two variables are computed combinatorially, [**72**].

The b-function of almost all irreducible and reduced plane curves with fixed Puiseux pairs is conjectured to be determined by a precise formula depending on the Puiseux pairs [**133**]. There are only partial results on determination of the b-function of plane curves, [**32, 14**].

The poles of motivic (p-adic, topological) zeta functions for any ideal of polynomials in two variables are determined in terms of the graph and the vanishing orders of the minimal log resolution, [**62, 125, 128, 32**].

The Strong Monodromy Conjecture is proven by Loeser for reduced plane curves, [**85**]. The Monodromy Conjecture is proven for ideals in two variables by Van Proeyen-Veys, [**126**].

Jet schemes of plane curves are considered in [**90**].

2.3. Nondegenerate polynomials. The monomials appearing in a polynomial f in n variables determine a set of points in $\mathbb{Z}^n_{\geq 0}$ whose convex hull is called the *Newton polytope* of f. The definition of *nondegenerate polynomial* is a condition involving the Newton polytope, which can differ in the literature. This is a condition that expresses in a precise way the fact that the polynomial is general and that it has an explicit log resolution. The following hold under nondegeneracy assumptions.

There are formulas in terms of the Newton polytope for: the Hodge spectrum [**118, 109**] and the monodromy Jordan normal form [**49**] when f has isolated singularities ; multiplier ideals and the jumping numbers in general [**81**];

Example. Let f be nondegenerate in the following sense: the form df_σ is nonzero on $(\mathbb{C}^*)^n \subset \mathbb{C}^n$, for every face σ of the Newton polytope, where f_σ is the polynomial composed of the terms of f which lie in σ. Then Howald showed that for $c < 1$ the multiplier ideals $\mathcal{J}(f^c)$ are the same as the multiplier ideals $\mathcal{J}(I_f^c)$, where I_f is the ideal generated by the terms of f. See next subsection for monomial ideals.

The poles of p-adic zeta functions are among a list determined explicitly by the Newton polytope, [**33, 135**]; the same holds for nondegenerate maps $f =$

$(f_1, \ldots, f_r)$, [**129**]. The motivic zeta function and the motivic Milnor fiber are considered in [**62**].

The Strong Monodromy Conjecture is proved for nondegenerate polynomials satisfying an additional condition, by displaying certain roots of the b-function, [**86**].

2.4. Monomial ideals. A *monomial ideal* is an ideal of polynomials generated by monomials. The *semigroup* of an ideal $I \subset \mathbb{C}[x_1, \ldots, x_n]$ is the set $\{u \mid x^u \in I\}$. The convex hull of this set is the *Newton polytope* $P(I)$ of the ideal. The Newton polytope of a monomial ideal equals the one of the integral closure of the ideal.

The Newton polytope of a monomial ideal determines explicitly the Hodge spectrum [**40**], the multiplier ideals and the jumping numbers, [**81**]; the test ideals and F-jumping numbers, [**68**]; and the p-adic zeta function, [**71**].

Example. Howald's formula for the multiplier ideals of a monomial ideal I is

$$\mathcal{J}(I^c) = \langle x^u \mid u + \mathbf{1} \in \text{Interior}(cP(I))\rangle.$$

The b-function of a monomial ideal has been computed in terms of the semigroup of the ideal. In general, the b-function cannot be determined by the Newton polytope alone, [**21, 22**].

The Strong Monodromy Conjecture is checked for monomial ideals, [**71**].

The geometry of the jet schemes of monomial ideals is described in [**55, 134**].

2.5. Hyperplane arrangements. Let K be a field. A *hyperplane arrangement* D in K^n is a possibly nonreduced union of hyperplanes of K^n. An invariant of D is *combinatorial* if it only depends on the lattice of intersections of the hyperplanes of D together with their codimensions.

Blowing up the intersections of hyperplanes gives an explicit log resolution. There is also a minimal resolution, [**28**].

Jet schemes of hyperplane arrangements are considered in [**93**]. Multiplier ideals are also considered here, see also [**121**].

A current major open problem in the theory of hyperplane arrangements is the combinatorial invariance of the Betti numbers of the cohomology of Milnor fiber, or stronger, of the dimensions of the Hodge pieces. The simplest unknown case is the cone over a planar line arrangement with at most triple points [**83**].

The jumping numbers and the Hodge spectrum of a hyperplane arrangement are explicitly determined combinatorial invariants, [**23**].

Example. Let $f \in \mathbb{C}[x, y, z]$ be a homogeneous reduced product of d linear forms. This plane arrangement is a cone over a line arrangement $D \in \mathbb{P}^2$. The Hodge spectrum multiplicities are:

$$n_{c,0}(f) = 0, \text{ if } cd \notin \mathbb{Z};$$

$$n_{c,0}(f) = \binom{i-1}{2} - \sum_{m\geq 3} \nu_m \binom{\lceil \frac{im}{d} \rceil \quad 1}{2}, \text{ if } c = \frac{i}{d}, i = 1, \ldots, d;$$

$$n_{c,0}(f) = (i-1)(d-i-1) - \sum_{m\geq 3} \nu_m \left(\left\lceil \frac{im}{d} \right\rceil - 1 \right) \left(m - \left\lceil \frac{im}{d} \right\rceil \right),$$
$$\text{if } c = \frac{i}{d} + 1, i = 1, \ldots, d;$$

$$n_{c,0}(f) = \binom{d-i-1}{2} - \sum_{m\geq 3} \nu_m \binom{m - \lceil \frac{im}{d} \rceil}{2} - \delta_{i,d},$$
$$\text{if } c = \frac{i}{d} + 2, i = 1, \ldots, d;$$

where $\nu_m = \#\{P \in D \mid mult_P D = m\}$, and $\delta_{i,d} = 1$ if $i = d$ and 0 otherwise.

The motivic, p-adic, and topological zeta functions also depend only on the combinatorics, [**25**].

The b-function is not a combinatorial invariant, according to a recent announcement of U. Walther. For computations of b-functions, by general properties already listed in this survey, it is enough to restrict to the case of so called "indecomposable central essential" complex arrangements. The *n/d-Conjecture* says that for such an arrangement of degree d, $-n/d$ is a root of the b-function. This is known only for reduced arrangements when $n \leq 3$, and when $n > 3$ for reduced arrangements with n and d coprime and one hyperplane in general position, [**25**]. For reduced arrangements as above, it is also known that: if $b_f(-c) = 0$ then $c \in (0, 2 - 1/d)$, and -1 is a root of multiplicity n of $b_f(s)$, [**111**].

Example. If f is a generic central hyperplane arrangement, then U. Walther [**130**], together with the information about the root -1 from above, showed that

$$b_f(s) = (s+1)^{n-1} \prod_{j=n}^{2d-2} \left(s + \frac{j}{d} \right).$$

The Monodromy Conjecture holds for all hyperplane arrangements; the Strong Monodromy Conjecture holds for a hyperplane arrangement $D \subset K^n$ if the n/d-Conjecture holds, [**24**]. In particular, the Strong Monodromy Conjecture holds for all reduced arrangements in ≤ 3 variables, and for the reduced arrangements in 4 variables of odd degree with one hyperplane in generic position, [**25**].

2.6. Discriminants of finite reflection groups. A *complex reflection group* is a group G, acting on a finite-dimensional complex vector space V, that is generated by elements that fix a hyperplane pointwise, i.e. by complex reflections. Weyl groups and Coxeter groups are complex reflection groups. The ring of invariants is a polynomial ring: $\mathbb{C}[V]^G = \mathbb{C}[f_1, \ldots, f_n]$. Here $n = \dim V$, and $f_1, \ldots, f_n$ are some algebraically independent invariant polynomials. The degrees $d_i = \deg f_i$ are determined uniquely. The finite irreducible complex reflection groups are classified by Shephard-Todd, [**15**] .

Let $D = \cup_i D_i$ be the union of the reflection hyperplanes, and let α_i denote a linear form defining D_i. Let e_i be the order of the subgroup fixing D_i. Consider the invariant polynomial

$$\delta = \prod_i \alpha_i^{e_i} \in \mathbb{C}[V]^G.$$

Viewed as a polynomial in the variables $f_1, \ldots, f_n$, it defines a regular map $\Delta : V/G \cong \mathbb{C}^n \to \mathbb{C}$, called the *discriminant.*

The b-functions of discriminants of the finite irreducible complex reflection groups have been determined in terms of the degrees d_i for Weyl groups in [**100**] and for Coxeter groups in [**101**]:

$$b_\Delta(s) = \prod_{i=1}^{n} \prod_{j=1}^{d_i - 1} \left(s + \frac{1}{2} + \frac{j}{d_i} \right).$$

In the remaining cases, the zeros of the b-functions are determined in [**35**].

The monodromy zeta function of Δ has also been determined in terms of the degrees d_i, [**35**].

2.7. Generic determinantal varieties. Let M be the space of all matrices of size $r \times s$, with $r \leq s$. The *k-th generic determinantal variety* is the subvariety D^k consisting of matrices of rank at most k.

The multiplier ideals $\mathcal{J}(M, c \cdot D^k)$ have been computed in [**75**]. In particular, the log canonical threshold is

$$lct\,(M, D^k) = \min_{i=0,\ldots,k} \frac{(r-i)(s-i)}{k+1-i}.$$

The topological zeta function is computed in [**42**]:

$$Z_{D^k}^{top}(s) = \prod_{c \in \Omega} \frac{1}{1 - sc^{-1}},$$

where

$$\Omega = \left\{ -\frac{r^2}{k+1}, -\frac{(r-1)^2}{k}, -\frac{(r-2)^2}{k-1}, \ldots, -(r-k)^2 \right\}.$$

The number of irreducible components of the n-th jet scheme D_n^k is 1 if $k = 0, r-1$, and is $n + 2 - \lceil (n+1)/(k+1) \rceil$ if $0 < k < r - 1$, [**42**].

It is not known in general how to compute the b-function of D^k.

Example. The oldest example of a nontrivial b-function is due to Cayley. Let $f = det(x_{ij})$ be the determinant of an $n \times n$ matrix of indeterminates. Then $b_f(s) = (s+1) \ldots (s+n)$ and the differential operator from the definition of the b-function is $P = det(\partial / \partial x_{ij})$.

2.8. Prehomogeneous vector spaces. A *prehomogeneous vector space (pvs)* is a vector space V together with a connected linear algebraic group G with a rational representation $G \to GL(V)$ such that V has a Zariski dense G-orbit. The complement of the dense orbit is called the *singular locus.* The pvs is *irreducible* if V is an irreducible G-module. The pvs is *regular* if the singular locus is a hypersurface $f = 0$. Irreducible regular pvs have been classified into 29 types by Kimura-Sato. The ones in a fixed type are related to each other via a so-called *castling transformation*, and within each type there is a "minimal" pvs called *reduced*, [**77**].

The b-functions $b_f(s)$ have been computed for irreducible regular pvs using microlocal calculus by Kimura (28 types) and Ozeki-Yano (1 type), [**77**]. For an introduction to microlocal calculus see [**76**].

The p-adic zeta functions of 24 types of irreducible regular pvs have been computed by Igusa. The Strong Monodromy Conjecture has been checked for irreducible regular pvs: 24 types by Igusa [**74, 77**], and the remaining types by Kimura-Sato-Zhu [**78**].

The castling transform for motivic zeta functions and for Hodge spectrum has been worked out in [**87**].

There are additional computations of b-functions for pvs beyond the case of irreducible and regular ones. We mention a few results. For the reducible pvs, an elementary method to calculate the b-functions of singular loci, which uses the known formula for b-functions of one variable, is presented in [**124**]. The decomposition formula for b-functions, which asserts that under certain conditions, the b-functions of reducible pvs have decompositions correlated to the decomposition of representations, was given in [**113**]. By using the decomposition formula, the b-functions of relative invariants arising from the quivers of type A have been determined in [**119**].

A *linear free divisor* $D \subset V$ is the singular locus of a particular type of pvs. One definition is that the sheaf of vector fields tangent to D is a free $\mathcal{O}_V$-module and has a basis consisting of vector fields of the type $\sum_j l_j \partial_{x_j}$, where l_j are linear forms. Another equivalent definition is that D is the singular locus of a pvs (G, V) with $\dim G = \dim V = \deg f$, where f is the equation defining D. To bridge the two definitions, one has that G is the connected component containing the identity of the group $\{A \in GL(V) |\ A(D) = D\}$. Quiver representations give often linear free divisors [**17**]. The b-functions for linear free divisors have been studied in [**57, 116**] and computed in some cases.

Example. Some interesting examples of b-functions, related to quivers of type A and to generic determinantal varieties with blocks of zeros inserted, are computed in [**119**]. For example, let X, Y, Z be matrices of three distinct sets of indeterminates of sizes (n_2, n_1), (n_2, n_3), (n_4, n_3), respectively, such that $n_1 + n_3 = n_2 + n_4$ and $n_1 < n_2$. Then for

$$f = \det \begin{bmatrix} X & Y \\ 0 & Z \end{bmatrix},$$

the b-function is

$$b_f(s) = (s+1)\dots(s+n_3) \cdot (s+n_2-n_1+1)\dots(s+n_2).$$

2.9. Quasi-ordinary hypersurface singularities. A germ of a hypersurface $(D, 0) \subset (\mathbb{C}^n, 0)$ is *quasi-ordinary* if there exists a finite morphism $(D, 0) \to (\mathbb{C}^{n-1}, 0)$ such that the discriminant locus is contained in a normal crossing divisor. In terms of equations, $f \in \mathbb{C}[x_1, \dots, x_n]$ has quasi-ordinary singularities if the discriminant of f with respect to $y = x_n$ equals $x_1^{u_1} \dots x_{n-1}^{u_{n-1}} \cdot h$, where $h(0) \neq 0$.

Quasi-ordinary hypersurface singularities generalize the case of plane curve in the sense that they are higher dimensional singularities with Puiseux expansions. The *characteristic exponents* $\lambda_1 < \dots < \lambda_g$ of an analytically irreducible quasi-ordinary hypersurface germ f are defined as follows. The roots of $f(y)$ are fractional power series $\zeta_i \in \mathbb{C}[[x_1^{1/e}, \dots, x_{n-1}^{1/e}]]$, where $e = \deg_y f$. The difference of two roots of f divides the discriminant, hence $\zeta_i - \zeta_j = x^{\lambda_{ij}} h_{ij}$, where h_{ij} is a unit. Then $\{\lambda_{ij}\}$ is the set of characteristic exponents which we order and relabel it $\{\lambda_k\}$. The set of characteristic exponents is an invariant of and equivalent to the embedded topological type of the germ, [**52**]. By a change of variable it can be assumed that the Newton polytope of f is determined canonically by the characteristic exponents. There is a canonical way to relabel the variables and to order the characteristic exponents, 2.9.1.

There are explicit embedded resolutions of quasi-ordinary singularities in terms of characteristic exponents, [**53**].

The monodromy zeta function has been computed in [**54**].

The Monodromy Conjecture holds for quasi-ordinary singularities [**6**].

Jet schemes of quasi-ordinary singularities have been analyzed in [**105, 56**]. For analytically irreducible quasi-ordinary hypersurface singularities the motivic zeta function, and hence the log canonical threshold, Hodge spectrum and the monodromy zeta function, have been computed in terms of the characteristic exponents in [**56**].

Example. A refined formula for the log canonical threshold of an analytically irreducible quasi-ordinary hypersurface singularity f was given in [**26**]. Let λ_i be ordered as in 2.9.1, and let $\lambda_{i,j}$ denote the j-th coordinate entry of the vector λ_i. Then:

(a) f is log canonical if and only if it is smooth, or $g = 1$ and the nonzero coordinates of λ_1 are $1/q$, or $g = 1$ and the nonzero coordinates of λ_1 are $1/2$ and 1.

(b) With i_1 and j_2 defined as in 2.9.1, if f is not log canonical, then:

(1) if $i_1 \neq n-1$,

$$lct\,(f) = \frac{1+\lambda_{1,n-1}}{e\lambda_{1,n-1}};$$

(2) if $i_1 = n-1, j_2 \neq 0$, and $\lambda_{2,j_2} \geq n_1(\lambda_{2,n-1} - \frac{1}{n_1} + 1)$,

$$lct\,(f) = \frac{1+\lambda_{2,j_2}}{\dfrac{e}{n_1}\cdot\lambda_{2,j_2}}$$

(3) if $i_1 = n-1$ and $j_2 = 0$; or if $i_1 = n-1$, $j_2 \neq 0$ and $\lambda_{2,j_2} < n_1(\lambda_{2,n-1} - \frac{1}{n_1} + 1)$,

$$lct\,(f) = \frac{1+\lambda_{2,n-1}}{\dfrac{e}{n_1}\cdot(\lambda_{2,n-1}+1-\dfrac{1}{n_1})}.$$

2.9.1. *Canonical order.* Let f be an analytically irreducible quasi-ordinary hypersurface. Let $0 = \lambda_0 < \lambda_1 < \ldots < \lambda_g \in \mathbb{Z}^{n-1}$ be the characteristic exponents of f. Set for $1 \leq j \leq g$,

$$n_j := \#\,(\mathbb{Z}^{n-1} + \lambda_1\mathbb{Z} + \ldots + \lambda_j\mathbb{Z})/(\mathbb{Z}^{n-1} + \lambda_1\mathbb{Z} + \ldots + \lambda_{j-1}\mathbb{Z}).$$

It is known that $n_1 \ldots n_g = \deg_y f$. We set $n_0 = 0$. One can permute the variables $x_1, \ldots, x_n$ such that for $j < j'$ we have $(\lambda_{1,j}, \ldots, \lambda_{g,j}) \leq (\lambda_{1,j'}, \ldots, \lambda_{g,j'})$ lexicographically.

This ordering defines $j_1 = n-1 > j_2 \geq j_3 \geq \ldots \geq j_g \geq 0$ such that $j_i = \max\{j \mid \lambda_{i-1,j} = 0\}$, where we set $j_i = 0$ if $\lambda_{i-1,j} \neq 0$ for all $j = 1, \ldots n-1$. For $j \leq j_i$, $\lambda_{i,j}$ can be written as a rational number with denominator n_i. Define

$i_k = \max\{j \leq j_k \mid \lambda_{k,j} = 1/n_k\}$ if $j_k > j_{k+1}$, and $i_k = j_{k+1}$ if $j_k = j_{k+1}$ or $\lambda_{k,i} \neq 1/n_k$ for all $i = 1, \ldots, j_k$. Whenever $j_k < j \leq j' \leq i_{k-1}$, we have $\lambda_{k,j} \leq \lambda_{k,j'}$ by our ordering. This defines the notation used in the previous example.

2.10. Computer programs. Most of the algorithms available for singularity invariants depend on Gröbner bases or resolution of singularities. Hence even for an input of small complexity one might not see the end of a computation. Nevertheless, the programs are very useful.

As a general rule, one should check the documentation of Singular [**31**], Macaulay2 [**58**], and Risa/Asir [**99**] for a list of packages pertaining to singularity theory, which can and hopefully does increase often. Singular has the most extensive packages dedicated to singularity theory; see [**61**], which contains the theory and subtleties in connection with local computations as opposed to global computations.

Packages involving $\mathcal{D}$-modules have been implemented extensively. For computation of b-functions, T. Oaku gave the first general algorithm. There are now algorithms implemented even for computation of generalized twisted b-functions $b_{f,g}(s)$ for ideals f [**1, 117, 9**].

Using that multiplier ideals can be written in terms of $\mathcal{D}$-modules [**20**], algorithms are now implemented that compute multiplier ideals and jumping numbers [**117, 9**].

Villamayor's algorithm for resolution of singularities has also been implemented. Some topological zeta functions can be computed using this, [**51**].

For computing the Hodge spectrum and b-functions of isolated hypersurface singularities, as well as other invariants, see [**114**].

For nondegenerate polynomials there are programs computing the p-adic, topological, monodromy zeta functions [**70**], and the Hodge spectrum [**48**].

Multiplier ideals for hyperplane arrangements can be computed with [**37**]. We are waiting for the authors to implement the combinatorial formulas of [**23**] for jumping numbers and Hodge spectra.

Currently there are no feasible algorithms available for computing test ideals or F-jumping numbers even for polynomials of very small complexity. This is due to appearance of high powers of the ideals involved in the definitions.

2.11. Questions.

Complete a general additive Thom-Sebastiani formula for b-functions. Are there additive Thom-Sebastiani rules for multiplier ideals and test ideals? Generalize [**34**] to obtain this rule for the motivic zeta function. This would recover the rule for the motivic Milnor fiber.

Is there a multiplicative Thom-Sebastiani for b-functions?

Let f and g be polynomials. Does there exist a Summation rule that allows one to "generate" the Verdier specialization functor $Sp_{(f,g)}$ of the ideal generated by f and g via the nearby cycles $\psi_{f^\lambda \cdot g^\mu}$ with $\lambda + \mu = 1$? Here λ and μ would represent eigenvalues of the monodromy along f and g, respectively. A similar question can be asked about motivic zeta functions.

Does the Hodge spectrum of a hypersurface germ satisfy a semicontinuity property similar to what happens in the isolated case?

For reduced and irreducible plane curves, determine the arithmetic invariants needed along with the Puiseux pairs to compute the test ideals and F-jumping numbers for a fixed reduction modulo p.

Prove or correct the formula conjectured by T. Yano for the b-function of a general reduced and irreducible plane curve among those with fixed Puiseux pairs.

Write the Hodge spectrum of nondegenerate polynomials, with non-necessarily isolated singularities, in terms of the Newton polytope.

Compute the F-jumping numbers of hyperplane arrangements, or of some other class of examples besides monomial ideals.

Solve the combinatorial problem that completes the proof of the n/d-Conjecture for hyperplane arrangements, and thus of the Strong Monodromy Conjecture for hyperplane arrangements.

We have already mentioned the problem of combinatorial invariance of the dimension of the (Hodge pieces of the) cohomology of the Milnor fibers for hyperplane arrangements. This is currently viewed as the "the holy-grail" in the theory of hyperplane arrangements. The problem fits between the combinatorial invariance of the fundamental group of the complement, which is not true [**106**], and that of the cohomology of the complement, which is true [**102**].

What can one say about the other zeta functions, besides the monodromy zeta function, for discriminants of irreducible finite reflection groups? Since the b-functions are already determined, maybe the Strong Monodromy Conjecture can be checked.

Compute the b-function of generic determinantal varieties. These varieties have certain analogies with monomial ideals, [**16**]. Maybe the strategy for computing b-functions of monomial ideals can be pushed to work for determinantal varieties.

Compute the Hodge spectrum of the 29 types of irreducible regular prehomogeneous vector spaces. By the castling transformation formula of Loeser [**87**], it is enough to compute the Hodge spectrum for the reduced ones.

Construct feasible algorithms for computing test ideals and F-jumping numbers. We are also lacking algorithms for Hodge spectra and p-adic zeta functions besides the cases mentioned in 2.10. However, due to their relation with $\mathcal{D}$-modules and resolution of singularities, it should be possible to give such algorithms.

References

[1] D. Andres, M. Brickenstein, V. Levandovskyy, J. Martín-Morales, and H. Schönemann. Constructive $\mathcal{D}$-module Theory with SINGULAR. arXiv:1005.3257. MR2775997

[2] M. Aprodu and D. Naie, Log-canonical threshold for curves on a smooth surface. arXiv:0707.0783.

[3] V.I. Arnold, S.M. Gusein-Zade, and A.N. Varchenko, *Singularities of differentiable maps, Vol. I and II*, Birkhäuser, 1985/1988. MR777682 (86f:58018)

[4] E. Artal Bartolo, P. Cassou-Noguès, I. Luengo, and A. Melle Hernández, On the log-canonical threshold for germs of plane curves. *Singularities I*, 1–14, Contemp. Math., 474, Amer. Math. Soc., Providence, RI, 2008. MR2454343 (2009m:32050)

[5] E. Artal Bartolo, P. Cassou-Noguès, I. Luengo, A. Melle Hernández, The Denef-Loeser zeta function is not a topological invariant. J. London Math. Soc. (2) 65 (2002), no. 1, 45–54. MR1875134 (2002k:14005)

[6] E. Artal Bartolo, P. Cassou-Noguès, I. Luengo, and A. Melle Hernández, *Quasi-ordinary power series and their zeta functions.* Mem. Amer. Math. Soc. 178 (2005), no. 841, vi+85 pp. MR2172403 (2007d:14005)

[7] T. Ashikaga, Normal two-dimensional hypersurface triple points and the Horikawa type resolution. Tohoku Math. J. (2) 44 (1992), no. 2, 177–200. MR1161610 (93e:14040)

[8] T. Ashikaga, Surface singularities on cyclic coverings and an inequality for the signature. J. Math. Soc. Japan 51 (1999), no. 2, 485–521. MR1674761 (2000b:14041)

[9] C. Berkesch and A. Leykin, Algorithms for Bernstein-Sato polynomials and multiplier ideals. arXiv:1002.1475.

[10] M. Blickle, M. Mustaţă, and K. Smith, Discreteness and rationality of F-thresholds. Special volume in honor of Melvin Hochster. Michigan Math. J. 57 (2008), 43–61. MR2492440 (2010c:13003)

[11] A. Borel, P.-P. Grivel, B. Kaup, A. Haefliger, B. Malgrange, and F. Ehlers, *Algebraic D-modules.* Perspectives in Mathematics, 2. Academic Press, Inc., Boston, MA, 1987. xii+355 pp. MR882000 (89g:32014)

[12] A. Bravo and K. Smith, Behavior of test ideals under smooth and étale homomorphisms. J. Algebra 247 (2002), no. 1, 78–94. MR1873384 (2002m:13007)

[13] T. Brélivet, Variance of the spectral numbers and Newton polygons. Bull. Sci. Math. 126 (2002), no. 4, 332–342. MR1909464 (2003f:14003)

[14] J. Briançon, P. Maisonobe, and T. Torrelli, Matrice magique associée à un germe de courbe plane et division par l'idéal jacobien. Ann. Inst. Fourier (Grenoble) 57 (2007), no. 3, 919–953. MR2336834 (2008i:32044)

[15] M. Broué, *Introduction to complex reflection groups and their braid groups.* Lecture Notes in Mathematics, 1988. Springer-Verlag, Berlin, 2010. xii+138 pp. MR2590895 (2011d:20078)

[16] W. Bruns and U. Vetter, *Determinantal rings.* Lecture Notes in Mathematics, 1327. Springer-Verlag, Berlin, 1988. viii+236 pp. MR953963 (89i:13001)

[17] R.-O. Buchweitz and D. Mond, Linear free divisors and quiver representations, *Singularities and computer algebra,* London Math. Soc. Lecture Note Ser., vol. 324, Cambridge Univ. Press, Cambridge, 2006, pp. 41–77. MR2228227 (2007d:16028)

[18] N. Budur, On Hodge spectrum and multiplier ideals. Math. Ann. 327 (2003), no. 2, 257–270. MR2015069 (2005b:14006)

[19] N. Budur, On the V-filtration of $\mathcal{D}$-modules. In *Geometric methods in algebra and number theory,* 59–70, Progr. Math., 235, Birkhäuser Boston, Boston, MA, 2005. MR2159377 (2007a:32010)

[20] N. Budur, M. Mustaţă, and M. Saito. Bernstein-Sato polynomials of arbitrary varieties. Compositio Math. 142, 779–797 (2006). MR2231202 (2007c:32036)

[21] N. Budur, M. Mustaţă, and M. Saito. Combinatorial description of the roots of the Bernstein-Sato polynomials for monomial ideals. Comm. Algebra 34 (2006), no. 11, 4103–4117. MR2267574 (2007h:32041)

[22] N. Budur, M. Mustaţă, and M. Saito. Roots of Bernstein-Sato polynomials for monomial ideals: a positive characteristic approach. Math. Res. Lett. 13 (2006), no. 1, 125–142. MR2200051 (2006k:14003)

[23] N. Budur and M. Saito, Jumping coefficients and spectrum of a hyperplane arrangement. Math. Ann. 347 (2010), no. 3, 545–579. MR2640043 (2011j:14042)

[24] N. Budur, M. Mustaţă, and Z. Teitler, The monodromy conjecture for hyperplane arrangements. To appear in Geom. Dedicata. arXiv:0906.1991. MR2819667

[25] N. Budur, M. Saito, and S. Yuzvinsky, On the local zeta functions and the b-functions of certain hyperplane arrangements. arXiv:1002.0629.

[26] N. Budur and M. González Villa, preprint 2010.

[27] E. Casas-Alvero, *Singularities of plane curves.* London Mathematical Society Lecture Note Series, 276. Cambridge University Press, Cambridge, 2000. xvi+345 pp. MR1782072 (2003b:14035)

[28] C. De Concini and C. Procesi, Wonderful models of subspace arrangements. Selecta Math. (N.S.) 1 (1995), no. 3, 459–494. MR1366622 (97k:14013)

[29] T. de Fernex, L. Ein, and M. Mustaţă, Bounds for log canonical thresholds with applications to birational rigidity. Math. Res. Lett. 10 (2003), no. 2-3, 219–236. MR1981899 (2004e:14060)
[30] T. de Fernex, L. Ein, and M. Mustaţă, Shokurov's ACC conjecture for log canonical thresholds on smooth varieties. Duke Math. J. 152 (2010), no. 1, 93–114. MR2643057 (2011c:14036)
[31] W. Decker, G.-M. Greuel, G. Pfister, and H. Schönemann: SINGULAR 3-1-1 — A computer algebra system for polynomial computations. http://www.singular.uni-kl.de (2010).
[32] J. Denef, Report on Igusa's local zeta function. Séminaire Bourbaki, Vol. 1990/91. Astérisque No. 201-203 (1991), Exp. No. 741, 359–386 (1992). MR1157848 (93g:11119)
[33] J. Denef and K. Hoornaert, Newton polyhedra and Igusa's local zeta function. J. Number Theory 89 (2001), no. 1, 31–64. MR1838703 (2002g:11170)
[34] J. Denef and W. Veys, On the holomorphy conjecture for Igusa's local zeta function. Proc. Amer. Math. Soc. 123 (1995), no. 10, 2981–2988. MR1283546 (95m:11132)
[35] J. Denef and F. Loeser, Regular elements and monodromy of discriminants of finite reflection groups. Indag. Math. (N.S.) 6 (1995), no. 2, 129–143. MR1338321 (96h:20077)
[36] J. Denef and F. Loeser, Geometry on arc spaces of algebraic varieties. *European Congress of Mathematics, Vol. I (Barcelona, 2000),* 327–348, Progr. Math., 201, Birkhäuser, Basel, 2001. MR1905328 (2004c:14037)
[37] G. Denham and G. Smith, `HyperplaneArrangements.m2`, v. 0.5. A Macaulay2 package. 2009.
[38] A. Dimca, Monodromy and Hodge theory of regular functions. New developments in singularity theory (Cambridge, 2000), 257–278, Kluwer Acad. Publ., Dordrecht, 2001. MR1849312 (2003a:32049)
[39] A. Dimca, *Sheaves in topology.* Universitext. Springer-Verlag, Berlin, 2004. xvi+236 pp. MR2050072 (2005j:55002)
[40] A. Dimca, P. Maisonobe, and M. Saito, Spectrum and multiplier ideals of arbitrary subvarieties. arXiv:0705.4197.
[41] A. Dimca, Tate properties, polynomial-count varieties, and monodromy of hyperplane arrangements. arXiv:1012.1437.
[42] R. Docampo, Arcs on determinantal varieties, arXiv:1011.1930. MR2713835
[43] A. Durfee, The signature of smoothings of complex surface singularities. Math. Ann. 232 (1978), no. 1, 85–98. MR0466620 (57:6497)
[44] L. Ein, R. Lazarsfeld, and M. Mustaţă, Contact loci in arc spaces. Compos. Math. 140 (2004), no. 5, 1229–1244. MR2081163 (2005f:14006)
[45] L. Ein, R. Lazarsfeld, K. Smith, and D. Varolin, D., Jumping coefficients of multiplier ideals, Duke Math. J. 123 (2004), no. 3, 469–506. MR2068967 (2005k:14004)
[46] L. Ein and M. Mustaţă, Inversion of adjunction for local complete intersection varieties. Amer. J. Math. 126 (2004), no. 6, 1355–1365. MR2102399 (2005j:14020)
[47] L. Ein and M. Mustaţă, Generically finite morphisms and formal neighborhoods of arcs. Geom. Dedicata 139 (2009), 331–335. MR2481855 (2009m:14002)
[48] S. Endrass, `spectrum.lib`. A SINGULAR 3-1-1 library for computing the singularity spectrum for nondegenerate singularities (2001).
[49] A. Esterov and K. Takeuchi, Motivic Milnor fibers over complete intersection varieties and their virtual Betti numbers, arXiv:1009.0230v4.
[50] C. Eyral, Zariski's multiplicity question - a survey. New Zealand J. of Math., 36 (2007), 253–276. MR2476643 (2010a:32040)
[51] A. Frühbis-Krüger and G. Pfister: `resolve.lib, reszeta.lib`. SINGULAR 3-1-1 libraries (2005).
[52] Y.-N. Gau, *Embedded topological classification of quasi-ordinary singularities,* Mem. Amer. Math. Soc., vol. 388 (1988). MR954948 (89m:14002)
[53] P.D. González Pérez, Toric embedded resolutions of quasi-ordinary hypersurface singularities. Ann. Inst. Fourier (Grenoble) 53 (2003), no. 6, 1819–1881. MR2038781 (2005b:32064)
[54] P.D. González Pérez, L.J. McEwan, and A. Némethi, The zeta-function of a quasi-ordinary singularity. II. *Topics in algebraic and noncommutative geometry (Luminy/Annapolis, MD, 2001),* 109–122, Contemp. Math., 324, Amer. Math. Soc., Providence, RI, 2003. MR1986117 (2004k:32050)
[55] R. Goward and K. Smith, The jet scheme of a monomial scheme. Comm. Algebra 34 (2006), no. 5, 1591–1598. MR2229478 (2007e:13037)
[56] M. González Villa, Ph.D. thesis, Universidad Complutense de Madrid, 2010.

[57] M. Granger and M. Schulze. On the symmetry of b-functions of linear free divisors. arXiv:0807.0560. MR2760735 (2011k:14014)

[58] D. Grayson and M. Stillman, Macaulay2, a software system for research in algebraic geometry. Available at http://www.math.uiuc.edu/Macaulay2/.

[59] G.-M. Greuel, Constant Milnor number implies constant multiplicity for quasihomogeneous singularities. Manuscr. Math. 56 (1986), 159–166. MR850367 (87j:32020)

[60] G.-M. Greuel, C. Lossen, and E. Shustin, *Introduction to Singularities and Deformations*. Springer Verlag, Berlin, Heidelberg, New York, 2007. xii+471 pp. MR2290112 (2008b:32013)

[61] G.-M. Greuel and G. Pfister: A SINGULAR Introduction to Commutative Algebra, 2nd Edition. Springer Verlag, Berlin, Heidelberg, New York, 2008. xx+689 pp. MR2363237 (2008j:13001)

[62] G. Guibert, Espaces d'arcs et invariants d'Alexander. Comment. Math. Helv. 77 (2002), no. 4, 783–820. MR1949114 (2003k:14021)

[63] G. Guibert, F. Loeser, and M. Merle, Nearby cycles and composition with a nondegenerate polynomial. Int. Math. Res. Not. 2005, no. 31, 1873–1888. MR2171196 (2006f:11145)

[64] G. Guibert, F. Loeser, and M. Merle, Iterated vanishing cycles, convolution, and a motivic analogue of a conjecture of Steenbrink. Duke Math. J. 132 (2006), no. 3, 409–457. MR2219263 (2007e:14011)

[65] G. Guibert, F. Loeser, and M. Merle, Composition with a two variable function. Math. Res. Lett. 16 (2009), no. 3, 439–448. MR2511624 (2010j:14021)

[66] L.H. Halle and J. Nicaise, Motivic zeta functions of abelian varieties, and the monodromy conjecture. arXiv:0902.3755v3. MR2782205

[67] N. Hara and K.-i. Watanabe, F-regular and F-pure rings vs. log terminal and log canonical singularities. J. Algebraic Geom. 11, 2 (2002), 363–392. MR1874118 (2002k:13009)

[68] N. Hara and K.-i. Yoshida, A generalization of tight closure and multiplier ideals. Trans. Amer. Math. Soc. 355 (2003), no. 8, 3143–3174. MR1974679 (2004i:13003)

[69] C. Hertling, Frobenius manifolds and variance of the spectral numbers. New developments in singularity theory (Cambridge, 2000), 235–255, Kluwer Acad. Publ., Dordrecht, 2001. MR1849311 (2002k:53174)

[70] K. Hoornaert and D. Loots, POLYGUSA v. 0.5: A computer program to calculate Igusa's local zeta function and the topological zeta function for non-degenerated polynomials. Available at: http://wis.kuleuven.be/algebra/kathleen/program.htm.

[71] J. Howald, M. Mustaţă, and C. Yuen, On Igusa zeta functions of monomial ideals. Proc. Amer. Math. Soc. 135 (2007), no. 11, 3425–3433. MR2336554 (2008j:11177)

[72] E. Hyry and T. Järvilehto, Jumping numbers and ordered tree structures on the dual graph. arXiv:1001.1220.

[73] J.-i. Igusa, Complex powers and asymptotic expansions I.; Complex powers and asymptotic expansions II.; J. Reine Angew. Math. 268/269 (1974), 110–130; and 278/279 (1975), 307–321.

[74] J.-i. Igusa, *An introduction to the theory of local zeta functions*. AMS/IP Studies in Advanced Mathematics, 14. American Mathematical Society, Providence, RI; International Press, Cambridge, MA, 2000. xii+232 pp. MR1743467 (2001j:11112)

[75] A. Johnson, Multiplier ideals of determinantal ideals, Ph.D. thesis, University of Michigan, 2003. MR2704808

[76] M. Kashiwara, *$\mathcal{D}$-modules and microlocal calculus*. Translated from the 2000 Japanese original by Mutsumi Saito. Translations of Mathematical Monographs, 217. Iwanami Series in Modern Mathematics. American Mathematical Society, Providence, RI, 2003. xvi+254 pp. MR1943036 (2003i:32018)

[77] T. Kimura, *Introduction to prehomogeneous vector spaces*. Translated from the 1998 Japanese original by Makoto Nagura and Tsuyoshi Niitani and revised by the author. Translations of Mathematical Monographs, 215. American Mathematical Society, Providence, RI, 2003. MR1944442 (2003k:11180)

[78] T. Kimura, F. Sato, and X.-W. Zhu, On the poles of p-adic complex powers and the b-functions of prehomogeneous vector spaces. Amer. J. Math. 112 (1990), no. 3, 423–437. MR1055652 (91i:11171)

[79] J. Kollár, Singularities of Pairs, Proc. Sympos. Pure Math. A.M.S. 62, Part 1 (1997), 221–287. MR1492525 (99m:14033)

[80] V. Kulikov, *Mixed Hodge structures and singularities.* Cambridge Univ. Press, Cambridge, 1998. xxii+186 pp. MR1621831 (99d:14009)

[81] R. Lazarsfeld, *Positivity in algebraic geometry. II. Positivity for vector bundles, and multiplier ideals.* Ergebnisse der Mathematik und ihrer Grenzgebiete. 3. Folge. A Series of Modern Surveys in Mathematics, 49. Springer-Verlag, Berlin, 2004. xviii+385 pp. MR2095472 (2005k:14001b)

[82] R. Lazarsfeld and K. Lee, Local syzygies of multiplier ideals, Invent. Math. 167 (2007), no. 2, 409–418. MR2270459 (2007h:13021)

[83] A. Libgober, On combinatorial invariance of the cohomology of Milnor fiber of arrangements and Catalan equation over function field. arXiv:1011.0191.

[84] J. Lipman and K.-i. Watanabe, Integrally closed ideals in two-dimensional regular local rings are multiplier ideals. Math. Res. Lett. 10 (2003), no. 4, 423–434. MR1995782 (2004m:13059)

[85] F. Loeser, Fonctions d'Igusa p-adiques et polynômes de Bernstein, Amer. J. Math. 110 (1988), no. 1, 1–21. MR926736 (89d:11110)

[86] F. Loeser, Fonctions d'Igusa p-adiques, polynômes de Bernstein, et polyèdres de Newton, J. reine angew. Math. 412 (1990), 75–96. MR1079002 (92c:11139)

[87] F. Loeser, Motivic zeta functions for prehomogeneous vector spaces and castling transformations. Nagoya Math. J. 171 (2003), 85–105. MR2002014 (2004h:11100)

[88] A. Melle-Hernández, Milnor numbers for surface singularities. Israel J. Math. 115 (2000), 29–50. MR1749672 (2001i:32044)

[89] J. Milnor, *Singular points of complex hypersurfaces*, Princeton Univ. Press, 1968. iii+122 pp. MR0239612 (39:969)

[90] H. Mourtada, Jet schemes of complex plane branches and equisingularity. arXiv:1009.5845.

[91] M. Mustaţă, Singularities of pairs via jet schemes, J. Amer. Math. Soc. 15, (2002), 599–615. MR1896234 (2003b:14005)

[92] M. Mustaţă, The multiplier ideals of a sum of ideals, Trans. Amer. Math. Soc. 354 (2002), no. 1, 205–217. MR1859032 (2002k:14006)

[93] M. Mustaţă, Multiplier ideals of hyperplane arrangements. Trans. Amer. Math. Soc. 358 (2006), no. 11, 5015–5023. MR2231883 (2007d:14007)

[94] M. Mustaţă, Bernstein-Sato polynomials in positive characteristic. arXiv:0711.3794. MR2469353 (2010b:13008)

[95] M. Mustaţă, S. Takagi, and K.-i. Watanabe, F-thresholds and Bernstein-Sato polynomials. In *European Congress of Mathematics.* Eur. Math. Soc., Zürich, 2005, pp. 341–364. MR2185754 (2007b:13010)

[96] M. Mustaţă and K. Yoshida, Test ideals vs. multiplier ideals , Nagoya Math. J. 193 (2009), 111–128. MR2502910 (2010a:13010)

[97] A. Némethi, Dedekind sums and the signature of $f(x,y)+z^N$. Selecta Math. (N.S.) 4 (1998), no. 2, 361–376. MR1669948 (99m:32050)

[98] J. Nicaise, An introduction to p-adic and motivic zeta functions and the monodromy conjecture. *Algebraic and analytic aspects of zeta functions and L-functions,* 141–166, MSJ Mem., 21, Math. Soc. Japan, Tokyo, 2010. MR2647606 (2011g:11223)

[99] Noro, et al., Risa/Asir: an open source general computer algebra system. 1994. Developed by Fujitsu Labs LTD, Kobe Distribution by Noro et al., see http://www.math.kobe-u.ac.jp/Asir/index.html.

[100] E. M. Opdam, Some applications of hypergeometric shift operators. Invent. Math. 98 (1989), no. 1, 1–18. MR1010152 (91h:33024)

[101] E. M. Opdam, Dunkl operators, Bessel functions and the discriminant of a finite Coxeter group. Compositio Math. 85 (1993), no. 3, 333–373. MR1214452 (95j:33044)

[102] P. Orlik, and L. Solomon, Combinatorics and topology of complements of hyperplanes. Invent. Math. 56 (1980), no. 2, 167–189. MR558866 (81e:32015)

[103] D. O'Shea, Topologically trivial deformations of isolated quasihomogeneous hypersurface singularities are equimultiple, Proc. Amer. Math. Soc. 101 (1987) 260–262. MR902538 (88k:32056)

[104] C. Peters and J. Steenbrink, *Mixed Hodge structures.* Ergebnisse der Mathematik und ihrer Grenzgebiete. 3. Folge. A Series of Modern Surveys in Mathematics, 52. Springer-Verlag, Berlin, 2008. xiv+470 pp MR2393625 (2009c:14018)

[105] G. Rond, Séries de Poincaré motiviques d'un germe d'hypersurface irréductible quasi-ordinaire. Astérisque No. 323 (2009), 371–396. MR2647979 (2011c:14039)

[106] G. Rybnikov. On the fundamental group of the complement of a complex hyperplane arrangement. DIMACS 94-13. math.AG/9805056.

[107] C. Sabbah, Modules d'Alexander et $\mathcal{D}$-modules. Duke Math. J. 60 (1990), no. 3, 729–814. MR1054533 (91i:32008)

[108] M. Saito, Introduction to mixed Hodge modules. *Actes du Colloque de Théorie de Hodge (Luminy, 1987)*. Astérisque No. 179-180 (1989), 10, 145–162. MR1042805 (91j:32041)

[109] M. Saito, Exponents and Newton polyhedra of isolated hypersurface singularities. Math. Ann. 281 (1988), no. 3, 411–417. MR954149 (89h:32027)

[110] M. Saito, Exponents of an irreducible plane curve singularity. math.AG/0009133.

[111] M. Saito, Bernstein-Sato polynomials of hyperplane arrangements. math.AG/0602527.

[112] M. Saito, On real log canonical thresholds. arXiv:0707.2308.

[113] F. Sato and K. Sugiyama, Multiplicity one property and the decomposition of b-functions. Internat. J. Math. 17 (2006), no. 2, 195–229. MR2205434 (2007c:11135)

[114] M. Schulze, `gaussman.lib`. A SINGULAR 3-1-1 library for computing invariants related to the the Gauss-Manin system of an isolated hypersurface singularity.

[115] K. Schwede and K. Tucker, On the behavior of test ideals under separable finite morphisms. arXiv:1003.4333.

[116] C. Sevenheck, Bernstein polynomials and spectral numbers for linear free divisors. arXiv:0905.0971. MR2828135

[117] T. Shibuta, An algorithm for computing multiplier ideals. arXiv:0807.4302. MR2811565

[118] J. Steenbrink, Mixed Hodge structures on the vanishing cohomology, in *Real and Complex Singularitites, Oslo 1976*, Alphen aan den Rijn, Oslo, (1977), 525–563. MR0485870 (58:5670)

[119] K. Sugiyama, b-Functions associated with quivers of type A. arXiv:1005.3596.

[120] S. Takagi, Formulas for multiplier ideals on singular varieties. Amer. J. Math. 128 (2006), no. 6, 1345–1362. MR2275023 (2007i:14006)

[121] Z. Teitler, A note on Mustaţă's computation of multiplier ideals of hyperplane arrangements. Proc. Amer. Math. Soc. 136 (2008), 1575-1579. MR2373586 (2008k:14005)

[122] L. V. Thành and J. Steenbrink, Le spectre d'une singularité d'un germe de courbe plane. Acta Math. Vietnam. 14 (1989), no. 1, 87–94. MR1058278 (91k:32042)

[123] M. Tomari, The inequality $8p_g < \mu$ for hypersurface two-dimensional isolated double points. Math. Nachr. 164 (1993), 37–48. MR1251454 (94j:32032)

[124] K. Ukai, b-functions of prehomogeneous vector spaces of Dynkin-Kostant type for exceptional groups. Compositio Math. 135 (2003), no. 1, 49–101. MR1955164 (2004c:11226)

[125] L. Van Proeyen and W. Veys, Poles of the topological zeta function associated to an ideal in dimension two. Math. Z. 260 (2008), no. 3, 615–627. MR2434472 (2009f:14025)

[126] L. Van Proeyen and W. Veys, The monodromy conjecture for zeta functions associated to ideals in dimension two. arXiv:0910.2179. MR2722244 (2011m:14036)

[127] J.-L. Verdier, Spécialisation de faisceaux et monodromie modérée. *Analysis and topology on singular spaces, II, III (Luminy, 1981)*, 332–364, Astérisque, 101–102, Soc. Math. France, Paris, 1983. MR737938 (86f:32010)

[128] W. Veys, Determination of the poles of the topological zeta function for curves, Manuscripta Math. 87 (1995), 435–448. MR1344599 (97a:11192)

[129] W. Veys and W.A. Zúñiga-Galindo, Zeta functions for analytic mappings, log-principalization of ideals, and Newton polyhedra. Trans. Amer. Math. Soc. 360 (2008), no. 4, 2205–2227. MR2366980 (2008i:11140)

[130] U. Walther, Bernstein-Sato polynomial versus cohomology of the Milnor fiber for generic hyperplane arrangements. Compos. Math. 141 (2005), no. 1, 121–145. MR2099772 (2005k:32030)

[131] Y.-J. Xu and S. S.-T. Yau, Durfee conjecture and coordinate free characterization of homogeneous singularities. J. Differential Geom. 37 (1993), no. 2, 375–396. MR1205449 (94f:32073)

[132] T. Yano, On the theory of b-functions. Publ. Res. Inst. Math. Sci. 14 (1978), no. 1, 111–202. MR499664 (80h:32026)

[133] T. Yano, Exponents of singularities of plane irreducible curves. Sci. Rep. Saitama Univ. Ser. A 10, no. 2, (1982), 21–28. MR662407 (84h:32009)

[134] C. Yuen, The multiplicity of jet schemes of a simple normal crossing divisor. Comm. Algebra 35 (2007), no. 12, 3909–3911. MR2372306 (2008i:13051)

[135] W.A. Zúñiga-Galindo, Local zeta functions and Newton polyhedra. Nagoya Math. J. 172 (2003), 31–58. MR2019519 (2004h:11098)

Department of Mathematics, University of Notre Dame, 255 Hurley Hall, Indiana 46556, USA

E-mail address: nbudur@nd.edu

Contemporary Mathematics
Volume **566**, 2012
http://dx.doi.org/10.1090/conm/566/11221

Finite families of plane valuations: value semigroup, graded algebra and Poincaré series

Carlos Galindo and Francisco Monserrat

1. Introduction

The formal definition of valuation was firstly given by the Hungarian mathematician J. Kürschák in 1912 supported with ideas of Hensel. Valuation theory, based on this concept, has been developed by a large number of contributors (some of them distinguished mathematicians as Krull or Zariski) and it has a wide range of applications in different context and research areas as, for instance, algebraic number theory or commutative algebra and its application to algebraic geometry or theory of diophantine equations.

In this paper, we are interested in some applications of valuation theory to algebraic geometry and, particularly, to singularity theory. Valuation theory was one of the main tools used by Zariski when he attempted to give a proof of resolution of singularities for algebraic schemes. In characteristic zero, resolution was proved by Hironaka without using that tool; however there is no general proof for positive characteristic and valuations seem to be suitable algebraic objects for this purpose. Valuations associated with irreducible curve singularities are one of the best known classes of valuations, especially the case corresponding to plane branches where valuations and desingularization process are very related. Germs of plane curves can contain several branches and, for this reason, it is useful to study their corresponding valuations, not only in an independent manner but as a whole [**6, 7, 8**]. Valuations of the fraction field of some 2-dimensional local regular Noetherian ring R centered at R, that we call *plane valuations*, are a very interesting class of valuations which includes the above mentioned family related with branches. These valuations were studied by Zariski and their study was revitalized by the paper [**46**]. Very little is known about valuations in higher dimension.

The aim of this paper is to provide a concise survey of some aspects of the theory of plane valuations, adding some comments upon more general valuations when it is possible. For those valuations, we describe value semigroup, graded algebra and Poincaré series emphasizing on the recent study of the same algebraic objects for finite families of valuations and their relation with the corresponding ones for reduced germs of plane curves.

1991 *Mathematics Subject Classification.* Primary 14B05, 13A18.

Supported by Spain Ministry of Education MTM2007-64704 and Bancaixa P1-1B2009-03.

Section 2 of the paper recalls the general notion of valuation and compiles the main known facts with respect to the value group and semigroup of a valuation. We show a new condition, Proposition 2.2, given in [**14**], that has to do with the number of generators of the value semigroups of Noetherian local domains (see [**17**] for a more general result). We also give in Proposition 2.4 a numerical condition, called combinatorially finiteness, that those value semigroups satisfy. The graded algebra of a valuation ν, $\mathrm{gr}_\nu R$, is introduced in Section 3. There, we explain how to construct a minimal free resolution of $\mathrm{gr}_\nu R$ as a module over a polynomial ring and, in Proposition 3.2, how to compute the dimension of its ith syzygy module. This graded algebra is the main ingredient in the Teissier's idea to prove resolution of singularities. When ν is plane $\mathrm{gr}_\nu R$ is Noetherian, notwithstanding this is not true for higher dimension (see Proposition 3.4). Section 4 is devoted to introduce plane valuations, their main invariants and to classify them by means of an algebraic device that allows us to get parametric equations of the valuations. The introduction and computation of the Poincaré series of plane valuations (with particular attention to the divisorial case) is given in Section 5.

Finite families of valuations whose value group is that of integer numbers, $\mathbb{Z}$, are considered in Section 6. For them we define the concepts of graded algebra, generating sequence and Poincaré series, explaining that this series is a rational function whenever one considers certain families of valuations which include the divisorial ones in the plane and those associated with a rational surface singularity. Following [**19**], and also in this section, semigroup of values, generating sequences and Poincaré series for finite families of plane divisorial valuations are explicitly computed. We also add some information given in [**10**] corresponding to families of any type of plane valuations. Finally, in Section 7 we provide an specific calculation of the Poincaré series of multiplier ideals of a plane divisorial valuation ν, Theorem 7.5. That series gathers information on the multiplier ideals and jumping numbers corresponding to the singularity that ν encodes and the proof of Theorem 7.5 uses techniques and results involving the family of plane divisorial valuations given by the exceptional divisors appearing in the blowing-up sequence determined by ν.

2. Valuations

2.1. Definition and a bit of history. Between 1940 and 1960, Zariski [**51, 52**] and Abhyankar [**2, 3**] developed the theory of valuations in the context of the theory of singularities with the aim of proving resolution for algebraic schemes. The concept of valuation is analogue to that of place. Places were introduced by Dedekind and Weber in the nineteenth century [**21**] with the purpose of constructing the Riemann surface associated with an affine curve from the field of functions of the curve. Also in that century, to study diophantine equations by using the Hensel's Lemma and solutions of the equations in the completions $\mathbb{Q}_p$, Hensel [**31**] considered p-adic valuations on the field of rational numbers, $\mathbb{Q}$, defined as $\nu_p(q) := \alpha$, whenever $\mathbb{Q} \setminus \{0\} \ni q = p^\alpha(r/s)$ and $\gcd(r,p) = \gcd(s,p) = 1$. The properties of ν_p give rise to the definition of valuation and its definition has to do with that of valuations centered at the completion of the local ring of a branch of a plane curve. In 1964, Hironaka [**34**] proved resolution of singularities in characteristic zero (some more recent references are [**49, 22**]) and valuations were forgotten for a large period. However, activity in valuation theory has been increased in the last two decades, probably due to the lack of success in proving resolution in positive characteristic. Next we define the concept of valuation and some related objects.

DEFINITION 2.1. A *valuation* of a commutative field K is a surjective map $\nu : K^*(:= K \setminus \{0\}) \to G$, where G is a totally ordered commutative group, such that for $f, g \in K^*$

- $\nu(fg) = \nu(f) + \nu(g)$.
- $\nu(f + g) \geq \min\{\nu(f), \nu(g)\}$ and the equality holds whenever $\nu(f) \neq \nu(g)$.

G is usually named the *value group* of ν and the set $R_\nu := \{f \in K^* | \nu(f) \geq 0\} \cup \{0\}$ is a local ring, called the *valuation ring* of ν, whose maximal ideal is $m_\nu := \{f \in K^* | \nu(f) > 0\} \cup \{0\}$. The *rank* of ν ($\mathrm{rk}(\nu)$) is the Krull dimension of the ring R_ν and the dimension of the $\mathbb{Q}$-vector space $G \otimes_{\mathbb{Z}} \mathbb{Q}$ is the *rational rank* of ν ($\mathrm{r.rk}(\nu)$).

2.2. Value group and value semigroup. Along this paper we shall consider a Noetherian local domain (R, m) whose field of fractions is K and we shall assume that each valuation ν dominates R, that is $R \subset R_\nu$ and $R \cap m_\nu = m$. In this case, in addition to the two previous numerical invariants associated with ν, we can consider the so-called *transcendence degree* of ν ($\mathrm{tr.deg}(\nu)$), which is the transcendence degree of the field k_ν over k, where $k_\nu := R_\nu / m_\nu$ and $k := R/m$. Unless otherwise stated, we shall assume that k is algebraically closed. The mentioned invariants are useful to classify valuations when $\dim R = 2$. The value groups G of valuations ν as above have been studied and classified [**41, 42, 52, 38**]. G can be embedded in $\mathbb{R}^n$ with lexicographical ordering, n being the dimension of R and $\mathbb{R}$ the real numbers.

An interesting object which is not well-understood in general is the *value semigroup* of a valuation ν associated with R. This one is defined as

$$S := \{\nu(f) | f \in R \setminus \{0\}\}.$$

Interesting data concerning ideal theory, singularities and topology are encoded by this semigroup. The two main facts which are known about it are:

1) The Abhyankar inequalities:

$$\mathrm{rk}(\nu) + \mathrm{tr.deg}(\nu) \leq \mathrm{r.rk}(\nu) + \mathrm{tr.deg}(\nu) \leq \dim(R).$$

Moreover, if $\mathrm{rk}(\nu) + \mathrm{tr.deg}(\nu) = \dim R$, then G is isomorphic to $\mathbb{Z}^{\mathrm{rk}(\nu)}$ with lexicographical ordering and whenever $\mathrm{r.rk}(\nu) + \mathrm{tr.deg}(\nu) = \dim R$, then G is isomorphic to $\mathbb{Z}^{\mathrm{r.rk}(\nu)}$.

2) S is a well-ordered subset of the positive part of the value group G of ordinal type at most $\omega^{rk(\nu)}$, ω being the ordinal type of the set $\mathbb{N}$ of non-negative integers.

When R is regular and $\dim R = 1$, the semigroups S are isomorphic to the natural numbers. The case $\dim R = 2$ is also known; later we shall give more information about it. For higher dimension, very little is known. The second inequality in condition 1) gives a constraint on the value semigroup and recently, Cutkosky [**14**] has proved that the mentioned inequality and condition 2) do not characterize value semigroups on equicharacteristic Noetherian local domains. To do it he proves the forthcoming Proposition 2.2, which gives a new necessary condition for a semigroup to be a value semigroup. This allows him to provide an example of a well ordered sub-semigroup of the positive rational numbers $\mathbb{Q}_+$ of ordinal type ω which is not a value semigroup of some equicharacteristic local domain.

PROPOSITION 2.2. *With the above notations, let assume that R is an equicharacteristic local domain and ν a valuation of K that dominates R. Set $s_0 := \min\{\nu(f)|f \in m \setminus \{0\}\}$, $n := \dim_k m/m^2$ and $S_\Psi := \nu(m \setminus \{0\}) \cap \Psi$, Ψ being the convex subgroup of real rank 1 of G. Then,*

$$\mathrm{card}\,(S_\Psi \cap [0,(d+1)s_0)) < \binom{n+d}{n},$$

for all nonnegative integer d, where we have set $[a,b) := \{c \in \Psi | a \leq c < b\}$, $a, b \in \Psi$.

Ideals in R which are contraction of ideals in the valuation ring R_ν are named valuation ideals or ν-ideals. The following result collects basic results on value semigroups and ν-ideals. Recall that an order $\leq$ in a semigroup is called cancellative if $\alpha+\beta = \alpha+\gamma$ implies $\beta = \gamma$ and it is admissible if $\alpha+\gamma \leq \beta+\gamma$ whenever $\gamma \geq 0$ and $\alpha \leq \beta$.

PROPOSITION 2.3. *The value semigroup S of a valuation ν of a field K, centered at R, is a cancellative, commutative, free of torsion, well-ordered semigroup with zero, where the associated order is admissible. Moreover, $F = \{P_\alpha\}_{\alpha\in S}$, where*

$$P_\alpha := \{f \in R \setminus \{0\} \,|\, \nu(f) \geq \alpha\} \cup \{0\}$$

is the family of ν-ideals (in R) of the valuation ν.

PROOF. We shall prove that S is free of torsion, F is the family of ν-ideals and, finally, that S is well-ordered. The remaining properties are clear. Assume that $\nu(u) \neq 0$, $u \in K \setminus \{0\}$, then either $\nu(u) > 0$ or $\nu(u^{-1}) > 0$, so either $u \in m_\nu$ or $u^{-1} \in m_\nu$ and therefore either $u^p \in m_\nu$ or $u^{-p} \in m_\nu$, p being a positive integer. Thus $\nu(u^p) \neq 0$ and the group spanned by S, $G(S)$ (which is G) is free of torsion. This proves that S is also. R is a Noetherian ring and then $\mathrm{rk}(\nu) < \infty$, so each ν-ideal $\mathcal{I}$ is finitely generated. Consider a finite set of generators for $\mathcal{I}$ and set α the minimum of the values (by ν) of these generators, then it is straightforward that $\mathcal{I} = P_\alpha$ and so $\mathcal{I} \in F$. Finally, S is well-ordered because the family of ν-ideals F is also [**52**, App. 3]. □

Let S be the value semigroup of a valuation ν. S satisfies that $(-S) \cap S = \{0\}$. This means that $\sum_{i=1}^m \alpha_i = 0$, $\alpha_i \in S$, implies $\alpha_i = 0$ for every index i. The *length function* of a semigroup S, $l : S \to \mathbb{N} \cup \{\infty\}$, is defined as $l(0) = 0$ and, for $\alpha \neq 0$,

$$l(\alpha) := \sup\{m \in \mathbb{N} | \alpha = \sum_{i=1}^m \alpha_i, \text{ where } \alpha_i \in S \setminus \{0\}\ \}.$$

In our case $l(\alpha) < \infty$ and therefore S is generated by its *irreducible elements*, that is those elements in S whose length is one. This is a consequence of the following result which can be deduced from the mentioned fact that G can be embedded in $\mathbb{R}^n$ with the lexicographical ordering, n being the dimension of R.

PROPOSITION 2.4. [**11**] *Let ν be a valuation and S its value semigroup. Then, for each $\alpha \in S$, it happens that $t(\alpha) < \infty$, where*

$$t(\alpha) := \mathrm{card}\,\{\{\alpha_i\}_{i=1}^m \textit{ finite subset of } S \setminus \{0\} | \alpha = \textstyle\sum_{i=1}^m \alpha_i\ \}.$$

Generally speaking, the semigroups S such that $t(\alpha) < \infty$ for all $\alpha \in S$ are called *combinatorially finite*.

3. Graded algebra of a valuation

3.1. Graded algebra and generators. Let ν be a valuation of the field K centered at the ring R. For each element α in the value semigroup S, consider the ν-ideals P_α and $P_\alpha^+ = \{f \in R | \nu(f) > \alpha\} \cup \{0\}$. The graded algebra of R relative to ν is defined to be as the graded k-algebra

$$\mathrm{gr}_\nu R := \bigoplus_{\alpha \in S} \frac{P_\alpha}{P_\alpha^+},$$

where the product of homogeneous elements is defined as follows: for $f \in P_\alpha$ and $g \in P_\beta$, f modulo P_α^+ times g modulo P_β^+ is the class fg modulo $P_{\alpha+\beta}^+$.

The field k_ν is an extension of the residue field of R, k. There is a canonical field embedding of k into k_ν and when this embedding is an isomorphism, one gets $\dim_k P_\alpha/P_\alpha^+ = 1$ for each $\alpha \in S$. In this case, if one fixes a nonzero element $[f_\beta] \in P_\beta/P_\beta^+$ for each $\beta \in \Lambda$, Λ being the set of irreducible elements in S, and consider the S-graded k-algebra, $k_\Lambda[S] := K[\{X_\beta\}_{\beta \in \Lambda}]$, where the X_β are indeterminates of degree β, then there exists an epimorphism of graded k-algebras $\psi : k_\Lambda[S] \to \mathrm{gr}_\nu R$, given by $\Psi(X_\beta) = [f_\beta]$, which is homogeneous of degree zero and allows us to regard $\mathrm{gr}_\nu R$ as $k_\Lambda[S]$-module, $\ker \psi$ being an ideal of $k_\Lambda[S]$ spanned by binomials.

Generally speaking k is not isomorphic to k_ν. In any case, the following property happens.

PROPOSITION 3.1. *For every $\alpha \in S$, P_α/P_α^+ is a finite dimensional k-vector space.*

PROOF. The inclusion $mP_\alpha \subset P_\alpha^+$ holds because s_0 $(:= \min\{\nu(f) | f \in m\}) > 0$ and therefore P_α/P_α^+ is a k-homomorphic image of P_α/mP_α which is a k-vector space of finite dimension because R is a Noetherian ring. □

This result allows us to get by a recursive procedure a minimal system of generators of $\mathrm{gr}_\nu R$, $M = \{[f_\gamma]\}_{\gamma \in \Gamma}$, and attach to it an S-graded polynomial algebra $A[\nu] := k[\{X_\gamma\}_{\gamma \in \Gamma}]$ that substitutes the former $k_\Lambda[S]$ for the general case. The procedure to obtain M works by recurrence on the length of the elements in S and it is based on the computation of certain bases of the vector spaces P_α/P_α^+ with $l(\alpha) = n$ from the knowledge of the vector spaces P_α/P_α^+ such that $l(\alpha) < n$. When $n = 1$, we pick an arbitrary basis of P_α/P_α^+. Otherwise set

$$\Omega_\alpha := \left\{ \{\alpha_1, \alpha_2, \ldots, \alpha_r\} \subseteq S \setminus \{0\}, r \geq 2 \mid \alpha = \sum_{i=1}^{r} \alpha_i \right\}$$

and since, by recurrence, we know a basis of each vector space $P_{\alpha_i}/P_{\alpha_i}^+$, we are able to compute the following vector subspace of P_α/P_α^+:

$$W_\alpha = \sum_{\{\alpha_1, \alpha_2, \ldots, \alpha_r\} \in \Omega_\alpha} \frac{P_{\alpha_1}}{P_{\alpha_1}^+} \cdot \frac{P_{\alpha_2}}{P_{\alpha_2}^+} \cdots \frac{P_{\alpha_r}}{P_{\alpha_r}^+},$$

where the products of elements are in the algebra $\mathrm{gr}_\nu R$. Now take an arbitrary linearly independent set of P_α/P_α^+ whose classes are a basis of the vector space $\frac{P_\alpha}{P_\alpha^+}/W_\alpha$. This set extends any basis of W_α to a basis of P_α/P_α^+. The set M is obtained joining the bases of the vector spaces P_α/P_α^+ such that $l(\alpha) = 1$ with the described sets extending bases of the spaces W_α such that $l(\alpha) > 1$. As a consequence, it holds that the elements $\gamma \in \Gamma$ are of the form $\gamma = (\beta, i_\beta)$ with

$\beta \in S$ and $1 \leq i_\beta \leq \dim\left(\frac{P_\beta}{P_\beta^+}/W_\beta\right)$, where we have set $W_\beta = 0$ when $l(\beta) = 1$, and $A[\nu]$ is S-graded by setting $\deg(X_\gamma) = \deg(\gamma) = \beta \in S$, [**11**].

3.2. Minimal free resolution of $\mathrm{gr}_\nu R$. Denote by $A[\nu]_\alpha$ the homogeneous component of degree α of the ring $A[\nu]$ and consider the map

$$\phi_0 : A[\nu] = \bigoplus_{\alpha \in S} A[\nu]_\alpha \longrightarrow \mathrm{gr}_\nu R$$

which maps X_γ to $[f_\gamma]$; it is a homogeneous k-algebra epimorphism. Also consider the graded ideals $m[\nu] := \bigoplus_{0 \neq \alpha \in S} A[\nu]_\alpha$ and $\mathcal{I}_0 := \ker(\phi_0)$, and a minimal homogeneous generating set of $\mathcal{I}_0$, $B = \cup_{\alpha \in S} B_\alpha$, B_α being the set of elements in B of degree α. By Nakayama's graded Lemma, the set of classes $[B_\alpha]$ of B_α in $\mathcal{I}_0/m[\nu]\mathcal{I}_0$ is a basis of the homogeneous component of degree α of $\mathcal{I}_0/m[\nu]\mathcal{I}_0$ and thus $[B_\alpha]$ and therefore B_α is finite since $A[\nu]_\alpha$ is a finite-dimensional vector space because S is a combinatorially finite semigroup. This allows us to provide a degree 0 homogeneous homomorphism $\phi_1 : L_1 := \oplus_{\alpha \in S}(A[\nu])^{l(\alpha)} \to A[\nu]$, $l(\alpha)$ being the cardinality of B_α and recursively a minimal free resolution of $\mathrm{gr}_\nu R$ as S-graded $A[\nu]$-module:

$$(A.) : \cdots \to L_i \stackrel{\phi_i}{\to} L_{i-1} \to \cdots \to L_1 \stackrel{\phi_1}{\to} A[\nu] \to \mathrm{gr}_\nu R \to 0.$$

Write $N_i := \ker(\phi_i)$, then the following result holds:

PROPOSITION 3.2. [**11**]

(1) *For every $i \geq 0$, there exists a homogeneous of degree 0 isomorphism of graded $A[\nu]$-modules between the ith* Tor *module* $\mathrm{Tor}_i^{A[\nu]}(\mathrm{gr}_\nu R, k)$ *and* $L_i \bigotimes_{A[\nu]} k$.

(2) *For each $\alpha \in S$, let denote the homogeneous component of degree α with the subindex α, then*

$$\dim_k \left(\mathrm{Tor}_{i+1}^{A[\nu]}(\mathrm{gr}_\nu R, k)\right)_\alpha = \dim_k \frac{(N_i)_\alpha}{(m[\nu]N_i)_\alpha}.$$

(3) *There exists an isomorphism of S-graded modules between* $\mathrm{Tor}_i^{A[\nu]}(k, \mathrm{gr}_\nu R)$ *and the ith homology $H_i(G[\nu])$ of an augmented Koszul complex of* $\mathrm{gr}_\nu R$*-modules.*

As a consequence of the commutative property of the Tor *functor and from item (2), the number of homogeneous elements of degree α in a minimal set of homogeneous generators of the ith syzygy module of $gr_\nu R$ as $A[\nu]$-module is* $\dim_k(H_i(G[\nu])_\alpha$.

The graded algebra relative to a valuation seems to be a useful tool to study the local uniformization problem. This consists of, given the local ring of an algebraic variety (assuming that it is an integral domain), finding, for each valuation ν centered at R, a regular local R-algebra R' essentially of finite type over R and contained in R_ν. In [**47**], Teissier proposes that R' might be obtained from an affine chart of a proper algebraic map $Z \to \mathrm{Spec}R$ which would be described as a proper and birational toric map with respect to some system of generators of the maximal ideal of R. An idea to do this would be to view R as a deformation of the graded ring $\mathrm{gr}_\nu R$ with respect to the filtration associated with the valuation and to obtain the uniformization of the valuation ν as a deformation of the valuation induced by

ν on $\mathrm{gr}_\nu R$; the motivating example is the case of complex plane branches which has been studied by Goldin and Teissier as deformations of monomial curves.

Without doubt, the most interesting valuations from a geometric point of view are the so-called divisorial valuations because they are attached to irreducible exceptional divisors of some birational map. Next we state the definition.

DEFINITION 3.3. Let us assume that $\dim R = n$. A valuation ν of K centered at R is called to be *divisorial* whenever its rank is 1 and its transcendence degree is $n-1$.

When $n = 2$, the graded algebra $\mathrm{gr}_\nu R$ of a divisorial valuation is Noetherian. Notwithstanding, this does not happen in higher dimension. For instance, let R be a 3-dimensional local regular ring and blow-up $X_0 = \mathrm{Spec} R$ at its maximal ideal m_0. Let X_1 be the obtained variety. Consider the cubic with equation $x^2z+xy^2+y^3=0$ on the obtained exceptional divisor $E_1 := \mathrm{Proj}(k[x,y,z])$ and a sequence of $n \geq 10$ point blowing-ups $X_n \to \cdots \to X_0$ centered at m_0 and at points m_i in X_i, $1 \leq i \leq n$, on the last obtained exceptional divisor E_i and on the strict transform of the cubic. Denote by ν the divisorial valuation given by the divisor E_n and set $R_i := \mathcal{O}_{X_i,m_i}$. It is not difficult to prove that $R_1 = k[a_1,b_1,c_1]_{(a_1,b_1,c_1)}$, where $a_1 = x$, $b_1 = y/x$ and $c_1 = (x/z)+(y/x)^2+(y/x)^3$. If A_1, B_1, C_1 are, respectively, the initial forms of a_1, b_1, c_1 on $\mathrm{gr}_\nu R_1 = k[A_1,B_1,C_1]$, then we can state

PROPOSITION 3.4. [**13**] *The family* A_1, A_1B_1, $A_1^3C_1$, $A_1B_1^2$, $A_1^2B_1^5$, $A_1^3B_1^8, \ldots,$ $A_1^iB_1^{3i-1}, \ldots$ *is a minimal system of generators of* $gr_\nu R \subset gr_\nu R_1$. *As a consequence* $\mathrm{gr}_\nu R$ *is not Noetherian.*

An interesting number associated with a divisorial valuation ν is the volume. In this case $\mathbb{Z}$ is the value group of ν and by definition, the *volume* of ν is

$$\mathrm{vol}(\nu) := \limsup_{\alpha\in\mathbb{N}} \frac{\mathrm{length}(R/P_\alpha)}{\alpha^n/n!}.$$

This definition corresponds to the analogue of the Samuel multiplicity for an m-primary ideal $p \subseteq R$:

$$e(p) := \limsup_{\alpha\in\mathbb{N}} \frac{\mathrm{length}(R/p^\alpha)}{\alpha^n/n!}.$$

It is known that the multiplicity is always an integer number and also [**23**] that

$$\mathrm{vol}(\nu) = \lim_{\alpha\to\infty} (e(P_\alpha)/\alpha^n).$$

However the volume of a divisorial valuation is not always an integer number although it is rational when its graded algebra is Noetherian. As a consequence valuations with irrational volume provide non-finitely generated attached graded algebras. For an example, see [**37**].

4. Plane valuations

4.1. Definition and geometric sense. From this section on we shall consider plane valuations, notwithstanding from time to time we shall speak about other types of valuations. We start this section with the definition.

DEFINITION 4.1. A *plane valuation* is a valuation of a field K which is the fraction field of a two-dimensional Noetherian local regular ring R and is centered at R.

Zariski in [**51**] classified plane valuations by attending invariants as the rank and the rational rank. By using previous results by Zariski, Spivakovsky [**46**] gives the following geometric view of plane valuations.

THEOREM 4.2. *There is a one to one correspondence between the set of plane valuations (of K centered at R) and the set of simple sequences of point blowing-ups of the scheme* Spec R.

The correspondence in Theorem 4.2 works as follows: each valuation ν is associated with the sequence

$$\pi : \cdots \longrightarrow X_{N+1} \xrightarrow{\pi_{N+1}} X_N \longrightarrow \cdots \longrightarrow X_1 \xrightarrow{\pi_1} X_0 = X = \text{Spec } R, \tag{4.1}$$

where π_{i+1} is the blowing-up of X_i at the unique closed point p_i of the exceptional divisor obtained after the blowing-up π_i, E_i, which satisfies that ν is centered at the local ring $\mathcal{O}_{X_i,p_i}$ $(:= R_i)$.

Theorem 4.2 allows Spivakovsky to give a classification of plane valuations which improves the Zariski's one and it is based in the form of the so-called dual graph of the sequence π. This graph is a (in general, infinite) tree whose vertices represent the strict transforms in X_l, l large enough, of the divisors E_i (also named E_i) and two vertices are joined by an edge whenever these strict transforms intersect. Set $\mathcal{C}_\nu = \{p_i\}_{i\geq 0}$ the configuration of infinitely near points determined by ν. We say that p_i is *proximate* to p_j (denoted by $p_i \to p_j$) whenever $i > j$ and p_i belongs either to E_{j+1} or to the strict transform of E_{j+1} at X_i and p_i is said to be *satellite* if there exists $j < i - 1$ such that $p_i \to p_j$. Valuations whose associated sequence (4.1) is finite are exactly the divisorial ones. The dual graph shape of a divisorial (plane) valuation is that of Figure 1.

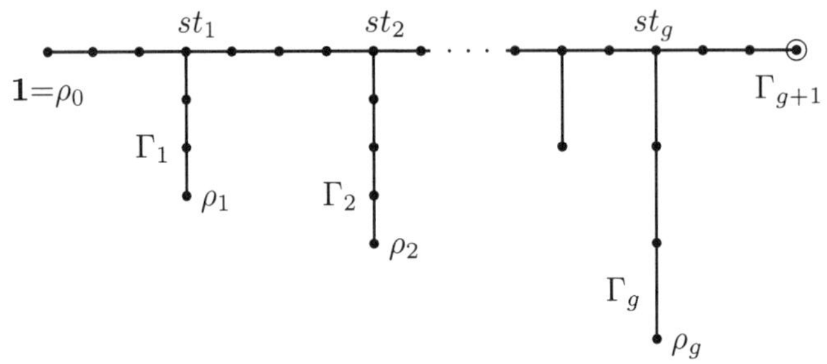

FIGURE 1. The dual graph of a divisorial valuation

The dual graph is not suitable when we desire to get parametric equations for computing valuations. Furthermore, the classical theory for curves uses, for this purpose, Puiseux exponents that only work for zero characteristic. Next, we recall the Spivakovsky's classification in terms of the so-called Hamburger-Noether expansions of valuations. These expansions provide parametric equations for plane valuations [**27**] and have been used in [**18**] to study saturation with respect to this type of valuations.

4.2. Hamburger-Noether expansions and classification of plane valuations. Let ν be a plane valuation and take $\{u, v\}$ a regular system of parameters for the ring R. Assume that $\nu(u) \leq \nu(v)$. This means that there exists an element $a_{01} \in k$ such that the set $\{u_1 = u, v_1 = (v/u) - a_{01}\}$ constitutes a regular system of parameters for the ring R_1. If, now, $\nu(u) \leq \nu(v_1)$ holds, then we repeat the above operation and we keep doing the same thing until we get $v = a_{01}u + a_{02}u^2 + \cdots + a_{0h}u^h + u^h v_h$, where either $\nu(u) > \nu(v_h)$ or $\nu(v_h) = 0$, or $v = a_{01}u + a_{02}u^2 + \cdots + a_{0h}u^h + \cdots$, with infinitely many steps. In the last two cases, we have got the Hamburger-Noether expansion for ν, obtaining $R_\nu = R_h$ when $\nu(v_h) = 0$. Otherwise, set $w_1 := v_h$ and reproduce the above procedure for the regular system of parameters $\{w_1, u\}$ of R_h. The procedure can continue indefinitely or we can obtain a last equality. In any case, we attach to ν a set of expressions called the *Hamburger-Noether expansion* of the valuation ν in the regular system of parameters $\{u, v\}$ of the ring R which provides a regular system of parameters for each local ring R_i given by the sequence π described in (4.1) and it has the form given in Figure 2.

$$
\begin{array}{rcl}
v & = & a_{01}u + a_{02}u^2 + \cdots + a_{0h_0}u^{h_0} + u^{h_0}w_1 \\
u & = & w_1^{h_1}w_2 \\
\vdots & & \vdots \\
w_{s_1-2} & = & w_{s_1-1}^{h_{s_1-1}}w_{s_1} \\
w_{s_1-1} & = & a_{s_1k_1}w_{s_1}^{k_1} + \cdots + a_{s_1h_{s_1}}w_{s_1}^{h_{s_1}} + w_{s_1}^{h_{s_1}}w_{s_1+1} \\
\vdots & & \vdots \\
w_{s_g-1} & = & a_{s_gk_g}w_{s_g}^{k_g} + \cdots + a_{s_gh_{s_g}}w_{s_g}^{h_{s_g}} + w_{s_g}^{h_{s_g}}w_{s_g+1} \\
\vdots & & \vdots \\
w_{i-1} & = & w_i^{h_i}w_{i+1} \\
\vdots & & \vdots \\
(w_{z-1} & = & w_z^{\infty}).
\end{array}
$$

FIGURE 2. Hamburger-Noether expansion of a plane valuation

The nonnegative integers $\{s_j\}_{j=0}^g$ correspond to rows with some nonzero a_{s_jl} (called free ones and that are those associated with the non-satellite blowing-up points), $g \in \mathbb{N} \cup \{\infty\}$ and $k_j = \min\{n \in \mathbb{N} \mid a_{s_j,n} \neq 0\}$. Thus, plane valuations can be classified in the following five types which we name with a letter or as in [**24**].

– *Type A or divisorial valuations.* Their Hamburger-Noether expansion is finite and their last row has the following shape

$$
w_{s_g-1} = a_{s_gk_g}w_{s_g}^{k_g} + \cdots + a_{s_gh_{s_g}}w_{s_g}^{h_{s_g}} + w_{s_g}^{h_{s_g}}w_{s_g+1}, \tag{4.2}
$$

where $g < \infty$, $h_{s_g} < \infty$, $w_{s_g+1} \in R_\nu$ and $\nu(w_{s_g+1}) = 0$.

– *Type B or curve valuations.* Their Hamburger-Noether expansion has a last equality associated with an infinite sum like this $w_{s_g-1} = \sum_{i=k_g}^{\infty} a_{s_gi}w_{s_g}^i$. Here $g < \infty$ and there exists a positive integer i_0 such that p_i is free for all $i > i_0$.

– *Type C or exceptional curve valuations.* Their Hamburger-Noether expansion has a last free row like (4.2) and, after, finitely many non-free rows with the shape

$$
\begin{array}{rcl}
w_{s_g} & = & w_{s_g+1}^{h_{s_g+1}}w_{s_g+2} \\
\vdots & & \vdots \\
w_{z-1} & = & w_z^{\infty}.
\end{array}
$$

type	subtype	rk	r.rk	tr.deg
A	—	1	1	1
B	I	1	1	0
	II	2	2	0
C	—	2	2	0
D	—	1	2	0
E	—	1	1	0

TABLE 1

In this case, $g < \infty$, $s_g < \infty$ and there exists a positive integer i_0 such that $p_i \to p_{i_0}$ for all $i > i_0$.

– *Type D or irrational valuations.* A plane valuation will be called of type D, whenever its Hamburger-Noether expansion has a last free row like (4.2) followed by infinitely many rows with the shape $w_{i-1} = w_i^{h_i} w_{i+1}$ $(i > s_g)$. Now $g < \infty$ and there exists a positive integer i_0 such that p_i is a satellite point for all $i \geq i_0$ but ν is not a type C valuation.

– *Type E or infinitely singular valuations.* When the Hamburger-Noether expansion of a plane valuation repeats indefinitely the basic structure, then the valuation is called to be of type *E*. This means that the sequence $\mathcal{C}_\nu$ alternates indefinitely blocks of 1 free and $(1 \leq)\, l\, (< \infty)$ non-free rows. Here $g = z = \infty$.

This classification does not depend on the regular system of parameters we choose on R. Table 1 relates our classification with the invariants of ν above defined. Notice that classical invariants provide a refinement of type B valuations. We also add that in [**24**] the real-valued class of plane valuations is interpreted in a rooted metric tree in such a way that the valuations are partially ordered and there is a unique path from any valuation to any other, being this path isometric to a real interval.

4.3. Other invariants of plane valuations. Let ν be a plane valuation and $\{m_i\}_{i\geq 0}$ the family of maximal ideals of the rings R_i of the sequence (4.1). We attach to ν the following data:

– The sequence $\{\min\{\nu(f) | f \in m_i \setminus \{0\}\}_{i \geq 0}$, that we call *sequence of values* of ν.

– The sequence $\{\beta'_j\}_{0 \leq j < g+2}$ that we name *Puiseux exponents* of ν and, with the convention $1/\infty = 0$, it is defined as $\beta'_0 := 1$ and

$$\beta'_{j+1} := h_{s_j} - k_j + 1 + \cfrac{1}{h_{s_j+1} + \cfrac{1}{\ddots}}.$$

– Set $\beta'_j = p_j/n_j$ with $\gcd(p_j, n_j) = 1$ and $r_i = \nu(w_i)$ and $e_j = \nu(w_{s_j})$ for $0 \leq j < g+1$ and $i \geq 0$. Define

$$\beta_{j+1} := \beta_j + (h_{s_j} - k_j)e_j + r_{s_j+1}$$

and

$$\bar{\beta}_{j+1} := n_j \bar{\beta}_j + (h_{s_j} - k_j)e_j + r_{s_j+1}.$$

Then the sequence $\{\beta_j\}_{j=0}^{g+1}$ is called to be the *characteristic sequence* and the sequence $\{\bar{\beta}_j\}_{j=0}^{g+1}$ the *sequence of maximal contact values* both of ν.

Last three sequences are infinite in case E, and in case B we only consider sub-indices from $j = 0$ to $j = g$ although in case B-II we add to $\{\bar{\beta}_j\}_{j=0}^{g}$ the minimum element in the value semigroup S with non-zero first coordinate, denoted by $\bar{\beta}_{g+1}$. The main result concerning maximal contact values is that they are a set of generators of S. Moreover, if we delete the last one $\bar{\beta}_{g+1}$ in type A valuations, we get a minimal set of generators for S. We can determine the type of a valuation if we know either its sequence of values or its characteristic exponents or its maximal contact values, but this does not happen with the Puiseux exponents or with the semigroup. When one knows the type of the valuation, the following result holds.

PROPOSITION 4.3. [**18**] *Assume that ν is a plane valuation and that we know which is its type. Then any of the following invariants can be computed from whichever of the others: sequence of values, Puiseux exponents, maximal contact values, characteristic exponents, and semigroup S of the valuation (or pair $(S, \bar{\beta}_{g+1})$).*

5. Poincaré series of the graded algebra

5.1. General case. For a while, we consider a non-necessarily plane valuation ν. The *Poincaré series* (of the graded algebra) of ν is the formal power series in the indeterminate t:

$$H_{\mathrm{gr}_\nu R}(t) := \sum_{\alpha \in S} \dim_k \frac{P_\alpha}{P_\alpha^+}\, t^\alpha,$$

which, according Proposition 3.1, belongs to the power series ring on S.

Let us assume that the value group of ν is isomorphic to the integer numbers. This is equivalent to say that the ring R_ν is Noetherian [**52**], however the algebra $\mathrm{gr}_\nu R$ may be non-Noetherian and its Poincaré series a non-rational function, even $\mathrm{gr}_\nu R$ might be non-Noetherian but $H_{\mathrm{gr}_\nu R}(t)$ a rational function. When the ring R is 2-dimensional and normal and ν is a divisorial valuation, this series is the generating function of a sequence of integers which is residually equal to the sum of a polynomial with a periodic function (see [**16**] and [**13**]). An explicit computation for the plane divisorial case can be found in [**25**]; as we shall see, the Poincaré series is very close to that attached to the semigroup of the valuation or to the Poincaré series of the analytically irreducible germ of curve provided by a general element of the valuation [**30**]. One can found many papers studying Poincaré series for singularities (which need not to correspond to the irreducible case), some of them are [**8, 9, 15, 28, 40, 45**].

5.2. The plane case. An important concept for studying plane valuations is that of generating sequence. This concept was introduced in [**46**] and the existence of those sequences is discussed in [**29**]. Notice that the hypothesis of 2-dimensionality of R is not necessary to define this concept.

DEFINITION 5.1. A sequence $\{r_j\}_{j\in\mathcal{J}}$ of elements in the maximal ideal m of R is called to be a *generating sequence* (relative to R) of a valuation ν if, for any element $\alpha \in S$, P_α is spanned by the set

$$\left\{ \prod_{j\in\mathcal{J}_0\subseteq\mathcal{J},\mathcal{J}_0 \text{ finite}} r_j^{a_j} \mid a_j \in \mathbb{N}, a_j > 0 \text{ and } \textstyle\sum_{j\in\mathcal{J}_0} a_j\nu(r_j) \geq \alpha \right\}. \tag{5.1}$$

Assume that the Hamburger-Noether expansion of ν is that given in Figure 2. Set $q_0 = u$, $q_1 = v$ and, for $1 < j < g+2$, let q_j be the defining equation of some analytically irreducible germ of curve on Spec R whose Hamburger-Noether expansion in the basis $\{\bar{u} = u + (q_j), \bar{v} = v + (q_j)\}$ of $\hat{R}/(q_j)$ [**5**], $\hat{R}$ being the m-adic completion of R, is

$$\begin{array}{rcl} \bar{v} & = & a_{01}\bar{u} + a_{02}\bar{u}^2 + \cdots + a_{0h_0}\bar{u}^{h_0} + \bar{u}^{h_0}\bar{w}_1 \\ \bar{u} & = & \bar{w}_1^{h_1}\bar{w}_2 \\ \vdots & & \vdots \\ \bar{w}_{s_{j-1}-1} & = & a_{s_{j-1}k_{j-1}}\bar{w}_{s_{j-1}}^{k_{j-1}} + \cdots + a_{s_{j-1}h_{s_{j-1}}}\bar{w}_{s_{j-1}}^{h_{s_{j-1}}} + \cdots . \end{array}$$

In [**46**] it is proved that any generating sequence of a *divisorial valuation* contains a subsequence $\{q_j\}_{j=0}^{g}$. Moreover, this set is a *minimal generating sequence* (no subset of it is a generating sequence) whenever the dual graph of ν (Figure 1) contains no subgraph Γ_{g+1} (or equivalently $h_{s_g} - k_g = 0$); otherwise, $\{q_j\}_{j=0}^{g+1}$ is a minimal generating sequence. Now, let ν be a *valuation of type C or D*. In both cases a minimal generating sequence of ν is of the form $\{q_j\}_{j=0}^{g+1}$. In the first type of valuations $\nu(q_j)$ $(0 \leq j < g+1)$ are data lying on the line that joins the origin to $\nu(q_0)$, but $\nu(q_{g+1})$ does not satisfy this property. With respect to the second type, $\nu(q_j) \in \mathbb{Q}$ whenever $0 \leq j < g+1$, but $\nu(q_{g+1}) \in \mathbb{R} \setminus \mathbb{Q}$. Whenever ν is a *type E valuation*, a minimal generating sequence of ν is an infinite sequence of the form $\{q_j\}_{0 \leq j}$. However neither all valuations have minimal generating sequences nor every element in a minimal generating sequence must be analytically irreducible. Valuations of type B-II which admit minimal generating sequences with this last condition are called of type B-II-a and the remaining ones will be of type B-II-b. Type B-I valuations do not admit minimal generating sequences. To understand this fact, we have to consider an element q_{g+1} which, in general, will be in $\hat{R}$. q_{g+1} will be the defining equation of ν. If q_{g+1}, up to multiplication by an unit, belongs to R, then we are speaking about a valuation of type B-II-a and $\{q_j\}_{j=0}^{g+1}$ is a minimal generating sequence of ν. When there exists an element in R which factorizes in $\hat{R}$ as a product which contains q_{g+1} as a non-trivial factor, ν is of type B-II-b and otherwise it is of type B-I [**46**, Section 9, case 4].

One important property for the generating sequences is given in the next

THEOREM 5.2. [**25, 27**] *Let ν be a type A, B-II-a, C or D plane valuation. Then a set $\{r_j\}_{j \in \mathcal{J}}$ of elements in the maximal ideal m of R is a generating sequence of ν if, and only if, the k-algebra $gr_\nu R$ is spanned by the classes defined by the elements r_j in $gr_\nu R$. In addition, when ν is of type E, it is also true that the classes defined by the elements r_j in $gr_\nu R$ span that algebra whenever $\{r_j\}_{j \in \mathcal{J}}$ is a generating sequence.*

5.2.1. *The divisorial case.* Assume that ν is a divisorial plane valuation and pick a generating sequence of $r+1$ elements. Recalling Section 3, consider the ring $A[\nu] = k[X_0, \ldots, X_r]$ and the exact sequence of graded algebras $0 \to \mathcal{I}_0 \to A[\nu] \overset{\phi_0}{\to} \mathrm{gr}_\nu R \to 0$ which gives rise to the following equality of Poincaré series of graded algebras $H_{\mathrm{gr}_\nu R}(t) = H_{A[\nu]}(t) - H_{\mathcal{I}_0}(t)$. With the help of generating sequences and making use of the Hamburger-Noether expansion properties, one can get the following result proved in [**25**].

THEOREM 5.3. *Let ν be a plane divisorial valuation, $\{\bar{\beta}_j\}_{j=0}^{g+1}$ its maximal contact values, $e_j = \gcd(\bar{\beta}_0, \bar{\beta}_1, \ldots, \bar{\beta}_j)$ and $n_j = e_{j-1}/e_j$. Then, $H_{\mathrm{gr}_\nu R}(t) = H_S(t)H'(t)$, where*

$$H_S(t) := \sum_{\alpha \in S} t^\alpha = \frac{1}{1-t^{\bar{\beta}_0}} \prod_{j=1}^{g} \frac{1-t^{n_j \bar{\beta}_j}}{1-t^{\bar{\beta}_j}}$$

is the Poincaré series of the value semigroup of the valuation and $H'(t) = \frac{1}{1-t^{\bar{\beta}_{g+1}}}$.

As a consequence the Poincaré series and the dual graph of a plane divisorial valuation are equivalent data.

The case when k is infinite but it needs not to be algebraically closed has been recently treated in [**36**] where it is also introduced a motivic Poincaré series.

5.2.2. *The remaining plane cases.* Assume now that ν is a non-divisorial plane valuation. Then, $\dim P_\alpha/P_\alpha^+ = 1$ for any $\alpha \in S$ and then $\mathrm{gr}_\nu R$ is a k-algebra isomorphic to the algebra of the semigroup S. Thus the Poincaré series for S (that is, the series $H_S(t)$ defined in Theorem 5.3) and for $\mathrm{gr}_\nu R$ coincide. With notations as in Section 4, from [**18**, 1.10.5] it is not difficult to prove that

$$H_{\mathrm{gr}_\nu R}(t) = \frac{1}{1-t^{\bar{\beta}_0}} \prod_{j=1}^{g} \frac{1-t^{n_j \bar{\beta}_j}}{1-t^{\bar{\beta}_j}} \frac{1}{1-t^{\bar{\beta}_{g+1}}},$$

except in cases B-I and E. In these cases

$$H_{\mathrm{gr}_\nu R}(t) = \frac{1}{1-t^{\bar{\beta}_0}} \prod_{j=1}^{g} \frac{1-t^{n_j \bar{\beta}_j}}{1-t^{\bar{\beta}_j}},$$

and $g = \infty$ whenever ν is of type E.

6. Graded algebra and Poincaré series of finite families valuations

6.1. Families of valuations whose value group is $\mathbb{Z}$. Throughout this sub-section, we consider a family $V = \{\nu_i\}_{i=1}^m$ of valuations of the quotient field K of a Noetherian local domain (R, m) centered at R such that $\mathbb{Z}$ is the value group of each ν_i, $i \leq i \leq m$. It is known that these valuations are of rank 1. For $\underline{\alpha}, \underline{\beta} \in \mathbb{N}^m$, we say $\underline{\alpha} \geq \underline{\beta}$ whenever $\underline{\alpha} - \underline{\beta} \in \mathbb{N}^m$. Write $\underline{\nu}(f) = (\nu_1(f), \nu_2(f), \ldots, \nu_m(f))$ for $f \in K$ and define the ideal in R,

$$P_{\underline{\alpha}}^V := \{f \in R | \underline{\nu}(f) \geq \underline{\alpha}\} \cup \{0\}.$$

Now we introduce the concepts of graded algebra and Poincaré series for our family V of valuations. Set $\underline{e}_i \in \mathbb{N}^m$, the m-tuple such that all its coordinates are zero but the ith one which is 1, furthermore $\underline{e}_{\leq i} := \sum_{j=0}^{i} \underline{e}_j$, $\underline{e}_{\leq 0} := 0 \in \mathbb{N}^m$ and $\underline{e} := \underline{e}_{\leq m}$.

DEFINITION 6.1. We define the graded algebra associated with the family V as the graded k-algebra

$$\mathrm{gr}_V R =:= \bigoplus_{\underline{\alpha} \in \mathbb{N}^m} \frac{P_{\underline{\alpha}}^V}{P_{\underline{\alpha}+\underline{e}}^V}.$$

Since $\frac{P^V_{\underline{\alpha}}}{P^V_{\underline{\alpha}+\underline{e}}} \cdot \frac{P^V_{\underline{\alpha}}}{P^V_{\underline{\beta}+\underline{e}}} \subseteq \frac{P^V_{\underline{\alpha}+\underline{\beta}}}{P^V_{\underline{\alpha}+\underline{\beta}+\underline{e}}}$ when $\underline{\alpha}, \underline{\beta} \in \mathbb{N}^m$, $\mathrm{gr}_V R$ is a well-defined $\mathbb{N}^m$-graded algebra. On the other hand, Nakayama's Lemma proves that, for each $\underline{\alpha} \in \mathbb{N}^m$, $\frac{P^V_{\underline{\alpha}}}{P^V_{\underline{\alpha}+\underline{e}}}$ is a finite dimensional k-vector space. Denote $\underline{t} = (t_1, t_2, \ldots, t_m)$ and $\underline{t}^{\underline{\alpha}} = t_1^{\alpha_1} t_2^{\alpha_2} \cdots t_m^{\alpha_m}$.

DEFINITION 6.2. The multi-graded (or multi-index) Poincaré series of the graded algebra $\mathrm{gr}_V R$ is defined to be

$$H_{\mathrm{gr}_V R}(t_1, t_2, \ldots, t_m) = H_{\mathrm{gr}_V R}(\underline{t}) := \sum_{\underline{\alpha} \in \mathbb{N}^m} \dim_k (P_{\underline{\alpha}} / P_{\underline{\alpha}+e}) \underline{t}^{\underline{\alpha}} \in \mathbf{Z}[[t_1, \ldots, t_m]],$$

where $\dim_k$ means dimension as k-vector space.

Definition 5.1 can be extended by stating that a family $\Lambda = \{r_j\}_{j \in \mathcal{J}}$ of elements in m is a *generating sequence* (or a *generating set*) of V whenever $P^V_{\underline{\alpha}}$ is spanned by the set given in (5.1) but replacing ν with $\underline{\nu}$ and α with $\underline{\alpha}$. This allows us to give the following definition for families of valuations $V = \{\nu_i\}_{i=1}^m$ as above.

DEFINITION 6.3. A finite family of valuations V is said to be *monomial* with respect to some system of generators $\Lambda = \{r_j\}_{j \in \mathcal{J}}$ of the maximal ideal m of R if Λ is a generating set of V.

In these conditions, we have the following extended version of Theorem 5.2:

PROPOSITION 6.4. *Let $V = \{\nu_i\}_{i=1}^m$ be a family of valuations of K centered at R whose value group is $\mathbb{Z}$. Assume that there exists a finite generating sequence for some valuation of V. Then, a system of generators $\Lambda = \{r_j\}_{j \in \mathcal{J}}$ of the maximal ideal m is a generating set of the family V if, and only if, the k-algebra $\mathrm{gr}_V R$ is generated by the set $\{[r_j]\}_{j \in \mathcal{J}}$, where $[r_j]$ denotes the coset that r_j defines in $\mathrm{gr}_V R$ and the meaning of the expression "coset defined by r_j" is clarified in the remark after the proof.*

PROOF. [**12**] Along this proof, we set P instead P^V, γ will denote elements in $\mathbb{N}^s$, $s \geq 0$, whose jth component is γ_j, r^γ will stand for $\prod_{j=1}^s r_j^{\gamma_j}$ and $[r]^\gamma$ will be $\prod_{j=1}^s [r_j]^{\gamma_j}$. Assume that Λ is a generating set for V. Let $f + P_{\underline{\alpha}+\underline{e}}$ be a nonzero element in $\mathrm{gr}_V R$, then $f \in P_{\underline{\alpha}}$ and so f is in the ideal generated by the set given in (5.1) –with $\underline{\alpha}$ and $\underline{\nu}$ instead of α and ν– which we denote by $P'_{\underline{\alpha}}$. Therefore,

$$f = \sum_{\gamma \in Q_0 \subseteq Q_{\underline{\alpha}}, Q_0 \text{ finite}} a_\gamma r^\gamma, \tag{6.1}$$

where $a_\gamma \in k$ and $Q_{\underline{\alpha}} = \{\gamma | s \in \mathbb{N}, \underline{\nu}(r^\gamma) \geq \underline{\alpha}\}$. As a consequence, $f + P_{\underline{\alpha}+\underline{e}} = \sum_{\gamma \in Q'_0} a_\gamma [r]^\gamma$, where $Q'_0 = \{\gamma \in Q_0 | \underline{\nu}(r^\gamma) \geq \underline{\alpha}$ and the equality holds for some component$\}$.

Conversely, consider $\underline{\alpha} \in \mathbb{N}^m$. We only need to prove that $P_{\underline{\alpha}} \subseteq P'_{\underline{\alpha}}$. Let $f \in P_{\underline{\alpha}}$ be such that $\underline{\nu}(f) = \beta^0 \geq \underline{\alpha}$. $\{[r_j]\}_{j \in \mathcal{J}}$ generates $\mathrm{gr}_V R$, therefore $f + P_{\beta^0+\underline{e}} = \sum_{\gamma \in Q'} a_\gamma [r]^\gamma$. Thus $f - \sum_{\gamma \in Q'} a_\gamma r^\gamma \in P_{\underline{\beta^0}+\underline{e}}$ and as a consequence, $f + f_0 \in P_{\beta^0+\underline{e}}$ for some $f_0 \in P'_{\beta^0}$. Analogously, we can get $\beta^1 \in \mathbb{N}^m$, such that $\beta^1 > \beta^0$ and $f + f_0 \in P'_{\beta^1} + P_{\beta^1+\underline{e}}$. Iterating, it holds that

$$f \in \bigcap_{j=0}^{\infty} \left(P'_{\beta^0} + P_{\beta^j+\underline{e}}\right),$$

where $\beta^0 < \beta^1 < \cdots < \beta^i < \cdots$ are elements in $\mathbb{N}^m$. Assume that there exists a finite generating sequence for the valuation ν_1. Then, the equality (6.1) for the set $\{\nu_1\}$ proves that $P_{\alpha_1}^{\{\nu_1\}} \subseteq m^{\mu_{\underline{\alpha}}}$ and that $\underline{\alpha}' > \underline{\alpha}$ implies $\mu_{\underline{\alpha}'} > \mu_{\underline{\alpha}}$, whenever $\mu_{\underline{\alpha}} := \min\left\{\sum_{j=1}^{s} \gamma_j | \gamma \in Q_{\underline{\alpha}}\right\}$. Thus, $P_{\beta^j+\underline{e}} \subseteq P^{\{\nu_1\}}_{\beta_1^j+1} \subseteq m^{\mu_{\beta_1^j}}$. So

$$\bigcap_{j=0}^{\infty} \left(P'_{\beta^0} + P_{\beta^j+\underline{e}}\right) \subseteq \bigcap_{j=0}^{\infty} \left(P'_{\beta^0} + m^j\right).$$

Furthermore, the opposite inclusion also happens because R is a Noetherian domain and P_β an m-primary ideal. Finally considering the ideal of the quotient ring R/P'_{β^0}, $m + P'_{\beta^0} = \bar{m}$, one gets

$$\bigcap_{j=0}^{\infty} \left(P'_{\beta^0} + P_{\beta^j+\underline{e}}\right) = \bigcap_{j=0}^{\infty} \bar{m}^j = P'_{\beta^0}.$$

Hence $f \in P'_{\underline{\alpha}}$ because $f \in P'_{\beta^0}$. □

REMARK 6.5. Notice that if $r \in m$ and $\underline{\alpha} = \underline{\nu}(r)$, then $r \in P^V_{\underline{\beta}}$ for any $\underline{\beta} \leq \underline{\alpha}$. Denote $[r]_{\underline{\beta}} := r + P_{\underline{\beta}+\underline{e}}$. So, $[r]_{\underline{\beta}} \neq 0$ if, and only if, $\underline{\beta} + \underline{e} \not\leq \underline{\alpha}$. That is $[r]$ in Proposition 6.4 means $[r] := \{[r]_{\underline{\beta}} \mid \underline{\beta} \leq \underline{\alpha}$ and $\underline{\beta} + \underline{e} \not\leq \underline{\alpha}\}$, although for simplicity's sake, in the above proof, it means $[r]_{\underline{\beta}}$ for suitable $\underline{\beta}$.

The main result for the Poincaré series of these families V is the following (see [**12**]).

THEOREM 6.6. *Let $V = \{v_i\}_{i=1}^m$ be a family of monomial valuations (of K centered at R) with respect to a finite system $\Lambda = \{r_j\}_{j=1}^n$ of generators of m. Then, the multi-graded Poincaré series of $\mathrm{gr}_V R$, $H_{\mathrm{gr}_V R}(\underline{t})$, is a rational function. Moreover, a denominator of $H_{\mathrm{gr}_V R}(\underline{t})$ is given by*

$$\prod \left(1 - (t_1^{\delta_1})^{\alpha_{j1}} (t_2^{\delta_2})^{\alpha_{j2}} \cdots (t_m^{\delta_m})^{\alpha_{jm}}\right),$$

where we have written $\nu_i(r_j) = \alpha_{ji}$, $(1 \leq i \leq m; 1 \leq j \leq n)$ and the product runs over all expressions $(1 - (t_1^{\delta_1})^{\alpha_{j1}} (t_2^{\delta_2})^{\alpha_{j2}} \cdots (t_m^{\delta_m})^{\alpha_{jm}})$ with $1 \leq j \leq n$, $\delta_i \in \{0, 1\}$ $(1 \leq i \leq m)$ and not all the δ_i's are equal to 0.

Theorem 6.6 can be proved taking into account that $H_{\mathrm{gr}_V R}(\underline{t}) = \sum_{i=0}^{m-1} h_i$, where

$$h_i = \sum_{\underline{\alpha} \in \mathbb{N}^m} \dim_k \left(P^V_{\underline{\alpha}+\underline{e}_{\leq i}} / P^V_{\underline{\alpha}+\underline{e}_{\leq i+1}}\right) \underline{t}^{\underline{\alpha}}$$

is the Poincaré series of the graded algebra $\oplus_{\underline{\alpha} \in \mathbb{N}^m} P^V_{\underline{\alpha}+\underline{e}_{\leq i}} / P^V_{\underline{\alpha}+\underline{e}_{\leq i+1}}$. Interesting families of valuations satisfy the requirements of Theorem 6.6 as one can see in the following result.

THEOREM 6.7. [**12**] *Let R be either a two-dimensional regular local ring or the local ring of a rational surface singularity. Let $V = \{\nu_i\}_{i=1}^m$ be a family of divisorial valuations of K centered at R. Then V has a finite generating set.*

PROOF. Let $\pi : Y \to \mathrm{Spec} R = X$ be a resolution of singularities of X such that if $\{E_j\}_{j-1}^q$ are the irreducible components of the exceptional divisor of π, then the center of each valuation ν_i, $i \leq i \leq m$, is some of the E_j' that we denote by E_i and π is minimal with that property. Let $E^\diamond := \oplus_{j=1}^q \mathbb{Z} E_j$ be the group of the divisors

and we shall say that $\Lambda \subset m$ is a *generating sequence of* C whenever the ideals $P^C_{\underline{\alpha}}$ are generated by the images in $\mathcal{O}_C$ of the monomials in Λ. For convenience, we set $C(\underline{\alpha}) := \frac{P^C_{\underline{\alpha}}}{P^C_{\underline{\alpha}+\underline{e}}}$ and $c(\underline{\alpha}) := \dim_k C(\underline{\alpha})$.

Let $\mathcal{G}$ denote the dual graph (defined as in the case of a unique valuation) attached to V. For each vertex $a \in \mathcal{G}$, Q_a denotes some irreducible element of m such that the strict transform of the associated germ of curve C_{Q_a} on X is smooth and meets E_a transversely. A *general curve* C of V is a reduced plane curve with m branches defined by m different equations given by general elements of each valuation ν_i. An element $\underline{\alpha} \in S_V$ is said to be *indecomposable* if we cannot write $\underline{\alpha} = \underline{\beta} + \underline{\gamma}$ with $\underline{\beta}, \underline{\gamma} \in S_V \setminus \{0\}$. In both cases ($V$ and C) $\mathcal{G}$ is a tree, $\mathbf{1}$ denotes the vertex corresponding to the first exceptional divisor, $\mathcal{E}$ the set of dead ends (those which have only one adjacent vertex, where, to count adjacency, arrows must also be taken into account) and $[a, b]$ the path joining the vertices a and b in $\mathcal{G}$. In the case of plane valuations, for $1 \leq i \leq m$, $a(i)$ denotes the vertex of $\mathcal{G}$ corresponding to the defining divisor of ν_i and otherwise the $a(i)$'s are the vertices with arrow of the dual graph of C; finally, for each vertex $r \in \mathcal{E}$, denote by b_r the nearest vertex to r in $\Omega = \bigcup_{i=1}^m [\mathbf{1}, a(i)]$. Define

$$\mathcal{H} := \{\mathbf{1}\} \cup \mathcal{E} \cup (\Omega \setminus \{\Gamma \cup \{b_r \mid r \in \mathcal{E}\}\}) ,$$

where $\Gamma = \bigcap_{i=1}^{r} [\mathbf{1}, a(i)]$. The following result, which holds for a a reduced germ of curve C as above, is proved in [**6**].

THEOREM 6.8. *The set of indecomposable elements of the semigroup* S_C *is*

$$\{\underline{v}(Q_a) \mid a \in \mathcal{H}\} \cup \{\underline{v}(Q_{a(i)}) + (0, \ldots, 0, l, 0, \ldots, 0) \mid i = 1, \ldots, m \quad l \geq 1\}.$$

This theorem allows us to prove the following one concerning the set V [**19**].

THEOREM 6.9. *The set of indecomposable elements of the semigroup of values* S_V *is the set* $\{\underline{\nu}(Q_a) \mid a \in \mathcal{H}\}$. *In particular,* S_V *is finitely generated.*

PROOF. If $C = \bigcup_{i=1}^{m} C_i$ is any general curve of V, then $S_V \subseteq S_C$, therefore, by Theorem 6.8, the elements in the set $\{\underline{\nu}(Q_a) | a \in \mathcal{H}\}$ are indecomposable. Conversely, given $h \in R$ such that $\underline{\nu}(h)$ is indecomposable in S_V, choose a general curve C of V such that the strict transforms of C and C_h by the minimal resolution of V do not intersect. Consider the map $\underline{v}$ given by the valuations associated with C, then $\underline{\nu}(h) = \underline{v}(h)$ and $\underline{\nu}(Q_a) = \underline{v}(Q_a)$ for any vertex a. h must be irreducible and by the proof of Theorem 6.8, $\underline{v}(h)$ decomposes in S_C as sum of elements $\underline{v}(Q_b)$ with $b \in \mathcal{H}$, which proves that $\underline{\nu}(h) = \underline{\nu}(Q_a)$ for some $a \in \mathcal{H}$. □

Now we can say that the semigroup S_V has no conductor whenever $m > 1$, that is, there is no element $\delta \in S_V$ such that $\delta + \mathbb{Z}^m_{\geq 0} \subseteq S_V$. However, the semigroup of values of a curve with m branches does have a conductor δ and thus, it cannot be finitely generated if $m > 1$. In particular, if C is any general curve of V, $S_V \neq S_C$ when $m > 1$ (recall that $S_V = S_C$ when $m = 1$).

In the sequel, we shall use the following notations: for $J \subset I := \{1, 2, \ldots, m\}$, $\underline{e}_J$ is the element of $\mathbb{Z}^m$ whose jth component is 1 whenever $j \in J$ and 0 otherwise, $D(\underline{\alpha}) = P^V_{\underline{\alpha}} / P^V_{\underline{\alpha}+\underline{e}}$, $D_i(\underline{\alpha}) = P^V_{\underline{\alpha}} / P^V_{\underline{\alpha}+\underline{e}_i}$, $d(\underline{\alpha}) = \dim_k D(\underline{\alpha})$ and $d_i(\underline{\alpha}) = \dim_k D_i(\underline{\alpha})$ when $1 \leq i \leq m$. Also, we shall write $B^i = \underline{\nu}(Q_{a(i)})$. We summarize in the following propositions some results concerning those vector spaces and dimensions. As we shall see, interesting results can be deduced from them. Firstly, we shall give a theorem containing an explicit description of the semigroup S_V (see [**19**] for proofs).

PROPOSITION 6.10. *With the above notations assume* $i \in I$ *and* $\underline{\alpha} \in \mathbb{Z}^m$, *then the following properties hold:*

(1) *The natural homomorphism* $D(B^i) \to D_i(B^i)$ *is an isomorphism.*
(2) $d_i(\underline{\alpha}) \geq 2$ *if and only if* $d_i(\underline{\alpha} - B^i) \geq 1$.
(3) *Assume that* $d_i(\underline{\alpha}) \neq 0$ *then* $d_i(\underline{\alpha} + B^i) = 1 + d_i(\underline{\alpha})$.

PROPOSITION 6.11. *In this proposition, we assume* $i \in I$ *and* $\underline{\alpha} \in S_V$, *then*

(1) $d_i(\underline{\alpha}) \geq 2$ *if and only if* $\underline{\alpha} - B^i \in S_V$.
(2) *If* $I \ni j \neq i$, *then* $d_i(\underline{\alpha} + B^j) = d_i(\underline{\alpha})$.

THEOREM 6.12. *Let* $\underline{\alpha} \in S_V$, *then there exist unique nonnegative integers* z_i, $1 \leq i \leq m$, *and a unique value* $\underline{\beta} \in S_V$ *such that*

- $\underline{\alpha} = \underline{\beta} + \sum_{i=1}^m z_i B^i$.
- $d_i(\underline{\beta}) = 1$ *for every* i.

Each value z_i *satisfies the following equality* $z_i = \max\{l \in \mathbb{Z}, l \geq 0 \mid \underline{\alpha} - lB^i \in S_V\} = d_i(\underline{\alpha}) - 1$.

PROOF. First, let us prove that there exist the values z_i and $\underline{\beta}$. Indeed, define $z_i = \max\{l \in \mathbb{Z}, l \geq 0 \mid \underline{\alpha} - lB^i \in S_V\}$ and $\underline{\beta} = \underline{\alpha} - \sum_{i=1}^m z_i B^i$. It suffices to show that $\underline{\alpha} - B^i \in S_V$ and $\underline{\alpha} - B^j \in S_V$ imply $\underline{\alpha} - B^i - B^j \in S_V$. Indeed, propositions 6.10 and 6.11 allows us to state that $d_j(\underline{\alpha} - B^i) = d_j(\underline{\alpha}) \geq 2$ and hence that $\underline{\alpha} - B^i - B^j \in S_V$. To finish we prove uniqueness: Proposition 6.11 proves $1 = d_i(\underline{\beta}) = d_i(\underline{\alpha} - z_i B^i) = d_i(\underline{\alpha}) - z_i$, and by Proposition 6.11 it holds that $\underline{\beta} - B^i \notin S_V$, thus $z_i = \max\{l \in \mathbb{Z}, l \geq 0 \mid \underline{\alpha} - lB^i \in S_V\} = d_i(\underline{\alpha}) - 1$. □

Proposition 6.4 proves that V has a finite minimal generating sequence. Next result, proved in [**19**], shows how minimal generating sequences for V and for general curves C of V are. As above $\mathcal{G}$ denotes the dual graph attached either to V or to C, consider $f_i \in R$ which gives an equation for C_i and fix an element $Q_r \in R$ for each $r \in \mathcal{E}$. Set

$$\Lambda_{\mathcal{E}} := \{Q_r \mid r \in \mathcal{E}\} \text{ and } \Lambda_{\overline{\mathcal{E}}} := \{Q_r \mid r \in \mathcal{E}\} \cup \{f_i\}_{i=1}^m,$$

where we do not include $f = f_1$ whenever $m = 1$, then,

THEOREM 6.13. *The set* $\Lambda_{\mathcal{E}}$ *(*$\Lambda_{\overline{\mathcal{E}}}$*, respectively) is a minimal generating sequence of* V *(*C*, respectively). Moreover, any minimal generating sequence for* V *and* C *is of the described form.*

6.2.2. *Poincaré series.* In this subsection, we shall introduce a Poincaré series for finite families V of plane divisorial valuations (and also for general elements attached to those families) that contains the same information provided for the Poincaré series attached to their corresponding graded algebras. In this form it is easier to compute those series. Assume $m > 1$ and set $\mathcal{L} := \mathbb{Z}[[t_1, t_1^{-1}, \ldots, t_m, t_m^{-1}]]$. As above $\underline{t} = (t_1, \ldots, t_m)$ and $\underline{t}^{\underline{\alpha}} := t_1^{\alpha_1} \cdots t_m^{\alpha_m}$, for $\underline{\alpha} = (\alpha_1, \ldots, \alpha_m) \in \mathbb{Z}^m$. Clearly $\mathcal{L}$ is a $\mathbb{Z}[t_1, \ldots, t_r]$–module and a $\mathbb{Z}[t_1, t_1^{-1}, \ldots, t_r, t_r^{-1}]$–module.

For a reduced plane curve C with m branches, the formal Laurent series $L_C(\underline{t}) := \sum_{\underline{\alpha}\in\mathbb{Z}^m} c(\underline{\alpha})\underline{t}^{\underline{\alpha}} \in \mathcal{L}$ was introduced in [**8**]. There, the authors showed that $P'_C(\underline{t}) = L_C(\underline{t})\prod_{i=1}^m(t_i-1)$ is a polynomial that is divisible by $t_1\cdots t_m - 1$. The *Poincaré series for the curve* C is defined as the polynomial with integer coefficients $P_C(\underline{t}) = P'_C(\underline{t})/(t_1\cdots t_m - 1)$. In our case, a finite family V of plane divisorial valuations, we define

$$L_V(t_1,\dots,t_m) = \sum_{\underline{\alpha}\in\mathbb{Z}^m} d(\underline{\alpha})\underline{t}^{\underline{\alpha}} \in \mathcal{L}.$$

The series L_V is a Laurent series, but, since $d(\underline{\alpha})$ can be positive even if $\underline{\alpha}$ have some negative component α_i, it is not a power series. It can be proved [**19**] that $P'_V(\underline{t}) := L_V(\underline{t})\prod_{i=1}^m(t_i-1) \in \mathbb{Z}[[t_1,\dots,t_m]]$. We define the **Poincaré series** of V as

$$P_V(t_1\dots,t_m) = \frac{P'_V(t_1,\dots,t_m)}{t_1\cdots t_m - 1},$$

which is also a formal power series. Write $\widetilde{P}'_V(\underline{t}) = H_{\mathrm{gr}_V R}(\underline{t})\ \prod_{i=1}^m(t_i-1)$, then

$$\widetilde{P}'_V(t_1,\dots,t_m) = \sum_{J\subset I}(-1)^{\mathrm{card}(J)}\ P'_V(\underline{t})|_{\{t_i=1 \text{ for } i\in J\}}.$$

So one can compute $H_{\mathrm{gr}_V R}(\underline{t})$ from $P'_V(\underline{t})$. $H_{\mathrm{gr}_V R}(\underline{t})$ determines the series $L_V(\underline{t})$ since $d(\underline{\alpha}) = d(\max(\alpha_1,0),\dots,\max(\alpha_m,0))$ for $\alpha \not\leq -1 = (-1,\dots,-1)$ and $d(\underline{\alpha}) = 0$ for $\underline{\alpha} \leq -1$. The next result shows the relation between the Poincaré series of V and a general curve for it.

THEOREM 6.14. [**19**] *Let $V = \{\nu_i\}_{i=1}^m$ be a finite family of plane divisorial valuations and C a general curve for V, then the following equality holds.*

$$P_V(t_1,\dots,t_m) = \frac{P_C(t_1,\dots,t_m)}{\prod_{i=1}^r(1-\underline{t}^{B^i})}.$$

For a vertex a of the dual graph $\mathcal{G}$ of a set of valuations V as above, we denote by $\overset{\bullet}{E}_a = E_a \setminus \overline{(E-E_a)}$ the smooth part of an irreducible component E_a in the exceptional divisor E of the minimal resolution of V and by $\chi(\overset{\bullet}{E}_a)$ its Euler characteristic. In addition, set $\underline{\nu}^a := \underline{\nu}(Q_a)$. When the field k is the field of complex numbers and $R = \mathcal{O}_{\mathbb{C}^2,O}$ is the local ring of germs of holomorphic functions at the origin of the complex plane, the following formula of A'Campo's type [**1**] holds. (See [**19, 20**]).

THEOREM 6.15.

$$P_V(t_1,\dots,t_m) = \prod_{E_a\subset E}\left(1-\underline{t}^{\underline{\nu}^a}\right)^{-\chi(\overset{\bullet}{E}_a)}.$$

6.2.3. *Families of plane valuations.* The Poincaré series for families of plane valuations of the fraction field of $R = \mathcal{O}_{\mathbb{C}^2,O}$, centered at R, has been treated in [**10**]. Consider a finite family $V = \{\nu_1,\dots,\nu_m\}$ of plane valuations, denote by S_i the value semigroup of ν_i, set $S := S_1\times\cdots\times S_m$ and, for any $\underline{\alpha}\in S$, define $P^V_{\underline{\alpha}}$ as above and $P^V_{\underline{\alpha}+} := \{f\in R|\underline{\nu}(f) > \underline{\alpha}\}$. The usual definition of Poincaré series has no sense for any type of family V, so the authors define the Poincaré series of V,

P_V, by means of the following expression that coincides with the usual definition whenever the valuations are integer valuated:

$$P_V(t_1,\ldots,t_m)=\sum_{\underline{\alpha}\in S}\left(\sum_{J\subseteq I}(-1)^{\mathrm{card}(J)}\dim\frac{P^V_{\underline{\alpha}}\cap P^{V_J}_{\underline{\alpha}_J+}}{P^V_{\underline{\alpha}+}}\right)\underline{t}^{\underline{\alpha}},$$

where for $J\subseteq I$ we have written $V_J:=\{\nu_j|j\in J\}$ and $\underline{\alpha}_J$ is the projection of $\underline{\alpha}$ preserving only the coordinates corresponding to J. With the help of projective limits and as in the case of a unique valuation, it is possible to introduce a notion of resolution $\pi: X\to\mathbb{C}^2$ of V. Assuming that the valuations of type B-II are exactly ν_i, $1\le i\le r$, and denoting by f_i the last element of a generating sequence of each one of these valuations ν_i, it happens the following result, proved in [**10**] with the help of integration with respect to the Euler characteristic over the projectivization $\mathbb{P}R$ of R.

PROPOSITION 6.16. *Let $V=\{\nu_i\}_{i=1}^m$ be a finite family of plane valuations ordered as we have said. Then the Poincaré series $P_V(\underline{t})$ determines the types of the involved valuations, the dual graph of its minimal resolution up to combinatorial equivalence and divisors and sequences of divisors corresponding to valuations. Furthermore, a formula of A'Campo's type for $P_V(\underline{t})$ is*

$$P_V(\underline{t})=\prod_{E_a\subset E}\left(1-\underline{t}^{\underline{\nu}^a}\right)^{-\chi(\overset{\bullet}{E}_a)}\times\prod_{i=1}^{r}\left(1-t_i^{(1,0)}\prod_{j\neq i}t_j^{\nu_j(f_i)}\right)^{-1}.$$

7. An application: Poincaré series of multiplier ideals of a plane divisorial valuation

An important tool in singularity theory and birational geometry is the concept of multiplier ideal. Multiplier ideals provide information on the type of singularity attached to an ideal, divisor or metric, see for instance [**39**]. Although this tool is very useful, explicit computations are hard (see [**4, 32, 33, 43**]). In this section, we summarize the results in [**26**] that provide an specific calculation of a Poincaré series containing the essential information corresponding to jumping numbers and dimensions of quotients of consecutive multiplier ideals of the primary simple complete ideal attached to a plane valuation in the complex case. So, with the above notation, assume that $k=\mathbb{C}$, $\mathbb{C}$ being the field of complex numbers, and let ν be a plane divisorial valuation of K centered at R. It is known [**46**] that ν determines (and it is determined by) a simple complete m-primary ideal of R, $\mathcal{I}_\nu$, and we define jumping numbers and multiplier ideals attached to ν as the same objects corresponding to $\mathcal{I}_\nu$. Consider the blowing-up sequence (4.1) given by ν, being $\pi_N: X=X_N\to X_{N-1}$ the last blowing-up, and set $D=\sum_{i=1}^N a_iE_i$ the effective divisor such that $\mathcal{I}_\nu\mathcal{O}_X=\mathcal{O}_X(-D)$, then for any positive rational number ι, the *multiplier ideal* of ν and ι is defined as $\mathcal{J}(\nu^\iota):=\pi_*\mathcal{O}_X(K_{X|X_0}-\lfloor\iota D\rfloor)$, where $K_{X|X_0}$ is the relative canonical divisor and $\lfloor\cdot\rfloor$ represents the round-down or the integral part of the corresponding divisor. The family of multiplier ideals is totally ordered by inclusion and parameterized by non-negative rational numbers. Furthermore, there is an increasing sequence $\iota_0<\iota_1<\cdots$ of positive rational numbers, called *jumping numbers*, such that $\mathcal{J}(\nu^\iota)=\mathcal{J}(\nu^{\iota_l})$ for $\iota_l\le\iota<\iota_{l+1}$ and $\mathcal{J}(\nu^{\iota_{l+1}})\subset\mathcal{J}(\nu^{\iota_l})$ for each $l\ge0$; ι_0, usually named the log-canonical threshold of $\mathcal{I}_\nu$, is the least positive rational number such that $\mathcal{J}(\nu^{\iota_0})\neq R$.

The *star vertices* of the dual graph (labelled with the symbols st_j in Figure 1) will be those whose associated exceptional divisors E_{st_j} meet three distinct prime exceptional divisors. From now on, we shall denote by g^* the number of star vertices. Write

$$\mathcal{H}_j := \left\{ \iota(j,p,q,r) := \frac{p}{e_{j-1}} + \frac{q}{\bar{\beta}_j} + \frac{r}{e_j} \mid \frac{p}{e_{j-1}} + \frac{q}{\bar{\beta}_j} \leq \frac{1}{e_j}; p,q \geq 1, r \geq 0 \right\}$$

whenever $1 \leq j \leq g^*$, and

$$\mathcal{H}_{g^*+1} := \left\{ \iota(g^*+1,p,q) := \frac{p}{e_{g^*}} + \frac{q}{\bar{\beta}_{g^*+1}} \mid p,q \geq 1 \right\},$$

p, q and r being integer numbers. In [**35**], it is proved that the set $\mathcal{H}$ of jumping numbers of ν can be computed as $\mathcal{H} = \cup_{i=j}^{g^*+1} \mathcal{H}_j$.

Assume $\iota \in \mathcal{H}$ and $\iota \neq \iota_0 = \min \mathcal{H}$. We denote by $\iota^<$ the largest jumping number which is less than ι. By convention we set $\mathcal{J}(\nu^{\iota_0^<}) = R$. Nakayama's Lemma proves that, for any $\iota \in \mathcal{H}$, $\mathcal{J}(\nu^{\iota^<})/\mathcal{J}(\nu^\iota)$ is a finitely generated $\mathbb{C}$-vector space. Thus, the Poincaré series we referred to will be defined as follows.

DEFINITION 7.1. Let ν be a plane divisorial valuation. The *Poincaré series of multiplier ideals of* ν is defined to be the following fractional power series:

$$P_{\mathcal{J},\nu}(t) := \sum_{\iota \in \mathcal{H}} \dim_{\mathbb{C}} \left(\frac{\mathcal{J}(\nu^{\iota^<})}{\mathcal{J}(\nu^\iota)} \right) t^\iota,$$

t being an indeterminate.

The main result of this section is to give an explicit computation of the series $P_{\mathcal{J},\nu}$ which also proves that it is a rational function in certain sense that we shall clarify. The proof is supported in three interesting facts. On the one hand, results and proofs of propositions 6.10 and 6.11, where the family V of involved plane divisorial valuations is given by the N exceptional divisors E_i appearing in (4.1), and, on the other hand, the next two propositions. To state the first one, we need the concept introduced in Definition 7.2, where π and $D = \sum_{i=1}^N a_i E_i$ are, respectively, the sequence of point blowing-ups and the divisor attached to ν.

DEFINITION 7.2. A candidate jumping number from a prime exceptional divisor E_i given by π is a positive rational number ι such that ιa_i is an integer number. We shall say that E_i *contributes* ι whenever ι is a candidate jumping number from E_i and $\mathcal{J}(\nu^\iota) \subset \pi_* \mathcal{O}_X(-\lfloor \iota D \rfloor + K_{X|X_0} + E_i)$.

PROPOSITION 7.3. *A jumping number ι of a plane divisorial valuation ν belongs to the set $\mathcal{H}_j$ $(1 \leq j \leq g^*+1)$ if and only if the prime exceptional divisor F_j contributes ι, where F_j is defined to be E_{st_j} if $1 \leq j \leq g^*$ and E_N (the last obtained exceptional divisor) whenever $j = g^*+1$.*

Jumping numbers and multiplier ideals can also be introduced for analytically irreducible plane curves and for them a similar result to Proposition 7.3 is proved in [**48**] and [**44**]. Our proof [**26**] and the previous ones are independent and use different arguments. Now, we state the second result.

PROPOSITION 7.4. *Let ι be a jumping number of a plane divisorial valuation ν. Then*

$$\pi_*\mathcal{O}_X\left(-\lfloor \iota D\rfloor + K_{X|X_0} + \sum_{l=1}^{s} F_{j_l}\right) = \mathcal{J}\left(\nu^{\iota^<}\right),$$

where $\{j_1, j_2, \ldots, j_s\}$ is the set of indexes j, $1 \leq j \leq g^*+1$, such that $\iota \in \mathcal{H}_j$.

We end this paper by stating the mentioned main result.

THEOREM 7.5. *The Poincaré series $P_{\mathcal{J},\nu}(t)$ can be expressed as*

$$P_{\mathcal{J},\nu}(t) = \frac{1}{1-t}\sum_{j=1}^{g^*}\sum_{\iota\in\mathcal{H}_j,\iota<1} t^{\iota} + \frac{1}{(1-t)^2}\sum_{\iota\in\Omega} t^{\iota},$$

where

$$\Omega := \{\iota \in \mathcal{H}_{g^*+1} \mid \iota \leq 2 \text{ and } \iota - 1 \notin \mathcal{H}_{g^*+1}\}.$$

Notice that if one considers the indeterminates $z_j = t^{\frac{1}{e_{j-1}\beta_j}}$, then $P_{\mathcal{J},\nu}(t)$ belongs to the field of rational functions $\mathbb{C}(z_1, z_2, \ldots, z_{g^*+1})$.

References

[1] N. A'Campo, La fonction zêta d'une monodromie, *Comment. Math. Helv.* **50** (1975), 233—248. MR0371889 (51:8106)

[2] S. S. Abhyankar, Local uniformization on algebraic surfaces over ground field of characteristic $p \neq 0$, *Ann. Math.* **63** (1956), 491—526. MR0078017 (17:1134d)

[3] S. S. Abhyankar, On the valuations centered in a local domain, *Amer. J. Math.* **78** (1956), 321—348. MR0082477 (18:556b)

[4] N. Budur, Jumping numbers of hyperplane arrangements, *Comm. Algebra* **38** (2010), 1122—1136. MR2650395 (2011d:14032)

[5] A. Campillo, "Algebroid curves in positive characteristic", Lecture Notes in Math. 613, Springer-Verlag, 1980. MR584440 (82h:14001)

[6] A. Campillo, F. Delgado, S.M. Gusein-Zade, The extended semigroup of a plane curve singularity, *J. London Math. Soc.* **60** (1999), 420—430. MR1724869 (2000m:14002)

[7] A. Campillo, F. Delgado, S.M. Gusein-Zade, The Alexander polynomial of a plane curve singularity and integrals with respect to the Euler characteristic, *Int. J. Math.* **14** (2003), 47—52. MR1955509 (2004a:14031)

[8] A. Campillo, F. Delgado, S.M. Gusein-Zade, The Alexander polynomial of a plane curve singularity via the ring of functions on it, *Duke Math. J.* **117** (2003), 125—156. MR1962784 (2004e:14047)

[9] A. Campillo, F. Delgado, S.M. Gusein-Zade, Poincaré series of a rational surface singularity, *Invent. Math.* **155** (2004), 41—53. MR2025300 (2004j:14042)

[10] A. Campillo, F. Delgado, S.M. Gusein-Zade, F. Hernando, Poincaré series of collections of plane valuations, *Int. J. Math.* **21** (2010), 1461—1473. MR2747738 (2012a:13006)

[11] A. Campillo, C. Galindo, On the graded algebra relative to a valuation, *Manuscripta Math.* **92** (1997), 173—189. MR1428647 (98a:13003)

[12] A. Campillo, C. Galindo, The Poincaré series associated with finitely many monomial valuations, *Math. Proc. Camb. Phil. Soc.* **134** (2003), 433—443. MR1981210 (2004d:13022)

[13] V. Cossart, C. Galindo, O. Piltant, Un exemple effectif de gradué non noethérien associé à une valuation divisorielle, *Ann. Inst. Fourier* **50** (2000), 105—112. MR1762339 (2001g:13005)

[14] S.D. Cutkosky, Semigroups of valuations dominating local domains (2008), arxiv:math/0801.0449v1.

[15] S.D. Cutkosky, J. Herzog, A. Reguera, Poincaré series of resolutions of surface singularities, *Trans. Amer. Math. Soc.* **356** (2004), 1833—1874. MR2031043 (2005a:14015)

[16] S.D. Cutkosky, V. Srinivas, On a problem of Zariski on dimensions of linear systems, *Ann. Math.* **137** (1993), 531—559. MR1217347 (94g:14001)

[17] S.D. Cutkosky, B. Teissier, Semigroups of valuations on local rings II (2008), to appear in *Amer. J. Math.* MR2732345 (2011k:13005)

[18] F. Delgado, C. Galindo, A. Nuñez, Saturation for valuations on two-dimensional regular local rings, *Math. Z.* **234** (2000), 519—550. MR1774096 (2001h:13003)

[19] F. Delgado, C. Galindo, A. Nuñez, Generating sequences and Poincaré series for a finite set of plane divisorial valuations, *Adv. Math.* **219** (2008), 1632—1655. MR2458149 (2009i:13005)

[20] F. Delgado, S.M. Gusein-Zade, Poincaré series for several divisorial valuations, *Proc. Edinb. Math. Soc* **46** (2003), 501—509. MR1998577 (2004i:14003)

[21] R. Dedekind, H. Weber, Theorie der algebraischen functionen einer veränderlichen, *J. für Math.* (1882).

[22] S. Encinas, O. Villamayor, Good points and constructive resolution of singularities, *Acta Math.* **181** (1998), 109—158. MR1654779 (99i:14020)

[23] L. Ein, R. Lazarsfeld, K. Smith, Uniform approximation of Abhyankar valuation ideals in smooth function fields, *Amer. J. Math.* **125** (2003), 409—440. MR1963690 (2003m:13004)

[24] C. Favre, M. Jonsson, "The valuative tree", Lecture Notes in Math. 1853, Springer-Verlag, Berlin, 2004. MR2097722 (2006a:13008)

[25] C. Galindo, On the Poincaré series for a plane divisorial computation, *Bull. Belg. Math. Soc.* **2** (1995), 65—74. MR1323928 (96e:13004)

[26] C. Galindo, F. Monserrat, The Poincaré series of multiplier ideals of a simple complete ideal in a local ring of a smooth surface, *Adv. Math.* **225** (2010), 1046–1068. MR2671187 (2012a:14039)

[27] C. Galindo, M. Sanchis, Evaluation codes and plane valuations, *Des. Codes Crypt.* **41** (2006), 199—219. MR2271689 (2007h:94086)

[28] P.D. González-Pérez, F. Hernando, Quasi-ordinary singularities, essential divisors and Poincaré series, *J. London Math. Soc.* **79** (2009), 780-802. MR2506698 (2010i:32024)

[29] S. Greco, K. Kiyek, "General elements in complete ideals and valuations centered at a two-dimensional regular local ring", in Algebra, Arithmetic, and Geometry, with Applications, Springer (2003), 381—455. MR2037102 (2005f:13002)

[30] S.M. Gusein-Zade, F. Delgado, A. Campillo, On the monodromy of a plane curve singularity and the Poincaré series of its ring of functions, *Funct. Anal. Appl.* **33** (1999), 56—57. MR1711890 (2000f:32042)

[31] K. Hensel, Über eine neue begründung der theorie der algebraischen zahlen, *Deutsch. Math.* **6** (1897).

[32] J. Howald, Multiplier ideals of monomial ideals, *Trans. Amer. Math. Soc.* **353** (2001), 2665—2671. MR1828466 (2002b:14061)

[33] J. Howald, Multiplier ideals for sufficiently general polynomials, arXiv:math/0303203v1.

[34] H. Hironaka, Resolution of singularities of an algebraic variety over a field of characteristic zero, *Ann. Math.* **79** (1964). MR0199184 (33:7333)

[35] T. Järvilehto,"Jumping numbers of a simple complete ideal in a two-dimensional regular local ring", Ph. D. thesis, University of Helsinky, 2007.

[36] K. Kiyek, J.J. Moyano-Fernández, The Poincaré series of a simple complete ideal of a two-dimensional regular local ring, *J. Pure Appl. Algebra* **213** (2009), 1777—1787. MR2518176 (2010j:13046)

[37] A. Küronya, A divisorial valuation with irrational volume, *J. Algebra* **262** (2003), 413—423. MR1971047 (2004d:13003)

[38] F. V. Kuhlmann, Value groups, residue fields and bad places of algebraic function fields, *Trans. Amer. Math. Soc.* **356** (2004), 4559—4600. MR2067134 (2005d:12010)

[39] R. Lazarsfeld, "Positivity in algebraic geometry. Vol. II", Springer, 2004.

[40] A. Lemahieu, Poincaré series of a toric variety, *J. Algebra* **315** (2007), 683—697. MR2351887 (2009e:14084)

[41] S. MacLane, A construction for absolute values in polynomial rings, *Trans. Amer. Math. Soc.* **40** (1936), 363—395. MR1501879

[42] S. MacLane, O. Schilling, Zero-dimensional brances of rank 1 on algebraic varieties, *Ann. Math.* **40** (1939), 507—520. MR0000158 (1:26c)

[43] M. Mustaţă, Multiplier ideals of hyperplane arrangements, *Trans. Amer. Math. Soc.* **358** (2006), 5015—5023. MR2231883 (2007d:14007)

[44] D. Naie, Jumping numbers of a unibranch curve on a smooth surface, *Manuscripta Math.* **128** (2009), 33—49. MR2470185 (2009j:14034)

[45] A. Nemethi, *Poincaré series associated with surface singularities.* Proceedings of the international conference "School and workshop on the geometry and topology of singularities" in honor of the 60th birthday of Lê Dung Tráng. *Contemporary Math.* **474** (2008), 271—297. MR2454352 (2010f:32025)

[46] M. Spivakovsky, Valuations in function fields of surfaces, *Amer. J. Math.* **112** (1990), 107—156. MR1037606 (91c:14037)

[47] B. Teissier, *Valuations, deformations and toric geometry.* Proceedings of the Saskatoon conference and workshop on valuation theory, *Fields Institute Comm.* **33** (2003), 361—459. MR2018565 (2005m:14021)

[48] K. Tucker, "Jumping numbers and multiplier ideals on algebraic surfaces" PhD Dissertation, University of Michigan, 2010. MR2736766

[49] O. Villamayor, Constructiveness of Hironaka's resolution, *Ann. Sc. Éc. Norm. Sup.* **22** (1989), 1—32. MR985852 (90b:14014)

[50] O. Zariski, Local uniformization on algebraic varieties, *Ann. Math.* **41** (1940), 852—896. MR0002864 (2:124a)

[51] O. Zariski, The reduction of singularities of an algebraic surface, *Ann. Math.* **40** (1939), 639—689. MR0000159 (1:26d)

[52] O. Zariski and P. Samuel, "Commutative Algebra. Vol II", Springer-Verlag, 1960. MR0120249 (22:11006)

Current address: Departamento de Matemáticas & Instituto Universitario de Matemáticas y Aplicaciones de Castellón (IMAC), Universitat Jaume I. Campus de Riu Sec, 12071 Castellón, Spain.

E-mail address: `galindo@mat.uji.es`

Current address: Instituto Universitario de Matemática Pura y Aplicada (IUMPA), Universidad Politécnica de Valencia, Camino de Vera s/n, 46022 Valencia, Spain.

E-mail address: `framonde@mat.upv.es`

Contemporary Mathematics
Volume **566**, 2012
http://dx.doi.org/10.1090/conm/566/11222

q,t-Catalan numbers and knot homology

E. Gorsky

Abstract. We propose an algebraic model of the conjectural triply graded homology of S. Gukov, N. Dunfield and J. Rasmussen for some torus knots. It turns out to be related to the q,t-Catalan numbers of A. Garsia and M. Haiman.

1. Introduction

In [**7**] A. Garsia and M. Haiman constructed a series of bivariate polynomials $C_n(q,t)$. In [**14**] M. Haiman proved that these polynomials have non-negative integer coefficients, and they generalize two known one-parametric deformations of the Catalan numbers, in particular, the value $C_n(1,1)$ equals to the n-th Catalan number. One of deformations can be expressed in terms of q-binomial coefficients, while the second one counts Dyck paths weighted by the area above them. M. Haiman also related these invariants to the geometry of the Hilbert scheme of points on $\mathbb{C}^2$.

Let $Hilb^n(\mathbb{C}^2)$ denote the Hilbert scheme of n points on $\mathbb{C}^2$, and let $Hilb^n(\mathbb{C}^2,0)$ parametrize 0-dimensional subschemes of length n supported at the origin. Let V be the tautological n-dimensional bundle over $Hilb^n(\mathbb{C}^2)$.

Theorem 1.1. *([**14**], Theorem 2) Consider the natural torus action on $\mathbb{C}^2$ and extend it to Hilbert schemes. Then*

$$C_n(q_1,q_2)=\chi^T(Hilb^n(\mathbb{C}^2,0),\Lambda^nV),$$

where q_1 and q_2 are equivariant parameters corresponding to the torus action.

We construct a sequence of the bigraded subspaces in the space of symmetric polynomials such that their Hilbert functions coincide with $C_n(t,q)$ for $n\le 4$.

Let Λ denote the ring of symmetric polynomials in the infinite number of variables. Let e_k denote the elementary symmetric polynomials and h_k denote the complete symmetric polynomials. One can equip Λ with the pair of gradings - one of them is the usual (homogeneous) degree, and the second one is the degree of a symmetric polynomial as a polynomial in variables e_k. In other words,

$$S(e_{\alpha_1}\ldots e_{\alpha_r})=\alpha_1+\ldots+\alpha_r,\quad b(e_{\alpha_1}\ldots e_{\alpha_r})=r.$$

1991 *Mathematics Subject Classification.* Primary 57M27, 05A19, 05A30.
Key words and phrases. Torus knots, Khovanov homology, q,t-Catalan numbers.
Partially supported by the grants RFBR-08-01-00110-a, RFBR-10-01-00678, NSh-8462.2010.1 and the Dynasty fellowship for young scientists.

We also define the sequence of spaces $\Lambda(n,r) \subset \Lambda$ which are generated by the monomials with b-grading less than or equal to r and S-grading equal to n.

DEFINITION 1.2. Let $L_n \subset \Lambda$ be the subspace generated by all monomials $h_{\alpha_1}h_{\alpha_2}\dots h_{\alpha_n}$ such that $\alpha_k \le k$ for all k.

THEOREM 1.3. *For $n \le 4$, the bivariate Hilbert function of L_n equals to*

$$\sum_{m,r=0}^{\infty} q^n t^r \dim[(L_n \cap \Lambda(m,r))/(L_n \cap \Lambda(m,r-1))] = q^{n(n-1)/2} C_n(q^{-1},t). \tag{1.1}$$

The construction of the spaces L_n is expected to be related to some constructions in knot theory.

DEFINITION 1.4. ([**6**]) The HOMFLY polynomial $\overline{P}$ is defined by the following skein relation:

$$a\overline{P}\langle \nwarrow\!\!\!\!\nearrow \rangle - a^{-1}\overline{P}\langle \nwarrow\!\!\!\!\nearrow \rangle = (q-q^{-1})\overline{P}\langle)(\rangle,$$

the multiplication property $\overline{P}(K_1 \sqcup K_2) = \overline{P}(K_1)\overline{P}(K_2)$ and its non-vanishing at the unknot. One can check that $\overline{P}(unknot) = (a-a^{-1})/(q-q^{-1})$, and we will also use the *reduced* HOMFLY polynomial P

$$P(K)(a,q) = \overline{P}(K)(a,q)/\overline{P}(unknot).$$

The HOMFLY polynomial unifies the quantum $sl(N)$ polynomial invariants of K

$$\overline{P}_N(K)(q) = \overline{P}(K)(a = q^N, q).$$

The original Jones polynomial $J(K)$ equals to $\overline{P}_2(K)$. The HOMFLY polynomial encodes the Alexander polynomial as well: $\Delta(q) = \overline{P}(K)(a=1,q)$.

The structure of the HOMFLY polynomial for torus knots was described by V. Jones in [**16**]. In particular, this result gives the answers for the Alexander, Jones and $sl(N)$ polynomials for all torus knots.

More recently, several knot homology theories had been developed: P. Ozsváth and Z. Szabó constructed ([**23**]) the Heegard-Floer knot homology theory categorifying the Alexander polynomial by the methods of the symplectic topology. For all algebraic (and hence torus) knots they managed ([**26**], see also [**15**]) to calculate explicitly the Heegard-Floer homology. It can be reconstructed by a certain combinatorial procedure from the Alexander polynomial.

M. Khovanov ([**17**]) constructed a homology theory categorifying the Jones polynomial. Later Khovanov and Rozansky gave a unified construction ([**19**]) of the homology theories categorifying $sl(N)$ Jones polynomials, and also another homology theory ([**20**]) categorifying the HOMFLY polynomial.

Although the complexes in the homology theories of Khovanov and Rozansky are defined combinatorially in terms of the knot diagrams, the explicit Poincaré polynomials for the corresponding homology groups of torus knots are known only in some particular cases.

To get all these theories together, Dunfield, Gukov and Rasmussen conjectured ([**4**]) that all these theories are parts, or specializations of a unified picture. Namely, for a given knot K they conjectured the existence of a triply-graded knot homology theory $\mathcal{H}_{i,j,k}(K)$ with the following properties:

- **Euler characteristic.** Consider the Poincaré polynomial
$$\mathcal{P}(K)(a,q,t)=\sum a^i q^j t^k \dim \mathcal{H}_{i,j,k}.$$
Its value at $t=-1$ equals to the value of the reduced HOMFLY polynomial of the knot K:
$$\mathcal{P}(K)(a,q,-1)=P(K)(a,q).$$

- **Differentials.** There exist a set of anti-commuting differentials d_j for $j\in\mathbb{Z}$ acting in $\mathcal{H}_*(K)$. For $N>0$, d_N has triple degree $(-2,2N,-1)$, d_0 has degree $(-2,0,-3)$ and for $N<0$ d_N has degree $(-2,2N,-1+2N)$.
- **Symmetry.** There exists a natural involution ϕ such that
$$\phi d_N = d_{-N}\phi$$
for all $N\in\mathbb{Z}$.

For $N\geq 0$, the homology of d_N are supposed to be tightly related to the $sl(N)$ Khovanov-Rozansky homology. Namely, let
$$\mathcal{H}^N_{p,k}(K)=\bigoplus_{iN+j=p}\mathcal{H}_{i,j,k}(K).$$

CONJECTURE 1.5. *([**4**]). There exists a homology theory with above properties such that for all $N>1$ the homology of $(\mathcal{H}^N_*(K),d_N)$ is isomorphic to the $sl(N)$ Khovanov-Rozansky homology. For $N=0$, $(\mathcal{H}^0_*(K),d_0)$ is isomorphic to the Heegard-Floer knot homology. The homology of d_1 are one-dimensional.*

In [**30**] J. Rasmussen proved a weaker version of this conjecture. Namely, for all $N>0$ he constructed explicit spectral sequences starting from the Khovanov-Rozansky categorification of HOMFLY polynomial and converging to $sl(N)$ homology. For the Heegard-Floer homology no relation to the other knot homology theories is known yet.

We propose a conjectural algebraic construction of vector spaces $\mathcal{H}(T_{n,n+1})$ associated with the $(n,n+1)$ torus knots for $n\leq 4$. In order to approach the Conjecture 1.5, we prove the following

THEOREM 1.6. *For $n\leq 4$ the Euler characteristic of $\mathcal{H}(n,n+1)$ coincides with the HOMFLY polynomial of the $(n,n+1)$ torus knot. One can define the differentials d_0, d_1 and d_2 such that the following properties hold:*

1. *The homology of $\mathcal{H}(n,n+1)$ with respect to the differential d_1 is one-dimensional.*
2. *The homology with respect to d_2 is isomorphic to the reduced Khovanov homology of the corresponding knot*
3. *The homology with respect to d_0 is isomorphic to the Heegard-Floer homology of the corresponding knot.*

Two latter statements are based on the tables from [**1**] and explicit description of the Heegard-Floer homology of algebraic knots proposed in [**23**] (see also [**15**],[**9**]).

The paper is organized as follows. Section 2 is devoted to the combinatorics of (q,t)-Catalan numbers and their polynomial "categorifications". In Subsection 2.1 we define these numbers and list some of their properties following A. Garsia and M. Haiman. In Subsection 2.2 we define the bounce statistic introduced by J. Haglund

([**11**]) and propose a "slicing" construction dividing a Young diagram into smaller "stable" subdiagrams. In the next subsection we associate a Schur polynomial to a stable Young diagram, and the product of such polynomials for "stable slices" to unstable one. This construction associates a symmetric polynomial to a Dyck path in the $n \times n$ square. It turns out that the subspace generated by these polynomials coincides with the space L_n (defined above), and the gradings of the polynomials are clearly expressed via the area and bounce statistics. This proves Theorem 1.3. In Subsection 2.4 we discuss a generalization of this construction applied to the (q,t)-deformation of Schröder numbers defined by J. Haglund.

Section 3 deals with the HOMFLY polynomials of torus knots and its conjectural categorification. Using the formula of V. Jones, we prove that the coefficients of the HOMFLY polynomial in the power expansion in the variable a can be expressed via certain products of the q-binomial coefficients. These coefficients are equal to the generalized Catalan and Schröder numbers. Moreover, the categorification procedure introduces one additional parameter t in the picture, so the resulting coefficients at given powers of a should be some bivariate deformations of the Catalan and Schröder numbers.

Therefore it is quite natural to relate them to the above constructions. Namely, we identify the space $\mathcal{H}(T_{n,m})$ corresponding to a torus knot with a certain subspace in a free polynomial algebra with even and odd generators. This space is equipped with the three gradings: two of them are defined on the ring of symmetric functions as above, and the third one equals to the degree in the skew variables. The differentials of Gukov-Dunfield-Rasmussen are supposed to be realized as certain differential operators acting on the skew variables. Moreover, we consider the bigger algebra $\mathcal{A}_{n,m}$ acting on $\mathcal{H}(T_{n,m})$. We suppose that for $(n, n+1)$ torus knots the space $\mathcal{H}(T_{n,n+1})$ is generated by the volume form and the action of the algebra $\mathcal{A}_{n,n+1}$. The generators of $\mathcal{A}_{n,n+1}$ can be naturally labelled by the diagonals of the $(n+2)$-gon.

In the Subsection 3.2 we also discuss the "stable limit" of the homology of (n,m)-torus knots at $m \to \infty$, following [**4**]. We identify this limit with the free supercommutative algebra $\mathcal{H}_n$ with $n-1$ even and $n-1$ odd generators, and show that the grading conditions define some differentials completely. We compare the resulting constructions and the homology with [**1**] and [**4**].

As a byproduct of the above conjectures, we propose an interesting combinatorial conjecture on the limit $q = 1$ in triply graded homology. It is well-known in the theory of Heegard-Floer homology that there is a spectral sequence starting from the homology of a given knot and converging to the one-dimensional Heegard-Floer homology of 3-sphere. This means that for any knot the value of the Poincaré polynomial for Heegard-Floer homology at $q = 1$ equals to 1.

For the triply graded theory, the limit of the Poincaré polynomial at $q = 1$ is a polynomial in a and t.

CONJECTURE 1.7. *Consider the $n \times m$ rectangle and the diagonal in it. Let $D_{n,m}(k)$ denote the set of lattice paths above the diagonal in this rectangle with k marked external corners. For a path $\pi \in D_{n,m}(k)$ let $S(\pi)$ denote the area above π. Let*

$$Q_{n,m}(a,t) = \sum_k \sum_{\pi \in D_{n,m}(k)} a^{2k} t^{k+2S(\pi)}.$$

Then the polynomial $Q_{n,m}$ coincides with the limit of the Poincaré polynomial for reduced triply graded homology of the torus (n,m)-knot at $q=1$.

One can say that the "homological grading" t is related to the area statistics. This conjecture seems to be coherent to some concepts in mathematical physics (e.g. [**10**]) relating knot homology theories to the geometry of Hilbert schemes and Donaldson-Thomas invariants.

In [**22**] A. Oblomkov and V. Shende developed a remarkable algebro-geometric description of the HOMFLY polynomial for algebraic knots as certain integrals with respect to the Euler characteristic. They proved their conjectures in full generality for all torus knots, and the reduction to the Alexander polynomial for all algebraic knots. We plan to find out the relation between their approach and the one presented here.

The author is grateful to S. Gusein-Zade, S. Gukov, B. Feigin, S. Loktev, A. Gorsky, M. Bershtein and especially to M. Gorsky for lots of useful discussions and remarks. The author also thanks the University of Kumamoto where the part of the work has been done and personally S. Tanabe for the hospitality. The research was partially benefited from the support of the "EADS Foundation Chair in mathematics".

The author is grateful to F. Bergeron for pointing that Theorem 1.3 is not true for $n\geq 5$.

2. Bivariate Catalan numbers

2.1. Properties.

DEFINITION 2.1. A Dyck path in the $n\times n$ square is a lattice path starting at the origin $(0,0)$ and ending at (n,n) consisting of North $N(0,1)$ and East $E(1,0)$ steps which never goes below the line $y=x$.

The number C_n of Dyck paths in the $n\times n$ square equals to the n-th Catalan number, i. e. $C_n=\frac{1}{n+1}\binom{2n}{n}$.

In [**7**] A. Garsia and M. Haiman introduced a remarkable two-parametric deformation of the Catalan numbers.

For a cell x in a Young diagram μ let $l(x),a(x),l'(x),a'(x)$ denote respectively leg, arm, co-leg and co-arm lengths of x. Let

$$n(\mu)=\sum_{x\in\mu}l(x),\quad n(\mu')=\sum_{x\in\mu}a(x).$$

DEFINITION 2.2. ([**7**]) We set

(2.1)

$$C_n(t,q) = \sum_{|\mu|=n}\frac{t^{n(\mu)}q^{n(\mu')}(1-t)(1-q)(\prod_{x\in\mu\setminus(0,0)}(1-t^{l'(x)}q^{a'(x)}))(\sum_{x\in\mu}t^{l'(x)}q^{a'(x)})}{\prod_{x\in\mu}(1-t^{1+l(x)}q^{-a(x)})(1-t^{-l(x)}q^{1+a(x)})}$$

Garsia and Haiman observed that $C_n(t,q)$ is a polynomial with the non-negative integer coefficients ([**8**]), and $C_n(1,1)$ equals to the Catalan number c_n. The geometric meaning of this bivariate deformation of Catalan numbers is described by the following theorem of Haiman.

Let $Hilb^n(\mathbb{C}^2)$ be the Hilbert scheme of n points on $\mathbb{C}^2$, and let $Hilb^n(\mathbb{C}^2, 0)$ parametrize 0-dimensional subschemes of length n supported at the origin. Let V be the tautological n-dimensional bundle over $Hilb^n(\mathbb{C}^2)$.

THEOREM 2.3. *([**14**]) Consider the diagonal action of the torus $(\mathbb{C}^*)^2 = T$ on $\mathbb{C}^2$ and extend it to Hilbert schemes. Then*

$$C_n(q_1, q_2) = \chi^T(Hilb^n(\mathbb{C}^2, 0), \Lambda^n V),$$

where q_1 and q_2 are equivariant parameters corresponding to the torus action.

As a corollary, $C_n(q_1, q_2)$ is a symmetric function of the parameters q_1 and q_2. Two different specializations of $C_n(q_1, q_2)$ are known.

We will use the standard notation

$$[k]_q = (1 - q^k)/(1 - q), [k]_q! = [1]_q[2]_q \cdots [k]_q, \binom{n}{k}_q = [n]_q!/[k]_q![n-k]_q!.$$

PROPOSITION 2.4. *([**7**],[**14**]) 1. The values $C_n(q^{-1}, q)$ are related to the deformation of Catalan numbers based on q-binomial coefficients:*

$$q^{\binom{n}{2}} C_n(q^{-1}, q) = \frac{1}{[n+1]_q} \binom{2n}{n}_q. \tag{2.2}$$

*2. The values $C_n(1, q)$ coincide with the Carlitz-Riordan ([**3**]) q-deformation of the Catalan numbers, which are defined by the recursive equation*

$$C_n(q) = \sum_{k=0}^{n-1} q^k C_k(q) C_{n-1-k}(q), C_0(q) = 1.$$

*It is also known (e. g. [**11**]) that $C_n(q) = \sum_\pi q^{S(\pi)}$, where the summation is done over the set of Dyck paths and $S(\pi)$ denotes the area above the path π.*

2.2. Bounce and area statistics.

DEFINITION 2.5. For a Dyck path π in the $n \times n$ square we define two statistics, following J. Haglund ([**11**]). First, $S(\pi)$ is the area above the path π.

The second one is called the *bounce* statistic. Consider a ball starting from the NE corner (n, n). A ball rolls west until it meets π, then turns south until it meets the diagonal, then reflects from the diagonal and moves west etc. It finishes at the last point $(0, 0)$. During its motion the ball touches the diagonal at points $(j_1, j_1), (j_2, j_2), \ldots$. We define

$$bounce(\pi) = j_1 + j_2 + \ldots.$$

The ball's path is called *bounce path.*

In the below picture the Dyck path is bold and its bounce path is dashed.

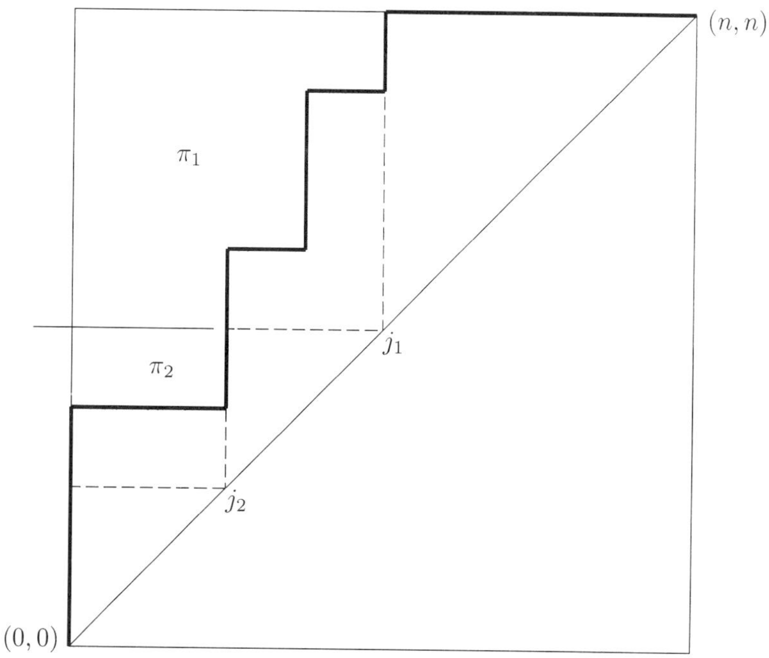

Bounce statistic for a Dyck path and slicing of a Young diagram

It turns out that the area and bounce statistics (which are defined here in a slightly different way than in [**11**]) are related to the (q,t)-deformation of the Catalan numbers.

THEOREM 2.6. *([**8**],[**11**]) The polynomial $C_n(q_1,q_2)$ can be presented as the following sum over Dyck paths:*

$$C_n(q_1,q_2)=\sum_{\pi}(q_1)^{\binom{n}{2}-S(\pi)}(q_2)^{bounce(\pi)}. \tag{2.3}$$

DEFINITION 2.7. Consider a Dyck path π in the square $n\times n$ and the bounce path for it. Let us continue horisontal bounce lines and cut the Young diagram above π along these lines. If bounce points are at $j_1>j_2>\ldots>j_r$, then we get r Young diagrams $\pi_1,\pi_2,\ldots,\pi_r$ corresponding to proper Dyck paths in the squares $n\times n, j_1\times j_1,\ldots,j_r\times j_r$. We will refer to the decomposition

$$\pi=\pi_1\sqcup\ldots\sqcup\pi_r$$

as to the **slicing** of a diagram T.

DEFINITION 2.8. A Dyck path π in the square $n\times n$ is **stable**, if

$$width(\pi)+height(\pi)<n.$$

The following properties of slicing and stable paths follows from the definitions.

PROPOSITIONS 2.9. *1. A path π is stable if and only if*

$$bounce(\pi)=width(\pi).$$

2. If $\pi = \pi_1 \sqcup \ldots \sqcup \pi_r$ is a slicing of a diagram π, then slices π_i are stable. Moreover, $width(\pi_m) = j_m$, so

$$bounce(\pi) = bounce(\pi_1) + \ldots + bounce(\pi_r). \tag{2.4}$$

We conclude that bounce and area statistics can be reconstructed from the slicing of the initial path.

2.3. Symmetric polynomials. Let Λ denote the ring of symmetric polynomials in the infinite number of variables. Let e_k denote the elementary symmetric polynomials and let h_k denote the complete symmetric polynomials. One can equip Λ with the pair of gradings - one of them is usual degree, and the second one is the degree of a symmetric polynomial as a polynomial in variables e_k. In other words,

$$a(e_{\alpha_1} \ldots e_{\alpha_r}) = \alpha_1 + \ldots + \alpha_r, \quad b(e_{\alpha_1} \ldots e_{\alpha_r}) = r.$$

We also define the sequence of spaces $\Lambda(n, r) \subset \Lambda$ which are generated by the monomials with b-grading less than or equal to r and a-grading equal to n.

We are ready to associate a symmetric polynomial from the ring Λ to a Dyck path π. For stable diagrams the result will not depend of n, while for unstable ones it depends on the slicing (and hence on n).

DEFINITION 2.10. Let π be a stable Young diagram in the square $n \times n$, let π^* be its transpose, $\pi = (\mu_1, \ldots, \mu_s), \pi^* = (\lambda_1, \ldots, \lambda_r)$. We define the corresponding symmetric polynomial as a Schur polynomial of π^*.

$$Z(\pi) = \det(e_{\lambda_i - i + j + 1}) = \det(h_{\mu_i - i + j + 1}). \tag{2.5}$$

It is clear that

$$a(Z(\pi)) = S(\pi), b(Z(\pi)) = width(\pi) = bounce(\pi).$$

DEFINITION 2.11. Let π be a Dyck path in the square $n \times n$, $\pi = \pi_1 \sqcup \ldots \sqcup \pi_r$ is its slicing. Then we define

$$Z(\pi) = Z(\pi_1) \cdot \ldots \cdot Z(\pi_r).$$

From the equation (2.4) it follows that the map Z respects both gradings:

$$a(Z(\pi)) = S(\pi), b(Z(\pi)) = bounce(\pi).$$

The subtle point is that the polynomials $Z(T)$ (as well as Schur polynomials) are homogeneous in the a-grading, but not homogeneous in the b-grading.

Proof of Theorem 1.3. First, let us remark that for a Dyck path π the polynomial $Z(\pi)$ belongs to the space L_n. Consider the slicing $\pi = \pi_1 \sqcup \ldots \sqcup \pi_r$. Using the second equation of (2.5), one can represent $Z(\pi_j)$ as a determinant in h's whose size is the number of rows in π_j. By the multiplication of such determinants we get a determinant of a block-diagonal matrix of size $n \times n$. From the definition of the bounce path one can prove that all monomials in the expansion of this determinant belong to the space L_n, therefore $Z(\pi) \in L_n$.

Second, all polynomials $Z(\pi)$ are linearly independent since their h-lex-minimal terms are different.

Since the dimension of the subspace generated by $Z(\pi)$ equals to the Catalan number, we conclude that $Z(\pi)$ form a basis in L_n.

Moreover, one can check that for $n \leq 4$ the b-maximal parts of $Z(\pi)$ are linearly independent.

We conclude that the images of the polynomials $Z(\pi)$ form a basis in all quotients $[(L_n \cap \Lambda(m,r))/(L_n \cap \Lambda(m,r-1))]$. Now the statement follows from the equation (2.3). □

REMARK 2.12. For $n = 5$ this proof fails, since the b-maximal parts of $Z(\pi)$ are no longer linearly independent for different T. Namely, one can check that

$$Z(4,2,2) = h_4(h_2^2 - h_1h_3), \quad Z(3,2,2,1) = (h_3h_2 - h_1h_4)h_2h_1.$$

Both polynomials have gradings $S = 8, b = 6$, but their b-maximal parts are proportional to $e_1^4(e_2^2 - e_1e_3)$.

2.4. Schröder numbers.

DEFINITION 2.13. A Schröder path is a lattice path starting at the origin $(0,0)$ and ending at (n,n) consisting of North $N(0,1)$, East $E(1,0)$ and Diagonal $D(1,1)$ steps which never goes below the line $y = x$.

We will denote by $S_{n,k}$ the number of Schröder paths in a square $n \times n$ with exactly k diagonal steps (large Schröder number), and by $R_{n,k}$ the number of such paths with no D steps on the diagonal $y = x$ (little Schröder number). It is well-known that $R_{n,k}$ equals to the number of ways to draw $n-k-1$ non-intersecting diagonals in a convex n-gon, that is, to the number of k-dimensional faces of the associahedron. The combinatorial formula for these numbers looks as ([**11**],[**13**])

$$S_{n,k} = \frac{(2n-k)!}{(n-k+1)!(n-k)!k!}, R_{n,k} = \frac{(2n-k)!}{n(n+1)\cdot k!(n-k)!(n-k-1)!}.$$

In [**11**] (see also [**2**]) a certain bivariate deformation of Schröder numbers was proposed. To any Schröder path π we associate a Dyck path $T(\pi)$ which is nothing but π with all D steps thrown away.

DEFINITION 2.14. Let $S(\pi)$ be the area above the path π.

Now we define the *bounce* statistic. First, consider the Dyck path $T(\pi)$ and the bounce path corresponding to it. Let us call the vertical lines of ball's motion *peak lines*. For a D-type step $x \in \pi$ let $nump(x)$ denote the number of peak lines to the east from x. Now let

$$b(\pi) = bounce(T(\pi)) + \sum_x nump(x),$$

where summation is done over all D-steps x.

DEFINITION 2.15. ([**5**],[**11**]) The (q,t)-Schröder polynomials are defined as

$$S_{n,k}(q,t) = \sum_{\pi} q^{\binom{n}{2}+\frac{k}{2}-S(\pi)}t^{b(\pi)}, \tag{2.6}$$

where the summation is done over all (n,k)-Schröder paths. The definition of $R_{n,k}(q,t)$ is analogous.

It is conjectured ([**11**]) that the polynomials $S_{n,k}(q,t)$ are symmetric in q and t. In what follows we will use the following

PROPOSITION 2.16. *(Corollary 4.8.1 in* [**11**]*) For* $0 \le k \le n$

(2.7)

$$q^{\binom{n}{2}-\binom{k}{2}}S_{n,k}(q,q^{-1}) = \frac{1}{[n-k+1]_q}\binom{2n-k}{n-k,n-k,k}_q = \frac{[2n-k]!_q}{[n-k+1]!_q[n-k]!_q[k]!_q}$$

REMARK 2.17. The equation (2.7) can be rewritten as

$$\sum_{\pi} q^{S(\pi)+b(\pi)} = q^{\frac{k^2}{2}} \frac{1}{[n-k+1]_q} \binom{2n-k}{n-k, n-k, k}_q . \tag{2.8}$$

We conjecture the following analogue of this identity for little Schröder numbers.

CONJECTURE 2.18.

$$q^{\binom{n}{2}-\binom{k}{2}} R_{n,k}(q, q^{-1}) = \frac{[2n-k]!_q}{[n]_q[n+1]_q[n-k-1]!_q[n-k]!_q[k]!_q} \tag{2.9}$$

Let us introduce the natural analogue for the Schröder paths.

DEFINITION 2.19. Let π be a Schröder path, and $T(\pi)$ is a Dyck path defined as above. Consider a slicing of $T(\pi)$, and lift the horisontal cuts of the slicing to the diagram of π. Now, take away all horisontal lines ending by D steps. We will receive a Young diagram sliced analogously to $T(\pi)$, which will be called sliced Young diagram of π.

We expect the existence some analogue map Z for the Schröder diagrams. Such a map is supposed to take values not in the ring Λ itself, but in its extension $\Lambda < \xi_1, \xi_2, \ldots >$, where ξ_j are some additional anti-commuting variables. The gradings S and b are extended to these skew variables by the formula

$$S(\xi_j) = j - \frac{1}{2}, b(\xi_j) = 0.$$

3. Homological knot invariants

3.1. HOMFLY polynomial for torus knots. Let $T_{n,m}$ be a torus knot of type (n,m), where n and m are coprime integers, $n < m$. The explicit expression for the HOMFLY polynomials $P(T_{n,m})$ was found by Jones in [**16**], we'll use it in a slightly rewritten form of [**4**]:

(3.1)

$$P(T_{n,m}) = (aq)^{(n-1)(m-1)} \frac{1-q^{-2}}{1-q^{-2n}} \sum_{b=0}^{n-1} q^{-2mb} (\prod_{i=1}^{b} \frac{a^2q^{2i}-1}{q^{2i}-1})(\prod_{j=1}^{n-1-b} \frac{a^2-q^{2j}}{1-q^{2j}}).$$

To compare the knots with different n and m, it is more convenient to get rid of negative powers and consider the rescaled version of this polynomial, namely

$$\begin{aligned} P_s(T_{n,m}) &= (a^{-1}q)^{(n-1)(m-1)} P(T_{n,m}) \\ &= \frac{1-q^2}{1-q^{2n}} \sum_{b=0}^{n-1} q^{2m(n-1-b)} (\prod_{i=1}^{b} \frac{a^2q^{2i}-1}{q^{2i}-1})(\prod_{j=1}^{n-1-b} \frac{a^2-q^{2j}}{1-q^{2j}}) \\ &= \frac{1-q^2}{1-q^{2n}} \sum_{b=0}^{n-1} q^{2mb} (\prod_{i=1}^{n-1-b} \frac{a^2q^{2i}-1}{q^{2i}-1})(\prod_{j=1}^{b} \frac{a^2-q^{2j}}{1-q^{2j}}). \end{aligned} \tag{3.2}$$

In [**4**] the expansion of P_s by the powers of a was carefully studied. For example, the following equation holds (as above, we have $n < m$):

$$P_s(T_{n,m}) = \sum_{J=0}^{n-1} a^{2J} P_s^J(T_{n,m}). \tag{3.3}$$

$\{E_j\}_{j=1}^q$ and T the set of m-primary complete ideals $\mathcal{I} \subset R$ such that $\mathcal{I}\mathcal{O}_Y$ is an invertible sheaf. For those ideals $\mathcal{I}$, denote by $D_{\mathcal{I}} \in E^{\diamond}$ the unique exceptional divisor such that $\mathcal{I}\mathcal{O}_Y = \mathcal{O}_Y(-D_{\mathcal{I}})$. T is a finitely generated semigroup because T is isomorphic to the sub-semigroup of $E^{\diamond}$ of lattice points D which are inside the rational polyhedral in $E^{\diamond} \otimes_{\mathbb{Z}} \mathbb{Q}$ given by the constrains $(-D)E_j \geq 0$ for all j.

Consider generators $\{\mathcal{I}_l\}_{l=1}^t$ of the semigroup T. For each l, pick a set of generators of $\mathcal{I}_l$ and denote by $\Lambda = \{r_s\}_{s=1}^n$ the set union of the above chosen sets of generators for all integers l. Λ is a generating set of the set V and to prove it we only need to check that every ideal $P_{\underline{\alpha}}^V$ is generated by the monomials in the r_s's. Consider the divisor $D'_{\underline{\alpha}} = \sum_{i=1}^m \alpha_i E_i$ and apply the Laufer algorithm to find another divisor $D_{\underline{\alpha}} \in E^{\diamond}$ with $(-D_{\underline{\alpha}})E_j \geq 0$ for all j and such that

$$P_{\underline{\alpha}}^V = \pi_*\left(\mathcal{O}_Y(-D'_{\underline{\alpha}})\right) = \pi_*\left(\mathcal{O}_Y(-D_{\underline{\alpha}})\right).$$

As a consequence, for suitable nonnegative integers a_l, $P_{\underline{\alpha}}^V = \prod_{l=1}^t \mathcal{I}_l^{a_l}$ and since each ideal I_j is spanned by monomials in the set $\{r_s\}_{s=1}^n$, $P_{\underline{\alpha}}^V$ is also generated by monomials in the r_s's. □

6.2. Families of plane divisorial valuations.

6.2.1. *Semigroup of values and graded algebra.* Along this section $V = \{\nu_i\}_{i=1}^m$ will be a finite family of plane divisorial valuations and we shall assume that R is complete; we know that its Poincaré series is a rational function and our goal is to compute this series and to give more information about its value semigroup. We also relate these data with the corresponding data for the close and rather studied families of valuations attached to plane curve singularities [**6, 8**].

The *semigroup of values* of V is defined to be the additive sub-semigroup S_V of $\mathbb{Z}^m$ given by

$$S_V = \{\underline{\nu}(f) := (\nu_1(f), \ldots, \nu_m(f) \mid f \in R \setminus \{0\}\}.$$

We also need to consider the *minimal resolution* of V, which is a modification $\pi : X \to \mathrm{Spec}R$ such that ν_i is the $E_{a(i)}$-valuation for an irreducible component of the exceptional divisor E given by π, $1 \leq i \leq m$, and π is minimal with this property.

On the other hand, let $C = \bigcup_{i=1}^m C_i$ be a reduced germ of curve, with irreducible components $C_1, \ldots, C_m$, defined by an element $f \in R$, and denote by $R/(f)^*$ the set of nonzero divisors of the ring $\mathcal{O}_C := R/(f)$. The *semigroup of values* S_C of C is the additive sub-semigroup of $\mathbb{Z}^m$ given by

$$S_C := \{\underline{v}(g) = (v_1(g), \ldots, v_m(g)) \mid g \in R/(f)^*\},$$

where each v_i is the valuation corresponding to C_i. The dual graph of C, denoted by $\mathcal{G}$, is the dual graph of its minimal embedded resolution, attaching an arrow, for each irreducible component C_i of C, to the vertex corresponding to the exceptional component which meets the strict transform on X of C_i. Here, we can also consider the valuation ideals $P_{\underline{\alpha}}^C := \{g \in \mathcal{O}_C | \underline{v}(g) \geq \underline{\alpha}\} \cup \{0\}$ and the corresponding graded algebra

$$\mathrm{gr}\mathcal{O}_C := \bigoplus_{\underline{\alpha} \in (\mathbb{Z}_{\geq 0})^m} \frac{P_{\underline{\alpha}}^C}{P_{\underline{\alpha}+\underline{e}}^C},$$

THEOREM 3.1. *The following equation for the coefficients P_s^J holds:*

$$P_s^k(T_{n,m}) = (-1)^k q^{2\binom{k+1}{2}} \frac{[m+n-k-1]_{q^2}!}{[n]_{q^2}[m]_{q^2}[k]_{q^2}![m-k-1]_{q^2}![n-k-1]_{q^2}!}. \tag{3.4}$$

The proof of this identity can be found in the Appendix.

REMARK 3.2. It is not clear from (3.2), that the right hand side is symmetric in m and n although it should be so. The coefficients (3.4) reveal this symmetry.

COROLLARY 3.3. *The terms of top and low degree have the q-binomial presentations:*

(3.5)

$$P_s^{(n-1)}(T_{n,m}) = \frac{(-1)^{n-1}q^{n(n-1)}}{[n]_{q^2}}\binom{m-1}{n-1}_{q^2}, P_s^0(T_{n,m}) = \frac{1}{[n]_{q^2}}\binom{m+n-1}{n-1}_{q^2}$$

At the limit $q=1$ we get

$$P_s^{(n-1)}(T_{n,m})(q=1) = \frac{(-1)^{n-1}}{n}\binom{m-1}{n-1}, P_s^0(T_{n,m})(q=1) = \frac{1}{n}\binom{m+n-1}{n-1}.$$

Also an interesting "blow-up" equation follows from (3.5):

(3.6)

$$P_s^0(T_{n,m}) = (-1)^{n-1}q^{-n(n-1)}P_s^{(n-1)}(T_{n,m+n}) = (-1)^{m-1}q^{-m(m-1)}P_s^{(m-1)}(T_{m,m+n}).$$

COROLLARY 3.4. *If we focus on the case $m=n+1$, we have*

$$P_s^k(T_{n,n+1}) = (-1)^k q^{k(k+1)} \frac{1}{[n-k]_{q^2}}\binom{n-1}{k}_{q^2}\binom{2n-k}{n+1}_{q^2}. \tag{3.7}$$

At the limit q=1 we have

$$(-1)^{k-1}P_s^k(T_{n,n+1}) = \frac{1}{n-k}\binom{n-1}{k}\binom{2n-k}{n+1},$$

what is equal to the little Schröder number $R_{n,k}$. At the lowest level $k=0$ we get the n-th Catalan number.

DEFINITION 3.5. We call a Dyck path marked if some of its external corners are marked.

THEOREM 3.6. *The number of marked Dyck paths in the rectangle $m \times n$ with k marks equals to*

$$\frac{(m+n-k-1)!}{m \cdot n \cdot (n-k-1)!(m-k-1)!k!}.$$

PROOF. Follows from the Lemmas 4.2 and 4.3 from the Appendix. □

COROLLARY 3.7. *The coefficient $P_s^k(T_{n,m})$ of the HOMFLY polynomial for (n,m) torus knot is a certain q-deformation of the number of marked Dyck paths in the rectangle $n \times m$ with k marks.*

The Corollary 3.7 means that the coefficients at a^{2k} in the Poincaré polynomial of the Gukov-Dunfield-Rasmussen homology of the (n,m)-torus knot should be certain (q,t)-deformations of the above combinatorial data. For example, we know that $P_s^{2k}(T_{n,n+1})$ is a q-deformation of the Schröder number $R_{n,k}$, and it is natural to assume that $\mathcal{P}_s^{2k}(T_{n,n+1})$ is related to the (q,t)-deformation of this number. By (2.9) we have

$$q^{2\binom{n}{2}-2\binom{k}{2}}R_{n,k}(q^2,q^{-2})=\frac{[2n-k]!_{q^2}}{[n]_{q^2}[n+1]_{q^2}[n-k-1]!_{q^2}[n-k]!_{q^2}[k]!_{q^2}},$$

$$q^{2\binom{n}{2}}R_{n,k}(q^2,q^{-2})=(-1)^k q^{-2k}P_s^{2k}(T_{n,n+1})=q^{-2k}\mathcal{P}_s^{2k}(T_{n,n+1})(q,-1),$$

what motivates the following

CONJECTURE 3.8. *The following equation holds:*

(3.8)
$$\mathcal{P}_s^{2k}(T_{n,n+1})(q,t)=q^{2\binom{n}{2}+2k}t^{2\binom{n}{2}+3k}R_{n,k}(q^{-2}t^{-2},q^2)=q^kt^{2k}\sum_{\pi}q^{2(S(\pi)+b(\pi))}t^{2S(\pi)},$$

where summation in the right hand side is done over all (n,k)-Schröder paths with no D steps on the diagonal.

COROLLARY 3.9. *The following equation holds:*

$$\mathcal{P}_s^0(T_{n,n+1})(q,t)=q^{2\binom{n}{2}}t^{2\binom{n}{2}}C_n(q^{-2}t^{-2},q^2)=\sum_{\pi}q^{2(S(\pi)+b(\pi))}t^{2S(\pi)}, \tag{3.9}$$

where summation in the right hand side is done over all Dyck paths in $n\times n$ square.

Since (q,t)-Schröder number are supposed to be symmetric in q and t, one can check the symmetry property for $\mathcal{P}$ which agrees with the properties of the involution ϕ from Conjecture 1.5.

The following equation is a corollary of (3.9):

$$\mathcal{P}_s^{2k}(T_{n,n+1})(1,t)=t^{2k}\sum_{\pi}t^{2S(\pi)},$$

where summation is over all (n,k)-Schröder paths with no D steps on the diagonal and $S(\pi)$ denotes the area above the path. We generalize this remark to the following

CONJECTURE 3.10. *The following equation holds:*

$$\mathcal{P}_s^{2k}(T_{n,m})(1,t)=t^k\sum_{\pi}t^{2S(\pi)}, \tag{3.10}$$

where summation is over all marked Dyck paths in the $m\times n$ rectangle with k marks.

3.2. Stable limit. The right hand side of (3.2) in the limit $m\to\infty$ tends to

$$P_s(T_n)=\lim_{m\to\infty}P_s(T_{n,m})=\prod_{k=1}^{n-1}\frac{(1-a^2q^{2k})}{(1-q^{2k+2})}.$$

It is natural to consider the behaviour of the Gukov-Dunfield-Rasmussen homology in this limit too. Following the discussions in Section 6 of [**4**], we conjecture that the limit homology $\mathcal{H}(T_n)=\lim_{m\to\infty}\mathcal{H}(T_{n,m})$ is a free polynomial algebra with $n-1$ even generators with gradings $(0,2k+2,2k)$ and $n-1$ odd generators with gradings $(2,2k,2k+1)$, and therefore

$$\mathcal{P}_s(T_n)=\prod_{k=1}^{n-1}\frac{(1+a^2q^{2k}t^{2k+1})}{(1-q^{2k+2}t^{2k})}.$$

We denote the odd generators by $\xi_1,\ldots,\xi_{n-1}$, and even generators by $e_1,\ldots,e_{n-1}$. The notation for even generators is motivated by the above constructions related with (q,t)-Catalan numbers. To be more precise, we identify e_k with the k-th elementary symmetric polynomial, and the even part of $\mathcal{H}(T_n)$ with the ring of symmetric polynomials in $n-1$ variables. Recall that we had two natural gradings on this ring defined by the equations

$$S(e_k)=k, b(e_k)=1.$$

Therefore the triple grading on the even part equals to $(0,2(S+b),2b)$.

Let us construct the action of the differentials on $\mathcal{H}(T_n)$. The differentials send ξ_k to some polynomials in e_m, and they are extended to the whole algebra by the Leibnitz rule. Taking into account the gradings, one can uniquely guess the equations

$$d_{-n}(\xi_k)=\delta_{k,n}, d_0(\xi_k)=e_{k-1}, d_1(\xi_k)=e_k.$$

Let us compute the homology of $\mathcal{H}(T_n)$ with respect to differentials d_N. From the properties of the Koszul complex one can deduce the following

PROPOSITIONS 3.11. *1. The complexes $(\mathcal{H}(T_n),d_{-N})$ are acyclic.*

2. The homology of $(\mathcal{H}(T_n),d_0)$ is the polynomial algebra generated by ξ_1 and e_{n-1}.

3. The homology of $(\mathcal{H}(T_n),d_1)$ is one-dimensional and generated by 1.

The construction of the higher differentials is less restricted by the grading, however for small degrees one has no choice but to define

$$d_2(\xi_2)=e_1^2, d_2(\xi_3)=e_1e_2, d_3(\xi_3)=e_1^3.$$

EXAMPLE 3.12. The homology of $(\mathcal{H}(T_3),d_2)$ is generated by ξ_1,e_2 and e_1 modulo relation $e_1^2=0$ since $d_2(\xi_2)=e_1^2$. These generators have gradings $(2,2,3)$, $(0,6,4)$ and $(0,4,2)$, so the Poincaré polynomial for these homology equals to

$$\frac{(1+q^4t^2)(1+a^2q^2t^3)}{(1-q^6t^4)}.$$

EXAMPLE 3.13. The homology of $(\mathcal{H}(T_4),d_2)$ is generated by the elements ξ_1,e_2,e_3,e_1 and $\mu=e_1\xi_3-e_2\xi_2$ modulo relation $e_1^2=0, e_1e_2=0, e_1\mu=0$, so it is isomorphic to

$$H(\mathcal{H}(T_4),d_2)=\mathbb{C}[\xi_1,e_3]\otimes(<e_1>\oplus\mathbb{C}[\mu,e_2]).$$

The Poincaré polynomial for this homology equals to

$$\frac{(1+a^2q^2t^3)}{(1-q^8t^6)}[q^4t^2+\frac{1+a^2q^{10}t^9}{1-q^6t^4}].$$

One can compare these answers with [**4**].

3.3. $(2,k)$ **and** $(3,k)$ **torus knots.** For the torus knots with the small number of strands (2 or 3) it is possible to give a clear algebraic description of the structure predicted in [**4**]: the triply graded homology and surviving differentials $d_0,d_{\pm1},d_{\pm2}$.

CONJECTURE 3.14. *The triply graded homology for the (n,k) torus knot for $n<3$ can be realized as a subspace in $\mathcal{H}(T_n)$. This subspace generated by its top level and the action of the differentials. The top level subspace is generated by the following monomials:*

$$e_1^i\xi_1, \quad 2i \leq k-3 \qquad \text{for} \quad n=2,$$
$$e_1^i e_2^j \xi_1\xi_2, \quad i+3j \leq k-4 \qquad \text{for} \quad n=3.$$

EXAMPLE 3.15. The homology of the trefoil knot $T_{2,3}$ is generated over $d_{\pm 1}$ by one element ξ_1. The differentials act as $d_{-1}(\xi_1) = 1$ and $d_1(\xi_1) = e_1$.

EXAMPLE 3.16. The homology of the trefoil knot $T_{3,4}$ is generated over the differentials $d_0, d_{\pm 1}, d_{\pm 2}$ by one element $\xi_1\xi_2$. This homology is presented in [**4**] by a diagram with three levels. On the level 2 we have one element $\xi_1\xi_2$. On the level 1 we have

$$d_{-2}(\xi_1\xi_2) = \xi_1, d_{-1}(\xi_1\xi_2) = \xi_2, d_0(\xi_1\xi_2) = e_1\xi_1, d_1(\xi_1\xi_2) = e_1\xi_2 - e_2\xi_1, d_2(\xi_1\xi_2) = e_1^2\xi_1,$$

and on the level 0 we have

$$d_{-1}(\xi_1) = d_{-2}(\xi_2) = 1,$$
$$d_1(\xi_1) = d_0(\xi_2) = d_{-1}(e_1\xi_1) = d_{-2}(e_1\xi_2 - e_2\xi_1) = e_1;$$
$$d_1(\xi_2) = -d_{-1}(e_1\xi_2 - e_2\xi_1) = e_2$$
$$d_2(\xi_2) = d_1(e_1\xi_1) = d_0(e_1\xi_2 - e_2\xi_1) = d_{-1}(e_1^2\xi_1) = e_1^2,$$
$$d_2(e_1\xi_2 - e_2\xi_1) = d_1(e_1^2\xi_1) = e_1^3.$$

These equations represent exactly the dot diagram for the homology of $T_{3,4}$ presented in [**4**]:

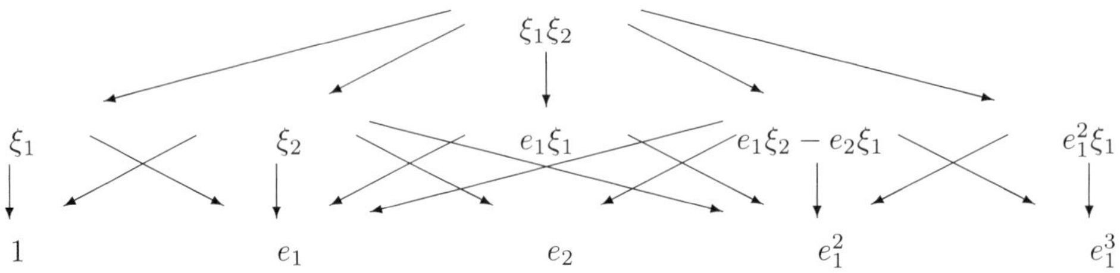

THEOREM 3.17. *Under the assumptions of the Conjecture* 3.14 *the Poincaré polynomial for the triply-graded homology has a form:*

$$\mathcal{P}_{2,2k+1} = \sum_{i=0}^{k} q^{4i}t^{2i} + a^2 \sum_{i=0}^{k-1} q^{4i+2}t^{2i+3}, \tag{3.11}$$

$$\mathcal{P}_{3,k} = 1 + \sum_{0<i+3j\leq k-1} q^{4i+6j}t^{2i+4j}(1+a^2q^{-2}t) + a^4q^6t^8 \sum_{i+3j\leq k-4} q^{4i+6j}t^{2i+4j}(1+a^{-2}q^2t^{-1}). \tag{3.12}$$

REMARK 3.18. The polynomials (3.11) and (3.12) coincide with the ones conjectured in [**4**].

PROOF. As a vector space, the homology of $T_{2,2k+1}$ is generated by

$$1, e_1, e_1^2 \ldots e_1^k \qquad \text{on lower level,}$$
$$\xi_1, e_1\xi_1, e_1^2\xi_1 \ldots e_1^{k-1}\xi_1 \qquad \text{on top level.}$$

This implies the equation (3.11).

By Conjecture 3.14, we know the structure of $\mathcal{H}(3,k)$ as a vector space at the level 2. Let us describe it on levels 1 and 0. Remark that $\mathcal{H}(3,k)$ is generated by its top level, and the differentials commute with the multiplication by even variables, so the levels in $\mathcal{H}(3,k)$ are just the products of the top level even monomials with the corresponding levels of $\mathcal{H}(3,4)$.

Therefore on the lower level we get

$$< 1, e_1, e_1^2, e_1^3, e_2 > \otimes < e_1^i e_2^j | i+3j \leq k-4 >=< e_1^i e_2^j | 3i+j \leq k-1 > .$$

It rests to know that the homology of d_1 is one-dimensional and generated by 1, so we can compute the homology at level 1 as well. □

COROLLARY 3.19. *The reduced $sl(2)$ homology of $(2,n)$ torus knot coincides with the triply-graded homology as $d_2 = 0$.*

COROLLARY 3.20. *The reduced $sl(2)$ homology of $(3,n)$ torus knot is spanned by the monomials*

$$1, e_2^i, e_1 e_2^j, e_2^j \xi_1, e_1 e_2^{i-1} \xi_1 \tag{3.13}$$

with $0 < 3i \leq n-1, 3j \leq n-2$.

PROOF. Since $d_2(\xi_1) = 0, d_2(\xi_2) = e_1^2$, to compute the homology of $\mathcal{H}(T_{3,n})$ with respect to d_2 one should throw away from $\mathcal{H}(T_{3,n})$ all monomials divisible by ξ_2 and e_1^2. Therefore it is spanned by the monomials of the form (3.13) belonging to $\mathcal{H}(T_{3,n})$. □

One can compare these results with [**1**].

EXAMPLE 3.21. The reduced $sl(2)$ homology of $(3,4)$ torus knot is spanned by the monomials $1, e_1, e_2, \xi_1, e_1\xi_1$, what gives the following Poincare polynomial:

$$\mathcal{P}_2(T_{3,4}) = 1 + q^4t^2 + q^6t^4 + q^6t^3 + q^{10}t^5.$$

3.4. $(4,5)$ **knot.** Following the above description of the triply graded homology for $(2,n)$ and $(3,n)$ torus knots, it is natural to assume that the homology of a torus knot is generated by its top level under the action of some algebra.

CONJECTURE 3.22. *There exists a sequence of algebras $\mathcal{A}_{n,m}$ together with their representations in the stable homology space $\mathcal{H}(T_{n,m})$ satisfying the following conditions:*

1) The action of $\mathcal{A}_{n,m}$ commutes with the multiplication by the polynomials in $e_1, \ldots e_n$

2) For every k ($|k| < n$) the differential d_k belongs to $\mathcal{A}_{n,m}$

3) The space $\mathcal{H}(T_{n,m})$ is generated by its top level under the action of $\mathcal{A}_{n,m}$.

4) The top level of homology for $(m, m+n)$ torus knot coincides with the zero level of the homology of (m,n) torus knot multiplied by the "volume form" $\xi_1\xi_2\ldots\xi_{m-1}$ (compare with (3.6)*).*

For $(2,n)$ and $(3,n)$ knots the corresponding algebras are supposed to be generated by the differentials $d_0, d_{\pm 1}, d_{\pm 2}$. Unfortunately, for bigger knots this is not true: for example, $\mathcal{H}_{n,n+1}$ has dimension 1 on the top level,so it is supposed to be generated by the "volume form" $\xi_1\xi_2\ldots\xi_{n-1}$. On the level $(n-2)$ it has dimension $R(n,n-2)=(n+2)(n-1)/2$ that is greater than the number $(2n-1)$ of available differentials.

For example, for $n=4$ we have $R(4,2)=9$ and 7 differentials at our disposal. Nevertheless, by grading reasons one can uniquely guess the algebraic description of the differentials as well as two missing operators. Presuming the action of differentials on ξ_1 and ξ_2 being the same as above, one can extend it to ξ_3 by the formula

$$d_{-1}(\xi_3)=d_{-2}(\xi_3)=0,\quad d_{-3}(\xi_3)=1,$$

$$d_0(\xi_3)=e_2,\quad d_1(\xi_3)=e_3,\quad d_2(\xi_3)=e_1e_2,\quad d_3(\xi_3)=e_1^3.$$

The two missing operators are

$$\alpha_1=e_1^{-1}d_3, \alpha_2=e_1^{-2}d_3.$$

Suppose that the space $\mathcal{H}(T_{4,5})$ is generated by the "volume form" $\xi_1\xi_2\xi_3$ under the action of the differentials and the operators α_1 and α_2. Let us describe the basis on each level in this space and indicate the gradings of the basis elements. For each basis element we also indicate the element in $\mathcal{A}_{4,5}$ producing it from $\xi_1\xi_2\xi_3$.

Level	Basis element	Element of $\mathcal{A}_{4,5}$	(a,q,t)
3	$\xi_1\xi_2\xi_3$	1	$(6,12,15)$
2	$\xi_1\xi_2$	d_{-3}	$(4,6,8)$
2	$\xi_1\xi_3$	d_{-2}	$(4,8,10)$
2	$\xi_2\xi_3$	d_{-1}	$(4,10,12)$
2	$e_1\xi_1\xi_2$	α_2	$(4,10,10)$
2	$e_1\xi_1\xi_3-e_2\xi_1\xi_2$	d_0	$(4,12,12)$
2	$e_1\xi_2\xi_3-e_2\xi_1\xi_3+e_3\xi_1\xi_2$	d_1	$(4,14,14)$
2	$e_1^2\xi_1\xi_2$	α_1	$(4,14,12)$
2	$e_1^2\xi_1\xi_3-e_1e_2\xi_1\xi_2$	d_2	$(4,16,14)$
2	$e_1^3\xi_1\xi_2$	d_3	$(4,18,14)$
1	ξ_1	$d_{-2}d_{-3}$	$(2,2,3)$
1	ξ_2	$d_{-1}d_{-3}$	$(2,4,5)$
1	ξ_3	$d_{-1}d_{-2}$	$(2,6,7)$
1	$e_1\xi_1$	d_0d_{-3}	$(2,6,5)$
1	$e_1\xi_2-e_2\xi_1$	d_1d_{-3}	$(2,8,7)$
1	$e_1\xi_3-e_3\xi_1$	d_1d_{-2}	$(2,10,9)$
1	$e_1\xi_3-e_2\xi_2$	d_0d_{-1}	$(2,10,9)$
1	$e_1^2\xi_1$	d_2d_{-3}	$(2,10,7)$
1	$e_1^3\xi_1$	$d_2\alpha_2$	$(2,14,9)$
1	$e_1^4\xi_1$	$d_2\alpha_1$	$(2,18,11)$
1	$e_1^5\xi_1$	d_2d_3	$(2,22,13)$
1	$e_1\xi_2$	$d_{-1}\alpha_2$	$(2,8,7)$
1	$e_1^2\xi_2$	$d_{-1}\alpha_1$	$(2,12,9)$
1	$e_1^3\xi_2$	$d_{-1}d_3$	$(2,16,11)$
1	$e_1^2\xi_2-e_1e_2\xi_1$	$d_1\alpha_2$	$(2,12,9)$
1	$e_1^3\xi_2-e_1^2e_2\xi_1$	$d_1\alpha_1$	$(2,16,11)$

Level	Basis element	Element of $\mathcal{A}_{4,5}$	(a,q,t)
1	$e_1^2\xi_3 - e_1e_2\xi_2$	d_2d_{-1}	$(2,14,11)$
1	$e_2\xi_3 - e_3\xi_2$	d_1d_{-1}	$(2,12,11)$
1	$e_1^3\xi_2 - e_1^2e_2\xi_1$	d_1d_3	$(2,18,13)$
1	$(e_2^2 - e_1e_3)\xi_1 - e_1e_2\xi_2 + e_1^2\xi_3$	d_1d_0	$(2,14,11)$
1	$(e_1e_2^2 - e_1^2e_3)\xi_1 - e_1^2e_2\xi_2 + e_1^3\xi_3$	d_1d_2	$(2,18,13)$
0	1	$d_{-1}d_{-2}d_{-3}$	$(0,0,0)$
0	e_1	$d_1d_{-2}d_{-3}$	$(0,4,2)$
0	e_2	$d_1d_{-1}d_{-3}$	$(0,6,4)$
0	e_3	$d_1d_{-1}d_{-2}$	$(0,8,6)$
0	e_1^2	$d_1d_0d_{-3}$	$(0,8,4)$
0	e_1^3	$d_1d_2d_{-3}$	$(0,12,6)$
0	e_1^4	$d_1d_2\alpha_2$	$(0,16,8)$
0	e_1^5	$d_1d_2\alpha_1$	$(0,20,10)$
0	e_1^6	$d_1d_2d_3$	$(0,24,12)$
0	e_1e_2	$d_1d_{-1}\alpha_2$	$(0,10,6)$
0	$e_1e_3 - e_2^2$	$d_1d_0d_{-1}$	$(0,12,8)$
0	$e_1^2e_3 - e_1e_2^2$	$d_1d_2d_{-1}$	$(0,16,10)$
0	$e_1^2e_2$	$d_1d_{-1}\alpha_1$	$(0,14,8)$
0	$e_1^3e_2$	$d_1d_{-1}d_3$	$(0,18,10)$

REMARK 3.23. The homology of d_1 is one-dimensional and spanned by 1.

REMARK 3.24. The homology of d_2 is spanned by

$$1, e_1, \xi_1, e_2, e_1\xi_1, e_3, e_2\xi_1, e_1e_3 - e_2^2, e_1\xi_3 - e_2\xi_2,$$

what agrees with the generating function for the reduced Khovanov homology of $T_{4,5}$ (compare with [**1**]):

$$\mathcal{P}_2(T_{4,5}) = 1 + q^4t^2 + q^6t^3 + q^6t^4 + q^{10}t^5 + q^8t^6 + q^{12}t^7 + q^{12}t^8 + q^{14}t^9.$$

REMARK 3.25. At the lower level of the triply-graded homology, we get the space L_4. In particular, this space has no monomial baisis.

4. Appendix

The following lemma is a well known q-analogue of the binomial identity.

LEMMA 4.1.

$$(1+z)(1+qz)\cdot\ldots\cdot(1+q^{n-1}z) = \sum_{j=0}^{n}\binom{n}{j}_q q^{\binom{j}{2}}z^j. \tag{4.1}$$

PROOF. Induction by n. □

Proof of the Theorem 3.1 First, by (4.1), we have

$$\prod_{i=1}^{n-1-b}(1-a^2q^{2i}) = \sum_{i=0}^{n-1-b}(-1)^i a^{2i}q^{2\binom{i+1}{2}}\binom{n-1-b}{i}_{q^2},$$

$$\prod_{j=1}^{b}(a^2-q^{2j}) = \sum_{j=0}^{b}(-1)^{b-j}a^{2j}q^{2\binom{b-j+1}{2}}\binom{b}{j}_{q^2}.$$

If we multiply these expressions and take the coefficient at a^{2k}, we get

$$\sum_{i=k-b}^{n-1-b} (-1)^{b-k} q^{2\binom{i+1}{2}+2\binom{b-k+i+1}{2}} \binom{n-1-b}{i}_{q^2} \binom{b}{k-i}_{q^2}.$$

Therefore

$$P_s^{2k}(T_{n,m}) = \frac{1}{[n]_{q^2}} \sum_{b=0}^{n-1} q^{2mb} \sum_{i=k-b}^{n-1-b} (-1)^{b-k} q^{2\binom{i+1}{2}+2\binom{b-k+i+1}{2}} \times$$

$$\frac{[b]_{q^2}![n-1-b]_{q^2}!}{[i]_{q^2}![n-1-b-i]_{q^2}![k-i]_{q^2}![b-k+i]_{q^2}!} \frac{(1-q^2)^{n-1}}{[b]_{q^2}![n-1-b]_{q^2}!} =$$

$$\frac{(1-q^2)^{n-1}}{[n]_{q^2}} \sum_{i=0}^{k} \frac{(-1)^i q^{2m(k-i)} \cdot q^{2\binom{i+1}{2}} [k]_{q^2}!}{[n-1-k]_{q^2}![k]_{q^2}![i]_{q^2}![k-i]_{q^2}!} \times$$

$$\sum_{b=k-i}^{n-1-i} (-1)^{b-k+i} q^{2m(b-k+i)} q^{2\binom{b-k+i+1}{2}} \frac{[n-1-k]_{q^2}!}{[n-1-b-i]_{q^2}![b-k+i]_{q^2}!}.$$

Now by (4.1) we simplify the inner sum, denoting $l = b-k+i$:

$$\sum_{l=0}^{n-1-k} (-1)^l q^{2ml} q^{2\binom{l+1}{2}} \binom{n-1-k}{l}_{q^2} = (1-q^{2m+2})(1-q^{2m+4}) \cdot \ldots \cdot (1-q^{2m+2n-2-2k})$$

$$= (1-q^2)^{n-1-k} \frac{[m+n-1-k]_{q^2}!}{[m]_{q^2}!},$$

And this sum does not depend on i. Analogously we have

$$\sum_{i=0}^{k} (-1)^i q^{2\binom{i+1}{2}} q^{2m(k-i)} \binom{k}{i}_{q^2} = (q^{2m}-q^2) \cdot \ldots \cdot (q^{2m}-q^{2k}) =$$

$$(-1)^k (1-q^2)^k q^{2\binom{k+1}{2}} \frac{[m-1]_{q^2}!}{[m-k-1]_{q^2}!}.$$

Finally,

$$P_s^{2k}(T_{n,m}) = (-1)^k q^{2\binom{k+1}{2}} \frac{[m-1]_{q^2}![m+n-1-k]_{q^2}!}{[n]_{q^2}[m]_{q^2}![m-k-1]_{q^2}![n-1-k]_{q^2}![k]_{q^2}!}.$$

□

To what follows we will need the the following sequence of "generalized Narayana numbers" (for given n, m):

$$N_k = \frac{(m-1)!(n-1)!}{k!(k+1)!(m-k-1)!(n-k-1)!} = \frac{1}{k+1}\binom{m-1}{k}\binom{n-1}{k}.$$

LEMMA 4.2. *N_k equals to the number of Dyck paths in $m \times n$ rectangle with k external corners.*

PROOF. First, let us remark that $\binom{m-1}{k}\binom{n-1}{k}$ equals to the number of lattice paths in the $m \times n$ rectangle with k external corners. Given a such path, let us continue it periodically to get an infinite path. This construction maps exactly $k+1$ different paths (we have $k+1$ corners, as we have a corner at the starting point) into one. On the other hand, exactly one of them is totally above the diagonal - it corresponds to the set of corners with the lowest value of the linear function $my - nx$. □

LEMMA 4.3.

$$\sum_{l=k}^{n-1}\binom{l}{k}N_l=\frac{(m+n-k-1)!}{m\cdot n\cdot(n-k-1)!(m-k-1)!k!}.$$

PROOF. Remark that

$$N_l=\frac{k!(k+1)!(m-k-1)!(n-k-1)!}{l!(l+1)!(m-l-1)!(n-l-1)!}N_k,$$

so

$$\binom{l}{k}N_l=\frac{(k+1)!(m-k-1)!(n-k-1)!}{(l-k)!(l+1)!(m-l-1)!(n-l-1)!}N_k=$$

$$\binom{m-k-1}{l-k}\binom{n}{n-l-1}\binom{n}{n-k-1}^{-1}N_k.$$

Now we have

$$\sum_{l=k}^{n-1}\binom{m-k-1}{l-k}\binom{n}{n-l-1}=\binom{m+n-k-1}{n-k-1},$$

so

$$\sum_{l=k}^{n-1}\binom{l}{k}N_l=\binom{m+n-k-1}{n-k-1}\binom{n}{n-k-1}^{-1}N_k=\frac{(m+n-k-1)!(k+1)!}{m!n!}N_k.$$

□

References

[1] D. Bar-Natan, S. Morrisson. The Knot Atlas. http://katlas.org.

[2] M. B. Can. Nested Hilbert schemes and the nested q,t-Catalan series. 20th Annual International Conference on Formal Power Series and Algebraic Combinatorics (FPSAC 2008), 61–69, Discrete Math. Theor. Comput. Sci. Proc. MR2721442 (2011k:14001)

[3] L. Carlitz, J. Riordan. Two element lattice permutation numbers and their q-generalization. Duke Math. J. **31** (1964), 371–388. MR0168490 (29:5752)

[4] N. Dunfield, S. Gukov, J. Rasmussen.The Superpolynomial for Knot Homologies. Experimental Math. **15** (2006), 129–159 MR2253002 (2008c:57020)

[5] E. Egge, J. Haglund, K. Killpatrick, and D. Kremer, A Schröder generalization of Haglund's statistic on Catalan paths, Electron. J. Combin. **10** (2003), Research Paper 16, 21 pp. (electronic). MR1975766 (2004c:05008)

[6] J. Hoste, A. Ocneanu, K. Millett, P. Freyd, W. B. R. Lickorish, D. Yetter. A New Polynomial Invariant of Knots and Links. Bulletin of the American Mathematical Society **12** (1985), 239–246. MR776477 (86e:57007)

[7] A. M. Garsia, M. Haiman. A remarkable q,t-Catalan sequence and q-Lagrange inversion. J. Algebraic Combin. **5** (1996), no. 3, 191–244. MR1394305 (97k:05208)

[8] A. M. Garsia, J. Haglund. A proof of the q,t–Catalan positivity conjecture. Discrete Math. **256** (2002), no. 3, 677–717. MR1935784 (2004c:05207)

[9] E. Gorsky. Combinatorial computation of the motivic Poincaré series. Journal of Singularities **3** (2011), 48–82. MR2785859

[10] S. Gukov, A. Iqbal, C. Kozcaz, C. Vafa. Link homologies and the refined topological vertex. Comm. Math. Phys. **298** (2010), no. 3, 757–785. MR2670927 (2011h:81223)

[11] J. Haglund. The q,t-Catalan Numbers and the Space of Diagonal Harmonics with an Appendix on the Combinatorics of Macdonald Polynomials. AMS University Lecture Series, Vol.41. MR2371044 (2009f:05261)

[12] J. Haglund. Conjectured statistics for the q,t–Catalan numbers. Adv. Math. **175** (2003), no. 2, 319–334. MR1972636 (2004c:05021)

[13] J. Haglund. A proof of the q,t–Schröder conjecture. Int. Math. Res. Not. 2004, no. 11, 525–560. MR2038776 (2005d:05147)

[14] M. Haiman. t, q-Catalan numbers and the Hilbert scheme. Discrete Math. **193** (1998), 201–224. MR1661369 (2000k:05264)
[15] M. Hedden. On knot Floer homology and cabling II. Int. Math. Res. Not. IMRN 2009, no. 12, 2248–2274. MR2511910 (2011f:57015)
[16] V. Jones. Hecke algebra representations of braid groups and link polynomials. Ann. of Math. **126** (1987), no.2, 335–388. MR908150 (89c:46092)
[17] M. Khovanov. A categorification of the Jones polynomial. Duke Math. J. **101** (2000), no. 3, 359–426 MR1740682 (2002j:57025)
[18] M. Khovanov. Triply-graded link homology and Hochschild homology of Soergel bimodules. Int. Journal of Math. **18** (2007), no. 8 869–885. MR2339573 (2008i:57013)
[19] M. Khovanov, L. Rozansky. Matrix factorizations and link homology I. Fund. Math. **199** (2008), no. 1, 1–91. MR2391017 (2010a:57011)
[20] M. Khovanov, L. Rozansky. Matrix factorizations and link homology II. Geom. Topol. **12** (2008), no. 3, 1387–1425. MR2421131 (2010g:57014)
[21] I. G. Macdonald. Symmetric functions and Hall polynomials. Oxford University Press, 1995. MR1354144 (96h:05207)
[22] A. Oblomkov, V. Shende. The Hilbert scheme of a plane curve singularity and the HOMFLY polynomial of its link. arXiv:1003.1568
[23] P. Ozsváth, Z. Szabó. Holomorphic discs and knot invariants. Adv. Math., **186** 2004, no.1, 58–116. MR2065507 (2005e:57044)
[24] P. Ozsváth, Z. Szabó.Holomorphic disks, link invariants and the multi-variable Alexander polynomial. Algebr. Geom. Topol. **8** (2008), no. 2, 615–692. MR2443092 (2010h:57023)
[25] P. Ozsváth, Z. Szabó. Holomorphic discs and topological invariants for closed three-manifolds. Ann. of Math. (2). **159** (2004), no. 3, 1027–1158. MR2113019 (2006b:57016)
[26] P. Ozsváth, Z. Szabó. On the Floer homology of plumbed three-manifolds. Geometry and Topology, **7** (2003), 185–224. MR1988284 (2004h:57039)
[27] P. Ozsváth, Z. Szabó. On knot Floer homology and lens space surgeries. Topology **44** (2005), no. 6, 1281–1300. MR2168576 (2006f:57034)
[28] J. Przytycki, P. Traczyk, Conway algebras and skein equivalence of links. Proc. Amer. Math. Soc. **100** (1987) 744–748. MR894448 (88m:57012)
[29] J. Rasmussen. Knot polynomials and knot homologies, in *Geometry and Topology of Manifolds*, Boden et. al. eds., Fields Institute Communications **47** (2005), 261–280. MR2189938 (2006i:57029)
[30] J. Rasmussen. Some differentials on Khovanov-Rozansky homology. arXiv: math.GT/0607544
[31] R. P. Stanley. Enumerative Combinatorics. Vol. 2, Cambridge Studies in Advanced Mathematics, vol. 62, Cambridge University Press, Cambridge, 1999. MR1676282 (2000k:05026)

Department of Mathematics, Stony Brook University, Stony Brook, New York 11794

E-mail address: egorsky@math.sunysb.edu

Contemporary Mathematics
Volume **566**, 2012
http://dx.doi.org/10.1090/conm/566/11223

Motivic zeta functions for degenerations of abelian varieties and Calabi-Yau varieties

Lars Halvard Halle and Johannes Nicaise

1. Introduction

Let $f \in \mathbb{Z}[x_1, \ldots, x_n]$ be a non-constant polynomial, and let p be a prime. Igusa's p-adic zeta function $Z_f^p(s)$ is a meromorphic function on the complex plane that encodes the number of solutions of the congruence $f \equiv 0$ modulo powers p^m of the prime p. Igusa's *p-adic monodromy conjecture* predicts in a precise way how the singularities of the complex hypersurface defined by the equation $f = 0$ influence the poles of $Z_f^p(s)$ and thus the asymptotic behaviour of this number of solutions as m tends to infinity. The conjecture states that, when p is sufficiently large, poles of $Z_f^p(s)$ should correspond to local monodromy eigenvalues of the polynomial map $\mathbb{C}^n \to \mathbb{C}$ defined by f. We refer to Section 3 for a precise formulation.

Starting in the mid-nineties, J. Denef and F. Loeser developed the theory of motivic integration, which had been introduced by M. Kontsevich in his famous lecture at Orsay in 1995. Denef and Loeser used this theory to construct a motivic object $Z_f^{mot}(s)$ that interpolates the p-adic zeta functions $Z_f^p(s)$ for $p \gg 0$ and captures their geometric essence. This object is called the *motivic zeta function* of f. Denef and Loeser also formulated a motivic upgrade of the monodromy conjecture (Conjecture 4.17). Its precise relation with the p-adic monodromy conjecture is explained in Section 4.7.

The aim of this paper is to present a *global* version of Denef and Loeser's motivic zeta functions. Let X be a Calabi-Yau variety over a complete discretely valued field K (i.e., a smooth, proper and geometrically connected variety with trivial canonical sheaf). We'll define the motivic zeta function $Z_X(T)$ of X. This is a formal power series with coeffients in a certain localized Grothendieck ring of varieties over the residue field k of K. We'll show that $Z_X(T)$ has properties analogous to Denef and Loeser's zeta function, and we'll prove a global version of the motivic monodromy conjecture when X is an abelian variety, under a certain tameness condition on X (Theorem 5.7).

The link between Denef and Loeser's motivic zeta function $Z_f^{mot}(s)$ and our global variant is an alternative interpretation of $Z_f^{mot}(s)$ in terms of non-archimedean geometry, due to J. Sebag and the second author [**NS07**]. This interpretation is

2010 *Mathematics Subject Classification.* Primary 14D06, 14G10; Secondary 11G10.

The second author was partially supported by the Fund for Scientific Research - Flanders (G.0415.10).

based on the theory of motivic integration on rigid varieties developed by F. Loeser and J. Sebag [**LS03**], which explains how one can associate a *motivic volume* to a gauge form on a smooth rigid variety over a complete discretely valued field. J. Sebag and the second author constructed the *analytic Milnor fiber* of a hypersurface singularity, a non-archimedean model for the classical Milnor fibration in the complex analytic setting. The analytic Milnor fiber is a smooth rigid variety over a field K of Laurent series. The motivic zeta function can be realized as a generating series whose coefficients are motivic volumes of a so-called *Gelfand-Leray* form on the analytic Milnor fiber over finite totally ramified extensions of the base field K. This is explained in detail in Section 4.

This interpretation of the motivic zeta function admits a natural generalization to the global case, where we replace the analytic Milnor fiber by a Calabi-Yau variety X over a complete discretely valued field K and the Gelfand-Leray form by a suitably normalized gauge form ω on X. The zeta function $Z_X(T)$ is studied in Section 5 when X is an abelian variety and in Section 6 in the general case. We raise the question whether there exists a relation between the poles of $Z_X(T)$ and the monodromy eigenvalues of X as predicted by the monodromy conjecture in the case of hypersurface singularities (Question 6.10).

We studied motivic zeta functions of abelian varieties in detail in the papers [**HN10a, HN10c, HN10d**]. Section 5 gives an overview of the results and methods used in those papers. A powerful and central tool is the Néron model of an abelian K-variety A, which is the "minimal" extension of A to a smooth group scheme over R. The Néron model $\mathcal{A}$ of A comes equipped with much interesting structure, such as the Chevalley decomposition of the identity component of its special fiber, the Lie algebra $\mathrm{Lie}(\mathcal{A})$ and the component group Φ_A. The key point in the study of $Z_A(T)$ is to understand how these objects change under ramified extensions of K.

Our main result is Theorem 5.7, which states that if A is a tamely ramified abelian K-variety, then $Z_A(\mathbb{L}^{-s})$ is rational with a unique pole at $s = c(A)$, where $c(A)$ denotes Chai's base change conductor of A [**Ch00**]. Moreover, for every embedding of $\mathbb{Q}_\ell$ in $\mathbb{C}$, the complex number $\exp(2\pi c(A)i)$ is an eigenvalue of the monodromy transformation on $H^g(A \times_K K^t, \mathbb{Q}_\ell)$, where K^t denotes a tame closure of K, and where g is the dimension of A. This shows that a global version of Denef and Loeser's motivic monodromy conjecture holds for tamely ramified abelian varieties.

The situation for general Calabi-Yau varieties is at the moment far less clear than in the abelian case. Our proofs for abelian varieties rely heavily on the theory of Néron models, and these methods do not extend to the general case. However, if we restrict ourselves to equal characteristic zero, there is still much that can be said, and we present some of our results under this assumption in Section 6.

A particular advantage in characteristic zero, and the basis for many applications, is that we can find an *sncd*-model of X, i.e., a regular proper R-model $\mathcal{X}$ whose special fiber $\mathcal{X}_s$ is a divisor with strict normal crossings. We explain in Section 6 how the results in [**NS07**] yield an explicit expression for $Z_X(T)$ in terms of the model $\mathcal{X}$. This expression shows that $Z_X(T)$ is rational, and yields a finite subset of $\mathbb{Q}$ that contains all the poles of $Z_X(\mathbb{L}^{-s})$. However, due to cancellations in the formula, it is often difficult to use this description to determine the precise set of poles.

The opposite of the largest pole of $Z_X(\mathbb{L}^{-s})$ turns out to be an interesting invariant of X, we call it the log canonical threshold $lct(X)$ of X. It can be easily computed on the model $\mathcal{X}$. We can show that $lct(X)$ corresponds to a monodromy eigenvalue on the degree $\dim(X)$ cohomology of X. The value $lct(X)$ is a global version of the log canonical threshold for complex hypersurface singularities, we explain the precise relationship in Section 6.3. Since we know that for an abelian K-variety A, the base change conductor $c(A)$ is the unique pole of $Z_A(\mathbb{L}^{-s})$, we find that $lct(A) = -c(A)$. This yields an interesting relation between the Néron model of A and the birational geometry of *sncd*-models of A. Our explicit expression for the zeta function allows to compute many other arithmetic invariants of A on an *sncd*-model, in particular the number of connected components of the Néron model. This generalizes the results that were known for elliptic curves. Conversely, we can use the zeta function to extend many interesting invariants of abelian K-varieties to arbitrary Calabi-Yau varieties.

2. Preliminaries

2.1. Notation. For every ring A, an A-variety is a reduced separated A-scheme of finite type. An algebraic group over a field F is a reduced group scheme of finite type over F. We denote by μ the profinite group scheme of roots of unity.

2.2. Local monodromy eigenvalues. Let k be a subfield of $\mathbb{C}$, let X be a k-variety, and let

$$f : X \to \mathbb{A}^1_k = \operatorname{Spec} k[t]$$

be a k-morphism. Let x be a point of $X(\mathbb{C})$ such that $f(x) = 0$. We denote by X^{an} the complex analytification of $X \times_k \mathbb{C}$, by $f^{\mathrm{an}} : X^{\mathrm{an}} \to \mathbb{C}$ the complex analytic map induced by f, and by X^{an}_s the zero locus of f^{an} in X. We say that a complex number α is a local monodromy eigenvalue of f at x if there exists an integer $j \geq 0$ such that α is an eigenvalue of the monodromy transformation on $R^j\psi_{f^{\mathrm{an}}}(\mathbb{C})_x$. Here

$$R\psi_{f^{\mathrm{an}}}(\mathbb{C}) \in D^b_c(X^{\mathrm{an}}_s, \mathbb{C})$$

denotes the complex of nearby cycles associated to f^{an} [**Di04**, §4.2]. If X is smooth at x, then the complex vector space $R^j\psi_{f^{\mathrm{an}}}(\mathbb{C})_x$ is isomorphic to the degree j singular cohomology space of the Milnor fiber of f^{an} at the point x.

If f is a polynomial in $k[x_1, \ldots, x_n]$, then we can speak of local monodromy eigenvalues of f by considering f as a morphism $\mathbb{A}^n_k \to \mathbb{A}^1_k$.

2.3. The Bernstein-Sato polynomial. Let k be a field of characteristic zero. Let X be a smooth irreducible k-variety of dimension n, endowed with a morphism

$$f : X \to \mathbb{A}^1_k = \operatorname{Spec} k[t].$$

Denote by X_s the fiber of f over the origin. For every closed point x of X_s, we denote by k_x the residue field at x and by $b_{f,x}(s)$ the Bernstein-Sato polynomial of the formal germ of f in $\widehat{\mathcal{O}}_{X,x} \cong k_x[[x_1, \ldots, x_n]]$ (see [**Bj79**, 3.3.6]). We call $b_{f,x}(s)$ the local Bernstein-Sato polynomial of f at x. If $k = \mathbb{C}$, then $b_{f,x}(s)$ coincides with the Bernstein polynomial of the analytic germ of f in $\mathcal{O}_{X^{\mathrm{an}},x}$, by [**MN91**, §4.2].

If $h : Y \to X$ is a morphism of smooth k-varieties and y is a closed point of Y such that $x = h(y)$ and h is étale at x, then the faithfully flat local homomorphism $\widehat{\mathcal{O}}_{X,x} \to \widehat{\mathcal{O}}_{Y,y}$ satisfies the conditions in [**MN91**, §4.2]. It follows that

$b_{f,x}(s) = b_{f\circ h,y}(s)$. The same argument shows that $b_{f,x}(s)$ is invariant under arbitrary extensions of the base field k. If $k = \mathbb{C}$, then Kashiwara has shown that the roots of $b_{f,x}(s)$ are rational numbers [**Ka76**]. By [**Sa94**], they lie in the interval $]-n, 0[$. Invoking the Lefschetz principle, we see that these properties hold for arbitrary k.

If $k = \mathbb{C}$, then it was proven by Malgrange [**Ma83**] that, for every root α of the local Bernstein-Sato polynomial $b_{f,x}(s)$, the value $\exp(2\pi i\alpha)$ is a local monodromy eigenvalue of f at some point of X_s^{an}. Moreover, if we allow x to vary in the zero locus of f, all local monodromy eigenvalues arise in this way. Since $b_{f,x}(s)$ is invariant under extension of the base field k, this property still holds over all subfields k of $\mathbb{C}$.

By constructibility of the nearby cycles complex, the local monodromy eigenvalues of f form a finite set $\mathrm{Eig}(f)$. Thus, as x runs through the set of closed points of X_s, the polynomials $b_{f,x}(s)$ form a finite set, since they are all monic polynomials whose roots belong to the finite set of rational numbers α in $[-n, 0[$ such that $\exp(2\pi i\alpha)$ lies in $\mathrm{Eig}(f)$. We call the least common multiple of the polynomials $b_{f,x}(s)$ the Bernstein-Sato polynomial of f, and we denote it by $b_f(s)$. If $X = \mathbb{A}^n_k$, then by [**MN91**, §4.2], this definition coincides with the usual definition of the Bernstein-Sato polynomial of an element f in $k[x_1, \ldots, x_n]$.

3. *P*-adic and motivic zeta functions

3.1. The Poincaré series. Let f be an element of $\mathbb{Z}[x_1, \ldots, x_n] \setminus \mathbb{Z}$, for some integer $n > 0$, and let p be a prime number. For every integer $m \geq 0$, we denote by S_m the set of solutions of the congruence $f \equiv 0$ modulo p^{m+1}, i.e.,

$$S_m = \{a \in (\mathbb{Z}/p^{m+1}\mathbb{Z})^n \mid f(a) \equiv 0 \mod p^{m+1}\}.$$

We put $N_m = \sharp S_m$.

DEFINITION 3.1. The Poincaré series associated to f and p is the generating series

$$P(T) = \sum_{m \geq 0} N_m T^m \quad \in \mathbb{Z}[[T]].$$

EXAMPLE 3.2. If the closed subscheme X of $\mathbb{A}^n_{\mathbb{Z}_p}$ defined by the equation $f = 0$ is smooth over $\mathbb{Z}_p$, then the Poincaré series $P(T)$ is easy to compute. For every integer $m \geq 0$, the set S_m is the set of $(\mathbb{Z}/p^{m+1}\mathbb{Z})$-valued points on X. Locally at every point, X admits an étale morphism to $\mathbb{A}^{n-1}_{\mathbb{Z}_p}$. The infinitesimal lifting criterion for étale morphisms implies that the map $S_{m+1} \to S_m$ is surjective, and that every fiber has cardinality p^{n-1}. In this way, we find that

$$P(T) = \frac{\sharp X(\mathbb{F}_p)}{1 - p^{n-1}T}. \tag{3.1}$$

If X is not smooth over $\mathbb{Z}_p$, then the behaviour of the values N_m is much harder to understand. The following conjecture was mentioned in [**BS66**], Chapter 1, Section 5, Problem 9.

CONJECTURE 3.3. *The Poincaré series $P(T)$ is rational, i.e., it belongs to the subring $\mathbb{Q}(T) \cap \mathbb{Z}[[T]]$ of $\mathbb{Q}((T))$.*

3.2. The p-adic zeta function.

DEFINITION 3.4. We denote by $|\cdot|_p$ the p-adic absolute value on $\mathbb{Q}_p$. The p-adic zeta function of f is defined by

$$Z_f^p(s) = \int_{\mathbb{Z}_p^n} |f(x)|_p^s \, dx$$

for every complex number s with $\Re(s) > 0$.

The p-adic zeta function Z_f^p is an analytic function on the complex right half plane $\Re(s) > 0$. It was introduced by Weil, and systematically studied by Igusa. It can be defined in a much more general set-up, starting from a p-adic field K, an analytic function f on K^n, a Schwartz-Bruhat function Φ on K^n and a character χ of $\mathcal{O}_K^\times$. Moreover, one can formulate analogous definitions over the archimedean local fields $\mathbb{R}$ and $\mathbb{C}$. For a survey, we refer to [**De91b**] or [**Ig00**].

We can write $Z_f^p(s)$ as a power series in p^{-s}, in the following way:

$$Z_f^p(s) = \sum_{m \geq 0} \mu_{\mathrm{Haar}}\{a \in (\mathbb{Z}_p)^n \mid v_p(f(a)) = m\} p^{-ms}$$

where v_p denotes the p-adic valuation on $\mathbb{Z}_p$. Direct computation shows that, if we set $T = p^{-s}$, then $Z_f^p(s)$ is related to the Poincaré series $P(T)$ by the formula

$$P(p^{-n}T) = \frac{p^n(1 - Z_f^p(s))}{1 - T} \tag{3.2}$$

Thus the zeta function $Z_f^p(s)$ contains exactly the same information as the Poincaré series $P(T)$, namely, the values N_m for all $m \geq 0$.

EXAMPLE 3.5. In the set-up of Example 3.2, we have

$$Z_f^p(s) = 1 - \sharp X(\mathbb{F}_p) p^{-(n-1)} \left(\frac{p^s - 1}{p^{s+1} - 1} \right).$$

THEOREM 3.6 (Igusa [**Ig74, Ig75**]). *The p-adic zeta function $Z_f^p(s)$ is rational in the variable p^{-s}. In particular, it admits a meromorphic continuation to $\mathbb{C}$.*

By (3.2), this gives an affirmative answer to Conjecture 3.3:

COROLLARY 3.7. *The Poincaré series $P(T)$ is rational.*

Igusa proved Theorem 3.6 by taking an embedded resolution of singularities for the zero locus of f in the p-adic manifold $\mathbb{Z}_p^n$, and applying the change of variables formula for p-adic integrals to compute the p-adic zeta function locally on the resolution space. This essentially reduces the problem to the case where f is a monomial, in which case one can make explicit computations.

The poles of $P(T)$, or equivalently, $Z_f^p(s)$, contain information about the asymptotic behaviour of N_m as $m \to \infty$. Igusa's proof shows that there exists a finite subset $\mathscr{S}^p$ of $\mathbb{Q}_{<0}$ such that the set of poles of $Z_f^p(s)$ is given by

$$\{\alpha + \frac{2\pi i}{\ln p}\beta \mid \alpha \in \mathscr{S}^p,\ \beta \in \mathbb{Z}\}.$$

By Denef's explicit formula for the p-adic zeta function in [**De91a**], one can associate to every embedded resolution for f over $\mathbb{Q}$ a finite subset $\mathscr{S}$ of $\mathbb{Q}_{<0}$ such that $\mathscr{S}^p \subset \mathscr{S}$ for $p \gg 0$. The set $\mathscr{S}$ is computed from the so-called *numerical* *data* of the resolution (in the notation of [**De91a**], $\mathscr{S}$ is the set of values $-\nu_i/N_i$ with i in T). In general, many of the elements in $\mathscr{S}$ are not poles of $Z_f^p(s)$, due to cancellations in the formula for the zeta function. This phenomenon is related to the *Monodromy Conjecture*.

3.3. Igusa's monodromy conjecture. Example 3.5 suggests that the poles of the zeta function $Z_f^p(s)$ should be related to the singularities of the polynomial f. The relation is made precise by Igusa's Monodromy Conjecture.

CONJECTURE 3.8 (Igusa's Monodromy Conjecture, strong form). *If we denote by $b_f(s)$ the Bernstein-Sato polynomial of f, then for $p \gg 0$, the function $b_f(s)Z_f^p(s)$ is holomorphic at every point of $\mathbb{R}$.*

In other words, the conjecture states that for every pole α of $Z_f^p(s)$, the real part $\Re(\alpha)$ is a root of the Bernstein-Sato polynomial $b_f(s)$, and the order of the pole is at most the multiplicity of the root. The Monodromy Conjecture describes in a precise way how the singularities of f influence the asymptotic behaviour of the values N_m as $m \to \infty$, for $p \gg 0$.

EXAMPLE 3.9. Assume that the closed subscheme of $\mathbb{A}^n_{\mathbb{Q}}$ defined by the equation $f = 0$ is smooth over $\mathbb{Q}$. Then the Bernstein-Sato polynomial $b_f(s)$ is equal to $s+1$. For $p \gg 0$, the closed subscheme of $\mathbb{A}^n_{\mathbb{Z}_p}$ defined by $f = 0$ is smooth, so that the zeta function $Z_f^p(s)$ has a unique real pole at $s = -1$, of order one, by Example 3.5.

Because of Kashiwara and Malgrange's result mentioned in Section 2.3, Conjecture 3.8 implies the following weaker statement.

CONJECTURE 3.10 (Igusa's Monodromy Conjecture, weak form). *For $p \gg 0$, the following holds: if α is a pole of $Z_f^p(s)$, then $\exp(2\pi\Re(\alpha)i)$ is an eigenvalue of the monodromy action on $R^j\psi_{f^{\mathrm{an}}}(\mathbb{C})_x$, for some integer $j \geq 0$ and some point x of $\mathbb{C}^n$ with $f^{\mathrm{an}}(x) = 0$.*

Several special cases of the Monodromy Conjecture have been proven, but the general case remains wide open. For a survey of known results and the relation with archimedean zeta functions over the local fields $\mathbb{R}$ and $\mathbb{C}$, we refer to [**Ni10**].

3.4. The motivic zeta function. In the nineties, Denef and Loeser defined a "motivic" object $Z_f^{\mathrm{mot}}(s)$ that interpolates the p-adic zeta functions for $p \gg 0$. It captures the geometric nature of the p-adic zeta functions and explains their uniform behaviour in p. Denef and Loeser called $Z_f^{\mathrm{mot}}(s)$ the *motivic zeta function* associated to f. They showed that it is rational over an appropriate ring of coefficients, and they conjectured that its poles correspond to roots of the Bernstein-Sato polynomial as in Conjecture 3.8. We will refer to this conjecture as the Motivic Monodromy Conjecture. It will be discussed in more detail in Section 4.6. For a survey on motivic integration and motivic zeta functions, and the precise relation with p-adic zeta functions, we refer to [**Ni10**].

Denef and Loeser defined the motivic zeta function by measuring spaces of the form

$$\{\psi \in (k[[t]]/t^{m+1})^n \mid f(\psi) \equiv 0 \mod t^{m+1}\} \tag{3.3}$$

with $m \geq 0$ and k a field of characteristic zero. In contrast with the p-adic case, the set (3.3) is no longer finite, because $k((t))$ is not a local field. Thus we cannot simply count points in (3.3). Instead, one shows that one can interpret (3.3) as the set of k-points on an algebraic $\mathbb{Q}$-variety, and one uses the *Grothendieck ring of varieties* to measure the size of an algebraic variety (see Section 4.1).

In the following section, we will explain an alternative interpretation of the motivic zeta function, due to J. Sebag and the second author [**NS07**][**Ni09a**], based on Loeser and Sebag's theory of motivic integration on non-archimedean analytic spaces [**LS03**]. This interpretation will eventually lead to the definition of the motivic zeta function of an abelian variety and, more generally, a Calabi-Yau variety over a complete discretely valued field.

4. Motivic integration on rigid varieties and the analytic Milnor fiber

4.1. The Grothendieck ring of varieties. Let F be a field. We denote by $K_0(\mathrm{Var}_F)$ the *Grothendieck ring of varieties over* F. As an abelian group, $K_0(\mathrm{Var}_F)$ is defined by the following presentation:

- *generators:* isomorphism classes $[X]$ of separated F-schemes of finite type X,
- *relations:* if X is a separated F-scheme of finite type and Y is a closed subscheme of X, then
$$[X] = [Y] + [X \setminus Y].$$
These relations are called *scissor relations.*

By the scissor relations, one has $[X] = [X_{\mathrm{red}}]$ for every separated F-scheme of finite type X, where X_{red} denotes the maximal reduced closed subscheme of X. We endow the group $K_0(\mathrm{Var}_F)$ with the unique ring structure such that
$$[X] \cdot [X'] = [X \times_F X']$$
for all F-varieties X and X'. The identity element for the multiplication is the class $[\operatorname{Spec} F]$ of the point. We denote by $\mathbb{L}$ the class $[\mathbb{A}^1_F]$ of the affine line, and by $\mathcal{M}_F$ the localization of $K_0(\mathrm{Var}_F)$ with respect to $\mathbb{L}$.

The scissor relations allow to cut an F-variety into subvarieties. For instance, we have
$$[\mathbb{P}^2_F] = \mathbb{L}^2 + \mathbb{L} + 1$$
in $K_0(\mathrm{Var}_F)$. Since these are the only relations that we impose on the isomorphism classes of F-varieties, taking the class of a variety in the Grothendieck ring should be viewed as the most general way to measure the size of the variety.

For technical reasons, we'll also need to consider the *modified Grothendieck ring of F-varieties* $K_0^{\mathrm{mod}}(\mathrm{Var}_F)$ [**NS11a**, §3.8]. This is the quotient of $K_0(\mathrm{Var}_F)$ by the ideal $\mathcal{I}_F$ generated by elements of the form $[X] - [Y]$ where X and Y are separated F-schemes of finite type such that there exists a finite, surjective, purely inseparable F-morphism $Y \to X$.

If F has characteristic zero, then it is easily seen that $\mathcal{I}_F$ is the zero ideal [**NS11a**, 3.11]. It is not known if $\mathcal{I}_F$ is non-zero if F has positive characteristic. In particular, if F' is a non-trivial finite purely inseparable extension of F, it is not known whether $[\operatorname{Spec} F'] \neq 1$ in $K_0(\mathrm{Var}_F)$. With slight abuse of notation, we'll again denote by $\mathbb{L}$ the class of $\mathbb{A}^1_F$ in $K_0^{\mathrm{mod}}(\mathrm{Var}_F)$. We denote by $\mathcal{M}_F^{\mathrm{mod}}$ the localization of $K_0^{\mathrm{mod}}(\mathrm{Var}_F)$ with respect to $\mathbb{L}$.

For a detailed survey on the Grothendieck ring of varieties and some intriguing open questions, we refer to [**NS11a**].

4.2. Motivic integration on rigid varieties. Let R be a complete discrete valuation ring, with quotient field K and perfect residue field k. We fix an absolute value on K by assigning a value $|\pi| \in]0,1[$ to a uniformizer π of R. If R has equal characteristic, then we set $\mathcal{M}_k^R = \mathcal{M}_k$. If R has mixed characteristic, we set $\mathcal{M}_k^R = \mathcal{M}_k^{\mathrm{mod}}$.

If $\mathfrak{X}$ is a formal R-scheme of finite type, then we denote by $\mathfrak{X}_s = \mathfrak{X} \times_R k$ its special fiber (this is a k-scheme of finite type) and by $\mathfrak{X}_\eta$ its generic fiber (this is a quasi-compact and quasi-separated rigid K-variety; see [**Ra74**] or [**BL93**]).

DEFINITION 4.1. A rigid K-variety X is called *bounded* if there exists a quasi-compact open subvariety U of X such that $U(K') = X(K')$ for all finite unramified extensions K' of K.

DEFINITION 4.2. Let X be a rigid K-variety. A weak Néron model for X is a smooth formal R-scheme of finite type $\mathfrak{X}$, endowed with an open immersion $\mathfrak{X}_\eta \to X$, such that $\mathfrak{X}_\eta(K') = X(K')$ for all finite unramified extensions K' of K.

Note that, if X is separated, then $\mathfrak{X}$ will be separated by [**BL93**, 4.7].

THEOREM 4.3 (Bosch-Schlöter). *A quasi-separated smooth rigid K-variety X is bounded if and only if X admits a weak Néron model.*

PROOF. Since the generic fiber of a formal R-scheme of finite type is quasi-compact, it is clear that the existence of a weak Néron model implies that X is bounded. The converse implication is [**BS95**, 3.3]. □

PROPOSITION 4.4. *Let X be a bounded quasi-separated smooth rigid K-variety, and let U be as in Definition* 4.1. *If $\mathfrak{X}$ is a regular formal R-model of U, then the R-smooth locus* $\mathrm{Sm}(\mathfrak{X})$ *(endowed with the open immersion* $\mathrm{Sm}(\mathfrak{X})_\eta \hookrightarrow \mathfrak{X}_\eta = U \hookrightarrow X$*) is a weak Néron model for X.*

PROOF. If R' is a finite unramified extension of R, with quotient field K', then the specialization map $\mathfrak{X}_\eta \to \mathfrak{X}$ induces a bijection $\mathfrak{X}_\eta(K') = \mathfrak{X}(R')$. Every R'-point on $\mathfrak{X}$ factors through $\mathrm{Sm}(\mathfrak{X})$, by [**NS11b**, 2.37]. □

A weak Néron model is far from unique, in general, as is illustrated by the following example.

EXAMPLE 4.5. Consider the open unit disc

$$B(0,1^-) = \{z \in \mathrm{Sp}\, K\{x\} \mid |x(z)| < 1\}.$$

Let π be a uniformizer in R and K' a finite unramified extension of K. Then all K'-points in $B(0,1^-)$ are contained in the closed disc

$$B(0,|\pi|) = \{z \in \mathrm{Sp}\, K\{x\} \mid |x(z)| \leq |\pi|\,\}$$

because $|\pi|$ is the largest element in the value group $|(K')^*| = |K^*| = |\pi|^{\mathbb{Z}}$ that is strictly smaller than one. It follows that $B(0,1^-)$ is bounded, and that $\mathfrak{X} = \mathrm{Spf}\, R\{u\}$ is a weak Néron model for $B(0,1^-)$ with respect to the open immersion

$$\mathfrak{X}_\eta = \mathrm{Sp}\, K\{u\} \to B(0,1^-)$$

defined by $x \mapsto \pi^{-1}u$. This weak Néron model is not unique: one could also remark that all the unramified points in $B(0,1^-)$ lie on the union of the circle $|x(z)| = |\pi|$ and the closed disc $B(0,|\pi|^2)$. In this way, we get a weak Néron model $\mathfrak{X}'$ that is the disjoint union of Spf $R\{u\}$ and Spf $R\{v, v^{-1}\}$. Note that $\mathfrak{X}'$ can be obtained by blowing up $\mathfrak{X}$ at the origin of $\mathfrak{X}_s$ and taking the R-smooth locus.

The open annulus

$$A(0; 0^+, 1^-) = \{z \in \operatorname{Sp} K\{x\} \,|\, 0 < |x(z)| < 1\}.$$

is not bounded, since K-points can lie arbitrarily close to zero.

Let X be a smooth rigid K-variety of pure dimension m, and assume that X admits a weak Néron model $\mathfrak{X}$. Let ω be a gauge form on X, i.e., a nowhere vanishing differential form of maximal degree. Then for every connected component C of $\mathfrak{X}_s$, we can consider the order $\mathrm{ord}_C\omega$ of ω along C. It is the unique integer γ such that $\pi^{-\gamma}\omega$ extends to a generator of $\Omega^m_{\mathfrak{X}/R}$ at the generic point of C. In geometric terms, it is the order of the zero or minus the order of the pole of the form ω along C.

THEOREM-DEFINITION 4.6 (Loeser-Sebag). *Let X be a separated, smooth and bounded rigid K-variety of pure dimension m, and let $\mathfrak{X}$ be a weak Néron model for X. Let ω be a gauge form on X. Then the expression*

$$\int_X |\omega| := \mathbb{L}^{-m} \sum_{C\in\pi_0(\mathfrak{X}_s)} [C]\mathbb{L}^{-\mathrm{ord}_C\omega} \quad \in \mathcal{M}_k^R \tag{4.1}$$

only depends on (X, ω), and not on the choice of weak Néron model $\mathfrak{X}$. We call it the motivic integral *or* motivic volume *of ω on X.*

PROOF. This is a slight generalization of the result in [**LS03**, 4.3.1]. A proof can be found in [**HN10c**, 2.3]. □

In this way, we can measure the space of unramified points on a bounded separated smooth rigid K-variety X with respect to a motivic measure defined by a gauge form ω on X. Intuitively, one can view the set of unramified points on X as a family of open balls parameterized by the special fiber of a weak Néron model. The gauge form ω renormalizes the volume of each ball in such a way that the total volume of the family is independent of the chosen model. We refer to [**NS11b**] for more background and further results.

4.3. The algebraic case. One can also define the notion of weak Néron model in the algebraic setting. Let K^s be a separable closure of K. Denote by R^{sh} the strict henselization of R in K^s, and by K^{sh} its quotient field. The residue field k^s of R^{sh} is a separable closure of k.

DEFINITION 4.7. Let X be a smooth algebraic K-variety. A weak Néron model is a smooth R-variety $\mathfrak{X}$ endowed with an isomorphism

$$\mathfrak{X} \times_R K \to X$$

such that the natural map

$$\mathfrak{X}(R^{sh}) \to \mathfrak{X}(K^{sh}) = X(K^{sh}) \tag{4.2}$$

is a bijection.

Note that any k^s-point on $\mathfrak{X}_s$ lifts to an R^{sh}-point on $\mathfrak{X}$, because $\mathfrak{X}$ is smooth and R^{sh} is henselian. Thus $\mathfrak{X}_s$ is empty if and only if $X(K^{sh})$ is empty.

REMARK 4.8. Since R^{sh} is the direct limit of all finite unramified extensions of R inside K^s, and $\mathfrak{X}$ is of finite type over R, we have that (4.2) is a bijection if and only if $\mathfrak{X}(R') \to X(K')$ is a bijection for every finite unramified extension R' of R. Here K' denotes the quotient field of R'.

PROPOSITION 4.9. *Let X be a smooth algebraic K-variety. Then X admits a weak Néron model $\mathfrak{X}$ iff the rigid analytification X^{rig} admits a weak Néron model, i.e., iff X^{rig} is bounded. In that case, the formal $\mathfrak{m}$-adic completion of $\mathfrak{X}$ is a weak Néron model for X^{rig}.*

PROOF. This follows from [**Ni11c**, 3.15, 4.3 and 4.9]. □

In particular, if X is proper over K, then X^{rig} is quasi-compact, so that X admits a weak Néron model.

If X is a smooth K-variety with weak Néron model $\mathfrak{X}$, and ω is a gauge form on X, then one can define the order $\mathrm{ord}_C\omega$ of ω along a connected component C of $\mathfrak{X}_s$ exactly as in the formal-rigid case.

DEFINITION 4.10. Let X be a smooth algebraic K-variety of pure dimension such that the rigid analytification X^{rig} of X is bounded. Let ω be a gauge form on X, and denote by ω^{rig} the induced gauge form on X^{rig}. Then we set

$$\int_X |\omega| = \int_{X^{\mathrm{rig}}} |\omega^{\mathrm{rig}}| \quad \in \mathcal{M}_k^R.$$

By Proposition 4.9, the motivic integral of ω on X can also be computed on a weak Néron model of X:

PROPOSITION 4.11. *Let X be a smooth algebraic K-variety of pure dimension m, and assume that X admits a weak Néron model $\mathfrak{X}$. For every gauge form ω on X, we have*

$$\int_X |\omega| = \mathbb{L}^{-m} \sum_{C\in\pi_0(\mathfrak{X}_s)} [C]\mathbb{L}^{-\mathrm{ord}_C\omega}$$

in $\mathcal{M}_k^R$.

4.4. The analytic Milnor fiber. Let k be any field, and set $R = k[[t]]$ and $K = k((t))$. We fix a t-adic absolute value on K by choosing a value $|t| \in]0,1[$.

Let X be a k-variety, endowed with a flat morphism

$$f : X \to \mathbb{A}_k^1 = \operatorname{Spec} k[t].$$

Let x be a closed point of the special fiber $X_s = f^{-1}(0)$ of f. Taking the completion of f at the point x, we obtain a morphism of formal schemes

$$\widehat{f}_x : \operatorname{Spf} \widehat{\mathcal{O}}_{X,x} \to \operatorname{Spf} R. \tag{4.3}$$

We consider the generic fiber $\mathscr{F}_x$ of $\widehat{f}_x$ in the sense of Berthelot [**Bert96**]. This is a separated rigid variety over the non-archimedean field K. It is bounded, by [**NS08**, 5.8]. If f is generically smooth (e.g., if k has characteristic zero and $X \setminus X_s$ is regular) then $\mathscr{F}_x$ is smooth over K.

DEFINITION 4.12. We call $\mathscr{F}_x$ the analytic Milnor fiber of f at the point x.

Note that the construction of the analytic Milnor fiber is a non-archimedean analog of the definition of the classical Milnor fibration in complex singularity theory. For an explicit dictionary, see [**NS11b**, 6.1].

EXAMPLE 4.13. Assume that x is k-rational and that X is smooth over k at x. By [**EGA4a**, 19.6.4], there exists an isomorphism of k-algebras

$$\widehat{\mathcal{O}}_{X,x} \cong k[[x_1,\ldots,x_n]]$$

where n is the dimension of X at x. Viewing f as an element of $k[[x_1,\ldots,x_n]]$, it defines an analytic function on the open unit polydisc

$$B^n(0,1^-) = \{z \in \operatorname{Sp} K\{x_1,\ldots,x_n\} \mid |x_i(z)| < 1 \text{ for all } i\}.$$

The analytic Milnor fiber $\mathscr{F}_x$ is the closed subvariety of $B^n(0,1^-)$ defined by the equation $f = t$. Note that, for every finite unramified extension K' of K, the K'-points on $\mathscr{F}_x$ are all contained in the closed polydisc

$$B^n(0,|\pi|) = \{z \in \operatorname{Sp} K\{x_1,\ldots,x_n\} \mid |x_i(z)| \leq |t| \text{ for all } i\}$$

by the same argument as in Example 4.5. This shows that $\mathscr{F}_x$ is bounded.

The following result follows immediately from [**Berk96**, 1.3 and 3.5].

THEOREM 4.14 (Berkovich). *Assume that k is algebraically closed. Let ℓ be a prime invertible in k, and denote by K^s a separable closure of K. Then for every integer $i \geq 0$, there exists a canonical $G(K^s/K)$-equivariant isomorphism*

$$H^i(\mathscr{F}_x \widehat{\times}_K \widehat{K^s}, \mathbb{Q}_\ell) \cong R^i\psi_f(\mathbb{Q}_\ell)_x. \tag{4.4}$$

In the left hand side of (4.4), we take Berkovich's étale cohomology for K-analytic spaces [**Berk93**]. In the right hand side, $R\psi_f(\mathbb{Q}_\ell)$ denotes the complex of étale ℓ-adic nearby cycles associated to f. If $k = \mathbb{C}$, then by Deligne's comparison theorem [**SGA7b**, Exp.XIV], there exists a canonical isomorphism

$$R^i\psi_f(\mathbb{Q}_\ell)_x \cong R^i\psi_{f^{\mathrm{an}}}(\mathbb{Q}_\ell)_x$$

where $f^{\mathrm{an}} : X^{\mathrm{an}} \to \mathbb{C}$ is the complex analytification of f and $\psi_{f^{\mathrm{an}}}$ is the complex analytic nearby cycles functor. Under this isomorphism, the action of the canonical generator of $G(K^s/K) = \mu(\mathbb{C})$ on $R^i\psi_f(\mathbb{Q}_\ell)_x$ corresponds to the monodromy transformation on $R^i\psi_{f^{\mathrm{an}}}(\mathbb{Q}_\ell)_x$. This means that we can read the local monodromy eigenvalues of f^{an} at x from the étale cohomology of the analytic Milnor fiber.

The following proposition shows that the analytic Milnor fiber $\mathscr{F}_x$ completely determines the formal germ $\widehat{f}_x$ of f at x, if X is normal at x. We denote by $\mathcal{O}(\mathscr{F}_x)$ the K-algebra of analytic functions on $\mathscr{F}_x$.

PROPOSITION 4.15 (de Jong [**dJ95**], Prop. 7.4.1; see also [**Ni09a**], Prop. 8.8). *If X is normal at x, then there exists a natural isomorphism of R-algebras*

$$\widehat{\mathcal{O}}_{X,x} \cong \{h \in \mathcal{O}(\mathscr{F}_x) \mid |h(z)| \leq 1 \textit{ for all } z \in \mathscr{F}_x\}.$$

There is an interesting relation between the Berkovich topology on $\mathscr{F}_x$ and the limit mixed Hodge structure on the nearby cohomology of f at x; see [**Ni11b**].

4.5. The Gelfand-Leray form. We keep the notations of Section 4.4, and we assume that k has characteristic zero and that X is smooth over k at the point x. Replacing X by an open neighbourhood of x, we can assume that f is smooth on $X^o = X \setminus X_s$, of relative dimension m, and that $\Omega^{m+1}_{X^o/k}$ is free, i.e., that X admits a gauge form ϕ. Then the exact complex

$$\Omega^{m-1}_{X^o/k} \xrightarrow{\wedge df} \Omega^m_{X^o/k} \xrightarrow{\wedge df} \Omega^{m+1}_{X^o/k} \longrightarrow 0$$

induces an isomorphism of sheaves

$$\Omega^m_{X^o/\mathbb{A}^1_k} \to \Omega^{m+1}_{X^o/k}$$

and thus an isomorphism of $\mathcal{O}(X^o)$-modules

$$\Omega^m_{X^o/\mathbb{A}^1_k}(X^o) \to \Omega^{m+1}_{X^o/k}(X^o).$$

The inverse image in $\Omega^m_{X^o/\mathbb{A}^1_k}(X^o)$ of the restriction of ϕ to X^o is called the *Gelfand-Leray form* on X^o associated to f and ϕ. It is denoted by ϕ/df. It induces a gauge form on the analytic Milnor fiber $\mathscr{F}_x$, since $\mathscr{F}_x$ is an open rigid sub-K-variety of the rigid analytification of $X \times_{k[t]} K$ [**Bert96**, 0.2.7 and 0.3.5]. We denote this gauge form again by ϕ/df. It can be constructed intrinsically on $\mathscr{F}_x$; see Proposition 4.15 and [**Ni09a**, § 7.3].

4.6. The motivic zeta function and the motivic monodromy conjecture. We keep the notations of Section 4.4. We assume that k has characteristic zero, and that X is smooth at x, of dimension n. For simplicity, we suppose that x is k-rational. Let ϕ be a gauge form on some open neighbourhood of x in X, and consider the Gelfand-Leray form ϕ/df on $\mathscr{F}_x$ constructed in Section 4.5. We set $\omega = t \cdot \phi/df$. This is a gauge form on $\mathscr{F}_x$. For every integer $d > 0$, we set $K(d) = k((\sqrt[d]{t}))$. This is a totally ramified extension of K of degree d. We set $\mathscr{F}_x(d) = \mathscr{F}_x \times_K K(d)$. For every differential form ω' on $\mathscr{F}_x$, we denote by $\omega'(d)$ the pullback of ω' to $\mathscr{F}_x(d)$.

We denote by $Z_{f,x}(T) \in \mathcal{M}_k[[T]]$ Denef and Loeser's local motivic zeta function of f at the point x (obtained from the zeta function in [**DL01**, 3.2.1] by taking the fiber at x and forgetting the μ-action). The reader who is unfamiliar with Denef and Loeser's definition may take the following theorem as a definition.

THEOREM 4.16 (Nicaise-Sebag). *We have*

$$Z_{f,x}(T) = \mathbb{L}^{n-1} \sum_{d>0} \left(\int_{\mathscr{F}_x(d)} |\omega(d)| \right) T^d \tag{4.5}$$

in $\mathcal{M}_k[[T]]$.

PROOF. Note that, for every d in $\mathbb{N}$, we have

$$\int_{\mathscr{F}_x(d)} |\omega(d)| = \mathbb{L}^{-d} \int_{\mathscr{F}_x(d)} |(\phi/df)(d)|$$

in $\mathcal{M}_k$, since t has valuation d in $K(d)$. Thus the theorem is a reformulation of [**Ni09a**, 9.7], which is a consequence of the comparison theorem in [**NS07**, 9.10]. □

The proof of Theorem 4.16 is based on an explicit construction of weak Néron models for the rigid varieties $\mathscr{F}_x(d)$, starting from an embedded resolution of singularities for (X, X_s). In this way, one obtains an explicit formula for the right hand side in (4.5) in terms of such a resolution, and one can compare this expression to the formula for $Z_{f,x}(T)$ obtained by Denef and Loeser [**DL01**, 3.3.1]. This formula implies in particular that $Z_{f,x}[T]$ is contained in the subring

$$\mathcal{M}_k\left[T, \frac{1}{1 - \mathbb{L}^a T^b}\right]_{a \in \mathbb{Z}_{<0},\, b \in \mathbb{Z}_{>0}}$$

of $\mathcal{M}_k[[T]]$.

If the residue field k_x of x is not k, one can adapt the construction as follows. Since k has characteristic zero, k_x is a separable extension of k so that the morphism $\widehat{f}_x$ from (4.3) factors through a morphism

$$\widehat{\mathcal{O}}_{X,x} \to \operatorname{Spf} k_x[[t]]$$

by [**EGA4a**, 19.6.2]. In this way, we can view $\mathscr{F}_x$ as a rigid variety over $K_x = k_x((t))$. Since K_x is separable over K, the natural morphism $\Omega^i_{\mathscr{F}_x/K_x} \to \Omega^i_{\mathscr{F}_x/K}$ is an isomorphism for all i, so that we can consider the Gelfand-Leray form ϕ/df as an element of $\Omega^m_{\mathscr{F}_x/K_x}$. Then the equality (4.5) holds in $\mathcal{M}_{k_x}[[T]]$.

CONJECTURE 4.17 (Motivic Monodromy Conjecture). *Assume that k is a subfield of $\mathbb{C}$. There exists a finite subset $\mathscr{S}$ of $\mathbb{Z}_{<0} \times \mathbb{Z}_{>0}$ such that $Z_{f,x}(T)$ belongs to the subring*

$$\mathcal{M}_{k_x}\left[T, \frac{1}{1-\mathbb{L}^a T^b}\right]_{(a,b)\in\mathscr{S}}$$

of $\mathcal{M}_{k_x}[[T]]$, and such that for every couple (a,b) in $\mathscr{S}$, the quotient a/b is a root of the Bernstein-Sato polynomial $b_f(s)$ of f. In particular, there exists a point y of $X(\mathbb{C})$ such that $f(y) = 0$ and such that $\exp(2\pi i a/b)$ is a local monodromy eigenvalue of f at the point y.

REMARK 4.18. One can drop the condition that k is a subfield of $\mathbb{C}$ by using ℓ-adic nearby cycles to define the notion of local monodromy eigenvalue. This does not yield a more general conjecture, since by the Lefschetz principle, one can always reduce to the case where k is a subfield of $\mathbb{C}$.

One needs to be careful when speaking about poles of the zeta function, since $\mathcal{M}_{k_x}$ is not a domain. A precise definition is given in [**RV03**]. The formulation in Conjecture 4.17 implies that, for any reasonable definition of pole in this context, the poles of $Z_{f,x}(\mathbb{L}^{-s})$ are of the form a/b, with $(a,b) \in \mathscr{S}$. With some additional work, one can define the order of a pole [**RV03**], and conjecture that the order of a pole of $Z_{f,x}(\mathbb{L}^{-s})$ is at most the multiplicity of the corresponding root of $b_f(s)$.

4.7. Relation with the p-adic monodromy conjecture. Let us explain the precise relation between Conjectures 4.17 and 3.8. In [**DL01**, 3.2.1], Denef and Loeser define the motivic zeta function $Z_f(T)$ (there denoted by $Z(T)$) associated to the morphism f. It carries more structure than we've considered so far: it is a formal power series with coefficients in the *equivariant* Grothendieck ring of X_s-varieties $\mathcal{M}^{\mu}_{X_s}$. The elements of this ring are virtual classes of X_s-varieties that carry an action of the profinite group scheme μ of roots of unity. The X_s-structure allows to consider various "motivic Schwartz-Bruhat functions" and the μ-action allows to twist the motivic zeta function by "motivic characters", like in the p-adic case; see [**DL98**]. The motivic zeta function that we've alluded to in Section 3.4 corresponds to the trivial character; it is the "naïve" motivic zeta function from [**DL01**, 3.2.1]. If $k = \mathbb{Q}$ and $X = \mathbb{A}^n_k$, then for $p \gg 0$, we can specialize the naïve motivic zeta function of f to the p-adic one, by counting rational points on the reductions of the coefficients modulo p. This is explained in [**Ni10**, §5.3].

The local zeta function $Z_{f,x}(T)$ that we've considered above is obtained from $Z_f(T)$ by applying the morphism

$$\mathcal{M}^{\mu}_{X_s} \to \mathcal{M}_{k_x}$$

(base change to x and forgetting the μ-structure) to the coefficients of $Z_f(T)$. In an unpublished manuscript, the second author has shown that it is possible to recover the μ-structure on $Z_{f,x}(T)$ by considering the Galois action on the extensions $K(d)$ of K.

One can formulate Conjecture 4.17 for $Z_f(T)$ instead of $Z_{f,x}(T)$, as follows:

CONJECTURE 4.19 (Motivic Monodromy Conjecture II). *Assume that k is a subfield of $\mathbb{C}$. There exists a finite subset $\mathscr{S}$ of $\mathbb{Z}_{<0} \times \mathbb{Z}_{>0}$ such that $Z_f(T)$ belongs to the subring*

$$\mathcal{M}^{\mu}_{X_s}\left[T, \frac{1}{1-\mathbb{L}^a T^b}\right]_{(a,b)\in\mathscr{S}}$$

of $\mathcal{M}^{\mu}_{X_s}[[T]]$, and such that for every couple (a,b) in $\mathscr{S}$, the quotient a/b is a root of the Bernstein-Sato polynomial $b_f(s)$ of f. In particular, there exists a point y of $X(\mathbb{C})$ such that $f(y)=0$ and such that $\exp(2\pi i a/b)$ is a local monodromy eigenvalue of f at the point y.

This conjecture implies

(1) Conjecture 4.17, since we can specialize $Z_f(T)$ to $Z_{f,x}(T)$ by taking fibers at x and forgetting the μ-structure,
(2) Conjecture 3.8, because $Z_f(T)$ can be specialized to the p-adic zeta function,
(3) more generally, the p-adic monodromy conjecture for zeta functions that are twisted by characters [**DL98**, §2.4].

Various weaker reformulations of Conjecture 4.19 have appeared in the literature (e.g. in [**DL98**, §2.4]). The formulation we use seems to be part of general folklore. We attribute it to Denef and Loeser.

5. The motivic zeta function of an abelian variety

5.1. Some notation. Let R be a complete discrete valuation ring with maximal ideal $\mathfrak{m}$, fraction field K and algebraically closed residue field k. We denote by p the characteristic exponent of k, and by $\mathbb{N}'$ the set of strictly positive integers that are prime to p. We fix a prime $\ell \neq p$ and a separable closure K^s of K. The Galois group $G(K^s/K)$ is called the *inertia group* of K.

A finite extension of K is called *tame* if its degree is prime to p. For every d in $\mathbb{N}'$, the field K admits a unique degree d extension $K(d)$ in K^s. It is obtained by joining a d-th root of a uniformizer to K. The extension $K(d)/K$ is Galois, with Galois group $\mu_d(k)$.

The union of the fields $K(d)$ is a subfield of K^s, called the *tame closure* K^t of K. The Galois group $G(K^t/K)$ is called the *tame inertia group* of K. It is canonically isomorphic to the procyclic group

$$\mu'(k) = \varprojlim_{d\in\mathbb{N}'} \mu_d(k)$$

where the elements in $\mathbb{N}'$ are ordered by divisibility and the transition morphisms in the projective system are given by

$$\mu_{de}(k) \to \mu_d(k) : x \mapsto x^e$$

for all d,e in $\mathbb{N}'$. We call every topological generator of $G(K^t/K)$ a *tame monodromy operator*. The Galois group $P = G(K^s/K^t)$ is a pro-p-group which is called the *wild inertia subgroup* of $G(K^s/K)$. We have a short exact sequence

$$1 \to P \to G(K^s/K) \to G(K^t/K) \to 1.$$

5.2. Néron models and semi-abelian reduction. Let A be an abelian K-variety of dimension g. It is not always possible to extend A to an abelian scheme over R. However, there exists a canonical way to extend A to a smooth commutative group scheme over R, the so-called *Néron model* of A.

DEFINITION 5.1. A Néron model of A is a smooth R-scheme of finite type $\mathcal{A}$, endowed with an isomorphism

$$\mathcal{A} \times_R K \to A,$$

such that the natural map

$$\mathrm{Hom}_R(T, \mathcal{A}) \to \mathrm{Hom}_K(T \times_R K, A) \tag{5.1}$$

is a bijection for every smooth R-scheme T.

Thus $\mathcal{A}$ is the minimal smooth R-model of A. The existence of a Néron model was first proved by A. Néron [**Ne64**]. For a modern scheme-theoretic treatment of the theory and an accessible proof of Néron's theorem, we refer to [**BLR90**]. The universal property of the Néron model implies that the Néron model $\mathcal{A}$ is unique up to unique isomorphism, and that the K-group structure on A extends uniquely to a commutative R-group structure on $\mathcal{A}$. Taking for T the spectrum of a finite unramified extension of R, we see that $\mathcal{A}$ is also a weak Néron model for A.

The special fiber $\mathcal{A}_s := \mathcal{A} \times_R k$ is a smooth commutative algebraic k-group. We denote by $\mathcal{A}_s^o$ the identity component of $\mathcal{A}_s$, i.e., the connected component containing the identity point for the group structure. The quotient $\Phi_A := \mathcal{A}_s/\mathcal{A}_s^o$ is called the *component group*. It is a finite étale group scheme over k whose group of k-points corresponds bijectively to the the set of connected components of $\mathcal{A}_s$. Since k is assumed to be algebraically closed, we will not distinguish between the group scheme Φ_A and the abstract group $\Phi_A(k)$.

The identity component $\mathcal{A}_s^o$ fits into a canonical short exact sequence of algebraic k-groups, the *Chevalley decomposition*,

$$0 \to T \times_k U \to \mathcal{A}_s^o \to B \to 0 \tag{5.2}$$

where B is an abelian variety, T is a torus and U is a unipotent group commonly referred to as the *unipotent radical* of $\mathcal{A}_s^o$. We call the dimension of T the *toric rank* of A, and the dimension of U the *unipotent rank* of A.

DEFINITION 5.2. We say that A has semi-abelian reduction if the unipotent rank of $\mathcal{A}_s^o$ is zero.

A celebrated result by A. Grothendieck, the Semi-Stable Reduction Theorem for abelian varieties [**SGA7a**, IX.3.6], asserts that there exists a finite separable extension K'/K such that $A \times_K K'$ has semi-abelian reduction over the integral closure R' of R in K'. Inside our fixed separable closure K^s of K, there exists a unique minimal extension L with this property, and it is Galois over K. By [**SGA7a**, IX.3.8], L is the fixed field of the subgroup of $G(K^s/K)$ consisting of the elements that act unipotently on the ℓ-adic Tate module $T_\ell A$ of A. If L is a tame extension of K then we say that A is *tamely ramified*. Since the P-action on $T_\ell A$ factors through a finite quotient of P [**LO85**, p.3], A is tamely ramified if and only if P acts trivially on $T_\ell A$. In that case, P acts trivially on the ℓ-adic cohomology of A, and the natural morphism

$$H^i(A \times_K K^t, \mathbb{Q}_\ell) \to H^i(A \times_K K^s, \mathbb{Q}_\ell)$$

is an isomorphism for every i in $\mathbb{N}$.

5.3. The base change conductor and the potential toric rank. In [**Ch00**], Chai introduced an invariant that measures how far the abelian K-variety A is from having semi-abelian reduction. He called it the *base change conductor* of A and denoted it by $c(A)$. It is defined as follows. Let K'/K be a finite separable extension such that A acquires semi-abelian reduction over K', and let $\mathcal{A}'$ be the Néron model of $A \times_K K'$. By the universal property of the Néron model $\mathcal{A}'$, there is a unique morphism

$$h : \mathcal{A} \times_R R' \to \mathcal{A}' \tag{5.3}$$

that extends the canonical isomorphism between the generic fibers. The induced map

$$\mathrm{Lie}(h) : \mathrm{Lie}(\mathcal{A} \times_R R') \to \mathrm{Lie}(\mathcal{A}')$$

is an injective homomorphism of free R'-modules of rank g, so that $\mathrm{coker}(\mathrm{Lie}(h))$ is an R'-module of finite length. The rational number

$$c(A) := [K' : K]^{-1} \cdot \mathrm{length}_{R'}(\mathrm{coker}(\mathrm{Lie}(h)))$$

is independent of the choice of K'.

The importance of the Semi-Stable Reduction Theorem lies in the fact that, if A has semi-abelian reduction, then h is an open immersion, so that it induces an isomorphism between the identity components of $\mathcal{A} \times_R R'$ and $\mathcal{A}'$ [**SGA7a**, IX.3.3] (the number of connected components of $\mathcal{A}$ might still change, though; this will be discussed below). Thus $c(A)$ vanishes if A has semi-abelian reduction. Conversely, if $c(A) = 0$ then h must be étale, and the fact that h restricts to an isomorphism between the generic fibers then implies that h is an open immersion. Thus $c(A)$ is zero if and only if A has semi-abelian reduction.

Another invariant that will be important for us is the toric rank of $A \times_K K'$. We call this value the *potential toric rank* of A, and denote it by $t_{\mathrm{pot}}(A)$. Again, it is independent of the choice of K'. Moreover, it is the maximum of the toric ranks of the abelian varieties $A \times_K K''$ as K'' ranges over all the finite separable extensions of K. The potential toric rank is a measure for the potential degree of degeneration of A over the closed point of $\mathrm{Spec}\, R$; it vanishes if and only if, after a finite separable extension of the base field K, the abelian variety A extends to an abelian scheme over R. In this case, we say that A has *potential good reduction.* If $t_{pot}(A)$ has the largest possible value, namely, the dimension of A, then we say that A has *potential purely multiplicative reduction.* Thus A has potential purely multiplicative reduction if and only if the identity component of $\mathcal{A}'_s$ is a torus.

5.4. The motivic zeta function of an abelian variety. For every d in $\mathbb{N}'$, we set $A(d) = A \times_K K(d)$ and we denote by $\mathcal{A}(d)$ the Néron model of $A(d)$. For every gauge form ω on A, we denote by $\omega(d)$ its pullback to $A(d)$. We define the order $\mathrm{ord}_{\mathcal{A}(d)^o_s}\omega(d)$ of $\omega(d)$ along $\mathcal{A}(d)^o_s$ as in Section 4.2.

DEFINITION 5.3. A gauge form ω on A is *distinguished* if it is the restriction to A of a generator of the free rank one $\mathcal{O}_{\mathcal{A}}$-module $\Omega^g_{\mathcal{A}/R}$.

It is clear from the definition that a distinguished gauge form always exists, and that it is unique up to multiplication with a unit in R. Note that a gauge form ω on A is distinguished if and only if $\mathrm{ord}_{\mathcal{A}_s^o}\omega = 0$.

In general, a distinguished gauge form on A does not remain distinguished under base change to a finite tame extension $K(d)$ of K. To measure the defect, we introduce the following definition.

DEFINITION 5.4. Let A be an abelian K-variety, and let ω be a distinguished gauge form on A. The order function of A is the function

$$\mathrm{ord}_A : \mathbb{N}' \to \mathbb{N} : d \mapsto -\mathrm{ord}_{\mathcal{A}(d)_s^o}\omega(d).$$

This definition does not depend on the choice of distinguished gauge form, since multiplying ω with a unit in R does not affect the order of $\omega(d)$ along $\mathcal{A}(d)_s^o$. The fact that ord_A takes its values in $\mathbb{N}$ follows easily from the existence of the morphism h in (5.3), for arbitrary finite extensions K' of K.

DEFINITION 5.5. Let A be an abelian K-variety, and let ω be a distinguished gauge form on A. We define the motivic zeta function $Z_A(T)$ of A as

$$Z_A(T) = \sum_{d\in\mathbb{N}'} [\mathcal{A}(d)_s]\mathbb{L}^{\mathrm{ord}_A(d)}T^d \in \mathcal{M}_k[[T]].$$

The following proposition gives an interpretation of $Z_A(T)$ in terms of the volumes of the "motivic Haar measures" $|\omega(d)|$.

PROPOSITION 5.6. *Let A be an abelian K-variety of dimension g, and let ω be a distinguished gauge form on A. The image of $Z_A(T)$ in the quotient ring $\mathcal{M}_k^R[[T]]$ of $\mathcal{M}_k[[T]]$ is equal to*

$$\mathbb{L}^g \sum_{d\in\mathbb{N}'} \left(\int_{A(d)} |\omega(d)| \right) T^d.$$

PROOF. We've already observed that every Néron model is also a weak Néron model. The gauge form ω is translation-invariant, so that

$$\mathrm{ord}_C\omega(d) = \mathrm{ord}_{\mathcal{A}(d)_s^o}\omega(d)$$

for every connected component C of $\mathcal{A}(d)_s$. Now the result follows immediately from the definition of the motivic integral, and the fact that

$$[\mathcal{A}(d)_s] = \sum_{C\in\pi_0(\mathcal{A}(d)_s)} [C]$$

by the scissor relations in $\mathcal{M}_k$. □

5.5. The monodromy conjecture. Now we come to the formulation of the main result of [**HN10c**], which is a variant of Conjecture 3.10 for abelian varieties. For every integer $d \geq 0$, we denote by $\Phi_d(t) \in \mathbb{Z}[t]$ the cyclotomic polynomial whose roots are the primitive d-th roots of unity. For every rational number q, we denote by $\tau(q)$ its order in the group $\mathbb{Q}/\mathbb{Z}$.

THEOREM 5.7 (Monodromy conjecture for abelian varieties). *Let A be a tamely ramified abelian variety of dimension g, and let σ be a tame monodromy operator in $G(K^t/K)$.*

(1) *The motivic zeta function* $Z_A(T)$ *belongs to the subring*

$$\mathcal{M}_k\left[T, \frac{1}{1-\mathbb{L}^a T^b}\right]_{(a,b)\in\mathbb{N}\times\mathbb{Z}_{>0}, a/b=c(A)}$$

of $\mathcal{M}_k[[T]]$. *The zeta function* $Z_A(\mathbb{L}^{-s})$ *has a unique pole at* $s=c(A)$, *of order* $t_{pot}(A)+1$.

(2) *The cyclotomic polynomial* $\Phi_{\tau(c(A))}(t)$ *divides the characteristic polynomial of the tame monodromy operator* σ *on* $H^g(A\times_K K^t, \mathbb{Q}_\ell)$. *Thus for every embedding* $\mathbb{Q}_\ell \hookrightarrow \mathbb{C}$, *the value* $\exp(2\pi c(A)i)$ *is an eigenvalue of* σ *on* $H^g(A\times_K K^t, \mathbb{Q}_\ell)$.

We'll briefly sketch the main ideas of the proof. The first step is to refine the expression for the motivic zeta function, as follows. For every d in $\mathbb{N}'$, we denote by $t_A(d)$ and $u_A(d)$ the toric, resp. unipotent rank of $A(d)$, and we denote by $B_A(d)$ the abelian quotient in the Chevalley decomposition of $\mathcal{A}(d)^o_s$. Moreover, we denote by $\phi_A(d)=|\Phi_{A(d)}|$ the number of connected components of the k-group $\mathcal{A}(d)_s$.

PROPOSITION 5.8. *For every abelian* K*-variety* A, *we have*

$$\begin{aligned} Z_A(T) &= \sum_{d\in\mathbb{N}'} [\mathcal{A}(d)_s]\mathbb{L}^{ord_A(d)}T^d \\ &= \sum_{d\in\mathbb{N}'} \left(\phi_A(d)\cdot(\mathbb{L}-1)^{t_A(d)}\cdot\mathbb{L}^{u_A(d)+ord_A(d)}\cdot[B_A(d)]T^d\right) \end{aligned}$$

in $\mathcal{M}_k[[T]]$.

PROOF. The first equality is simply the definition of the zeta function. For every $d\in\mathbb{N}'$, the connected components of $\mathcal{A}(d)_s$ are all isomorphic to $\mathcal{A}(d)^o_s$, because k is algebraically closed. Thus by the scissor relations in the Grothendieck ring, one has

$$[\mathcal{A}(d)_s] = \phi_A(d)\cdot[\mathcal{A}(d)^o_s].$$

Now consider the Chevalley decomposition

$$0\to T_A(d)\times_k U_A(d)\to \mathcal{A}(d)^o_s \to B_A(d)\to 0$$

of $\mathcal{A}(d)^o_s$. The torus $T_A(d)$ is isomorphic to $\mathbb{G}_{m,k}^{t_A(d)}$, and $U_A(d)$ is a successive extension of additive groups $\mathbb{G}_{a,k}$. It follows that $\mathcal{A}(d)^o_s\to B_A(d)$ is a Zariski-locally trivial fibration. Moreover, as a k-variety, $U_A(d)$ is isomorphic to $\mathbb{A}_k^{u_A(d)}$. Thus

$$[\mathcal{A}(d)^o_s] = (\mathbb{L}-1)^{t_A(d)}\cdot\mathbb{L}^{u_A(d)}\cdot[B_A(d)]$$

in $\mathcal{M}_k$. □

Thus the study of $Z_A(T)$ can be split up into the following subproblems:

(1) How do $t_A(d)$, $u_A(d)$ and $B_A(d)$ vary with d?
(2) What is the shape of the order function ord_A?
(3) How does $\phi_A(d)$ vary with d?

Our main tool in the study of (1) was a theorem due to B. Edixhoven [**Ed92**], which says that for every $d\in\mathbb{N}'$, the Néron model $\mathcal{A}$ is canonically isomorphic to the fixed locus of the $G(K(d)/K)$-action on the Weil restriction of $\mathcal{A}(d)$ to R. This result enabled us to show that $t_A(d)$, $u_A(d)$ and $[B_A(d)]$ only depend on the residue class of d modulo e, where e is the degree of the minimal extension of K where A acquires semi-abelian reduction. In the same paper, Edixhoven constructs a filtration on $\mathcal{A}_s$ by closed algebraic subgroups, indexed by $\mathbb{Q}\cap[0,1]$. This filtration measures the behaviour of the Néron model of A under tamely ramified base change. The *jumps* of A are the indices where the subgroup changes. Edixhoven related these jumps to the Galois action of $G(K(e)/K) = \mu_e(k)$ on the k-vector space $\mathrm{Lie}(\mathcal{A}(e)_s^o)$. We deduced from Edixhoven's theory that $c(A)$ is the sum of the jumps of A and that on every residue class of $\mathbb{N}'$ modulo e, the function ord_A is affine with slope $c(A)$. The function ord_A is responsable for the pole of $Z_A(\mathbb{L}^{-s})$ at $s = c(A)$.

To control the behaviour of $\phi_A(d)$ turned out to be rather involved, we treated this in the separate paper [**HN10a**]. There, we used rigid uniformization of A in the sense of [**BX96**] to reduce to the case of tori and abelian varieties with potential good reduction, where more explicit methods could be used to describe the change in the component groups under ramified base extensions. In this way, we obtained the following result.

THEOREM 5.9. *Let A be an abelian K-variety, and let e be the degree of the minimal extension of K where A acquires semi-abelian reduction. Denote by $t(A)$ the toric rank of A and by $\phi(A)$ the number of connected components of the Néron model of A. Assume either that A is tamely ramified or that A has potential purely multiplicative reduction. Then for every element d of $\mathbb{N}'$ that is prime to e, we have*

$$\phi_A(d) = \phi(A) \cdot d^{t(A)}.$$

This result was sufficient for our purposes. The behaviour of $\phi_A(d)$ is responsible for the order $t_{\mathrm{pot}}(A) + 1$ of the unique pole of the zeta function.

It remains to prove the relation between the base change conductor and the tame monodromy action on $H^g(A \times_K K^t, \mathbb{Q}_\ell)$. Here we again used Edixhoven's theory and we showed how to compute the eigenvalues of σ on $H^1(A \times_K K^t, \mathbb{Q}_\ell)$ from the Galois action of $\mu_e(k)$ on $\mathrm{Lie}(\mathcal{A}(e)_s^o)$.

5.6. Strong version of the monodromy conjecture. It is natural to ask for an analog of Conjecture 3.8 for abelian varieties. There is no good notion of Bernstein polynomials in this setting. However, the multiplicities of the roots of the Bernstein polynomial of a complex hypersurface singularity are closely related to the sizes of the Jordan blocks of the monodromy action on the cohomology of the Milnor fiber, so one may ask if the order of the pole of $Z_A(T)$ is related to Jordan blocks of the tame monodromy action on $H^g(A \times_K K^t, \mathbb{Q}_\ell)$. We've shown in [**HN10d**] that this is indeed the case.

THEOREM 5.10. *Let A be a tamely ramified abelian K-variety of dimension g. For every tame monodromy operator σ in $G(K^t/K)$ and every embedding of $\mathbb{Q}_\ell$ in $\mathbb{C}$, the value $\alpha = \exp(2\pi c(A)i)$ is an eigenvalue of σ on $H^g(A \times_K K^t, \mathbb{Q}_\ell)$. Each Jordan block of σ on $H^g(A \times_K K^t, \mathbb{Q}_\ell)$ has size at most $t_{pot}(A) + 1$, and σ has a Jordan block with eigenvalue α on $H^g(A \times_K K^t, \mathbb{Q}_\ell)$ with size $t_{pot}(A) + 1$.*

In the case $K = \mathbb{C}((t))$, we also gave in [**HN10d**] a Hodge-theoretic interpretation of the jumps in Edixhoven's filtration, in terms of the limit mixed Hodge structure associated to A.

5.7. Cohomological interpretation. The motivic zeta function of an abelian K-variety admits a cohomological interpretation, by [**Ni09b**]. We consider the unique ring morphism

$$\chi : \mathcal{M}_k \to \mathbb{Z}$$

that sends the class of a k-variety X to the ℓ-adic Euler characteristic

$$\chi(X) = \sum_{i\geq 0}(-1)^i \dim H^i_c(X, \mathbb{Q}_\ell).$$

Since χ sends $\mathbb{L}$ to 1, the image of $Z_A(T)$ under the morphism $\mathcal{M}_k[[T]] \to \mathbb{Z}[[T]]$ induced by χ is equal to

$$\chi(Z_A(T)) = \sum_{d\in\mathbb{N}'} \chi(\mathcal{A}(d)_s)T^d.$$

THEOREM 5.11. *Let A be a tamely ramified abelian K-variety. For every d in $\mathbb{N}'$, we have*

$$\sum_{i\geq 0}(-1)^i \mathrm{Trace}(\sigma^d \mid H^i(A\times_K K^t, \mathbb{Q}_\ell)) = \chi(\mathcal{A}(d)_s).$$

This value equals $\phi_A(d)$ if $A(d)$ has purely additive reduction (i.e, if $\mathcal{A}(d)^o_s$ is unipotent) and it equals zero in all other cases.

This result can be seen as a particular case of a more general theory that expresses a certain motivic measure for the number of rational points on a K-variety X in terms of the Galois action on the ℓ-adic cohomology of X; see [**Ni09a, Ni11c, Ni11d**]. For a similar formula for the zeta function of a hypersurface singularity, see [**DL02**, 1.1], [**NS07**, 9.12] and [**Ni09a**, 9.9].

6. Degenerations of Calabi-Yau varieties

We keep the notations from Section 5.1. To simplify the presentation, we assume that k has characteristic zero. Part of the theory below can be developed also in the case where k has positive characteristic; see [**HN10b**]. In particular, the definition of the zeta function remains valid.

6.1. Motivic zeta functions of Calabi-Yau varieties.

DEFINITION 6.1. A Calabi-Yau variety over a field F is a smooth, proper, geometrically connected F-variety with trivial canonical sheaf.

For instance, every abelian variety is Calabi-Yau. By definition, every Calabi-Yau variety admits a gauge form. In the definition of a Calabi-Yau variety X, one often includes the additional condition that $h^{i,0}(X)$ vanishes for $0 < i < \dim X$. We do not impose this condition.

Let X be a Calabi-Yau variety over K. We will now define the motivic zeta function $Z_X(T)$ of X, in a way that generalizes our construction for abelian varieties. There is no canonical weak Néron model as in the abelian case, but we can generalize the expression for the zeta function in Proposition 5.6 in terms of the motivic volume of an appropriate gauge form on X. We first show how the notion of distinguished gauge form extends to Calabi-Yau varieties.

PROPOSITION 6.2. *Let X be a Calabi-Yau variety over K, and let ω be a gauge form on X. Then for every weak Néron model $\mathfrak{X}$ of X, the value*

$$\mathrm{ord}(X,\omega) \quad := \quad \min\{ord_C(\omega) \mid C \in \pi_0(\mathfrak{X}_s)\} \quad \in \mathbb{Z}\cup\{-\infty\}$$

only depends on the pair (X, ω), and not on $\mathfrak{X}$. By convention, we set $\min\emptyset = -\infty$.

PROOF. Since every connected component of $\mathfrak{X}_s$ has the same dimension as X, the value $\mathrm{ord}(X,\omega)$ is precisely minus the *virtual dimension* of the motivic integral

$$\int_X |\omega| = \mathbb{L}^{-\dim(X)} \sum_{C\in\pi_0(\mathcal{X}_s)} [C]\mathbb{L}^{-\mathrm{ord}_C\omega} \in \mathcal{M}_k.$$

The virtual dimension of an element α in $\mathcal{M}_k$ can be defined, for instance, as half of the degree of the *Poincaré polynomial* of α [**Ni11c**, §8]. □

Note that $\mathrm{ord}(X,\omega) = -\infty$ if and only if $\mathfrak{X}_s$ is empty, i.e., if and only if $X(K)$ is empty.

DEFINITION 6.3. Let X be a Calabi-Yau variety over K. A distinguished gauge form on X is a gauge form ω such that $\mathrm{ord}(X,\omega)=0$.

Thus, a distinguished gauge form on X extends to a relative differential form on every weak Néron model, in a "minimal" way. It is clear that X admits a distinguished gauge form iff X has a K-rational point, and that a distinguished gauge form is unique up to multiplication with a unit in R.

DEFINITION 6.4. Let X be a Calabi-Yau variety over K, and assume that X has a K-rational point. Let ω be a distinguished gauge form on X. We define the motivic zeta function $Z_X(T)$ of X by

$$Z_X(T) = \mathbb{L}^{\dim(X)}\sum_{d\in\mathbb{N}}\left(\int_{X(d)}|\omega(d)|\right)T^d \quad \in \mathcal{M}_k[[T]].$$

This definition only depends on X, and not on the choice of distinguished gauge form ω, since multiplying ω with a unit in R does not affect the motivic integral of ω on X. It follows from Proposition 5.6 that, when X is an abelian variety, Definition 6.4 is equivalent to Definition 5.5.

By embedded resolution of singularities, we can find an *sncd*-model $\mathcal{X}$ for X, i.e., a regular proper R-model such that $\mathcal{X}_s = \sum_{i\in I} N_i E_i$ is a divisor with strict normal crossings. For every $i\in I$, we define the order $\mu_i = \mathrm{ord}_{E_i}\omega$ of ω along E_i as in [**NS07**, 6.8]. These values do not depend on the choice of distinguished gauge form ω. For every non-empty subset J of I, we set

$$\begin{aligned} E_J &= \cap_{j\in J}E_j, \\ E_J^o &= E_J\setminus(\cup_{i\in I\setminus J}E_i). \end{aligned}$$

These are locally closed subsets of $\mathcal{X}_s$, and we endow them with the induced reduced structure. As J runs through the set of non-empty subsets of I, the subvarieties E_J^o form a partition of $\mathcal{X}_s$.

It follows from [**NS07**, 7.7] that the motivic zeta function $Z_X(T)$ can be expressed in the form

$$Z_X(T) = \sum_{\emptyset\neq J\subset I}(\mathbb{L}-1)^{|J|-1}[\widetilde{E}_J^o]\prod_{j\in J}\frac{\mathbb{L}^{-\mu_j}T^{N_j}}{1-\mathbb{L}^{-\mu_j}T^{N_j}} \quad \in \mathcal{M}_k[[T]] \tag{6.1}$$

where $\widetilde{E}_J^o$ is a certain finite étale cover of E_J. By [**Ni11d**, 2.2.2], one can construct $\widetilde{E}_J^o$ as follows: set

$$N_J = \gcd\{N_j\,|\,j\in J\},$$

choose a uniformizer π in R, and denote by $\mathcal{Y}$ the normalization of

$$\mathcal{X} \times_R (R[x]/(x^{N_J} - \pi)).$$

Then there is an isomorphism of E_J^o-schemes

$$\widetilde{E}_J^o \cong E_J^o \times_{\mathcal{X}} \mathcal{Y}.$$

In particular, one sees from (6.1) that $Z_X(T)$ is a rational function and that every pole of $Z_X(\mathbb{L}^{-s})$ is of the form $s = -\mu_i/N_i$ for some $i \in I$. Every irreducible component E_i of the special fiber yields in this way a "candidate pole" $-\mu_i/N_i$ of the zeta function. Since the expression in (6.1) is independent of the chosen normal crossings model $\mathcal{X}$, one expects in general that not all of these candidate poles are actual poles of $Z_X(T)$. But even candidate poles that appear in *every* model will not always be actual poles. To explain this phenomenon, we will propose in Section 6.4 a version of the Monodromy Conjecture for Calabi-Yau varieties.

EXAMPLE 6.5. If X is an elliptic curve, then X admits a unique *minimal* regular model with strict normal crossings $\mathcal{X}$. It is not the case that all irreducible components of $\mathcal{X}_s$ give actual poles of the motivic zeta function. This can be seen by combining Theorem 5.7 with the Kodaira-Néron classification.

6.2. Log canonical threshold. Let X be a Calabi-Yau K-variety such that $X(K) \neq \emptyset$. From the formula in (6.1), we see that the poles of $Z_X(\mathbb{L}^{-s})$ form a finite subset of $\mathbb{Q}$. It turns out that the *largest* pole of $Z_X(T)$ is an interesting invariant for X, which can be read off from the numerical data associated to any *sncd*-model of X.

Choose a regular proper R-model $\mathcal{X}$ of X such that $\mathcal{X}_s$ is a strict normal crossings divisor $\mathcal{X}_s = \sum_{i\in I} N_i E_i$ and define the values μ_i, $i \in I$ as in Section 6.1. We put

$$\begin{aligned} lct(X) &= \min\{\mu_i/N_i \,|\, i \in I\}, \\ \delta(X) &= \max\{\,|J|\,|\,\emptyset \neq J \subset I,\ E_J \neq \emptyset,\ \mu_j/N_j = lct(X) \text{ for all } j \in J\} - 1. \end{aligned}$$

DEFINITION 6.6. We call $lct(X)$ the log canonical threshold of X, and $\delta(X)$ the degeneracy index of X.

The following theorem shows that these values do not depend on the chosen model $\mathcal{X}$.

THEOREM 6.7. *Let X be a Calabi-Yau variety with $X(K) \neq \emptyset$.*

(1) *The value $s = -lct(X)$ is the largest pole of the motivic zeta function $Z_X(\mathbb{L}^{-s})$, and its order equals $\delta(X) + 1$. In particular, $lct(X)$ and $\delta(X)$ are independent of the model $\mathcal{X}$. For every integer $d > 0$, we have*

$$\begin{aligned} \delta(X \times_K K(d)) &= \delta(X), \\ lct(X \times_K K(d)) &= d \cdot lct(X). \end{aligned}$$

(2) *Assume moreover that $K = \mathbb{C}((t))$ and that X admits a projective model $\mathcal{Y}$ over the ring $\mathbb{C}\{t\}$ of germs of analytic functions at the origin of the complex plane. If we put $\alpha = lct(X)$, then $\exp(-2\pi i\alpha)$ is an eigenvalue of the action of the semi-simple part of monodromy on*

$$\mathrm{Gr}_F^m H^m(\mathcal{Y}_\infty, \mathbb{C}) := \mathrm{Gr}_F^m \mathbb{H}^m(\mathcal{Y}_s^{\mathrm{an}}, R\psi_{\mathcal{Y}}(\mathbb{C}))$$

where $m = \dim(X)$. *In particular, for every embedding of* $\mathbb{Q}_\ell$ *in* $\mathbb{C}$, $\exp(-2\pi i\alpha)$ *is an eigenvalue of every tame monodromy operator* σ *on* $H^m(X \times_K K^t, \mathbb{Q}_\ell)$.

In part (2) of Theorem 6.7 above, $H^m(\mathcal{Y}_\infty, \mathbb{C})$ denotes the limit cohomology at $t = 0$ associated to any projective model for $\mathcal{Y}$ over a small open disc around the origin of $\mathbb{C}$. It carries a natural mixed Hodge structure [**St76, Na87**], and $F^\bullet$ denotes the Hodge filtration. Note that $K^t = K^s$ since k has characteristic zero.

Comparing Theorem 5.7 and Theorem 6.7, we find:

COROLLARY 6.8. *If* A *is an abelian* K*-variety, then* $lct(A) = -c(A)$ *and* $\delta(A) = t_{\mathrm{pot}}(A)$.

The degeneracy index of a Calabi-Yau variety X over K is a measure for the potential degree of degeneration of X over the closed point of $\operatorname{Spec} R$. If A is an abelian variety, then by Corollary 6.8, the degeneracy index $\delta(A)$ is zero if and only if A has potential good reduction, and $\delta(A)$ reaches its maximal value $\dim(A)$ if and only if A has potential purely multiplicative reduction.

Looking at the expression for the zeta function of an abelian variety in Proposition 5.8, one sees that the zeta function of an abelian variety encodes many other interesting invariants of the abelian variety, such as the order function ord_A and the number of components $\phi_A(d)$ for every d in $\mathbb{N}$. Our motivic zeta function allows to generalize these invariants to Calabi-Yau varieties. Using the expression (6.1) for the zeta function in terms of an *sncd*-model, all these invariants can be explicitly computed on such a model. See [**HN10b**, §5].

6.3. Comparison with the case of a hypersurface singularity. Let us return for a moment to the set-up of Section 4.4, still assuming that k has characteristic zero. We can also apply Definition 6.6 to this situation, replacing X by the analytic Milnor fiber $\mathscr{F}_x$ of f at x and taking for ω the gauge form $t \cdot \phi/df$ on $\mathscr{F}_x$, where ϕ/df is a Gelfand-Leray form. In this way, we define the log-canonical threshold $lct_x(f)$ of f at x and the degeneracy index $\delta_x(f)$ of f at x. One can deduce from [**Ni09a**, 7.30] that $lct_x(f)$ coincides with the usual log-canonical threshold of f at x as it is defined in birational geometry. The results in Theorem 6.7 remain valid; in particular, using Theorem 4.16, we see that $s = -lct_x(f)$ is the largest pole of the motivic zeta function $Z_{f,x}(\mathbb{L}^{-s})$ of f at x. We refer to [**HN10b**] for details.

6.4. Global Monodromy Property. In the light of our results for abelian varieties, it is natural to wonder if there is a relation between poles of $Z_X(T)$ and monodromy eigenvalues for Calabi-Yau varieties X, similar to the one predicted by the motivic monodromy conjecture for hypersurface singularities (Conjecture 4.19).

DEFINITION 6.9. Let X be a Calabi-Yau variety with $X(K) \neq \emptyset$, and let σ be a topological generator of $G(K^t/K) = G(K^s/K)$. We say that X satisfies the Global Monodromy Property (GMP) if there exists a finite subset $\mathcal{S}$ of $\mathbb{Z} \times \mathbb{Z}_{>0}$ such that

$$Z_X(T) \in \mathcal{M}_k\left[T, \frac{1}{1-\mathbb{L}^a T^b}\right]_{(a,b)\in\mathcal{S}}$$

and such that for each $(a,b) \in \mathcal{S}$, the cyclotomic polynomial $\Phi_{\tau(a/b)}(t)$ divides the characteristic polynomial of the monodromy operator σ on $H^i(X \times_K K^t, \mathbb{Q}_\ell)$ for some $i \in \mathbb{N}$.

Recall that $\tau(a/b)$ denotes the order of a/b in $\mathbb{Q}/\mathbb{Z}$. By Theorem 5.7, every abelian K-variety satisfies the Global Monodromy Property.

QUESTION 6.10. Is there a natural condition on X that guarantees that X satisfies the Global Monodromy Property (GMP)?

We don't know any example of a Calabi-Yau variety over K that does not satisfy the GMP. We would like to mention some work in progress where we can show that the GMP holds for certain types of varieties "beyond" abelian varieties.

Semi-abelian varieties. As a direct generalization of abelian varieties, it is natural to consider *semi-abelian* varieties, i.e., algebraic K-groups that are extensions of abelian varieties by tori. Néron models exist also for semi-abelian varieties, we refer to [**BLR90**] and [**HN10c**] for more details (the Néron model we consider is the maximal quasi-compact open subgroup scheme of the Néron lft-model from [**BLR90**]). We have generalized Theorem 5.7 to tamely ramified semi-abelian K-varieties, in arbitrary characteristic. The main complication is that one has to control the behaviour of the *torsion* part of the component group of the Néron lft-model under ramified base change.

K3-surfaces. Let X be a Calabi-Yau variety over K that admits a K-rational point. To show that X satisfies the Global Monodromy Property, one strategy would be to consider a regular proper model $\mathcal{X}$ whose special fiber has strict normal crossings. In principle, using the expression in (6.1), one can then determine the poles of $Z_X(T)$. The next step is to use A'Campo's formula (in the form of [**Ni11d**]) to compute the monodromy zeta function of X on the model $\mathcal{X}$ (the monodromy zeta function is the alternate product of the characteristic polynomials of the monodromy action on the cohomology spaces of X). In this way, one tries to show that the poles of $Z_X(T)$ correspond to monodromy eigenvalues. In practice, this kind of argumentation can be quite complicated. For one thing, when the dimension of X is greater than one, there is usually no distinguished *sncd*-model to work with, like the minimal *sncd*-model in the case of elliptic curves. And even when one has some more or less explicitly given model, the combinatorial and geometric complexity of the special fiber often make computations very hard: one needs to analyze the model in a very precise way to eliminate fake candidate poles and to find a sufficiently large list of monodromy eigenvalues. Worse, the monodromy zeta function might contain too little information to find all the necessary monodromy eigenvalues, due to cancellations in the alternate product.

There do however exist cases where this procedure leads to results. For instance, assume that X has dimension two and that it allows a triple-point-free degeneration. By this we mean that X has a proper regular model $\mathcal{X}/R$ where the special fiber $\mathcal{X}_s$ is a strict normal crossings divisor such that three distinct irreducible components of $\mathcal{X}_s$ never meet in one point. Such degenerate triple-point-free fibers have been classified by B. Crauder and D. Morrison [**CM83**]. In an ongoing project we use their classification to study the motivic zeta function of X, and we have been able to verify in almost all cases that the Global Monodromy Property holds.

References

[EGA4a] A. Grothendieck and J. Dieudonné. Eléments de Géométrie Algébrique, IV, Première partie. *Publ. Math., Inst. Hautes Étud. Sci.*, 20:5-259, 1964. MR0173675 (30:3885)

[SGA7a] *Groupes de monodromie en géométrie algébrique. I.* Séminaire de Géométrie Algébrique du Bois-Marie 1967-1969 (SGA 7 I), Dirigé par A. Grothendieck. Avec la collaboration de

M. Raynaud et D.S. Rim. Volume 288 of *Lecture Notes in Mathematics*. Springer-Verlag, Berlin, 1972. MR0354656 (50:7134)

[SGA7b] *Groupes de monodromie en géométrie algébrique. II.* Séminaire de Géométrie Algébrique du Bois-Marie 1967-1969 (SGA 7 II), Dirigé par P. Deligne et N. Katz. Volume 340 of *Lecture Notes in Mathematics*. Springer-Verlag, Berlin, 1973. MR0354657 (50:7135)

[Berk93] V.G. Berkovich. Étale cohomology for non-Archimedean analytic spaces. *Publ. Math., Inst. Hautes Étud. Sci.*, 78:5-171, 1993. MR1259429 (95c:14017)

[Berk96] V.G. Berkovich. Vanishing cycles for formal schemes, II. *Invent. Math.*, 125(2):367-390, 1996. MR1395723 (98k:14031)

[Bert96] P. Berthelot. Cohomologie rigide et cohomologie rigide à supports propres. *Prepublication, Inst. Math. de Rennes*, 1996.

[Bj79] J.-E. Björk. *Rings of differential operators*. Volume 21 of *North-Holland Mathematical Library*. North-Holland Publishing Co., Amsterdam-New York, 1979. MR549189 (82g:32013)

[BS66] A.I. Borevich and I.R. Shafarevich. *Number theory*. Volume 20 of *Pure and Applied Mathematics*. Academic Press, New York-London, 1966. MR0195803 (33:4001)

[BL93] S. Bosch and W. Lütkebohmert. Formal and rigid geometry. I. Rigid spaces. *Math. Ann.*, 295(2):291-317, 1993. MR1202394 (94a:11090)

[BLR90] S. Bosch, W. Lütkebohmert and M. Raynaud. *Néron models*, volume 21 of *Ergebnisse der Mathematik und ihrer Grenzgebiete*. Springer-Verlag, 1990. MR1045822 (91i:14034)

[BS95] S. Bosch and K. Schlöter. Néron models in the setting of formal and rigid geometry. *Math. Ann.*, 301(2):339-362, 1995. MR1314591 (96h:14035)

[BX96] S. Bosch and X. Xarles. Component groups of Néron models via rigid uniformization. *Math. Ann.*, 306:459-486, 1996. MR1415074 (97f:14022)

[Ch00] C.-L. Chai. Néron models for semiabelian varieties: congruence and change of base field. *Asian J. Math.*, 4(4):715-736, 2000. MR1870655 (2002i:14025)

[CM83] B. Crauder and D.R. Morrison. *Triple-point-free degenerations of surfaces with Kodaira number zero.* in: R. Friedman and D.R. Morrison (eds). *The birational geometry of degenerations* Volume 29 of *Progress in Mathematics*. Birkhäuser Boston, Boston, MA, pages 353-386, 1983. MR690270 (85g:14040)

[dJ95] A.J. de Jong. Crystalline Dieudonné module theory via formal and rigid geometry. *Publ. Math., Inst. Hautes Étud. Sci.*, 82:5-96, 1995.

[De91a] J. Denef. Local zeta functions and Euler characteristics. *Duke Math. J.* 63(3):713-721, 1991. MR1121152 (93a:11100)

[De91b] J. Denef. Report on Igusa's local zeta function. In *Séminaire Bourbaki, Vol. 1990/91, Exp. No.730-744*, volume 201-203, pages 359-386, 1991. MR1157848 (93g:11119)

[DL98] J. Denef and F. Loeser. Motivic Igusa zeta functions. *J. Algebraic Geom.*, 7:505-537, 1998. MR1618144 (99j:14021)

[DL01] J. Denef and F. Loeser. Geometry on arc spaces of algebraic varieties. *Progr. Math.*, 201:327-348, 2001. MR1905328 (2004c:14037)

[DL02] J. Denef and F. Loeser. Lefschetz numbers of iterates of the monodromy and truncated arcs. *Topology*, 41(5):1031-1040, 2002. MR1923998 (2003f:14004)

[Di04] A. Dimca. *Sheaves in topology*. Universitext. Springer-Verlag, Berlin, 2004. MR2050072 (2005j:55002)

[Ed92] B. Edixhoven. Néron models and tame ramification. *Compos. Math.*, 81:291-306, 1992. MR1149171 (93a:14041)

[HN10a] L.H. Halle and J. Nicaise. The Néron component series of an abelian variety. *Math. Ann.* 348(3):749-778, 2010. MR2677903 (2011j:11105)

[HN10b] L.H. Halle and J. Nicaise. Motivic zeta functions of abelian varieties, and the monodromy conjecture. *preprint*, arXiv:0902.3755v3. MR2782205

[HN10c] L.H. Halle and J. Nicaise. Motivic zeta functions of abelian varieties, and the monodromy conjecture. *Adv. Math.* 227, No. 1:610-653, 2011. MR2782205

[HN10d] L.H. Halle and J. Nicaise. Jumps and monodromy of abelian varieties. *submitted*, arXiv:1009.3777.

[Ig74] J. Igusa. Complex powers and asymptotic expansions. I. *J. Reine Angew. Math.* 268/269:110-130, 1974. MR0347753 (50:254)

[Ig75] J. Igusa. Complex powers and asymptotic expansions. II. *J. Reine Angew. Math.* 278/279: 307-321, 1975. MR0404215 (53:8018)

[Ig00] J. Igusa. *An introduction to the theory of local zeta functions.* AMS/IP Studies in Advanced Mathematics. Providence, RI; International Press, Cambridge, MA, 2000. MR1743467 (2001j:11112)

[Ka76] M. Kashiwara. B-functions and holonomic systems. Rationality of roots of B-functions. *Invent. Math.*, 38:33-53, 1976. MR0430304 (55:3309)

[LO85] H.W. Lenstra and F. Oort. Abelian varieties having purely additive reduction. *J. Pure Appl. Algebra*, 36:281-298, 1985. MR790619 (86e:14020)

[LS03] F. Loeser and J. Sebag. Motivic integration on smooth rigid varieties and invariants of degenerations. *Duke Math. J.*, 119:315-344, 2003. MR1997948 (2004g:14026)

[Ma83] B. Malgrange. Polynômes de Bernstein-Sato et cohomologie évanescente. *Astérisque*, 101/102:243-267, 1983. MR737934 (86f:58148)

[MN91] Z. Mebkhout and L. Narváez-Macarro. La théorie du polynôme de Bernstein-Sato pour les algèbres de Tate et de Dwork-Monsky-Washnitzer. *Ann. Sci. École Norm. Sup. (4)* 24(2): 227-256, 1991. MR1097693 (93a:32014)

[Na87] V. Navarro Aznar. Sur la théorie de Hodge-Deligne. *Invent. Math.*, 90:11-76, 1987. MR906579 (88j:32037)

[Ne64] A. Néron. Modèles minimaux des variétés abéliennes sur les corps locaux et globaux. *Inst. Hautes Études Sci. Publ.Math.* 21:5-128, 1964. MR0179172 (31:3423)

[Ni09a] J. Nicaise. A trace formula for rigid varieties, and motivic Weil generating series for formal schemes. *Math. Ann.*, 343(2):285-349, 2009. MR2461256 (2010b:14043)

[Ni09b] J. Nicaise. Trace formula for component groups of Néron models. *preprint*, arXiv:0901.1809v2.

[Ni10] J. Nicaise. An introduction to p-adic and motivic zeta functions and the monodromy conjecture. in: G. Bhowmik, K. Matsumoto and H. Tsumura (eds.), *Algebraic and analytic aspects of zeta functions and L-functions.* Volume 21 of *MSJ Memoirs*, Mathematical Society of Japan, pages 115-140, 2010. MR2647606 (2011g:11223)

[Ni11a] J. Nicaise. Motivic invariants of algebraic tori. *Proc. Amer. Math. Soc.*, 139:1163-1174, 2011. MR2748411

[Ni11b] J. Nicaise. Singular cohomology of the analytic Milnor fiber, and mixed Hodge structure on the nearby cohomology. *J. Algebraic Geom.*, 20:199-237, 2011. MR2762990

[Ni11c] J. Nicaise. A trace formula for varieties over a discretely valued field. *J. Reine Angew. Math.*, 650:193-238, 2011. MR2770561

[Ni11d] J. Nicaise. Geometric criteria for tame ramification. *submitted*, arXiv:0910.3812.

[NS07] J. Nicaise and J. Sebag. The motivic Serre invariant, ramification, and the analytic Milnor fiber. *Invent. Math.*, 168(1):133-173, 2007. MR2285749 (2009c:14040)

[NS08] J. Nicaise and J. Sebag. Motivic Serre invariants and Weil restriction. *J. Algebra*, 319(4):1585-1610, 2008. MR2383059 (2009e:14041)

[NS11a] J. Nicaise and J. Sebag. The Grothendieck ring of varieties. In: R. Cluckers, J. Nicaise and J. Sebag (eds). *Motivic integration and its interactions with model theory and non-archimedean geometry.* London Mathematical Society Lecture Notes Series, Cambridge University Press, pages 145-188, 2011.

[NS11b] J. Nicaise and J. Sebag. Motivic invariants of rigid varieties, and applications to complex singularities. In: R. Cluckers, J. Nicaise and J. Sebag (editors). *Motivic integration and its interactions with model theory and non-archimedean geometry.* London Mathematical Society Lecture Notes Series, Cambridge University Press, pages 244-304, 2011.

[Ra74] M. Raynaud. Géométrie analytique rigide d'après Tate, Kiehl, ... *Mémoires de la S.M.F.*, 39-40:319-327, 1974. MR0470254 (57:10012)

[Sa94] M. Saito. On microlocal b-function. *Bull. Soc. Math. France*, 122(2):163-184, 1994. MR1273899 (95c:32040)

[RV03] B. Rodrigues and W. Veys. Poles of Zeta functions on normal surfaces. *Proc. London Math. Soc.*, 87(3):164-196, 2003. MR1978573 (2004c:14022)

[St76] J.H.M. Steenbrink. Limits of Hodge structures. *Invent. Math.*, 31:229-257, 1976. MR0429885 (55:2894)

MATEMATISK INSTITUTT, UNIVERSITETET I OSLO, POSTBOKS 1053, BLINDERN, 0316 OSLO, NORWAY

E-mail address: larshhal@math.uio.no

KULEUVEN, DEPARTMENT OF MATHEMATICS, CELESTIJNENLAAN 200B, 3001 HEVERLEE, BELGIUM

E-mail address: johannes.nicaise@wis.kuleuven.be

Contemporary Mathematics
Volume **566**, 2012
http://dx.doi.org/10.1090/conm/566/11224

The lattice cohomology of $S^3_{-d}(K)$

András Némethi and Fernando Román

1. Introduction

Let $\{K_\ell\}_{\ell=1}^{\nu}$ be a collection of algebraic knots determined by certain irreducible plane curve singularities, and define their connected sum $K := K_1\#\cdots\#K_\nu \subset S^3$. The aim of the present note is the computation of the lattice cohomology of the 3–manifold $M := S^3_{-d}(K)$ obtained by $(-d)$–surgery of the 3–sphere S^3 along K. The final version describes the lattice cohomology $\mathbb{H}^*(M)$ in terms of the subtle distribution properties of the semigroups $\{\Gamma_\ell\}_{1\leq\ell\leq\nu}$.

Lattice cohomology of negative definite plumbed 3–manifolds was introduced in [**13**] and [**16**]. Since surgery manifolds constitute a key family of 3–manifolds, several articles target their study focusing on the case when K is an algebraic knot: [**14**] treats the negative integer surgery case, [**15**] the negative rational surgery case (this second preprint was published as part of [**18**]).

Any such surgery 3–manifold $M = S^3_{-d}(K)$ is the link of a certain normal surface singularity. In this way, the lattice cohomology makes a bridge between the analytic invariants of this singularity and topological–combinatorial invariants of M. For the analytic aspects and connections the reader is invited to inspect [**13**] and [**16**]. The present note emphasizes mostly the combinatorial aspects of $\mathbb{H}^*(M)$. This includes the connection with the Seiberg–Witten and Heegaard–Floer theories.

The Heegaard–Floer theory was developed by Oszváth and Szabó, see [**23**] and their long list of publication. The theory associates to any oriented 3–manifold several invariants, the most important one is $HF^+(M)$. Conjecturally, the lattice cohomology contains all the information about the Heegaard–Floer homology of M, provided that M is a negative definite plumbed manifold. The starting point of the correspondence appeared in [**25**], later it was developed in [**13, 16**]. For the corresponding connections, conjectures and partial results see also section 8. They include a precise correspondence for ν small. If $\nu = 1$ (in this case the corresponding graph is 'almost rational', hence the result follows already from [**13**])

2000 *Mathematics Subject Classification.* Primary 14E15, 57M27, 32Sxx; Secondary 14Bxx, 57R57, 58Kxx.

Key words and phrases. 3–manifolds, $\mathbb{Q}$–homology spheres, lattice cohomology, Ozsváth–Szabó invariant, Heegaard Floer homology, Seiberg–Witten invariant, knot surgeries, plane curve singularities, algebraic semigroups.

The first author is partially supported by OTKA Grants. The second author is partially supported by FPI Grant.

then $\mathbb{H}^0(M) = HF^+_{even}(M)$, and $HF^+_{odd}(M) = \mathbb{H}^q(M) = 0$ for $q \geq 1$. Theorem (8.2.1) of the present work treats the case $\nu = 2$ as well: $\mathbb{H}^0(M) = HF^+_{even}(-M)$, $\mathbb{H}^1(M) = FH^+_{odd}(-M)$, and $\mathbb{H}^q(M) = 0$ for $q \geq 2$. (For the Heegaard–Floer invariants of surgery manifolds in terms of the filtered chain homotopy type of the Heegaard–Floer complex associated with the pair (S^3, K) see also [**26**].)

More generally, for arbitrary ν, in (4.2.11) we prove that $\mathbb{H}^q(M) = 0$ whenever $q \geq \nu$. Moreover, in section 9 we show that this vanishing result is sharp. Indeed, we determine explicitly the lattice cohomology when each $\{K_\ell\}_{\ell=1}^{\nu}$ is a torus knots of type $(2,3)$, and we show that $\mathbb{H}^{\nu-1}(M) = \mathbb{Z}$.

The connection with the Seiberg–Witten theory is described in section 7.1. We prove that the normalized Euler–characteristic of the lattice cohomology coincides with the normalized Seiberg–Witten invariants of M. In particular, this provides a purely combinatorial formula for the Seiberg–Witten invariants. (This was generalized recently by the first author in [**20**].)

Although in this article we do not emphasize the special case provided by the hypersurface superisolated singularities, this constitutes a main motivation for us. For the corresponding connections with the classification of the rational cuspidal projective plane curves, see [**14, 15**].

Usually, the direct computation of the lattice cohomology is very hard. Our computation here is based on a crucial 'reduction theorem', the main result of [**8**], which reduces the rank of the plumbing lattice to ν, the number of 'bad' vertices in the plumbing graph of M.

A few words about the organization of the sections. Section 2 contains the necessary material about the plumbing graph of M, its connection with the classical invariants of the algebraic knots (Alexander polynomial, semigroups). It also presents some generalities regarding the *spin*^{c}–structures and Seiberg–Witten invariants of M. The next section is a review of the lattice cohomology. It also contains a 'new' Mayer–Vietoris type formula for the normalized Euler–characteristic (which will be used later). Section 4 contains the lattice reduction theorem and its consequences applied for this situation. This is continued in the next two section, where all the invariants are described in terms of the distribution properties of the semigroups. Section 7.1 determines the normalized Euler–characteristic in terms of the Seiberg–Witten invariants, while section 8 contains the connection with Heegaard–Floer theory. In the last two sections we list some examples.

2. The 3–manifold $S^3_{-d}(K)$

2.1. Let us fix a positive integer d and a collection of algebraic knots $\{K_\ell\}_{\ell=1}^{\nu}$; these are knots determined by isolated irreducible plane curve singularities. In this section we describe the oriented 3-manifold $M = S^3_{-d}(K)$, obtained by $(-d)$-surgery along the connected sum $K = K_1\#\cdots\#K_\nu \subset S^3$ of the knots K_ℓ. In (2.2) we review some of the complete invariants of algebraic knots which codify the isotopy types $K_\ell \subset S^3$. Then we provide the negative definite plumbing representation of $S^3_{-d}(K)$, and we determine some invariants of this manifold, including its Seiberg–Witten invariants.

2.2. Review of algebraic knots. Let $K \subset S^3$ be an algebraic knot, i.e. the link of an irreducible plane curve singularity f; see [**4**] and [**5**] as general references for their invariants. Then K is an iterated torus knot. We will assume that f is not smooth, i.e. K is not an unknot. The isotopy type of $K \subset S^3$ is completely characterized by any of the following invariants listed below: Newton pairs, linking pairs, Alexander polynomial, semigroup, embedded resolution graph (or, equivalently, plumbing graph of the knot).

2.2.1. The *Newton pairs* of f consist of $r \geq 1$ pairs of integers $\{(p_i, q_i)\}_{i=1}^r$, where $p_i \geq 2$, $q_i \geq 1$, $q_1 > p_1$ and $\gcd(p_i, q_i) = 1$. They appear naturally in the 'normal form' of the equation of f. On the other hand, in topological discussions one usually uses the *'linking pairs'* (or, the decorations of the splice diagram, cf. [**5**]) $(p_i, a_i)_{i=1}^r$, where

$$a_1 = q_1 \quad \text{and } a_{i+1} = q_{i+1} + p_{i+1}p_i a_i \quad \text{for} \quad i \geq 1.$$

2.2.2. The *Alexander polynomial* $\Delta(t)$ of $K_f \subset S^3$, normalized by $\Delta(1) = 1$, in terms of the pairs $(p_i, a_i)_i$ is given by

$$\Delta(t) = \frac{(t^{a_1p_1p_2\cdots p_r} - 1)(t^{a_2p_2\cdots p_r} - 1)\cdots(t^{a_rp_r} - 1)(t-1)}{(t^{a_1p_2\cdots p_r} - 1)(t^{a_2p_3\cdots p_r} - 1)\cdots(t^{a_r} - 1)(t^{p_1\cdots p_r} - 1)}.$$

The degree of Δ is the Milnor number μ of f.

2.2.3. The original definition of the *semigroup* Γ is analytic: it is the semigroup of the intersection multiplicities of f with all possible analytic germs. This is equivalent with the following combinatorial descriptions. Γ is the sub-semigroup of $\mathbb{Z}_{\geq 0}$ (with $0 \in \Gamma$) generated by the integers $\bar\beta_0 = p_1p_2\cdots p_r$, $\bar\beta_k = a_kp_{k+1}\cdots p_r$ for $1 \leq k \leq r-1$, and $\bar\beta_r = a_r$. This is also equivalent with the next identity (cf. [**7**]), which will be used in the sequel:

$$\frac{\Delta(t)}{1-t} = \sum_{k\in\Gamma} t^k. \tag{2.2.4}$$

In fact, $\#(\mathbb{Z}_{\geq 0} \setminus \Gamma) = \mu/2$, cf. [**11**]. The largest element of $\mathbb{Z}_{\geq 0} \setminus \Gamma$ is $\mu - 1$, and for $0 \leq k \leq \mu - 1$ one has the symmetry: $k \in \Gamma$ if and only if $\mu - 1 - k \notin \Gamma$.

The integer $\delta := \#(\mathbb{Z}_{\geq 0} \setminus \Gamma) = \mu/2$ is called the *delta-invariant* of f, which also equals the minimal Seifert genus of $K \subset S^3$.

2.2.5. The *embedded minimal good dual resolution graph* $G(f)$ of the pair $(\mathbb{C}^2, f^{-1}(0))$, or equivalently, the minimal (negative definite) plumbing graph of the pair (S^3, K) will be denoted schematically by

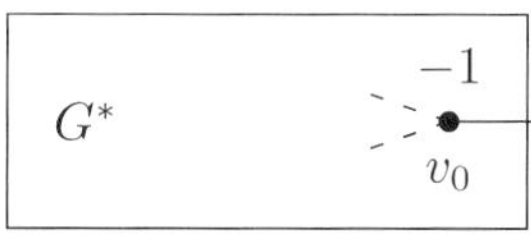

In fact, this is a tree of the following shape

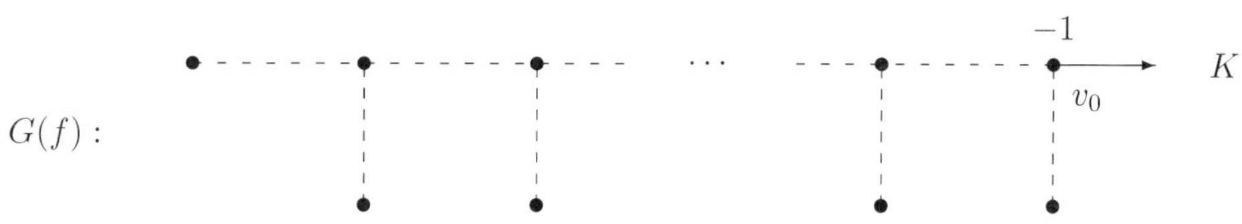

where the dash-lines represent strings. Here we emphasize only the r nodes (vertices of valency 3) and end–vertices. The graph has some additional decorations: each vertex has a self–intersection (Euler number) and a multiplicity decoration (the vanishing order of the pull–back of f along the corresponding exceptional divisor), while the arrowhead has the multiplicity decoration 1. For example, the multiplicity of the unique (-1)–vertex v_0 is $m = a_r p_r$, cf. [**5**]. One can prove that $m > \mu$. For more details, see [**4, 5**] or [**12**], section 4.I.

If we disregard the arrowhead, the graph G^* transforms into a plumbing graph of S^3, hence can be blown down completely.

2.3. The plumbing graph of $S^3_{-d}(K)$. In the sequel we consider ν algebraic knots $\{K_\ell\}_{\ell=1}^{\nu}$. Associated with K_ℓ, let r_ℓ be the number of linking pairs, $(p_i^{(\ell)}, a_i^{(\ell)})_{i=1}^{r_\ell}$ the corresponding linking pairs, Δ_ℓ, Γ_ℓ, μ_ℓ and δ_ℓ the Alexander polynomial, semigroup, Milnor number and delta–invariant respectively, while G_ℓ the graph of (S^3, K_ℓ), i.e. the embedded resolution graph of the pair $(\mathbb{C}^2, f_\ell^{-1}(0))$. Let Z_ℓ be the divisor of (the pullback of) f on G_ℓ, and let its coefficient on the (-1)–vertex be denoted by m_ℓ, which equals $m_\ell = a_{r_\ell}^{(\ell)} p_{r_\ell}^{(\ell)}$ (see above). Then, similarly as in [**14, 15**], the plumbing graph $G(M)$ of the oriented 3–manifold $M = S^3_{-d}(K)$ is obtained from the graphs $\{G_\ell\}_\ell$ (by deleting their multiplicities but unmodified otherwise) by adding a new vertex v_+ with self–intersection $-d - \sum_\ell m_\ell$ connected to the vertices $v_{0,\ell}$ of G_ℓ. (This statement can be checked by Kirby calculus).

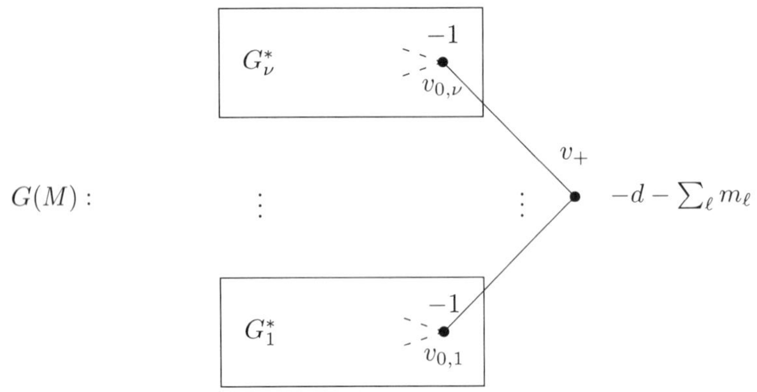

In the sequel we will fix this plumbing representation of M, and we assume that $d > 0$. In particular, the above plumbing graph has a negative definite intersection form.

Our goal is to compute the graded roots, lattice cohomologies and the Seiberg–Witten invariants associated with M.

2.4. Invariants of plumbing graphs/plumbed manifolds. First, we recall some notations and generalities regarding plumbing graphs.

2.4.1. **Generalities.** Let us assume that G is a connected negative definite plumbing graph. G can be realized as the resolution graph of some normal surface singularity $(X, 0)$, and the link M of $(X, 0)$ can be considered as the plumbed 3–manifold associated with G. In the sequel we assume that M is a rational homology sphere, or, equivalently, that G is a tree and all the genera decorations are zero. For more details regarding this section, see e.g. [**13, 14, 15, 16**].

Let $\tilde{X}$ be the smooth 4-manifold with boundary M obtained either by plumbing disc bundles along G, or via the resolution $\pi : \tilde{X} \to X$ of $(X, 0)$ with resolution graph G. Then $L = H_2(\tilde{X}, \mathbb{Z})$ is generated by $\{E_j\}_{j \in \mathcal{J}}$, the cores of the plumbing construction (or the irreducible components of the exceptional divisor $E := \pi^{-1}(0)$ of π). L is a lattice via the (negative definite) intersection form $I := \{(E_j, E_i)\}_{j,i}$. Let L' be the dual lattice $\{l' \in L \otimes \mathbb{Q} : (l', L) \subset \mathbb{Z}\}$. L' is generated by the (anti)dual elements E^*_j defined via $(E^*_j, E_k) = -\delta_{jk}$ (the negative of the Kronecker symbol).

Set $H := L'/L$. Then $H_1(M, \mathbb{Z}) = H$. The set of characteristic elements are defined by

$$Char := \{k \in L' : (k, x) + (x, x) \in 2\mathbb{Z} \text{ for any } x \in L\}.$$

The unique rational cycle $k_{can} \in L'$ which satisfies the system of adjunction relations $(k_{can}, E_j) = -(E_j, E_j) - 2$ for all j is called the *canonical cycle*. Then $Char = k_{can} + 2L'$. There is a natural action of L on $Char$ by $l * k := k + 2l$ whose orbits are of type $k + 2L$. Obviously, H acts freely and transitively on the set of orbits by $[l'] * (k + 2L) := k + 2l' + 2L$.

The first Chern class realizes an identification between the *spin*^{c}–structures $Spin^c(\tilde{X})$ on $\tilde{X}$ and $Char \subset L'$. $Spin^c(\tilde{X})$ is an L' torsor compatible with the above action of L' on $Char$.

All the *spin*^{c}–structures on M are obtained by restriction $Spin^c(\tilde{X}) \to Spin^c(M)$, $Spin^c(M)$ is an H torsor, and the actions are compatible with the factorization $L' \to H$. Hence, one has an identification of $Spin^c(M)$ with the set of L–orbits of $Char$, and this identification is compatible with the action of H on both sets. In this way, any *spin*^{c}-structure of M will be represented by an orbit $[k] := k + 2L \subset Char$. The *canonical* *spin*^{c}–structure corresponds to $[-k_{can}]$.

Recall that the Seiberg–Witten invariant associates with any $[k] \in Spin^c(M)$ is a rational number $\mathbf{sw}(M, [k]) \in \mathbb{Q}$. Moreover,

$$\mathbf{sw}(M, [-k]) = \mathbf{sw}(M, [k]) = -\mathbf{sw}(-M, [k]), \tag{2.4.2}$$

where $-M$ denotes M with opposite orientation.

2.4.3. **The Seiberg–Witten invariants of $S^3_{-d}(K)$.** Assume that $M = S^3_{-d}(K)$ with plumbing graph G as in (2.3). The base elements E_+, respectively $\{E_{0,\ell}\}_{\ell=1}^{\nu}$, correspond to the vertices v_+ and $\{v_{0,\ell}\}_{\ell=0}^{\nu}$.

We define $\Delta(t) := \prod_\ell \Delta_\ell(t)$. Its degree is $\mu := \sum_\ell \mu_\ell$; and we also set $\delta := \sum_\ell \delta_\ell = \mu/2$.

Since $\Delta(1) = 1$ and $\Delta'(1) = \delta$ (use e.g. the formula of (2.2.2)), one gets

$$\Delta(t) = 1 + \delta(t-1) + (t-1)^2 \cdot Q(t)$$

for some polynomial $Q(t) = \sum_{i=0}^{\mu-2} \alpha_i t^i$ of degree $\mu - 2$ with integral coefficients.

If $\nu = 1$, then all the coefficients $\{\alpha_i\}_{i=0}^{\mu-2}$ are strict positive. In fact, $\alpha_i = \#\{k \notin \Gamma_1 : k > i\}$ [**14**], hence

$$\delta = \alpha_0 \geq \alpha_1 \geq \cdots \geq \alpha_{\mu-2} = 1 \qquad (\text{for } \nu = 1).$$

If $\nu = 2$ then still $\alpha_i \geq 0$ for all i, but the monoteneity is lost; while for $\nu \geq 3$ some of the coefficients might be even negative. Nevertheless, in any situation (see e.g. the proof of (7.1.3) part (b)):

$$\alpha_0 = \delta \ \text{ and } \ \alpha_{\mu-2} = 1 \qquad (\text{for any } \nu). \tag{2.4.4}$$

Similarly as in [**14**, **15**], or [**3**, (8.1)] (where the case $\nu = 1$ is treated), one obtains that L'/L is the cyclic group with d elements generated by $[E^*_+]$. (This can be verified as follows: first one checks that $\det(-I) = d$, hence $|H| = d$. Since all

G^*_ℓ are unimodular, the coefficient of E_+ in E^*_+ is $1/d$, hence $[E^*_+]$ has order d in $L'/L = H$.)

THEOREM 2.4.5. *For any $a \in \{0, 1, \dots, d-1\}$ one has*

$$\mathbf{sw}(S^3_{-d}(K), [k_{can} + 2aE^*_+]) = -\sum_{l \in \mathbb{Z}_{\geq 0}} \alpha_{a+ld} + \frac{(\mu - 2 + d - 2a)^2}{8d} - \frac{1}{8}.$$

$$(k_{can} + 2aE^*_+)^2 + |\mathcal{J}| = -\frac{(\mu - 2 + d - 2a)^2}{d} + 1.$$

PROOF. The first statement is proved for $\nu = 1$ in [**3**, (8.1)] (where E^*_j is used with opposite sign). The proof of the general case runs in the same way based on [**3**, (5.0.3)], and on the fact that any component of $G \setminus v_+$ represent S^3. Similarly, the second statement follows from [**3**, (5.0.2)]. □

3. Review of lattice cohomologies

3.1. Preliminaries, $\mathbb{Z}[U]$–modules. One can associate lattice cohomologies to different objects at different levels (see below). All the time, the output is a graded $\mathbb{Z}[U]$–module. Some special $\mathbb{Z}[U]$–modules will stay as building blocks of the cohomology, we review their definitions first.

Consider the graded $\mathbb{Z}[U]$–module $\mathbb{Z}[U, U^{-1}]$, and (following [**25**]) denote by $\mathcal{T}^+_0$ its quotient by the submodule $U \cdot \mathbb{Z}[U]$. This has a grading in such a way that $\deg(U^{-d}) = 2d$ ($d \geq 0$). Similarly, for any $n \geq 1$, the quotient of $\mathbb{Z}\langle U^{-(n-1)}, U^{-(n-2)}, \dots, 1, U, \dots\rangle$ by $U \cdot \mathbb{Z}[U]$ (with the same grading) defines the graded module $\mathcal{T}_0(n)$. Hence, $\mathcal{T}_0(n)$, as a $\mathbb{Z}$–module, is freely generated by $1, U^{-1}, \dots, U^{-(n-1)}$, and has finite $\mathbb{Z}$-rank n.

More generally, for any graded $\mathbb{Z}[U]$–module P with d–homogeneous elements P_d, and for any $r \in \mathbb{Q}$, we denote by $P[r]$ the same module graded (by $\mathbb{Q}$) in such a way that $P[r]_{d+r} = P_d$. Then set $\mathcal{T}^+_r := \mathcal{T}^+_0[r]$ and $\mathcal{T}_r(n) := \mathcal{T}_0(n)[r]$. (Hence, for $m \in \mathbb{Z}$, $\mathcal{T}^+_{2m} = \mathbb{Z}\langle U^{-m}, U^{-m-1}, \dots\rangle$.)

3.2. Lattice cohomology associated with $\mathbb{Z}^s$ and a system of weights. [**16**] We fix a free $\mathbb{Z}$–module, with a fixed basis $\{E_j\}_j$, denoted by $\mathbb{Z}^s$. It is also convenient to fix a total ordering of the index set $\mathcal{J}$, which in the sequel will be denoted by $\{1, \dots, s\}$. Using the pair $(\mathbb{Z}^s, \{E_j\}_j)$ and a system of weights we determine a cochain complex whose cohomology is our central object.

3.2.1. **The cochain complex.** $\mathbb{Z}^s \otimes \mathbb{R}$ has a natural cellular decomposition into cubes. The set of zero–dimensional cubes is provided by the lattice points $\mathbb{Z}^s$. Any $l \in \mathbb{Z}^s$ and subset $I \subset \mathcal{J}$ of cardinality q defines a q–dimensional cube, which has its vertices in the lattice points $(l + \sum_{j \in I'} E_j)_{I'}$, where I' runs over all subsets of I. On each such cube we fix an orientation. This can be determined, e.g., by the order $(E_{j_1}, \dots, E_{j_q})$, where $j_1 < \dots < j_q$, of the involved base elements $\{E_j\}_{j \in I}$. The set of oriented q–dimensional cubes defined in this way is denoted by $\mathcal{Q}_q$ ($0 \leq q \leq s$).

Let $\mathcal{C}_q$ be the free $\mathbb{Z}$–module generated by oriented cubes $\square_q \in \mathcal{Q}_q$. Clearly, for each $\square_q \in \mathcal{Q}_q$, the oriented boundary $\partial \square_q$ has the form $\sum_k \varepsilon_k \square^k_{q-1}$ for some $\varepsilon_k \in \{-1, +1\}$. Here, in this sum, we write only those $(q-1)$–cubes which appear with non–zero coefficient. These are called *faces* of $\square_q$.

Although $\partial\circ\partial = 0$, the homology of the chain complex $(\mathcal{C}_*, \partial)$ (or, of the cochain complex $(\mathrm{Hom}_{\mathbb{Z}}(\mathcal{C}_*, \mathbb{Z}), \delta)$) is not very interesting: it is just the (co)homology of $\mathbb{R}^s$. In order to get a more interesting object, we consider a set of *weight functions* $w_q : \mathcal{Q}_q \to \mathbb{Z}$ $(0 \le q \le s)$.

3.2.2. **Definition.** A set of functions $w_q : \mathcal{Q}_q \to \mathbb{Z}$ $(0 \le q \le s)$ is called a *set of compatible weight functions* if the following hold:

(a) For any integer $k \in \mathbb{Z}$, the set $w_0^{-1}(\,(-\infty, k]\,)$ is finite;

(b) for any $\square_q \in \mathcal{Q}_q$ and for any face $\square_{q-1} \in \mathcal{Q}_{q-1}$ of it one has $w_q(\square_q) \ge w_{q-1}(\square_{q-1})$.

EXAMPLE 3.2.3. Assume that some $w_0 : \mathcal{Q}_0 \to \mathbb{Z}$ satisfies (a) for all $k \in \mathbb{Z}$. For any $q \ge 1$ set

$$w_q(\square_q) := \max\{w_0(v) \,:\, v \text{ is a vertex of } \square_q\}.$$

Then $\{w_q\}_q$ is a set of compatible weight functions.

In the presence of a compatible weight functions $\{w_q\}_q$, one sets $\mathcal{F}^q := \mathrm{Hom}_{\mathbb{Z}}(\mathcal{C}_q, \mathcal{T}_0^+)$. Then $\mathcal{F}^q$ is, in fact, a $\mathbb{Z}[U]$–module by $(p * \phi)(\square_q) := p(\phi(\square_q))$ $(p \in \mathbb{Z}[U])$, and $\mathcal{F}^q$ has a $\mathbb{Z}$–grading: $\phi \in \mathcal{F}^q$ is homogeneous of degree $d \in \mathbb{Z}$ if for each $\square_q \in \mathcal{Q}_q$ with $\phi(\square_q) \ne 0$, $\phi(\square_q)$ is a homogeneous element of $\mathcal{T}_0^+$ of degree $d - 2\cdot w(\square_q)$. (In the sequel sometimes we will omit the index q of w_q.)

Next, one defines $\delta_w : \mathcal{F}^q \to \mathcal{F}^{q+1}$. For this, fix $\phi \in \mathcal{F}^q$ and we show how $\delta_w\phi$ acts on a cube $\square_{q+1} \in \mathcal{Q}_{q+1}$. First write $\partial\square_{q+1} = \sum_k \varepsilon_k \square_q^k$, then set

$$(\delta_w\phi)(\square_{q+1}) := \sum_k \varepsilon_k\, U^{w(\square_{q+1}) - w(\square_q^k)}\, \phi(\square_q^k).$$

Then $\delta_w \circ \delta_w = 0$, hence $(\mathcal{F}^*, \delta_w)$ is a cochain complex. Moreover, $(\mathcal{F}^*, \delta_w)$ has an augmentation. Set $\mathfrak{m}_w := \min_{l\in\mathbb{Z}^s} w_0(l)$ and choose $l_w \in \mathbb{Z}^s$ such that $w_0(l_w) = \mathfrak{m}_w$. Then one defines the $\mathbb{Z}[U]$–linear map $\epsilon_w : \mathcal{T}^+_{2\mathfrak{m}_w} \longrightarrow \mathcal{F}^0$ such that $\epsilon_w(U^{-\mathfrak{m}_w - n})(l)$ is the class of $U^{-\mathfrak{m}_w + w_0(l) - n}$ in $\mathcal{T}_0^+$ for any $n \in \mathbb{Z}_{\ge 0}$. Then, ϵ_w is injective, and $\delta_w \circ \epsilon_w = 0$. Both ϵ_w and δ_w are homogeneous (degree zero) morphisms of $\mathbb{Z}[U]$–modules.

3.2.4. **Definitions.** The homology of the cochain complex $(\mathcal{F}^*, \delta_w)$ is called the *lattice cohomology* of the pair $(\mathbb{R}^s, w)$, and it is denoted by $\mathbb{H}^*(\mathbb{R}^s, w)$. The homology of the augmented cochain complex

$$0 \longrightarrow \mathcal{T}^+_{2\mathfrak{m}_w} \xrightarrow{\epsilon_w} \mathcal{F}^0 \xrightarrow{\delta_w} \mathcal{F}^1 \xrightarrow{\delta_w} \dots$$

is called the *reduced lattice cohomology* of the pair $(\mathbb{R}^s, w)$, and it is denoted by $\mathbb{H}^*_{red}(\mathbb{R}^s, w)$. For any $q \ge 0$, both $\mathbb{H}^q$ and $\mathbb{H}^q_{red}$ admit an induced graded $\mathbb{Z}[U]$–module structure, and one has graded $\mathbb{Z}[U]$–module isomorphisms $\mathbb{H}^q = \mathbb{H}^q_{red}$ (for $q > 0$) and $\mathbb{H}^0 = \mathcal{T}^+_{2\mathfrak{m}_w} \oplus \mathbb{H}^0_{red}$. Moreover, the $\mathbb{Z}$–grading of $\mathcal{F}^q$ induces a $\mathbb{Z}$–grading on $\mathbb{H}^q$ and $\mathbb{H}^q_{red}$; the d–homogeneous part is denoted by $\mathbb{H}^q_d$, or $\mathbb{H}^q_{red,d}$.

If each $\mathbb{H}^q_{red}$ has finite $\mathbb{Z}$–rank, then one can define the 'normalized Euler characteristic'

$$eu(\mathbb{H}^*(\mathbb{R}^s, w)) := -\mathfrak{m}_w + \textstyle\sum_q (-1)^q \,\mathrm{rank}_{\mathbb{Z}}(\mathbb{H}^q_{red}(\mathbb{R}^s, w)). \tag{3.2.5}$$

3.2.6. **Modification.** Clearly, instead of all the cubes of $\mathbb{R}^s$ we can consider only those ones which sit in $[0,\infty)^s$, or only in the rectangle $R := [0, T_1] \times \cdots \times [0, T_s]$ (for some $T_i \in \mathbb{Z}_{\ge 0}$). In such a case, we write $\mathbb{H}^*([0,\infty)^s, w)$ or $\mathbb{H}^*(R, w)$ for the corresponding lattice cohomologies.

3.3. The $\mathbb{S}^*$–representation. Next, we present a more geometric realization of the modules $\mathbb{H}^*$.

3.3.1. **Definitions.** For each $n \in \mathbb{Z}$, define $S_n = S_n(w) \subset \mathbb{R}^s$ as the union of all the cubes $\square_q$ (of any dimension) with $w(\square_q) \leq n$. Clearly, $S_n = \emptyset$, whenever $n < \mathfrak{m}_w$. For any $q \geq 0$, set

$$\mathbb{S}^q(\mathbb{R}^s, w) := \oplus_{n \geq \mathfrak{m}_w} H^q(S_n, \mathbb{Z}).$$

Then $\mathbb{S}^q$ is $\mathbb{Z}$ (in fact, $2\mathbb{Z}$)–graded, the $d = 2n$–homogeneous elements $\mathbb{S}^q_d$ consists of $H^q(S_n, \mathbb{Z})$. Also, $\mathbb{S}^q$ is a $\mathbb{Z}[U]$–module. The U–action is given by the restriction map $r_{n+1} : H^q(S_{n+1}, \mathbb{Z}) \longrightarrow H^q(S_n, \mathbb{Z})$, namely, $U * (\alpha_n)_n := (r_{n+1}\alpha_{n+1})_n$. Moreover, for $q = 0$ one has an augmentation $H^0(S_n, \mathbb{Z}) = \mathbb{Z} \oplus \widetilde{H}^0(S_n, \mathbb{Z})$, hence an augmentation of the graded $\mathbb{Z}[U]$–modules

$$\mathbb{S}^0 = (\oplus_{n \geq \mathfrak{m}_w} \mathbb{Z}) \oplus (\oplus_{n \geq \mathfrak{m}_w} \widetilde{H}^0(S_n, \mathbb{Z})) = \mathcal{T}^+_{2\mathfrak{m}_w} \oplus \mathbb{S}^0_{red}.$$

The point is that this $\mathbb{Z}[U]$–module $\mathbb{S}^*$ coincides with the lattice cohomology $\mathbb{H}^*$. More precisely, we have the following theorem whose proof (together with many other properties of the lattice cohomology) can be found in [**16**].

THEOREM 3.3.2. *There exists a graded $\mathbb{Z}[U]$–module isomorphism, compatible with the augmentations, between $\mathbb{H}^*(\mathbb{R}^s, w)$ and $\mathbb{S}^*(\mathbb{R}^s, w)$. Similar statement is valid for $\mathbb{H}^*([0, \infty)^s, w)$, or for $\mathbb{H}^*(\prod_i [0, T_i], w)$ too.*

From now on we will denote both modules with the same symbol $\mathbb{H}^*$, no matter which representation we use.

3.3.3. The above description allows us to reinterpret classical homological exact sequences in lattice cohomology. For example, the Mayer–Vietoris exact sequences will have the following consequence. For any integer T consider the half spaces $\mathbb{R}^s_{x_s \leq T} = \{\mathbf{x} \in \mathbb{R}^s : x_s \leq T\}$ and $\mathbb{R}^s_{x_s \geq T} = \{\mathbf{x} \in \mathbb{R}^s : x_s \geq T\}$ separated by the hyperplane $\mathbb{R}^s_{x_s = T} = \{\mathbf{x} \in \mathbb{R}^s : x_s = T\}$. The restriction of the weight function on these spaces determines lattice cohomologies.

PROPOSITION 3.3.4. **(Mayer Vietoris formula for** eu**)** *With the above notations one has:*

$$eu(\mathbb{H}^*(\mathbb{R}^s, w)) = eu(\mathbb{H}^*(\mathbb{R}^s_{x_s \leq T}, w) + eu(\mathbb{H}^*(\mathbb{R}^s_{x_s \geq T}, w) - eu(\mathbb{H}^*(\mathbb{R}^s_{x_s = T}, w).$$

Similar statement is valid for different rectangles instead of the whole space $\mathbb{R}^s$.

PROOF. Use the classical Mayer–Vietoris exact sequence and (3.3.2). □

3.4. The lattice cohomology associated with a plumbing graph. Let G be a negative definite plumbing graph as in (2.4.1). Let s be the number of vertices. Then we can associate to $L = \mathbb{Z}^s$ the the free $\mathbb{Z}$–module $\mathcal{C}_q$ generated by oriented cubes $\square_q \in \mathcal{Q}_q$, as in (3.2.1).

To any $k \in Char$ we associate weight functions $\{w_q\}_q$ as follows. First, we define $\chi_k : L \to \mathbb{Z}$ by

$$\chi_k(l) = -(l, l + k)/2.$$

We also write $\mathfrak{m}_k := \min \{ \chi_k(l) : l \in L\}$. Then the weight functions are defined as in (3.2.3) by:

$$w_q(\square_q) = \max\{\chi_k(v) : v \text{ is a vertex of } \square_q\}.$$

The associated lattice cohomology (resp. reduced lattice cohomology) will be denoted by $\mathbb{H}^*(G, k)$ (resp. $\mathbb{H}^*_{red}(G, k)$). It is proved (cf. [**16**]) that $\mathbb{H}^*_{red}(G, k)$ is finitely generated over $\mathbb{Z}$, hence $eu(\mathbb{H}^*(G,k)) := eu(\mathbb{H}^*(\mathbb{R}^s,w))$ is well–defined, cf. (3.2.5).

REMARK 3.4.1. Although each k provides a different cohomology module, there are only $|L'/L|$ essentially different ones. Indeed, assume that $[k]=[k']$, hence $k'=k+2l$ for some $l\in L$. Then

$$\chi_{k'}(x-l)=\chi_k(x)-\chi_k(l) \quad \text{for any } x\in L. \tag{3.4.2}$$

Therefore, the transformation $x\mapsto x':=x-l$ realizes the following identification:

$$\mathbb{H}^*(G,k')=\mathbb{H}^*(G,k)[-2\chi_k(l)].$$

3.5. The distinguished representatives k_r. We fix a *spin^c*–structure $[k]$. Recall, see (2.4.1), that $[k]$ has the form $k_{can}+2(l'+L)$ for some $l'\in L'$. Among all the characteristic elements in $[k]$ we will choose a very special one. Consider the (Lipman, or anti–nef) cone

$$\mathcal{S}':=\{l'\in L' : (l',E_v)\leq 0 \text{ for any vertex } v\}.$$

3.5.1. **Definition.** [**13**, (5.4–5.5)] We denote by $l'_{[k]}\in L'$ the unique minimal element of $(l'+L)\cap\mathcal{S}'$ and we call $k_r:=k_{can}+2l'_{[k]}$ the *distinguished representative* of the class $[k]$.

For example, since the minimal element of $L\cap\mathcal{S}'$ is the zero cycle, we get $l'_{[k_{can}]}=0$, and the distinguished representative in $[k_{can}]$ is the canonical cycle k_{can} itself.

The classes k_r generalize the canonical cycle for different *spin^c*–structures. For some applications (showing their importance) see [**13, 16, 18**]. The next properties are proved in [**16**]:

LEMMA 3.5.2. *(a)* $\mathbb{H}^*(G,k_r)\cong\mathbb{H}^*([0,\infty)^s,k_r)$.

(b) The set $\{\mathbb{H}^*(G,k_r)\}_{[k_r]}$ *is independent on the plumbing representation G of the 3–manifold M, hence it associates a $\mathbb{Z}[U]$–module to any pair (M,k_r), where $[k_r]\in Spin^c(M)$.*

4. The lattice reduction

4.1. The strategy. Our goal is the computation of the lattice cohomologies of the 3–manifold $S^3_{-d}(K)$. First, we need some preparations. In the first step we reduce the problem from the level of the lattice of rank $|\mathcal{J}|$ associated with G to another lattice of rank ν. This reduction is based on a general principle proved in [**8**]. Then, the lattice cohomologies of the reduced lattice will be described in terms of the semigroups $\{\Gamma_\ell\}_{\ell=1}^{\nu}$. In this way, the cohomology modules will codify the very subtle distribution properties of the semigroup elements of Γ_ℓ (as subsemigroups of $\mathbb{Z}_{\geq 0}$).

4.2. The lattice reduction. The general statement. In [**16**] is proved that the lattice cohomologies of rational graphs (for definition see below) are trivial. On the other hand, any non–rational graph can be transformed into a rational graph by surgery — decreasing some of the decorations — along some of its vertices. It turns out that all the lattice cohomology information is 'concentrated in these vertices'. They are called *bad vertices* (cf. [**25, 13, 16, 17**]).

4.2.1. **Definitions.** We say that a negative definite graph is *rational* if it is the resolution graph of a rational singularity. They were characterized combinatorially by Artin, for more details see [**2, 19, 13**], or (4.2.6) below. Next, we say that a subset of vertices $\{\overline{v}_j\}_{j=1}^{\nu}$ of a negative definite plumbing graph are *bad vertices* if replacing their decorations $e_{\overline{v}_j} := (E_{\overline{v}_j}, E_{\overline{v}_j})$ by some more negative integers $e'_{\overline{v}_j} \leq e_{\overline{v}_j}$ we get a rational graph.

For any graph there is a family of bad vertices with smallest cardinality. If this cardinality is less than or equal to one, then the graph is called *almost rational* (cf. [**13**]).

The idea of the reduction theorem is present already in [**13**]. The new lattice of rank ν will be associated with the bad vertices, the lattice points are associated with some important cycles of L (as distinguished members of Laufer–type computational sequences of L). In the next paragraphs we recall first the definitions of these cycles.

4.2.2. **The definition of the lattice points** $x(i_1, \ldots, i_\nu)$**.** Suppose we have a (minimal) family of bad vertices $\overline{\mathcal{J}} := \{\overline{v}_j\}_{j=1}^{\nu} \subset \mathcal{J}$. Then split the set of vertices $\mathcal{J}$ into the disjoint union $\overline{\mathcal{J}} \sqcup \mathcal{J}^*$. Furthermore, for any given vertex v and any (rational) cycle x, we write $m_v(x)$ for the v–coefficient $-(x, E_v^*)$ of x. (In other words, $x = \sum_v m_v(x) E_v$).

Then the cycles $x(i_1, \ldots, i_\nu)$ are defined via the next lemma.

LEMMA 4.2.3. *Fix* $[k]$ *as above. For any integer coordinates* $(i_1, \ldots, i_\nu) \in (\mathbb{Z}_{\geq 0})^\nu$, *there exists a unique cycle* $x(i_1, \ldots, i_\nu) \in L$ *satisfying the following properties:*

(a) $m_{\overline{v}_j}(x(i_1, \ldots, i_\nu)) = i_j$ *for any bad vertex* $\overline{v}_j$*;*

(b) $(x(i_1, \ldots, i_\nu) + l'_{[k]}, E_v) \leq 0$ *for every 'non-bad vertex'* $v \in \mathcal{J}^*$*;*

(c) $x(i_1, \ldots, i_\nu)$ *is minimal with the two previous properties.*

Moreover, $x(0, \ldots, 0) = 0$ *and* $x(i_1, \ldots, i_\nu) \geq 0$.

The proof is almost word–by–word the proof of Lemma 7.6. of [**13**], valid for $\nu = 1$, see also [**8**].

REMARK 4.2.4. Since $\chi_{k_r}(x + E_v) - \chi_{k_r}(x) = \chi_{k_r}(E_v) - (x, E_v) = 1 - (l'_{[k]} + x, E_v)$, property (*b*) is equivalent to $\chi_{k_r}(x(i_1, \ldots, i_\nu) + E_v) > \chi_{k_r}(x(i_1, \ldots, i_\nu))$ for every $v \in \mathcal{J}^*$. In fact (see [**13**, (9.1)] for $\nu = 1$, or [**8**]), the smallest χ_{k_r}–value on all the lattice points $\{x \in L : m_{\overline{v}_j}(x) = i_j \text{ for every } \overline{v}_j\}$ is realized by $x(i_1, \ldots, i_\nu)$.

The point is that in our main applications, we do not really need the cycles $x(i_1, \ldots, i_\nu)$ themselves, but only the values $\chi_{k_r}(x(i_1, \ldots, i_\nu))$. These can be computed inductively thanks to the following proposition.

In order to simplify the notation we set $\mathbf{i} := (i_1, \ldots, i_j, \ldots, i_\nu)$ and $\mathbf{i} + 1_j := (i_1, \ldots, i_j + 1, \ldots, i_\nu)$.

PROPOSITION 4.2.5. *For any distinguished representative* $k_r \in Char$, *any* $\mathbf{i} = (i_1, \ldots, i_\nu) \in (\mathbb{Z}_{\geq 0})^\nu$ *and* $j \in \overline{\mathcal{J}}$ *one has*

$$\chi_{k_r}(x(\mathbf{i} + 1_j)) = \chi_{k_r}(x(\mathbf{i})) + 1 - (x(\mathbf{i}) + l'_{[k]}, E_{\overline{v}_j}).$$

Moreover, $\chi_{k_r}(x(0, \ldots, 0)) = 0$.

PROOF. We construct the following computation sequence $\{x_n\}_{n=1}^N$ (as a generalization of Laufer's computation sequence) connecting $x(\mathbf{i})+E_{\overline{v}_j}$ and $x(\mathbf{i}+1_j)$, that is $x_1 = x(\mathbf{i})+E_{\overline{v}_j}$ and $x_N = x(\mathbf{i}+1_j)$ and $x_{n+1} = x_n + E_{v_{j(n)}}$ for some $j(n)\in \mathcal{J}^*$. The construction runs as follows: one start with $x_1 := x(\mathbf{i})+E_{\overline{v}_j}$. Then, assume that x_n is already constructed. If x_n satisfies the property (b) of (4.2.3) then we stop and we take $n=N$. Otherwise, there is at least one $j(n)\in\mathcal{J}^*$ such that $(x_n+l'_{[k]},E_{v_{j(n)}})>0$. Then write $x_{n+1}:=x_n+E_{v_{j(n)}}$.

We claim that this procedure is finite and its final term x_N is precisely $x(\mathbf{i}+1_j)$.

To prove this, it is enough to show that $x_n\leq x(\mathbf{i}+1_j)$ for any $1\leq n\leq N$; the minimality property (c) of $x(\mathbf{i}+1_j)$ will do the rest. For $n=1$ it is clear, so assume it is true for x_n and take $x_{n+1}=x_n+E_{v_{j(n)}}$. Then we have to verify that $m_{v_{j(n)}}(x_{n+1})\leq m_{v_{j(n)}}(x(\mathbf{i}+1_j))$ or equivalently $m_{v_{j(n)}}(x_n)<m_{v_{j(n)}}(x(\mathbf{i}+1_j))$. Suppose that this is not true, i.e. $m_{v_{j(n)}}(x(\mathbf{i}+1_j)-x_n)=0$. Then $(x_n+l'_{[k]},E_{v_{j(n)}}) = (x(\mathbf{i}+1_j)+l'_{[k]},E_{v_{j(n)}}) - (x(\mathbf{i}+1_j)-x_n,E_{v_{j(n)}})\leq 0$, a contradiction.

Since in every step of the construction of the sequence we imposed that $(x_n + l'_{[k]},E_{v_{j(n)}})>0$, we have $\chi_{k_r}(x_{n+1})\leq\chi_{k_r}(x_n)$ for any $1\leq n<N$. What we will prove next is that $(x_n+l'_{[k]},E_{v_{j(n)}})$ is exactly 1 for any $1\leq n<N$, hence $\chi_k(x_{n+1})=\chi_k(x_n)$, which proves the proposition.

The last claim is based on the following well–know fact proved by Laufer [**9**]

4.2.6. **[Laufer's Criterion]** *Let $\{z_n\}_{n=1}^M$ be a computation sequence provided by Laufer's algorithm* [**9**] *(similar as above with $[k]=[k_{can}]$) connecting $z_1=E_{v_j}$ (for some $j\in\mathcal{J}$) and the Artin's fundamental cycle $z_M=z_{min}$, the minimal non–zero cycle of $\mathcal{S}'\cap L$. (This means that $z_{n+1}=z_n+E_{v_n}$ for some v_n, where $(z_n,E_{v_n})>0$.) Then the graph is rational if and only if in all steps $1\leq n<M$ one has $(z_{n+1}-z_n,z_n)=1$.*

Now we can finish the proof; we just have to take $z_n:=x_n-x(\mathbf{i})$ for $1\leq n\leq N$. It is easy to see that $\{z_n\}_{n=1}^N$ is the beginning of a Laufer sequence $\{z_n\}_{n=1}^M$ connecting $E_{\overline{v}_j}$ with z_{min}. Moreover, the values $(z_n,E_{v_j(n)})$ will stay unmodified for every n if we replace our graph G with the rational graph $\widetilde{G}$ by decreasing the decorations of the bad vertices. Therefore, by Laufer's Criterion, $(z_n,E_{v_{j(n)}})=1$ in $\widetilde{G}$, and consequently in G too. This shows that

$$1=(x_n-x(\mathbf{i}),E_{v_{j(n)}})=(x_n+l'_{[k]},E_{v_{j(n)}})-(x(\mathbf{i})+l'_{[k]},E_{v_{j(n)}})\geq(x_n+l'_{[k]},E_{v_{j(n)}}).$$

Since $(x_n+l'_{[k]},E_{v_{j(n)}})>0$, it must equal 1. □

4.2.7. **Definition of the new lattice $\overline{L}$.** Let us fix $[k]$ and assume that the graph G admits ν bad vertices as above. Then define $\overline{L}=(\mathbb{Z}_{\geq 0})^\nu$, and the function $\overline{w}_0:(\mathbb{Z}_{\geq 0})^\nu\to\mathbb{Z}$ by

$$\overline{w}_0(i_1,\ldots,i_\nu):=\chi_{k_r}(x(i_1,\ldots,i_\nu)). \tag{4.2.8}$$

Then $\overline{w}_0$ defines a set $\{\overline{w}_q\}_{q=0}^\nu$ of compatible weight functions depending on $[k]$, defined similarly as in (3.2.3), denoted by $\overline{w}[k]$.

THEOREM 4.2.9 (Reduction Theorem [**8**]). *Let G be a negative definite connected graph and let k_r be the distinguished representative of a characteristic class. Suppose $\overline{\mathcal{J}}=\{\overline{v}_j\}_{j=1}^\nu$ is a (minimal) family of bad vertices and $(\overline{L},\overline{w}[k])$ is the* *new weighted lattice associated with $\overline{\mathcal{J}}$ and k_r. Then there is a graded $\mathbb{Z}[U]$–module isomorphism*

$$\mathbb{H}^*(G, k_r) \cong \mathbb{H}^*(\overline{L}, \overline{w}[k]). \tag{4.2.10}$$

This and (3.3.2) imply (see also [**17**]) the following:

COROLLARY 4.2.11. *Fix $\nu \geq 1$. If a graph G has a family of ν bad vertices then $\mathbb{H}^q(G, k) = 0$ for $q \geq \nu$ and any $k \in Char$.*

5. The lattice reduction for $S^3_{-d}(K)$

5.1. In this subsection we consider the graph $G = G(M)$ determined in (2.3) and we apply the general machinery of the previous subsection (4.2).

We first identify the bad vertices of the graph $G = G(M)$.

PROPOSITION 5.1.1. *The smallest family of bad vertices of G is exactly the family of (-1)–vertices $\overline{\mathcal{J}} = \{v_{0,\ell}\}_{\ell=1}^{\nu}$. In fact, if we replace all the (-1) decorations by (-2), then we get a rational graph.*

PROOF. It is not difficult to see (using e.g. Laufer's Criterion (4.2.6)) that a rational graph can never have a vertex v with decoration e_v greater than or equal to $2 - \partial_v$, where ∂_v is the *valency* of v (the number of adjacent vertices to v). Therefore, all the (-1)–vertices $v_{0,\ell}$ of G must be included in any family of bad vertices. We just have to show that decreasing their decoration we obtain a rational graph. In order to prove this we will use Laufer's Criterion (4.2.6).

Let us replace all the -1 decorations by -2. In this way the graphs G^*_ℓ $(1 \leq \ell \leq \nu)$ and G are replaced by $\tilde{G}^*_\ell$ and $\tilde{G}$ respectively. Let $\bar{G}^*_\ell$ be the subgraph of $\tilde{G}$ consisting of $\tilde{G}^*_\ell$ and E_+ and the connecting edge. In [**15**, (2.4.2)] is proved that $\bar{G}^*_\ell$ is a sandwiched graph, hence a rational graph. We wish to show that $\tilde{G}$ is rational too.

Let Z be the cycle on $\tilde{G}$ whose restriction on each $\tilde{G}^*_\ell$ is Z_ℓ (for its definition see (2.3)), and the multiplicity of E_+ is 1. Let $\bar{z}^*_{\ell,min}$ and $\tilde{z}_{min}$ be the Artin's fundamental cycle in $\bar{G}^*_\ell$ and $\tilde{G}$ respectively. Notice that for any E_v one has $(E_v, Z) \leq 0$, hence $Z \geq \tilde{z}_{min}$, hence $m_{v_+}(\tilde{z}_{min}) = m_{v_+}(\bar{z}^*_{\ell,min}) = 1$. In particular, the Laufer computation sequence associated with the fundamental cycle $\tilde{z}_{min}$ breaks into the computation sequences associated with the graphs $\bar{G}^*_\ell$. Since this graphs are rational, at any step $(z_{n+1} - z_n, z_n) = 1$, cf. (4.2.6). Hence, this is true for the graph $\tilde{G}$ too, which by Laufer's Criterion shows that $\tilde{G}$ is rational too. $\square$

5.2. Next, we wish to determine the cycles $l'_{[k]}$. As we already mentioned in (2.4.3), L'/L is the cyclic group of order d generated by $[E^*_+]$.

PROPOSITION 5.2.1. *Fix $a \in \{0, 1, \ldots, d-1\}$, and set $[k] = [k_{can} + 2aE^*_+]$. Then $l'_{[k]} = aE^*_+$.*

PROOF. Note that $\mathcal{S}'$ is generated over $\mathbb{Z}_{\geq 0}$ by the cycles E^*_j. Write $l'_{[k]}$ as $\sum_j a_j E^*_j$ for some $a_j \in \mathbb{Z}_{\geq 0}$. Then $aE^*_+ - \sum_j a_j E^*_j = l \in L$ with $l \geq 0$. We have to show that $l = 0$.

Let l_+ be the coefficient of E_+ in l; hence $l_+ \geq 0$. Since all the graphs G^*_ℓ are unimodular, the coefficient of E_+ in E^*_+ is $1/d$. Hence, l can be written as $l_+ dE^*_+ + \bar{l}$ for some $\bar{l} \in L$ supported on $\cup_\ell G^*_\ell$. Note that for any $j \neq j_+$ one has

$0 \leq a_j = (l, E_j) = (\bar{l}, E_j)$, hence, by the negative definiteness of the form we get (†) $\bar{l} \leq 0$.

Next, for $j = j_+$ we get $0 \leq a_{j_+} = (l - aE^*_+, E_+) = (l_+ dE^*_+ - aE^*_+ + \bar{l}, E_+) = -l_+ d + a + (\bar{l}, E_+)$. Since $\bar{l}$ is supported on $\cup_\ell G^*_\ell$, (†) implies that $(\bar{l}, E_+) \leq 0$. Hence $a \geq l_+ d$, which is possible (since $l_+ \geq 0$ and $a \in \{0, 1, \ldots, d-1\}$) only if $l_+ = 0$. But then $\bar{l} = l$, and since $l \geq 0$ we get $\bar{l} \geq 0$. This with (†) implies $\bar{l} = 0$, hence $l = 0$ too. This shows the minimality of aE^*_+. □

5.3. The cycles $x(\mathbf{i})$. In the sequel we fix a and $[k]$ as in (5.2.1), and we start to determine the cycles $x(\mathbf{i})$ associated with $[k]$, or rather their χ_{k_r}–values (using the recursive identities of (4.2.5)). According to this, we need to determine $(x(\mathbf{i}) + aE^*_+, E_{0,\ell})$ for any of the bad vertices $v_{0,\ell} \in \overline{\mathcal{J}}$.

Obviously, the part of the cycle $x(i_1, \ldots, i_\nu)$ contained in G^*_ℓ only depends on i_ℓ, and we will denote it by $x_\ell(i_\ell)$. Hence, for some $m_+(\mathbf{i}) \in \mathbb{Z}_{\geq 0}$ one has

$$x(i_1, \ldots, i_\nu) = \textstyle\sum_\ell x_\ell(i_\ell) + m_+(\mathbf{i})E_+. \tag{5.3.1}$$

The cycle $x_\ell(i_\ell)$ is the same as for the $\nu = 1$ case, hence we can read all its properties from [**18**].

For any ℓ, let Z_ℓ be an integral cycle supported on G^*_ℓ defined by $Z_\ell := E^*_{0,\ell}(G^*_\ell)$ (i.e. the dual cycle of $E_{0,\ell}$ *considered in the graph* G^*_ℓ embedded into L); it is also the compact part of the pull–back divisor of the germ f_ℓ, cf. (2.3).

LEMMA 5.3.2. [**18**] *Write $i_\ell = \alpha_\ell m_\ell + \beta_\ell$ with $\alpha_\ell \in \mathbb{Z}_{\geq 0}$ and $0 \leq \beta_\ell \leq m_\ell - 1$. Then*

$$x_\ell(i_\ell) = \alpha_\ell Z_\ell + x_\ell(\beta_\ell).$$

Moreover,

$$(x_\ell(i_\ell), E_{0,\ell}) = -\alpha_\ell + (x_\ell(\beta_\ell), E_{0,\ell}) = -\alpha_\ell + \begin{cases} 1 & \text{if } \beta_\ell \notin \Gamma_\ell, \\ 0 & \text{if } \beta_\ell \in \Gamma_\ell. \end{cases}$$

Next, we compute $m_+(\mathbf{i})$. By definition, $x(\mathbf{i})$ is the minimal cycle satisfying properties (a) and (b) of (4.2.3), therefore $m_+(\mathbf{i})$ is the smallest integer satisfying:

$$0 \overset{(b)}{\geq} (x(\mathbf{i}) + aE^*_+, E_+) \overset{(a)}{=} \sum_\ell i_\ell - a - (d + \sum_\ell m_\ell) \cdot m_+(\mathbf{i}).$$

Hence, with the notation $i := \sum_\ell i_\ell$ and $M := \sum_\ell m_\ell$:

$$m_+(\mathbf{i}) = \left\lceil \frac{i-a}{d+M} \right\rceil, \tag{5.3.3}$$

where $\lceil x \rceil$ represents the smallest integer greater than or equal to x.

Summarized all this together, and using (4.2.5) and (4.2.9), we obtain the following facts.

PROPOSITION 5.3.4. *For $M = S^3_{-d}(K)$ and any fixed $a = 0, \ldots, d-1$, the weights of the (first quadrant) lattice $\overline{L} = (\mathbb{Z}_{\geq 0})^\nu$ associated with the bad vertices $\{v_{0,\ell}\}_{\ell=1}^\nu$ of the graph $G(M)$ are recursively determined by $\overline{w}(\mathbf{0}) = 0$ and*

$$\overline{w}(\mathbf{i} + 1_\ell) - \overline{w}(\mathbf{i}) = 1 + \alpha_\ell - \left\lceil \frac{i-a}{d+M} \right\rceil - \begin{cases} 1 & \text{if } \beta_\ell \notin \Gamma_\ell, \\ 0 & \text{if } \beta_\ell \in \Gamma_\ell \end{cases} \tag{5.3.5}$$

for any $\mathbf{i}=(i_1,\dots,i_\ell,\dots,i_\nu)\in(\mathbb{Z}_{\geq 0})^\nu$. *Using this weight, one has a* $\mathbb{Z}[U]$*–module isomorphism:*

$$\mathbb{H}^*(G,k_r)\cong\mathbb{H}^*(\overline{L},\overline{w}[k]). \tag{5.3.6}$$

In fact, even $\overline{L}$ can be reduced more. First, we reduce it to a finite multi–rectangle:

COROLLARY 5.3.7. *Set* $\min:=\min_\ell\{m_\ell-\mu_\ell\}$. *Then the following facts hold.*
(a) Set

$$\alpha_0:=\left\lceil\frac{M-\min-a-\nu}{d}\right\rceil.$$

Let $\alpha\geq\alpha_0$ *be any integer, and consider* $\mathbf{i}$ *such that*

$$(\alpha m_1,\dots,\alpha m_\nu)\leq\mathbf{i}<((\alpha+1)m_1,\dots,(\alpha+1)m_\nu). \tag{5.3.8}$$

Then $\overline{w}(\mathbf{i}+1_\ell)\geq\overline{w}(\mathbf{i})$ *for any* $\ell\in\{1,\dots,\nu\}$.
(b)

$$\mathbb{H}^*(G,k_r)\cong\mathbb{H}^*([0,\alpha_0m_1]\times\cdots\times[0,\alpha_0m_\nu],\overline{w}).$$

PROOF. (a) Assume that $i\leq(\alpha+1)M-\min-\nu$ (†). Since $\alpha\geq\alpha_0$ we get $d\alpha\geq M-\min-a-\nu$. This together with (†) implies $(i-a)/(d+M)\leq\alpha$, hence (a) follows from (5.3.5). Similarly, if $i>(\alpha+1)M-\min-\nu$, that is if $\sum_\ell\beta_\ell>M-\min-\nu$, then (since $\beta_\ell\leq m_\ell-1$) necessarily $\beta_\ell\in\Gamma_\ell$ for all ℓ, hence the last contribution in (5.3.5) is zero. On the other hand, $i\leq(\alpha+1)M-1$, hence $(i-a)/(d+M)\leq\alpha+1$.

(b) Let $[0,\mathbf{i}]$ be the multi–rectangle in $\overline{L}$ consisting of the lattice points $\mathbf{j}$ with $0\leq\mathbf{j}\leq\mathbf{i}$ and all the cubes with these vertices. We show that if $\mathbf{i}$ is as in (5.3.8), then for any ℓ one has:

$$\mathbb{H}^*([0,\mathbf{i}],\overline{w}[k])=\mathbb{H}^*([0,\mathbf{i}+1_\ell],\overline{w}[k]).$$

In order to see this consider the inclusion $\iota:[0,\mathbf{i}]\to[0,\mathbf{i}+1_\ell]$ and its retract $\rho:[0,\mathbf{i}+1_\ell]\to[0,\mathbf{i}]$ (i.e. $\rho|[0,\mathbf{i}]$=identity) with $\rho(\mathbf{j}+1_\ell)=\mathbf{j}$ for any $\mathbf{j}$ satisfying $0\leq\mathbf{j}\leq\mathbf{i}$ and $j_\ell=i_\ell$. Using (4.2.5) for both $\mathbf{j}$ and $\mathbf{i}$, we get for any $\mathbf{j}$ satisfying $0\leq\mathbf{j}\leq\mathbf{i}$ and $j_\ell=i_\ell$ that

$$\overline{w}(\mathbf{j}+1_\ell)-\overline{w}(\mathbf{j})=\overline{w}(\mathbf{i}+1_\ell)-\overline{w}(\mathbf{i})+(x(\mathbf{i})-x(\mathbf{j}),E_{0,v_\ell}).$$

Since $\mathbf{i}\geq\mathbf{j}$ and $i_\ell=j_\ell$, the cycle $x(\mathbf{i})-x(\mathbf{j})$ is effective and it is supported on the complement of E_{0,v_ℓ}, hence $(x(\mathbf{i})-x(\mathbf{j}),E_{0,v_\ell})\geq 0$. This together with the inequality from part (a) provides $\overline{w}(\mathbf{j}+1_\ell)\geq\overline{w}(\mathbf{j})$ for any $\mathbf{j}$.

Hence, using the notation of (3.3.1), for any n the inclusion $\iota:S_n\cap[0,\mathbf{i}]\to S_n\cap[0,\mathbf{i}+1_\ell]$ and the retract $\rho:S_n\cap[0,\mathbf{i}+1_\ell]\to S_n\cap[0,\mathbf{i}]$ induce isomorphisms at the level of (simplicial) cohomology, hence the result follows from (3.3.2).

Finally, by induction and part (a), we get $\mathbb{H}^*([0,\alpha_0m_1]\times\cdots\times[0,\alpha_0m_\nu],\overline{w})\cong\mathbb{H}^*(\overline{L},\overline{w})$. □

The multi–rectangle $[0,\alpha_0m_1]\times\cdots\times[0,\alpha_0m_\nu]$ can be divided further in smaller parts: 'small rectangles' and 'stripes'. A 'small rectangle' has the form

$$R_{\alpha_1,\dots,\alpha_\nu}:=\left\{\ \mathbf{x}\in\mathbb{R}^\nu\ :\ \alpha_\ell m_\ell\leq x_\ell\leq(\alpha_\ell+1)m_\ell\ \text{ for any }\ell\ \right\}.$$

For simplicity, we also write $R_\alpha := R_{\alpha,\dots,\alpha}$. Clearly, $[0,\alpha_0 m_1]\times\cdots\times[0,\alpha_0 m_\nu] = \cup R_{\alpha_1,\dots,\alpha_\nu}$, where the union is over $0\le \alpha_\ell<\alpha_0$ for all ℓ. On the other hand, we also consider the following subsets for any $\alpha\in\mathbb{Z}_{\ge 0}$ and $\mathbf{x}\in\mathbb{R}^\nu$ with $x:=\sum x_\ell$:

$$B_\alpha := \{\ \mathbf{x}\ge 0\ :\ \left\lceil\frac{x-a}{d+M}\right\rceil = \alpha\},$$

$$T^+_\alpha := \{\ \mathbf{x}\ge 0\ :\ \frac{x-a}{d+M} = \alpha\},$$

$$T^-_\alpha := \{\ \mathbf{x}\ge 0\ :\ \frac{x-a-1}{d+M} = \alpha-1\}.$$

Obviously, $T^-_\alpha\cup T^+_\alpha\subset B_\alpha$ are the two limiting planes of the α–stripe B_α (with $T^-_0=\emptyset$). For any $\mathbf{i}\in B_\alpha$ the coefficient $m_+(\mathbf{i})$ takes constant value $m_+(\mathbf{i})=\alpha$. A direct computation shows:

LEMMA 5.3.9. *For any* $\alpha\ge 0$ *one has:*
(a) $\alpha\mathbf{m}\in\cup_{\alpha'\le\alpha} B_{\alpha'}$*;*
(b) if $(\alpha+1)\mathbf{m}\in\cup_{\alpha'\le\alpha}B_{\alpha'}$*, then* $\alpha\ge\alpha_0$*;*
(c) if $\alpha\mathbf{m}\in T^-_\alpha$ *then* $\alpha>\alpha_0$*.*

In particular, if $0\le\alpha\le\alpha_0$*, then* $\alpha\mathbf{m}\in B_\alpha\setminus T^-_\alpha$*; and* $\alpha\mathbf{m}\in T^+_\alpha$ *if and only if* $a=\alpha=0$*. Moreover, for* $0\le\alpha,\alpha'\le\alpha_0$ *one has* $T^+_\alpha\cap R_{\alpha'}\ne\emptyset$ *if and only if* $\alpha=\alpha'$*.*

COROLLARY 5.3.10. *Consider a small rectangle* $R=R_{\alpha_1,\dots,\alpha_\nu}$ *such that* $\alpha:=\max_\ell\{\alpha_\ell\}<\alpha_0$*. Then:*
(i) if $\mathbf{i}\in R$ *then* $\mathbf{i}\le(\alpha+1)\mathbf{m}$*, hence* $R\subset\cup_{\alpha'\le\alpha+1}B_{\alpha'}$*;*
(ii) if $R\cap B_{\alpha+1}\ne\emptyset$*, then* $R\subset B_\alpha\cup B_{\alpha+1}$*;*
(iii) if $R\cap B_{\alpha+1}=\emptyset$*, then* $R\subset\cup_{\alpha'\le\alpha}B_{\alpha'}$*.*

PROOF. Use (5.3.9) and direct verification. Notice that the difference between the two i–values of the two opposite corners of R is M, which is less than $d+M$, showing (ii). □

The next proposition shows that in the right hand side of (5.3.7)(b) one can omit all the small rectangles except those of type R_α.

PROPOSITION 5.3.11. $\mathbb{H}^*(G,k_r)\cong\mathbb{H}^*(\cup_{0\le\alpha<\alpha_0}R_\alpha,\overline{w})$.

PROOF. We proceed by induction. Assume that we already reduced the rectangle $[0,\alpha_0m_1]\times\cdots\times[0,\alpha_0m_\nu]$ by suitable deformation retracts to

$$(\cup_{\alpha_\ell\le\alpha}R_{\alpha_1,\dots,\alpha_\nu})\cup(\cup_{\alpha<\alpha'<\alpha_0}R_{\alpha'}).$$

Next, we wish to eliminate all the small rectangles not in $\cup_{\alpha_\ell<\alpha}R_{\alpha_1,\dots,\alpha_\nu}$, and not equal to R_α. Consider such a small rectangle. Then $\max_\ell\{\alpha_\ell\}=\alpha$, and $\bar{\alpha}=\min_\ell\{\alpha_\ell\}<\alpha$. We consider the different cases of (5.3.10).

In the case of (ii), $\lceil(i-a)/(d+M)\rceil\ge\alpha$ for all $\mathbf{i}\in R$, hence for any $\mathbf{i}\in R$ and any index j with $\alpha_j<\alpha$ one has $\overline{w}(\mathbf{i}+1_j)\le\overline{w}(\mathbf{i})$. Hence R can be contracted in the (increasing) direction of the j–coordinate similarly as in the proof of (5.3.7)(b).

In the case of (iii), $\lceil(i-a)/(d+M)\rceil\le\alpha$ for all $\mathbf{i}\in R$, hence for any $\mathbf{i}\in R$ and any index j with $\alpha_j=\alpha$ one has $\overline{w}(\mathbf{i}+1_j)\ge\overline{w}(\mathbf{i})$. Hence R can be contracted in the (decreasing) direction of the j–coordinate similarly as above.

The order how we eliminate the small rectangles is the following. In the first series we eliminate those one which satisfy (iii) (i.e. $R\cap B_{\alpha+1}=\emptyset$) in the following order: first those with $\alpha_1 = \alpha$ (in decreasing direction of the first coordinate), then those from the remaining ones with $\alpha_2 = \alpha$, and continuing in this way, finally all those remaining rectangles satisfying $R \cap B_{\alpha+1} = \emptyset$ with $\alpha_\nu = \alpha$.

After this series of contractions, all those which still have to be contracted satisfy (ii), hence any coordinate j with $\alpha_j < \alpha$ is a good candidate for a contraction in increasing direction. Then we contract in increasing direction of 1_1 all the rectangles with $\alpha_1 < \alpha$, then all the remaining ones with $\alpha_2 < \alpha$ in direction 1_2, and continuing in this way, at the end all those remaining one with $\alpha_\nu < \alpha$ in direction 1_ν. □

6. The lattice cohomology of $S^3_{-d}(K)$ via the semigroups Γ_ℓ

6.1. In this section we determine the weights of lattice points sitting in $\cup_{0\leq\alpha<\alpha_0} R_\alpha$ in terms of the semigroups $\{\Gamma_\ell\}_\ell$. As a consequence, we get a description of the lattice cohomology of $M := S^3_{-d}(K)$ in terms of these semigroups.

LEMMA 6.1.1. *For any* $0 \leq \alpha \leq \alpha_0$ *one has*

$$\overline{w}(\alpha\mathbf{m}) = \alpha(1 + a - \delta) + d\alpha(\alpha - 1)/2.$$

In particular, these values depend only on δ *(and* a*), but otherwise are independent on the structure of the semigroups* Γ_ℓ.

PROOF. First we verify that

$$\chi_{can}(Z_\ell) = m_\ell - \delta_\ell. \tag{6.1.2}$$

In the graph G_ℓ let K be the canonical cycle, E the union of all exceptional divisors, and ∂_v the valency of a vertex v (in G_ℓ with arrowhead). Then, by adjunction relations, $K + E = E^*_{0,\ell} + \sum_{v\in\mathcal{V}(G_\ell)}(2 - \partial_v)E^*_v$. Hence $-1 + (K, E^*_{0,\ell}) = (K + E, E^*_{0,\ell}) = (E^*_{0,\ell}, E^*_{0,\ell}) + \sum_v(2 - \partial_v)(E^*_v, E^*_{0,\ell})$. Note that $(E^*_{0,\ell}, E^*_{0,\ell}) = (Z_\ell, Z_\ell) = -m_\ell$ and $(E^*_v, E^*_{0,\ell}) = -m_v(Z_\ell)$. Therefore, by A'Campo's theorem [**1**] one gets $(K, E^*_{0,\ell}) = 1 - m_\ell - \sum_v(2 - \partial_v)m_v(Z_\ell) = -m_\ell + \mu_\ell$. Hence $\chi_{can}(Z_\ell) = -(E^*_{0,\ell}, E^*_{0,\ell} + K)/2 = m_\ell/2 - (-m_\ell + \mu_\ell)/2 = m_\ell - \delta_\ell$.

Next, we prove the statement of the lemma.

By (5.3.1) and (5.3.2) we have $\overline{w}(\alpha\mathbf{m}) = \chi_{k_r}(\alpha\sum_\ell Z_\ell + m_+E_+)$. By (5.3.9) $m_+ = \alpha$. Then the result follows by induction via the identities $\chi_{k_r}(A) = \chi_{can}(A) - a(A, E^*_+)$, $\chi(A+B) = \chi(A)+\chi(B)-(A, B)$, $\chi_{can}(Z_\ell) = m_\ell-\delta_\ell$ and $Z^2_\ell = -m_\ell$. □

LEMMA 6.1.3. *Fix* $0 \leq \alpha < \alpha_0$.
(a) Assume that $\mathbf{i} \in R_\alpha \cap B_\alpha$. *Then for all* ℓ *one has*

$$\overline{w}(\mathbf{i} + 1_\ell) - \overline{w}(\mathbf{i}) = \begin{cases} 0 & \text{if } \beta_\ell \notin \Gamma_\ell, \\ 1 & \text{if } \beta_\ell \in \Gamma_\ell. \end{cases}$$

Therefore, for any $\mathbf{i} \in R_\alpha$, $i_\ell = \alpha m_\ell + \beta_\ell$, *with* $i \leq \alpha(d + M) + a + 1$ *(in particular for any* $\mathbf{i} \in T^-_{\alpha+1} \cap R_\alpha$ *too), one has:*

$$\overline{w}(\mathbf{i}) - \overline{w}(\alpha\mathbf{m}) = \sum_\ell \#\{\gamma \in \Gamma_\ell : \gamma \leq \beta_\ell - 1\}. \tag{6.1.4}$$

(b) Assume that $\mathbf{i} \in R_\alpha \cap B_{\alpha+1}$. *Then for all* ℓ *with* $\mathbf{i} + 1_\ell \in B_{\alpha+1}$ *one has*

$$\overline{w}(\mathbf{i} + 1_\ell) - \overline{w}(\mathbf{i}) = -\begin{cases} 1 & \text{if } \beta_\ell \notin \Gamma_\ell, \\ 0 & \text{if } \beta_\ell \in \Gamma_\ell. \end{cases}$$

Therefore, for any $\mathbf{i} \in R_\alpha$, $i_\ell = \alpha m_\ell + \beta_\ell$, *with* $i \geq \alpha(d+M)+a+1$ *(in particular for any* $\mathbf{i} \in T^-_{\alpha+1} \cap R_\alpha$ *too), one has:*

$$\overline{w}(\mathbf{i}) - \overline{w}((\alpha+1)\mathbf{m}) = \sum_\ell \#\{\gamma \notin \Gamma_\ell \,:\, \gamma \geq \beta_\ell\}. \tag{6.1.5}$$

All these and the results of the previous section summarized provide the following statements. In this final version, we decrease even the bound α_0 (notice that $M - \min -\nu \geq \mu - 1$).

THEOREM 6.1.6. **(Final reduction theorem)** *Set*

$$\tilde{\alpha}_0 := \left\lceil \frac{\mu - a - 1}{d} \right\rceil,$$

and for any $0 \leq \alpha < \tilde{\alpha}_0$ *set* $\min T^-_{\alpha+1} := \min\{\, \overline{w}(\mathbf{i}) \,:\, \mathbf{i} \in T^-_{\alpha+1} \cap R_\alpha\}$. *Then the following facts hold:*

(a) $\overline{w}(\alpha\mathbf{m}) \leq \min T^-_{\alpha+1}$, $\overline{w}((\alpha+1)\mathbf{m}) \leq \min T^-_{\alpha+1}$.

(b) $\mathfrak{m}_{k_r} = \min \chi_{k_r} = \min_{0\leq\alpha\leq\tilde{\alpha}_0}\{\, \overline{w}(\alpha\mathbf{m})\,\}$.

(c) Let $\alpha_{\mathfrak{m}}$ *be the smallest integer satisfying* $\overline{w}(\alpha_{\mathfrak{m}}\mathbf{m}) = \mathfrak{m}_{k_r}$. *Then*

$$\begin{aligned}\mathbb{H}^0_{red}(G, k_r) = &\bigoplus_{0\leq\alpha<\alpha_{\mathfrak{m}}} \mathcal{T}_{2\overline{w}(\alpha\mathbf{m})}\big(\min T^-_{\alpha+1} - \overline{w}(\alpha\mathbf{m})\big)\\ &\oplus \bigoplus_{\alpha_{\mathfrak{m}}\leq\alpha<\tilde{\alpha}_0} \mathcal{T}_{2\overline{w}((\alpha+1)\mathbf{m})}\big(\min T^-_{\alpha+1} - \overline{w}((\alpha+1)\mathbf{m})\big).\end{aligned}$$

(d)

$$\begin{aligned}rank_{\mathbb{Z}}\, \mathbb{H}^0_{red}(G, k_r) = &\sum_{0\leq\alpha<\alpha_{\mathfrak{m}}} \big(\min T^-_{\alpha+1} - \overline{w}(\alpha\mathbf{m})\big)\\ &+ \sum_{\alpha_{\mathfrak{m}}\leq\alpha<\tilde{\alpha}_0} \big(\min T^-_{\alpha+1} - \overline{w}((\alpha+1)\mathbf{m})\big),\end{aligned}$$

or

$$rank_{\mathbb{Z}}\, \mathbb{H}^0_{red}(G, k_r) - \mathfrak{m}_{k_r} = \sum_{0\leq\alpha<\tilde{\alpha}_0} \big(\min T^-_{\alpha+1} - \overline{w}((\alpha+1)\mathbf{m})\big).$$

(e) For any $q > 0$ *one has*

$$\mathbb{H}^q(G, k_r) = \bigoplus_{0\leq\alpha<\tilde{\alpha}_0} \mathbb{H}^q(R_\alpha, \overline{w}).$$

PROOF. First we prove statements (a)–(e) for α_0 instead of $\tilde{\alpha}_0$.

(a) and (b) follow from (6.1.3). (c) follows by a direct verification (or, using the general statement [**13**, (3.6)]). The first part of (d) is based on (c), while the second one uses the general formula [**13**, (3.8)], or can be determined from the first identity and from the fact: $\overline{w}(\alpha_1\mathbf{m}) > \overline{w}(\alpha_2\mathbf{m})$ for all $\alpha_1 > \alpha_2 \geq \alpha_{\mathfrak{m}}$. For (e) use (3.3.2) and a Mayer–Vietoris argument.

Next, we prove that all the small rectangles R_α for $\alpha \geq \tilde{\alpha}_0$ can also be eliminated. We prove this by increasing induction: we verify that for any $\alpha'_0 > \tilde{\alpha}_0$ one has

$$\bigoplus_{0\leq\alpha<\alpha'_0} \mathbb{H}^q(R_\alpha, \overline{w}) = \bigoplus_{0\leq\alpha<\alpha'_0-1} \mathbb{H}^q(R_\alpha, \overline{w}), \tag{6.1.7}$$

i.e., the last small rectangle can be contracted into $(\alpha'_0-1)\mathbf{m}$ compatibly with the weights.

By a direct verification,

$$\alpha\geq\tilde{\alpha}_0 \text{ if and only if for all } \mathbf{i}\in T^-_{\alpha+1}\cap R_\alpha \text{ one has } \textstyle\sum_\ell\beta_\ell\geq\mu, \tag{6.1.8}$$

or, if there is an $\mathbf{i}\in T^-_{\alpha+1}\cap R_\alpha$ with $\beta_\ell\geq\mu_\ell$ for any ℓ. Since the largest element of $\mathbb{Z}\setminus\Gamma_\ell$ is $\mu_\ell-1$, such an $\mathbf{i}$ satisfies $\overline{w}(\mathbf{i})=\overline{w}((\alpha+1)\mathbf{m})$ by (6.1.5). Hence, in (c), proved for the bound α_0, all the contribution in $\mathbb{H}^0_{red}(G,k_r)$ corresponding to $\alpha\geq\tilde{\alpha}_0$ are zero, since $\min T^-_{\alpha+1}=\overline{w}((\alpha+1)\mathbf{m})$.

For $q>0$ we need a better argument, a sequence of contraction (which reproves the $q=0$ case too). For any $0\leq\beta_\ell\leq m_\ell$ set the rectangle

$$R^{\beta_1,\dots,\beta_\nu}_\alpha:=\{\mathbf{x}\in\mathbb{R}^\nu\,:\,\alpha m_\ell\leq x_\ell\leq\alpha m_\ell+\beta_\ell\}.$$

E.g., $R^{m_1,\dots,m_\nu}_\alpha=R_\alpha$ and $R^{0,\dots,0}_\alpha=\{\,\alpha\mathbf{m}\}$.

Fix $\beta^o_\ell\geq\mu_\ell$ for all ℓ such that $\sum_\ell\beta^o_\ell=\alpha d+a+1$, i.e. the corresponding $\mathbf{i}$ is on $T^-_{\alpha+1}$. Then we show that R_α contracts onto $R^{\beta^o_1,\dots,\beta^o_\nu}_\alpha$. Assume that R_α is already contracted onto $R^{\bar\beta_1,\dots,\bar\beta_\nu}_\alpha$ with $\bar\beta_\ell\geq\beta^o_\ell$ for all ℓ and $\bar\beta_\ell>\beta^o_\ell$ for some ℓ. Fix such an ℓ. We show that for any $\mathbf{i}\in R^{\bar\beta_1,\dots,\bar\beta_\nu}_\alpha\setminus R^{\bar\beta_1,\dots,\bar\beta_\ell-1,\dots,\bar\beta_\nu}_\alpha$ the inequality $\overline{w}(\mathbf{i})\geq\overline{w}(\mathbf{i}-1_\ell)$ holds. This defines a contraction of $R^{\bar\beta_1,\dots,\bar\beta_\nu}_\alpha$ onto $R^{\bar\beta_1,\dots,\bar\beta_\ell-1,\dots,\bar\beta_\nu}_\alpha$ compatibly with the weights.

Indeed, if $i>\alpha(d+M)+a+1$ then $\overline{w}(\mathbf{i})=\overline{w}(\mathbf{i}-1_\ell)$ from (6.1.3)(b). If $i\leq\alpha(d+M)+a+1$ then $\overline{w}(\mathbf{i})=\overline{w}(\mathbf{i}-1_\ell)+1$ from (6.1.3)(a). Hence R_α contracts onto $R^{\beta^o_1,\dots,\beta^o_\nu}_\alpha$.

Finally $R^{\beta^o_1,\dots,\beta^o_\nu}_\alpha$ contracts onto the point $\alpha\mathbf{m}$ using same type of contraction (but now in arbitrary order of the coordinates) via (6.1.3)(a) (repeating the case $i\leq\alpha(d+M)+a+1$). □

REMARK 6.1.9.

(a) The bound $\tilde{\alpha}_0$ in Theorem 6.1.6 is optimal. This can be seen already at the $\mathbb{H}^0$–level, cf. part (d), the statement (6.1.8) from the proof of the 'final reduction' and the comments after it.

(b) The cohomology $\mathbb{H}^q(R_\alpha,\overline{w}[k])$ depends only on the $\overline{w}$–values on $\alpha\mathbf{m}$, on $(\alpha+1)\mathbf{m}$ and along $T^-_{\alpha+1}$.

Indeed, for any n consider S_n as in (3.3.1). Then for $n<\min T^-_{\alpha+1}$ the space $S_n\cap R_\alpha$ has the same homotopy type as $S_n\cap\{\alpha\mathbf{m},(\alpha+1)\mathbf{m}\}$; while for $n\geq\min T^-_{\alpha+1}$ it has the homotopy type of the suspension of $S_n\cap T^-_{\alpha+1}$. In particular, all the nontrivial homogeneous elements of $\mathbb{H}^q(R_\alpha,\overline{w}[k])$ $(q>0)$ have degree $\geq\min T^-_{\alpha+1}$.

In fact, one can define on each T^-_α a 'generalized lattice structure' and the corresponding lattice cohomology. This can be done, similarly as in (3.3.1) in two different ways. The simplest definition is via the $\mathbb{S}^*$–representation: one defines

$$\mathbb{H}^*(T^-_\alpha,\overline{w}):=\oplus_{n\geq\min T^-_\alpha}H^*(S_n\cap T^-_\alpha,\mathbb{Z}). \tag{6.1.10}$$

Equivalently, one can define a lattice decomposition of T^-_α (or, just of $T^-_\alpha\cap R_{\alpha-1}$) as follows: the 'generalized q–cubes' of T^-_α are the intersection of $(q+1)$–cubes (as in (3.2.1)) with T^-_α. Lemma (6.1.3) guarantees that if $\square_{q+1}\cap T^-_\alpha\neq\emptyset$, then

$$\max\{w(v)\,:\,\text{vertex } v \text{ of } \square_{q+1}\}=\max\{w(v)\,:\,\text{vertex } v \text{ of } \square_{q+1}\cap T^-_\alpha\}.$$

Therefore, the right hand side of this identity defines an intrinsic set of weight functions on the generalized cubes of T^-_α, hence a cochain complex too determining the corresponding lattice cohomology.

With this 'generalized lattice structure' of T^-_α, one has

$$\mathbb{H}^q(R_\alpha,\overline{w})=\mathbb{H}^{q-1}_{red}(T^-_{\alpha+1},\overline{w}) \quad \text{for } q>0. \tag{6.1.11}$$

(c) In most of the notations above, we have omitted the symbol a codifying the characteristic element k_r. In fact, for any $\alpha\geq 0$ and $a\in\{0,\dots,d-1\}$, $T^-_{\alpha+1}$ is

$$T^-_{\alpha+1,a}:=\{\mathbf{i}\,:\,i_\ell=\alpha m_\ell+\beta_\ell;\ \sum_\ell\beta_\ell=\alpha d+a+1\}.$$

Note that when α runs over $\mathbb{Z}_{\geq 0}$ and $a\in\{0,\dots,d-1\}$, the integer $n=\alpha d+a$ runs over $\mathbb{Z}_{\geq 0}$. This motivates to consider for any $n\in\mathbb{Z}_{\geq 0}$

$$\mathbb{T}_n:=\{(\beta_1,\dots,\beta_\nu)\in[0,m_1]\times\cdots\times[0,m_\nu]\,:\,\sum_\ell\beta_\ell=n+1\}. \tag{6.1.12}$$

Then for d and a fixed $\mathbb{T}_{\alpha d+a}=T^-_{\alpha+1,a}$. Up to the shift $\overline{w}(\alpha\mathfrak{m})$ (whose values is given in (6.1.1)), and which is constant on each $\mathbb{T}_n$, but otherwise depends on $\alpha=\lfloor n/d\rfloor$, the weights on $\mathbb{T}_n\cap\mathbb{Z}^\nu$ are given by the right hand side of (6.1.4), or (6.1.5). The second one, namely (6.1.5), in which case the shift is $\overline{w}((\alpha+1)\mathbf{m})$, gives weights:

$$\overline{W}((\beta_1,\dots,\beta_\nu))=\sum_\ell\#\{\gamma\notin\Gamma_\ell\,:\,\gamma\geq\beta_\ell\}. \tag{6.1.13}$$

Then the weight function $\overline{W}$ restricted on all the level sets $\{\mathbb{T}_n\}_{n\geq 0}$ of $(\mathbb{Z}_{\geq 0})^\nu$ measures the very subtle distribution properties of the semigroups $\{\Gamma_\ell\}_\ell$, and they provide, as a universal object all the lattice cohomologies $\mathbb{H}^q(G(d),k_r)$ for all $q>0$ (and with the additional data of the shifts $\overline{w}((\alpha+1)\mathfrak{m})$, all the lattice cohomologies $\mathbb{H}^0(G(d),k_r)$ too) for all the possible values d and a. Here, and in the next discussion, we denote the dependence of G on d by $G(d)$.

More precisely, for any d and $a\in\{0,\dots,d-1\}$, one has:

$$\mathbb{H}^q(G(d),k_{can}+2aE^*_+)=\bigoplus_{\substack{n\equiv a\ (mod\ d)\\ n\geq 0}}\mathbb{H}^{q-1}_{red}(\mathbb{T}_n,\overline{W})[s_{n,d}] \qquad (q>0), \tag{6.1.14}$$

where $s_{n,d}$ indicates a value of the shift $2\overline{w}((\alpha+1)\mathbf{m})=2(\alpha+1)(1+a-\delta)+d(\alpha+1)\alpha$ (with $\alpha=\lfloor n/d\rfloor$). Moreover, the values $\{\min\overline{W}\,|\mathbb{T}_n\}_n$ and $s_{n,d}$ determine all the cohomology groups $\mathbb{H}^0(G(d),k_r)$ too. The second identity of (6.1.6)(d) together with (6.1.5) reads as:

(6.1.15)

$$-\mathfrak{m}_{k_r}+\text{rank }\mathbb{H}^0_{red}(G(d),k_{can}+2aE^*_+)=\sum_{\substack{n\equiv a\ (mod\ d)\\ n\geq 0}}\min\{\overline{W}\text{ restricted to }\mathbb{T}_n\}.$$

In particular (see (3.2.5) for the definition of $eu(\mathbb{H}^*)$), for any fixed $d>0$ and $a\in\{0,\dots,d-1\}$ one has:

$$-eu(\mathbb{H}^*(G(d),k_{can}+2aE^*_+))=\bigoplus_{\substack{n\equiv a\ (mod\ d)\\ n\geq 0}}eu(\mathbb{H}^*(\mathbb{T}_n,\overline{W})). \tag{6.1.16}$$

(d) Summing up (6.1.14) over a, for any fixed $d > 0$ and $q > 0$,

$$\mathbb{H}^q(G(d)) = \bigoplus_{a=0}^{d-1} \mathbb{H}^q(G(d), k_{can} + 2aE_+^*) = \bigoplus_{n\geq 0} \mathbb{H}^{q-1}_{red}(\mathbb{T}_n, \overline{W})[s_{n,d}]. \tag{6.1.17}$$

On the right hand side of (6.1.17) the numbers $s_{n,d}$ depend on d, but the rank of the right hand side is independent of d. In particular, up to shifts of different direct sum blocks, $\oplus_{q>0}\mathbb{H}^q(G(d))$ is independent on the choice of the integer d. This can also be deduced from the surgery exact sequences (Theorem A and Theorem B) of [**17**] (applied for the special vertex v_+, and using the fact that $G \setminus v_+$ is rational).

EXAMPLE 6.1.18.

(a) Assume that for some d and a one gets $\tilde{\alpha}_0 = 0$. Then $\mathbb{H}^*_{red}(G, k_r) = 0$, and $\mathbb{H}^0(G, k_r) = \mathcal{T}_0^+$.

(b) Assume that for some d and a one gets $\tilde{\alpha}_0 = 1$. Then $\mathbb{H}^*(G, k_r) = \mathbb{H}^*(R_0, \overline{w})$, hence everything is determined by $T^-_{1,a}$. Indeed,

$$\begin{aligned} \min T^-_{1,a} &= \min\Big\{\sum_\ell \#\{\gamma\in\Gamma_\ell\,:\,\gamma\leq i_\ell - 1\},\text{ where }\textstyle\sum_\ell i_\ell = a+1\Big\} \\ &= \min\Big\{\sum_\ell \#\{\gamma\notin\Gamma_\ell\,:\,\gamma\geq i_\ell\},\text{ where }\textstyle\sum_\ell i_\ell = a+1\Big\} + 1 + a - \delta, \end{aligned}$$

$\mathfrak{m}_{k_r} = \min\{0, 1+a-\delta\}$, $\mathbb{H}^0_{red}(G, k_r)$ is generated by one element of degree $2\max\{0, 1+a-\delta\}$, rank $\mathbb{H}^0_{red}(G, k_r) = \min T^-_{1,a} - \max\{0, 1+a-\delta\}$, and finally for $q > 0$ one has $\mathbb{H}^q(G, k_r) = \mathbb{H}^{q-1}_{red}(T^-_{1,a}, \overline{w})$.

(c) If $d \geq \mu - 1$ then $\tilde{\alpha}_0 = 1$ for $a < \mu - 1$, and $\tilde{\alpha}_0 = 0$ for $a \geq \mu - 1$.

REMARK 6.1.19. Assume that we know all the cohomology groups $\{\mathbb{H}^*(G(d), k_r)\}_{k_r}$ for some specific, very large d. Then using them, and also the values $\overline{w}(\alpha\mathbf{m}) = \alpha(1 + a - \delta) + d\alpha(\alpha - 1)/2$ for all $d > 0$, we can recover all the lattice cohomologies $\{\mathbb{H}^*(G(d), k_r)\}_{k_r}$ for any $d > 0$. Indeed, e.g. for any $n \geq 0$ with the choice of $d > \max\{n, \nu\}$ one has

$$\mathbb{H}^{q-1}_{red}(\mathbb{T}_n, \overline{W}) = \mathbb{H}^q(G(d), k_{can} + 2nE_+^*)[-s_{n,d}] \quad (q > 0);$$

and

$$\min\{\overline{W}\,|\mathbb{T}_n\} = rank_{\mathbb{Z}}\, \mathbb{H}^0_{red}(G(d), k_{can} + 2nE_+^*) + \max\{0, 1+n-\delta\}.$$

REMARK 6.1.20. In the above discussion, e.g. in (6.1.9), the space $\mathbb{T}_n$ (intersection of a simplex with a rectangle) can be replaced by the supporting simplex. Indeed, set

$$\Sigma_n := \{(\beta_1, \ldots, \beta_\nu) \in (\mathbb{R}_{\geq 0})^\nu\,:\, \sum \beta_\ell = n+1\}. \tag{6.1.21}$$

The simplex Σ_n has a natural 'generalized cube–decomposition' similarly as $\mathbb{T}_n$, cf. (6.1.9)(b), where the 0–cubes are the lattice points $\Sigma_n \cap \mathbb{Z}^\nu$. Moreover, we take the same weight function on $\Sigma_n \cap \mathbb{Z}^\nu$ as for $\mathbb{T}_n \cap \mathbb{Z}^\nu$ given in (6.1.13).

First assume that $n + 1 \leq M$, i.e. $\mathbb{T}_n = \Sigma_n \cap [0, m_1] \times \cdots \times [0, m_\nu]$ is non–empty. (In fact, if $\alpha < \tilde{\alpha}_0$ then $n = \alpha d + a < \mu - 1 < M$.) We claim that the inclusion $(\mathbb{T}_n, \overline{W}) \hookrightarrow (\Sigma_n, \overline{W})$ induces an isomorphism of the corresponding lattice cohomologies $\mathbb{H}^*(\Sigma_n, \overline{W}) \to \mathbb{H}^*(\mathbb{T}_n, \overline{W})$.

This can be verify as follows. Let us define an application $\varphi : \Sigma_n \cap \mathbb{Z}^\nu \to \Sigma_n \cap \mathbb{Z}^\nu$ as follows. If $\beta = (\beta_1, \ldots, \beta_\nu) \in \mathbb{T}_n \cap \mathbb{Z}^\nu$ then $\varphi(\beta) = \beta$. Otherwise, let i be the smallest index for which $\beta_i > M_i$, and j the largest index for which $\beta_j < M_j$. Then $\varphi(\beta) := (\beta_1, \dots, \beta_i - 1, \dots, \beta_j + 1, \dots, \beta_\nu)$. One can check that $\overline{W}(\varphi(\beta)) \le \overline{W}(\beta)$, repeated application of φ maps $\Sigma_n \cap \mathbb{Z}^\nu$ into $\mathbb{T}_n \cap \mathbb{Z}^\nu$, φ extends to the level of arbitrary dimensional cubes, hence induces the desired isomorphism at the level of lattice cohomology.

If $n + 1 \ge \mu$, then for a certain β with all $\beta_\ell \ge \mu_\ell$ one has $\overline{W}(\beta) = 0$, hence $\min \overline{W} = 0$. Moreover, by similar retraction as above one can show that $\mathbb{H}^*_{red}(\Sigma_n, \overline{W}) = 0$.

In particular, in all the discussion above, $H^*_{red}(\mathbb{T}_n, \overline{W})$ can be replaced by $H^*_{red}(\Sigma_n, \overline{W})$ for all $n \ge 0$.

REMARK 6.1.22. Let $\mathbb{H}^q_r$ denote the r–degree part of $\mathbb{H}^q$. We expect that the generating function $\sum_{n,q,r} \operatorname{rank} \mathbb{H}^q(\mathbb{T}_n)_r h^q t^n s^r \in \mathbb{Z}[[h,t,s]]$ can be organized into a 'nice' expression in terms of the semigroups (and similar result is valid for $\sum_{q,r} \operatorname{rank} \mathbb{H}^q(G(d))_r h^q s^r \in \mathbb{Z}[[h,s]]$ too).

7. The normalized Euler–characteristic and the Seiberg–Witten invariant

7.1. The Euler–characteristic formula. Consider the graphs $\{G^*_\ell\}_{\ell=1}^\nu$, $d > 0$ and the graph $G(d)$ as in (2.3). Let s be the number of vertices of $G(d)$. The main result of this section identified the 'normalized Euler characteristic' of the lattice cohomology with the 'normalized Seiberg–Witten invariant':

THEOREM 7.1.1. *For any $a \in \{0, \dots, d-1\}$ one has:*
(7.1.2)

$$\mathbf{sw}(S^3_{-d}(K), [k_{can} + 2aE^*_+]) + \frac{(k_{can} + 2aE^*_+)^2 + s}{8} = -eu(\mathbb{H}^*(G(d), k_{can} + 2aE^*_+)).$$

PROOF. Consider the polynomial $Q(t) = \sum_i \alpha_i t^i$ as in (2.4.3). Then by (2.4.5)

$$\mathbf{sw}(S^3_{-d}(K), [k_{can} + 2aE^*_+]) + \frac{(k_{can} + 2aE^*_+)^2 + s}{8} = - \sum_{\substack{n \equiv a \ (mod\ d) \\ n \ge 0}} \alpha_n.$$

This compared with (6.1.16) shows that it is enough to prove $\alpha_n = -eu(\mathbb{H}^*(\mathbb{T}_n, \overline{W}))$ for any $n \ge 0$. In particular, the identity (7.1.2) is independent of d, it is enough to prove it for d very large compared with μ (determined by the semigroups Γ_ℓ), and any $0 \le a < d$. In that case, we are in the situation of Example 6.1.18.

The point is that the wanted identity depends merely on the semigroups $\{\Gamma_\ell\}_\ell$, and in fact, it is not even important that they are semigroups of $\mathbb{Z}_{\ge 0}$. Indeed, the statement follows from the following combinatorial property of finite subsets of $\mathbb{Z}_{>0}$. $\square$

PROPOSITION 7.1.3. *Consider ν finite subsets $\Gamma^c_\ell \subset \mathbb{Z}_{>0}$ $(1 \le \ell \le \nu)$, and define $\Gamma_\ell := \mathbb{Z}_{\ge 0} \setminus \Gamma^c_\ell$. (Note that $0 \in \Gamma_\ell$, and even $\Gamma^c_\ell = \emptyset$ is allowed). Let δ_ℓ be the cardinality of Γ^c_ℓ, and set $\delta := \sum_\ell \delta_\ell$. Consider the polynomials $\Delta_l(t)$ defined by $(1-t)\sum_{\gamma \in \Gamma_\ell} t^\gamma$, and set $\Delta(t) := \prod_\ell \Delta_\ell(t)$. Then $\Delta(t)$ can be written as*

$$\Delta(t) = 1 + \delta(t-1) + (t-1)^2 Q(t)$$

for some polynomial $Q(t)$. Denote its coefficients by $Q(t) = \sum_{n \ge 0} \alpha_n t^n$.

On the other hand, one fixes for any ℓ an integer $\mu_\ell > \max \Gamma^c_\ell$ (if $\Gamma^c_\ell = \emptyset$, then one takes an arbitrary $\mu_\ell \geq 0$). Next, for any integer $a \geq 0$ one can consider the lattice cohomology $\mathbb{H}^*(R, w)$ of $R = [0, \mu_1] \times \cdots \times [0, \mu_\nu]$ associated with the weight

$$\begin{aligned} w_a(\mathbf{i}) = w(\mathbf{i}) &:= \sum_\ell \#\{\gamma \in \Gamma^c_\ell \,:\, \gamma \geq i_l\} - \delta + \min\{i, 1+a\} \\ &= \sum_\ell \#\{\gamma \in \Gamma_\ell \,:\, \gamma \leq i_l - 1\} + \min\{0, -i+1+a\}. \end{aligned}$$

where $\mathbf{i} = (i_1, \ldots, i_\nu) \in R \cap \mathbb{Z}^\nu$ and $i = \sum_\ell i_\ell$. Then its normalized Euler characteristic satisfies

$$eu((\mathbb{H}^*(R, w_a)) = \alpha_a.$$

(In particular, it is independent of the choice of the integers μ_ℓ, which can be taken even infinity.)

PROOF. It is convenient to denote the coefficient of t^a of a polynomial $P(t)$ by $P|_a$. Let us start with some remarks.

(a) If we write $\chi_{\Gamma^c_\ell}(t) := \sum_{\gamma \in \Gamma^c_\ell} t^\gamma$ then

$$\Delta_\ell = 1 + (t-1)\chi_{\Gamma^c_\ell} = 1 + (t-1)\delta_\ell + (t-1)^2 Q_\ell(t), \tag{7.1.4}$$

where $Q_\ell = \sum_{\gamma \in \Gamma^c_\ell}(1+t+\ldots+t^{\gamma-1})$. Hence $Q_\ell|_a = \#\{\gamma \in \Gamma^c_\ell \,:\, \gamma > a\}$. $Q_\ell|_0 = \delta_\ell$ since $0 \notin \Gamma^c_\ell$.

(b) Since $\chi_{\Gamma^c_\ell}|_0 = 0$ and $Q = \sum_\ell Q_\ell + \sum_t \tilde{Q}_t$, where each $\tilde{Q}_t$ is multiple of at least one $\chi_{\Gamma^c_\ell}$, one also gets that $Q|_0 = \delta$.

(c) The proof runs over double induction: we will assume that the property is true for $\nu - 1$ finite subsets, and also for any ν finite subsets with total cardinality $\delta - 1$, and we will verify that it is true for ν subsets with total cardinality δ as well. In order to run the induction, we will extend the property to $a = -1$ as well. The weight function w_{-1} will be defined by the same formula as in the theorem, while $Q|_{-1}$, by definition, is δ.

(d) Let us verify the extended property for $\nu = 1$ and any $a \geq -1$. In this case w_a is (non necessarily strictly) increasing on $[0, 1+a]$ and decreasing on $[1+a, \mu_1]$, with values 0 at 0, $-\delta + 1 + a$ at μ_1, and $Q_1|_a - \delta + 1 + a$ at $1 + a$ (cf. part (a)). Hence the identity $eu(\mathbb{H}^*(R, w_a)) = Q_1|_a$ follows.

(e) Next, we verify the other 'starting case' too, namely $\delta = 0$, that is, all the sets Γ^c_ℓ are empty. Then $w_a(\mathbf{i}) = \min\{i, 1+a\}$, hence $\mathbb{H}^*_{red} = 0$, and $\mathbb{H}^0 = \mathcal{T}^+_0$, hence $eu(\mathbb{H}^*) = 0$. But $Q = 0$ too.

(f) Now we will verify the inductive step. We start with ν finite sets as in the statement of the theorem. We assume that Γ^c_ν is non–empty, and set $\gamma_0 := \max \Gamma^c_\nu$. This situation (called 'old' situation) will be compared with that case (called 'new' situation), when all Γ^c_ℓ for $\ell < \nu$ are preserved, but Γ^c_ν is replaced by $\Gamma^c_\nu \setminus \{\gamma_0\}$.

The corresponding 'old' and 'new' invariants (with obvious notation) can be compared. Indeed, by (a), $\Delta^{old}_\nu - \Delta^{new}_\nu = (t-1)t^{\gamma_0}$. Hence, $\Delta^{old} - \Delta^{new} = (t-1)t^{\gamma_0} \cdot \Delta_{<\nu}(t)$ (where the index $< \nu$ indicated the corresponding invariant associated with the first $\nu - 1$ subsets), or

$$\Delta^{old} - \Delta^{new} = t^{\gamma_0}(t-1) \cdot \big(1 + (t-1)(\delta - \delta_\nu) + (t-1)^2 Q_{<\nu}\big). \tag{7.1.5}$$

Since in the 'new' case the total cardinality is $\delta - 1$, from the definition of the Q–polynomials we get

$$\Delta^{old} - \Delta^{new} = t - 1 + (t-1)^2(Q^{old} - Q^{new}). \tag{7.1.6}$$

The identities (7.1.5) and (7.1.6) combined provides

$$Q^{old} - Q^{new} = \frac{t^{\gamma_0} - 1}{t-1} + t^{\gamma_0}(\delta - \delta_\nu) + t^{\gamma_0}(t-1)Q_{<\nu}. \tag{7.1.7}$$

Next, we fix $a \geq -1$, and we will treat the lattice cohomologies associated with the two situtions: the corresponding weigh functions will be denoted by w^{old}, respectively w^{new}. In order to simplify the notation, we write $eu(R,w)$ instead of $eu(\mathbb{H}^*(R,w))$. For both weight functions, by Mayer–Vietoris argument (cf. (3.3.4)) we have

$$\begin{aligned} eu(R,w) = eu(R_{i_\nu \leq \gamma_0}, w) + eu(R_{\gamma_0 \leq i_\nu \leq \gamma_0+1}, w) + eu(R_{i_\nu \geq \gamma_0}, w) \\ -eu(R_{i_\nu = \gamma_0}, w) - eu(R_{i_\nu = \gamma_0+1}, w). \end{aligned} \tag{7.1.8}$$

Now we have to notice several facts. First, $w^{new}(\mathbf{i}) = w^{old}(\mathbf{i})$ if $i_\nu \leq \gamma_0$, and $w^{new}(\mathbf{i}) = w^{old}(\mathbf{i}) + 1$ if $i_\nu \geq \gamma_0 + 1$. Note also that if on a rectangle the weight function increases by N then the euler characteristic decreases by N (because of the $-\mathfrak{m}$–contribution). Also, the choice of γ_0 implies that for any $\mathbf{i}$ with $i_\nu = \gamma_0$ one has $w^{old}(\mathbf{i}) \geq w^{old}(\mathbf{i}+1_\nu)$. Hence $(R_{\gamma_0 \leq i_\nu \leq \gamma_0+1}, w^{old})$ contracts to $(R_{i_\nu=\gamma_0+1}, w^{old})$, showing that in (7.1.8) for w^{old} one has the cancelation $eu(R_{\gamma_0 \leq i_\nu \leq \gamma_0+1}, w^{old}) = eu(R_{i_\nu=\gamma_0+1}, w^{old})$.

The computation will be separated in several subcases.

Case $a < \gamma_0$. In this case for any $\mathbf{i}$ with $i_\nu = \gamma_0$ one has $w^{old}(\mathbf{i}) = w^{old}(\mathbf{i}+1_\nu) + 1$, hence $w^{new}(\mathbf{i}) = w^{new}(\mathbf{i}+1_\nu)$, showing by the same contraction argument that $eu(R_{\gamma_0 \leq i_\nu \leq \gamma_0+1}, w^{new}) = eu(R_{i_\nu=\gamma_0+1}, w^{new})$. Hence considering (7.1.8) for w^{old} and w^{new}, their difference provides

$$eu(R, w^{old}) - eu(R, w^{new}) = 1.$$

This should be compared with (7.1.7). Since by assumption the statement of the theorem is true for the 'new' case, we will have the same fact for the 'old' case provided that $(Q^{old} - Q^{new})|_a = 1$. But this is clear from the right hand side of (7.1.7) since $a < \gamma_0$.

Case $a \geq \gamma_0$. In this case for any $\mathbf{i}$ with $i_\nu = \gamma_0$ one has $w^{new}(\mathbf{i}) \leq w^{new}(\mathbf{i}+1_\nu)$, hence $eu(R_{\gamma_0 \leq i_\nu \leq \gamma_0+1}, w^{new}) = eu(R_{i_\nu=\gamma_0}, w^{new})$. Hence we get

$$eu(R, w^{old}) - eu(R, w^{new}) = eu(R_{i_\nu=\gamma_0+1}, w^{old}) - eu(R_{i_\nu=\gamma_0}, w^{old}). \tag{7.1.9}$$

In order to determine the right hand side, we will analyze $eu(R_{i_\nu=k}, w)$ for any integer $k \leq \gamma_0 + 1$ and $w = w^{old}$. For any $\mathbf{i}$ with $i_\nu = k$ one has the identity $w(\mathbf{i}) = w_{i_\nu=k}(\mathbf{i}) + N(k)$, where

$$w_{i_\nu=k}(\mathbf{i}) = \sum_{\ell<\nu} \#\{\gamma \in \Gamma_\ell^c \,:\, \gamma \geq i_\ell\} - (\delta - \delta_\nu) + \min\{i-k, 1+a-k\},$$

an expression which is the weight function associated with the first $\nu-1$ subsets and $a-k$, and $N(k)$ is a constant independent on $\mathbf{i}$, namely $\#\{\gamma \in \Gamma_\nu^c \,:\, \gamma \geq k\} - \delta_\nu + k$. Since $(R_{i_\nu=k}, w)$ satisfies the statement of the theorem by the inductive assumption, we have:

$$eu(R_{i_\nu=k}, w) = Q_{<\nu}|_{a-k} - N(k) = Q_{<\nu}|_{a-k} - \#\{\gamma \in \Gamma_\nu^c \,:\, \gamma \geq k\} + \delta_\nu - k.$$

Hence the right hand side of (7.1.9) is

$$eu(R_{i_\nu=\gamma_0+1}, w^{old}) - eu(R_{i_\nu=\gamma_0}, w^{old}) = Q_{<\nu}|_{a-\gamma_0-1} - Q_{<\nu}|_{a-\gamma_0}. \tag{7.1.10}$$

Since for $a \geq \gamma_0$ the a–th coefficient of the right hand side of (7.1.7) equals the right hand side of (7.1.10), the inductive step follows again. □

8. Connection with Heegaard–Floer homology

8.1. Review of Heegaard–Floer homology. Let M be an oriented 3–manifold; for simplicity we assume again that M as a rational homology sphere. The Heegaard–Floer homology $HF^+(M)$ was introduced by Ozsváth and Szabó in [**23**]. It is a $\mathbb{Z}[U]$-module with a $\mathbb{Q}$-grading compatible with the $\mathbb{Z}[U]$-action, where $\deg(U) = -2$. Additionally, $HF^+(M)$ also has an (absolute) $\mathbb{Z}_2$-grading; $HF^+_{even}(M)$, respectively $HF^+_{odd}(M)$, denote the part of $HF^+(M)$ with the corresponding parity. Moreover, $HF^+(M)$ has a natural direct sum decomposition of $\mathbb{Z}[U]$-modules (compatible with all the gradings) corresponding to the *spin*^{c}-structures of M:

$$HF^+(M) = \oplus_{[k]\in Spin^c(M)} HF^+(M,[k]).$$

For any *spin*^{c}-structure $[k]$, one has a graded $\mathbb{Z}[U]$-module isomorphism

$$HF^+(M,[k]) = \mathcal{T}^+_{d(M,[k])} \oplus HF^+_{red}(M,[k]),$$

where $HF^+_{red}(M,[k])$ has a finite $\mathbb{Z}$-rank and an induced $\mathbb{Z}_2$-grading. One also considers

$$\chi(HF^+(M,[k])) := \operatorname{rank}_{\mathbb{Z}} HF^+_{red,even}(M,[k]) - \operatorname{rank}_{\mathbb{Z}} HF^+_{red,odd}(M,[k]).$$

Then one recovers the Seiberg-Witten topological invariant of $(M,[k])$ (see [**27**]) via

$$\mathbf{sw}(M,[k]) := \chi(HF^+(M,[k])) - d(M,[k])/2. \tag{8.1.1}$$

With respect to the change of orientation the above invariants behave as follows: $d(M,[k]) = -d(-M,[k])$ and $\chi(HF^+(M,[k])) = -\chi(HF^+(-M,[k]))$.

If M is an integral homology sphere then for the unique *spin*^{c}–structure $\mathbf{sw}(M)$ equals the Casson invariant $\lambda(M)$ (normalized as in [**10**] (4.7)).

8.2. Lattice homology and Heegaard–Floer homology. In the sequel we set again $M = S^3_{-d}(K)$. First, assume that $\nu = 1$. Then G is almost rational, hence $\mathbb{H}^q(-M) = 0$ for any $q > 0$, cf. (4.2.11). Moreover, by [**25, 13, 15**], we have:

THEOREM 8.2.1. *Assume that $\nu = 1$. For any $[k] \in Spin^c(M)$*

$$HF^+_{odd}(-M,[k_r]) = 0,$$

and

$$HF^+_{even}(-M,[k_r]) = \mathbb{H}^0(G,k_r)\Big[-\frac{k_r^2+|\mathcal{J}|}{4}\Big].$$

In particular

$$d(M,[k_r]) = \max_{k'\in[k_r]} \frac{(k')^2+|\mathcal{J}|}{4} = \frac{k_r^2+|\mathcal{J}|}{4} - 2\mathfrak{m}_{k_r}. \tag{8.2.2}$$

The next theorem generalizes this fact for $\nu = 2$. Again, by (4.2.11) one gets $\mathbb{H}^q(-M) = 0$ for $q > 1$. Additionally:

THEOREM 8.2.3. *Assume that $\nu = 2$. Then for any $[k_r] \in Spin^c(M)$*

$$HF^+_{even}(-M, [k_r]) = \mathbb{H}^0(G, k_r)[-\overline{d}]$$

and

$$HF^+_{odd}(-M, [k_r]) = \mathbb{H}^1(G, k_r)[-\overline{d}],$$

where $\overline{d} := (k_r^2 + |\mathcal{J}|)/4$. In particular, (8.2.2) is valid in this case too.

PROOF. Let G be the plumbing graph of $S^3_{-d}(K)$ as above, G^- the graph obtained from G via replacing the decoration -1 of $v_{0,1}$ with -2, and finally, $G^\dagger$ the graph obtained from G by removing the vertex $v_{0,1}$ and all adjacent edges. From (5.1.1) follows that

(8.2.4) both graphs G^- and $G^\dagger$ are almost rational.

Let $HF^+_{e/o}(-\Gamma)$ be the even/odd Heegaard–Floer homology of the $-M(\Gamma)$, where $M(\Gamma)$ is the plumbed 3–manifold associated with Γ. Then one has the following commutative diagram:

$$\begin{array}{ccccccccc} 0 \longrightarrow HF^+_e(-G) & \longrightarrow & HF^+_e(-G^-) & \longrightarrow & HF^+_e(-G^\dagger) & \longrightarrow & HF^+_o(-G) \longrightarrow 0 \\ \downarrow T_G & & \downarrow T_{G^-} & & \downarrow T_{G^\dagger} & & \\ 0 \longrightarrow \mathbb{H}^0(G) & \longrightarrow & \mathbb{H}^0(G^-) & \longrightarrow & \mathbb{H}^0(G^\dagger) & \longrightarrow & \mathbb{H}^1(G) \longrightarrow 0 \end{array}$$

The first line is exact by [**25**, §2], and using the fact that $HF^+_o(-G^-) = HF^+_o(-G^\dagger) = 0$. This type of vanishing is proved in [**13**] for almost rational graphs, cf. (8.2.4).

The second line is exact by [**17**], and from the vanishing $\mathbb{H}^1(G^-) = \mathbb{H}^1(G^\dagger)$ (here we use again (8.2.4) and (4.2.11)). The natural maps T_G, T_{G^-} and $T_{G^\dagger}$ are constructed in [**25**]. Since G^- and $G^\dagger$ are almost rational, the maps T_{G^-} and $T_{G^\dagger}$, are isomorphisms by [**13**]. Hence, T_G is also an isomorphisms, and and $T_{G^\dagger}$ induces an isomorphism $HF^+_o(-G) \to \mathbb{H}^1(G)$ as well. □

REMARK 8.2.5.

(a) For any $\nu \geq 1$ the identities (7.1.1) and (8.1.1) imply that

(8.2.6)

$$\sum_q (-1)^q \operatorname{rank}_{\mathbb{Z}} \mathbb{H}^q_{red}(G, k_r) = \chi(HF^+(-M, [k_r])) + \frac{d(M, [k_r])}{2} - \frac{k_r^2 + |\mathcal{J}|}{8} + \mathfrak{m}_{k_r}.$$

We predict, cf. Conjecture 5.2.4 in [**16**], that

$$HF^+_{red,even}(-M, [k_r]) = \bigoplus_{q\ even} \mathbb{H}^q_{red}(G, k_r)[-\overline{d}],$$

and

$$HF^+_{red,odd}(M, [k_r]) = \bigoplus_{q\ odd} \mathbb{H}^q_{red}(G, k_r)[-\overline{d}]$$

and (8.2.2) is valid for any $\nu \geq 1$. This is compatible with the above theorems, and with the identity (8.2.6) already proved.

(b) The main obstruction to prove the conjectured isomorphisms connecting the Heegaard–Floer homologies with lattice cohomologies is that at this moment we know no natural application connecting them, except level $q = 0$ (as in the proof of (8.2.3)). Both isomorphisms obtained in the proof of Theorem 8.2.3 are induced by 0–level morphisms.

(c) Although Theorem 8.2.3 does not answer the arbitrary case, it is still important: it is the first family which identifies completely a non–zero $HF^+_{odd}(-M,[k])$ in terms of the combinatorial lattice cohomology.

9. Example. The case of ν cusps of type $(2,3)$

9.1. The setup. Assume that for all ℓ the knot K_ℓ is the $(2,3)$ torus knot, i.e. $\Gamma_\ell=\{0,2,3,\cdots\}$, $\Delta_\ell=t^2-t+1$, $\delta_\ell=1$. Hence $\delta=\nu$.

Our goal is to analyze the lattice cohomology $\mathbb{H}^*_{red}(\Sigma_n,\overline{W})$ for any $n\geq 0$, cf. (6.1.20). Recall that Σ_n is the *real* simplex $\{\beta\in\mathbb{R}^\nu_{\geq 0}\,:\,\sum\beta_\ell=n+1\}$ with lattice points $L\Sigma_n:=\Sigma_n\cap\mathbb{Z}^\nu$.

In the next discussion it is convenient to set $f_2(\beta):=\#\{\ell\,:\,\beta_\ell\geq 2\}$ for any $\beta\in L\Sigma_n$. Then it is easy to see that $\overline{W}(\beta)=\nu-f_2(\beta)$. Note that $f_2(\beta)\leq\nu$ and also $f_2(\beta)\leq\lfloor(n+1)/2\rfloor$, hence $\max f_2=\min\{\nu,\lfloor(n+1)/2\rfloor$. Therefore,

$$\min\overline{W}|\Sigma_n=\max\Big\{0,\nu-\Big\lfloor\frac{n+1}{2}\Big\rfloor\Big\}. \tag{9.1.1}$$

The function $\beta\mapsto\#\{\ell\,:\,\beta_\ell\geq 2\}$ provides a stratification of Σ_n, which is finer than the standard skeleton decomposition of the simplex Σ_n. Moreover, the associated (generalized) cube decomposition of Σ_n also depends on the integer n (i.e. the structure of the lattice points on the simplex).

Let us denote by $Sk_r=Sk_r(\Sigma_n)$ the real r–skeleton of the $\nu-1$–dimensional Σ_n, namely

$$Sk_r:=\{\beta\in\Sigma_n\,:\,\text{at most for } r+1 \text{ indices } \ell \text{ one has } \beta_\ell>0\}.$$

Clearly, $Sk_0\subset\cdots\subset Sk_{\nu-2}\subset Sk_{\nu-1}=\Sigma_n$, $Sk_r=\emptyset$ for $r<0$ and $Sk_r=\Sigma_n$ for $r\geq\nu-1$.

For $0\leq r\leq\nu-1$, each Sk_r is a simplicial complex of dimension r, and $Sk_{\nu-2}$ is homeomorphic with the sphere $S^{\nu-2}$. The lattice point distribution of $L\Sigma_n$ in the skeletons is reflected in the next definition:

DEFINITION 9.1.2. Set $LSk_r:=Sk_r\cap\mathbb{Z}^\nu=Sk_r\cap L\Sigma_n$.

(a) For any r define $\phi(r)=\phi_n(r)$ as the largest integer $r'\leq\max\{r,\nu-1\}$ such that $Sk_{r'}$ contains all the lattice points $L\Sigma_n$ sitting in Sk_r, but no other lattice points.

(b) For any $n\geq 0$ and $r\in\mathbb{Z}$ define

$$K^n_r:=Sk_{\phi_n(r)}.$$

LEMMA 9.1.3. *(a)*

$$K^n_r=\begin{cases}\emptyset & \text{if } r<0,\\ \Sigma_n & \text{if } r\geq n \text{ or } r\geq\nu-1,\\ Sk_r & \text{otherwise.}\end{cases}$$

(b) Let $\tilde{K}^n_r$ be the union of all generalized cubes with vertices in $LSk_r(\Sigma_n)$. Then K^n_r is the largest skeleton included in $\tilde{K}^n_r$, and $K^n_r\subset\tilde{K}^n_r$ is a homotopy equivalence.

PROOF. (a) If $r<0$ then $\phi(r)=-1$, hence $Sk_{\phi(r)}=\emptyset$. If $r\geq\nu-1$, then $\phi(r)=r$, and $Sk_{\phi(r)}=\Sigma_n$. Next assume that $0\leq r\leq\nu-2$. Then $LSk_{r+1}\setminus LSk_r\neq\emptyset$ if and only if $r\leq n-1$. Therefore, if $r\leq n-1$, then $\phi(r)=r$, otherwise $\phi(r)=\nu-1$. Part (b) follows from the definitions. □

EXAMPLE 9.1.4. Both situations $K^n_r=\tilde{K}^n_r$ and $K^n_r\subsetneq\tilde{K}^n_r$ might happen. First assume that $\nu=2$ and $n=1$. Then $K^1_1=\tilde{K}^1_1=\Sigma_1=Sk_2(\Sigma_1)$. But, if $\nu=2$ and $n=2$, then $K^2_1=Sk_1(\Sigma_2)\subsetneq\tilde{K}^2_1$. Indeed, note that Σ_2 can be cut into nine small triangles, these are the generalized cubes of Σ_2. Then $\tilde{K}^2_1$ contains additionally to K^2_1 the three little triangles, which are neighbors of $Sk_0(\Sigma_2)$.

Recall, that the lattice cohomology $\mathbb{H}^*_{red}(\Sigma_n,\overline{W})$ can also be computed via the usual (simplicial) cohomology, cf. (3.3.1), via the identity

$$\mathbb{H}^*_{red}(\Sigma_n,\overline{W})=\sum_{k\geq\min\overline{W}|\Sigma_n}\tilde{H}^*(S_k(\Sigma_n),\mathbb{Z}),\tag{9.1.5}$$

where $S_k(\Sigma_n)$ is the union of all generalized cubes in Σ_n whose vertices have weight $\leq k$. The next proposition provides (the homotopy type of) this simplicial complex in terms of the skeletons of Σ_n.

PROPOSITION 9.1.6. *For any k one has the homotopy equivalence*

$$S_k(\Sigma_n)\sim K^n_{n+k-\nu}\setminus K^n_{\nu-k-2}.\tag{9.1.7}$$

The right hand side written in this form covers the cases $k<\min\overline{W}|\Sigma_n$ too. Indeed,

$$K^n_{n+k-\nu}\setminus K^n_{\nu-k-2}=\begin{cases}\emptyset & \textit{if}\ \ k<\min\overline{W}\\ K^n_{n+k-\nu} & \textit{if}\ \ k>\nu-2\\ Sk_{n+k-\nu}\setminus Sk_{\nu-k-2} & \textit{if}\ \ \min\overline{W}\leq k\leq\nu-2.\end{cases}\tag{9.1.8}$$

PROOF. First we verify (9.1.8). If $k<0$ then $\nu-k-2\geq\nu-1$, hence $K^n_{\nu-k-2}=\Sigma_n$. Similarly, if $k<\nu-\lfloor(n+1)/2\rfloor$, then $n+k-\nu\leq\nu-k-2$ hence $K^n_{n+k-\nu}\setminus K^n_{\nu-k-2}=\emptyset$. This proves the first row. If $k>\nu-2$ then $\nu-k-2<0$ giving the second row. Finally, assume the third case. Then $k\geq\min\overline{W}$ implies $\nu-k-2\leq\min\{\nu-2,n-1\}$, hence $K^n_{\nu-k-2}=Sk_{\nu-k-2}$, cf. (9.1.3)(a). Furthermore, $k\leq\nu-2$ implies $n+k-\nu\leq n-2$, hence $K^n_{n+k-\nu}=Sk_{n+k-\nu}$ too by (9.1.3)(a).

Now we start to prove (9.1.7). It is convenient to write sometimes Σ_n as $\Sigma^{\nu-1}_n$ in order to emphasize both its dimension and its lattice point structure.

For any $\beta\in L\Sigma_n$ write $f_1(\beta):=\#\{\ell:\beta_\ell\geq1\}$. One verifies that $f_1\leq\min\{n+1,\nu\}$ and

$$f_2\leq\min\{f_1,n+1-f_1\}\leq n+1.\tag{9.1.9}$$

Our goal is to describe $S_k(\Sigma_n)$. For this first we have to list those lattice points $\beta\in L\Sigma_n$ with $\overline{W}(\beta)\leq k$, or $f_2(\beta)\geq\nu-k$.

Assume that for some $\beta\in L\Sigma_n$ one has $f_1(\beta)=p+1$. This means exactly that $\beta\in LSk_p(\Sigma_n)\setminus LSk_{p-1}(\Sigma_n)$. Hence β is a lattice point in the interior of one of the open p–faces. The convex closure of such points on a fixed p–face is a p–simplex, and its lattice points have the structure of $L\Sigma^p_{n-1-p}$ (if $p>0$ then on any edge there are $n-p$ lattice points). Their union is denoted by $Sk^o_p(\Sigma_n)$ with lattice points $LSk^o_p(\Sigma_n)$.

Moreover, for such points $f_2(\beta)\geq\nu-k$ if and only if

$$\beta\in L\Sigma^p_{n-1-p}\setminus LSk_{\nu-k-2}(\Sigma^p_{n-1-p}).$$

Next, we wish to list the possible values p. For this we use (9.1.9). If $f_1(\beta) \leq n+1-f_1(\beta)$, i.e. $p+1 \leq (n+1)/2$, then $f_2(\beta) \leq f_1(\beta)$, thus $\nu - k \leq p+1$. If $p+1 \geq (n+1)/2$, then $f_2(\beta) \leq n+1-f_1(\beta)$, or $\nu - k \leq n+1-p-1$. Hence the possible p–values are $\nu - k - 1 \leq p \leq n+k-\nu$. In fact, the first inequality as restriction is superfluous; indeed if $p \leq \nu - k - 2$ then $L\Sigma^p_{n-1-p} \setminus LSk_{\nu-k-2}(\Sigma^p_{n-1-p}) = \emptyset$. Summed up, we get

$$\begin{aligned} S_k(\Sigma_n) \cap \mathbb{Z}^\nu &= \bigcup_{p \leq n+k-\nu} LSk^o_p(\Sigma_n) \setminus LSk_{\nu-k-2}(Sk^o_p(\Sigma_n)) \\ &= LSk_{n+k-\nu}(\Sigma_n) \setminus \bigcup_{p \leq n+k-\nu} LSk_{\nu-k-2}(Sk^o_p(\Sigma_n)). \end{aligned} \tag{9.1.10}$$

Now we are ready to verify (9.1.7). If $k < \min \overline{W}$ then both sides are empty, cf. (9.1.8). If $k > \nu - 2$, then (9.1.10) reads as $S_k(\Sigma_n) \cap \mathbb{Z}^\nu = LSk_{n-k-\nu}$, hence $S_k(\Sigma_n) = \tilde{K}^n_{n-k-\nu}$, which by (9.1.3)(b) has the same homotopy type as $K^n_{n-k-\nu}$ and which agree with (9.1.8). Finally, assume that $\min \overline{W} \leq k \leq \nu - 2$. Recall from the first paragraph of the proof (the proof of the third row of (9.1.8), $\phi(n+k-\nu) = n+k-\nu$, i.e. the cubes filling $LSk_{n+k-\nu}$ provide exactly $\tilde{K}^n_{n+k-\nu}$ whose homotopy type is $K^n_{n+k-\nu} = Sk_{n+k-\nu}$.

The point is that in the range $p \leq n+k-\nu$, both inequalities $\nu-k-2 \leq p-1$ and $\nu - k - 2 < n-1-p$ hold (compare with (9.1.3)(a)), hence each p–simplex Σ^p_{n-1-p} contains additional lattice points which are not in $LSk_{\nu-k-2}(\Sigma^p_{n-1-p})$. Hence the union of the cubes missing from Σ^p_{n-1-p} due to the lattice points $\cup_p LSk_{\nu-k-2}(\Sigma^p_{n-1-p})$ forms a regular tubular neighborhood of $Sk_{\nu-k-2}$. □

We expect that the space $Sk_k \setminus Sk_l$ was already studied in the literature, nevertheless we were not able to find a reference for its homotopy type or homology; therefore we present the corresponding statement in the next proposition.

PROPOSITION 9.1.11. *Let Sk_k be the k–skeleton of the $(\nu-1)$–dimensional simplex $\Sigma = \Sigma^{\nu-1}$. Assume that $-1 \leq l < k \leq \nu - 1$. Then $Sk_k \setminus Sk_l$ has the homotopy type of a bouquet of spheres of dimension $k-l-1$:*

$$Sk_k \setminus Sk_l \sim \bigvee_{c_{k,l}} S^{k-l-1},$$

where

$$c_{k,l} = (-1)^{k-l-1}\Big[-1 + \sum_{i=l+1}^{k} (-1)^{i-l-1}\binom{\nu}{i+1}\binom{i}{l+1}\Big].$$

REMARK 9.1.12. Before we start to prove (9.1.11), let us list some particular cases (which are essential steps in the proof too). Note also that a bouquet of c ($c \geq 0$) copies of S^0 consists of $c+1$ points. Furthermore, the next combinatorial identity sometimes is applicable:

$$\sum_{i\in\mathbb{Z}} (-1)^{i-l-1}\binom{\nu}{i+1}\binom{i}{l+1} = 1. \tag{9.1.13}$$

It can be proved as follows. Expand $(1+z)^\nu$ into $\sum_k \binom{\nu}{k} z^k$, and $(1+z)^{-(l+2)}$ into $\sum_k (-1)^k \binom{l+1+k}{l+1} z^k$. Then the coefficient of $z^{\nu-(l+2)}$ in the product $(1+z)^\nu(1+z)^{-(l+2)}$ is 1.

(a) Assume that $0 \leq k \leq \nu - 1$ and $l = -1$. Then the dimension of the spheres is k and their number is $\binom{\nu-1}{k+1}$. (Use $\sum_{i=0}^{k+1}(-1)^{k+i-1}\binom{\nu}{i} = \binom{\nu-1}{k+1}$.)

(b) Assume that $0 \leq l \leq \nu - 2$ and $k = \nu - 1$. Then $\Sigma \setminus Sk_l$ is contractible (use (9.1.13)).

(c) Assume that $0 \leq l \leq \nu - 3$ and $k = \nu - 2$. Then the dimension of the spheres is $\nu - l - 3$ and their number is $\binom{\nu-1}{l+1}$ (use again (9.1.13)).

PROOF. The $(\nu - 1)$–dimensional simplex Σ has a natural cell–decomposition providing the skeleton decomposition too. Let C_k be the free $\mathbb{Z}$–module generated by the k–cells; it has rank $\binom{\nu}{k+1}$, $0 \leq k \leq \nu - 1$. Let $\partial_{k+1} : C_{k+1} \to C_k$ be the boundary map, $C_{-1} = \mathbb{Z}$, and $\partial_0 : C_0 \to C_{-1}$ the augmentation. The augmented complex $C_\bullet$ is exact. By induction on k one shows that the rank of $im\, \partial_k$ is $\binom{\nu-1}{k}$. The augmented chain–complex associated with Sk_k is $0 \to C_k \to \cdots \to C_{-1} \to 0$, which is exact except at place k, where the homology is $im\, \partial_{k+1}$. Since Σ is simply connected , Sk_2 is simply connected too, hence the homotopy type follows by Whitehead theorem. This provides the case $l = -1$ corresponding to (9.1.12)(a).

If $k = \nu - 1$ then $\Sigma \setminus Sk_l$ contracts to the barycenter of Σ, cf. (9.1.12)(b).

Assume that $k = \nu - 2$ and $l \geq 0$. Then $Sk_{\nu-2} = S^{\nu-2}$, hence the homological statement follows from Alexander duality. For the homotopy statement, or for an independent proof one can use the dual cell decomposition of the simplex. This shows the case (9.1.12)(c). The last two cases also prove the following statement:

Fact. *Let Σ^r be an r–dimensional simplex with boundary $\partial\Sigma^r$. Let $l < r-1$. Then the pair $(\Sigma^r, \partial\Sigma^r \setminus Sk_l)$ has the homotopy type of the pair $(Cone(\partial\Sigma^r \setminus Sk_l)), \partial\Sigma^r \setminus Sk_l)$. Since $\partial\Sigma^r \setminus Sk_l$ has the homotopy type of a bouquet of $(r-l-2)$–dimensional spheres, gluing Σ^r along $\partial\Sigma^r \setminus Sk_l$ is equivalent by gluing $(r - l - 1)$–cells. Their number is $\binom{r}{l+1}$.*

Now we prove the general case by increasing induction on k. Set $X_k := Sk_k \setminus Sk_l$. Start with $k = l + 1$. Then X_{l+1} is the collection of open $l + 1$ faces, hence it has the homotopy type of $\binom{\nu}{l+2}$ points. Next, we take X_{l+2}, it contains the space X_{l+1}, and we glue to it the $(l + 2)$ faces Σ^{l+2} along their partial boundaries $\partial\Sigma^{l+2} \setminus Sk_l$. Applied Fact for $r = l + 2$, each face contributes with $\binom{l+2}{l+1}$ copies of 1–cells. The total number of $l + 2$–faces is $\binom{\nu}{l+3}$. Hence we have to glue $\binom{\nu}{l+3}\binom{l+2}{l+1}$ 1–cells.

By similar argument, X_{l+3} is obtained from X_{l+2} by gluing $\binom{\nu}{l+4}\binom{l+3}{l+1}$ 2–cells. If we continue in this way finally we get $X_{\nu-1}$, which is contractible, hence when we glue in this inductive procedure the r cells we have to kill totally the r homology (and for $r = 2$, the fundamental group). Hence at each step we get a bouquet of spheres, and their number can be determined by an Euler characteristic argument. This is exactly the formula of the statement. $\square$

9.2. Particular cases.

EXAMPLE 9.2.1. Since $\tilde{H}^q(S_k(\Sigma_n), \mathbb{Z}) = 0$ for $q \geq \nu - 1$, we get that $\mathbb{H}^q_{red}(\Sigma_n, \overline{W}) = 0$ for $q \geq \nu - 1$; this also shows that $\mathbb{H}^q(S^3_{-d}(K)) = 0$ for any $q \geq \nu$, a fact compatible with (4.2.11).

In this example we wish to investigate $\tilde{H}^{\nu-2}(S_k(\Sigma_n))$ for any fixed $\nu \geq 2$. This group can be non–zero only if in (9.1.6) we have $Sk_k(\Sigma_n) = Sk_{\nu-2}$, or if the following three facts hold simultaneously: $k - \nu + n = \nu - 2$, $\nu - k - 2 < 0$ and

$k - \nu + n < n$. This system has a unique solution, namely $\nu = k + 1 = n + 1$. Therefore,

$$\operatorname{rank} \mathbb{H}^{\nu-1}(S^3_{-d}(K)) = \operatorname{rank} \mathbb{H}^{\nu-2}_{red}(\Sigma_{\nu-1}, \overline{W}) = 1.$$

(Compare also with (4.4.1) and (4.4.2) of [**16**].)

EXAMPLE 9.2.2. Assume that $\nu = 3$ and $d = 2$. For $a = 0$ one has $\mathfrak{m}_{k_r} = -2$, $\mathbb{H}^0_{red}(G(d), k_r) = \mathcal{T}_0(1)^2 \oplus \mathcal{T}_{-4}(2)$ of rank 4, $\mathbb{H}^1(G(d), k_r) = 0$ and $rank\, \mathbb{H}^2(G(d), k_r) = 1$. Hence $eu(\mathbb{H}^*(G(d), k_r)) = 7$.

On the other hand, for $a = 1$, one has $\mathfrak{m}_{k_r} = -1$, $\mathbb{H}^0_{red}(G(d), k_r) = \mathcal{T}_0(1)^2$ of rank 2, and rank $\mathbb{H}^1(G(d), k_r) = 4$, while $\mathbb{H}^2(G(d), k_r) = 0$. Hence $eu(\mathbb{H}^*(G(d), k_r)) = -1$. This shows that $eu(\mathbb{H}^*(G(d), k_r)) < 0$ can be realized.

The above data can be compared with the coefficients of the polynomial Q in the light of (7.1.1) and (7.1.3): by an easy computation $Q(t) = t^4 - t^3 + 3t^2 + 3$. Note that the sum of the coefficients of the monomials with even (respectively odd) exponent is 7 (resp. -1).

10. Another example

Assume that $\nu = 2$, both knots are torus knots, one of them of type $(3, 4)$ the other $(2, 7)$. Hence $\mu_1 = \mu_2 = 6$ and $\Gamma^c_1 = \{1, 2, 5\}$ and $\Gamma^c_2 = \{1, 3, 5\}$. The Alexander polynomials are $\Delta_1(t) = t^6 - t^5 + t^3 - t + 1$ and $\Delta_2(t) = t^6 - t^5 + t^4 - t^3 + t^2 - t + 1$. By a computation one gets $Q(t) = \sum_i \alpha_i t^i = t^{10} + t^8 + t^7 + 2t^6 + t^5 + 3t^4 + 3t^3 + 4t^2 + 4t + 6$. In the next example we take $d = 3$ and $a = 0$. The plumbing graph G is:

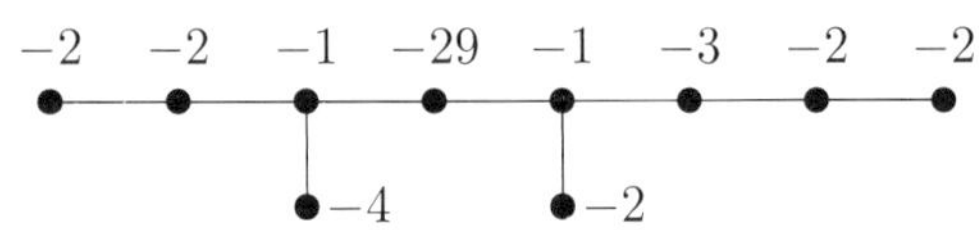

The bound $\tilde{\alpha}_0$ is 4, so we have to analyze four little rectangle. The values $\overline{w}(\alpha\mathbf{m})$ for $0 \leq \alpha \leq 4$ are 0, -5, -7, -6 and -2 respectively. The next table contains all the information regarding these values the weights along $T^-_{\alpha+1}$ and the ranks of $\mathbb{H}^0_{red}(\mathbb{T}_{3\alpha})$ (note that $\mathbb{H}^q_{red}(\mathbb{T}_{3\alpha}) = 0$ for $q \geq 1$).

	$\alpha = 0$	$\alpha = 1$	$\alpha = 2$	$\alpha = 3$
$\min \overline{w}\vert T^-_{\alpha+1}$	1	-3	-4	-1
$\min \overline{W}\vert \mathbb{T}^-_{3\alpha}$	6	4	2	1
$\overline{w}((\alpha+1)\mathbf{m})$	-5	-7	-6	-2
$rank\, \mathbb{H}^0_{red}(\mathbb{T}_{3\alpha})$	0	1	0	1
$-eu(\mathbb{H}^*(\mathbb{T}_{3\alpha}, \overline{W})$	6	3	2	0
$\alpha_{3\alpha}$	6	3	2	0

Hence the interested reader might exemplify using these entries the corresponding theoretical results from the body of the paper.

In particular, for $G = G(3)$ and $a = 0$ we get: $\min \overline{w} = -7$, rank $\mathbb{H}^0_{red}(G, can) = 6$, while rank $\mathbb{H}^1(G, can) = 2$, hence $eu\, \mathbb{H}^*(G, can) = 11$.

References

[1] A'Campo, N.: La fonction zeta d'une monodromy, *Com. Math. Helvetici*, **50** (1975), 233-248. MR0371889 (51:8106)

[2] Artin, M.: Some numerical criteria for contractibility of curves on algebraic surfaces. *Amer. J. of Math.*, **84** (1962), 485-496. MR0146182 (26:3704)

[3] Braun, G. and Némethi, A.: Surgery formula for the Seiberg-Witten invariants of negative definite plumbed 3-manifolds, *Journal für die Reine und angewandte Mathematik*, **638** (2010), 189-208. MR2595340 (2011c:57033)

[4] Brieskorn, E. and Knörrer, H.: *Plane Algebraic Curves*, Birkhäuser, Boston, 1986. MR886476 (88a:14001)

[5] Eisenbud, D. and Neumann, W.: *Three-dimensional link theory and invariants of plane curve singularities*, Ann. of Math. Studies, **110**, Princeton University Press, 1985. MR817982 (87g:57007)

[6] Gompf, R.E. and Stipsicz, I.A.: An Introduction to 4-Manifolds and Kirby Calculus, *Graduate Studies in Mathematics*, vol. **20**, Amer. Math. Soc., 1999. MR1707327 (2000h:57038)

[7] Gusein-Zade, S.M., Delgado, F. and Campillo, A.: On the monodromy of a plane curve singularity and the Poincaré series of the ring of functions on the curve, *Funct. Analysis and its Applications*, **33**(1) (1999), 56-67. MR1711890 (2000f:32042)

[8] László, T. and Némethi, A.: Reduction theorem for lattice cohomology, manuscript in preparation.

[9] Laufer, H.B.: On rational singularities, *Amer. J. of Math.*, **94** (1972), 597-608. MR0330500 (48:8837)

[10] Lescop, C.: *Global Surgery Formula for the Casson-Walker Invariant*, Ann. of Math. Studies, **140**, Princeton Univ. Press, 1996. MR1372947 (97c:57017)

[11] Milnor, J.: *Singular points of complex hypersurfaces*, Ann. of Math. Studies, **61**, Princeton Univ. Press, 1968. MR0239612 (39:969)

[12] Némethi, A.: Dedekind sums and the signature of $f(x,y)+z^N$, II, *Selecta Math.*, New series, **5** (1999), 161-179. MR1694898 (2000f:32039)

[13] Némethi, A.: On the Ozsváth-Szabó invariant of negative definite plumbed 3-manifolds, *Geometry and Topology* **9** (2005), 991-1042. MR2140997 (2006c:57011)

[14] Némethi, A.: On the Heegaard Floer homology of $S^3_{-d}(K)$ and unicuspidal rational plane curves, *Fields Institute Communications*, Vol. **47**, 2005, 219-234; "Geometry and Topology of Manifolds", Editors: H.U. Boden, I. Hambleton, A.J. Nicas and B.D. Park, (Proceedings of the Conference at McMaster University, May 2004). MR2189934 (2007g:57026)

[15] Némethi, A.: On the Heegaard Floer homology of $S^3_{-p/q}(K)$, math.GT/0410570, publishes as part of [**18**].

[16] Némethi, A.: Lattice cohomology of normal surface singularities, *Publ. RIMS. Kyoto Univ.*, **44** (2008), 507-543. MR2426357 (2009m:32053)

[17] Némethi, A.: Two exact sequences for lattice cohomology, Proceedings of the conference organized to honor H. Moscovici's 65th birthday, *Contemporary Math.* **546** (2011), 249–269. MR2815139

[18] Némethi, A.: Graded roots and singularities, Proceedings *Advanced School and Workshop* on *Singularities in Geometry and Topology* ICTP (Trieste, Italy), World Sci. Publ., Hackensack, NJ, 2007, 394–463. MR2311495 (2009c:14069)

[19] Némethi, A.: Five lectures on normal surface singularities, lectures at the Summer School in *Low dimensional topology* Budapest, Hungary, 1998; Bolyai Society Math. Studies **8** (1999), 269-351. MR1747271 (2001g:32066)

[20] Némethi, A.: The Seiberg–Witten invariants of negative definite plumbed 3–manifolds, *Journal of EMS* **13**(4) (2011), 959–974. MR2800481

[21] Némethi, A. and Nicolaescu, L.I.: Seiberg-Witten invariants and surface singularities, *Geometry and Topology*, Volume **6** (2002), 269-328. MR1914570 (2003i:14048)

[22] Neumann, W.D.: A calculus for plumbing applied to the topology of complex surface singularities and degenerating complex curves. *Transactions of the AMS*, **268** (2) (1981), 299-344. MR632532 (84a:32015)

[23] Ozsváth, P.S. and Szabó, Z.: Holomorphic disks and topological invariants for closed three-manifolds, *Ann. of Math.*, **(2) 159** (2004), no. 3, 1027–1158. MR2113019 (2006b:57016)

[24] Ozsváth, P.S. and Szabó, Z.: Holomorphic discs and three-manifold invariants: properties and applications, *Annals of Math.*, **159** (2004), 1159–1245. MR2113020 (2006b:57017)
[25] Ozsváth, P.S. and Szabó, Z.: On the Floer homology of plumbed three-manifolds, *Geom. Topol.* **7** (2003), 185-224. MR1988284 (2004h:57039)
[26] Ozsváth, P. and Szabó, Z.: Knot Floer homology and integer surgeries, math.GT/0410300.
[27] Rustamov, R.: A surgery formula for renormalized Euler characteristic of Heegaard Floer homology, math.GT/0409294.

A. Rényi Institute of Mathematics, 1053 Budapest, Reáltanoda u. 13-15, Hungary
E-mail address: nemethi@renyi.hu

Depart. de Álgebra, Universidad Complutense de Madrid, Plaza de Ciencias s/n, E-28040 Madrid, Spain
E-mail address: ferchu@mat.ucm.es

Part IV: Zeta functions for groups and representations

Contemporary Mathematics
Volume **566**, 2012
http://dx.doi.org/10.1090/conm/566/11226

Representation zeta functions of some compact p-adic analytic groups

Nir Avni, Benjamin Klopsch, Uri Onn, and Christopher Voll

ABSTRACT. Using the Kirillov orbit method, novel methods from $\mathfrak{p}$-adic integration and Clifford theory, we study representation zeta functions associated to compact p-adic analytic groups. In particular, we give general estimates for the abscissae of convergence of such zeta functions.

We compute explicit formulae for the representation zeta functions of some compact p-adic analytic groups, defined over a compact discrete valuation ring $\mathfrak{o}$ of characteristic 0. These include principal congruence subgroups of $\mathrm{SL}_2(\mathfrak{o})$, without any restrictions on the residue field characteristic of $\mathfrak{o}$, as well as the norm one group $\mathrm{SL}_1(\mathfrak{D})$ of a non-split quaternion algebra $\mathfrak{D}$ over the field of fractions of $\mathfrak{o}$ and its principal congruence subgroups. We also determine the representation zeta functions of principal congruence subgroups of $\mathrm{SL}_3(\mathfrak{o})$ in the case that $\mathfrak{o}$ has residue field characteristic 3 and is unramified over $\mathbb{Z}_3$.

1. Introduction

Let G be a group. For $n \in \mathbb{N}$, we denote by $r_n(G)$ the number of isomorphism classes of n-dimensional irreducible complex representations of G. If G is a topological group, we tacitly assume that representations are continuous. We call G (representation) *rigid* if, for every $n \in \mathbb{N}$, the number $r_n(G)$ is finite. It is known that a profinite group is rigid if and only if it is FAb, i.e. if each of its open subgroups has finite abelianisation. All the groups studied in this paper are rigid.

We say that G has *polynomial representation growth* if the sequence $R_N(G) := \sum_{n=1}^{N} r_n(G)$ is bounded by a polynomial in N. The *representation zeta function* of a group G with polynomial representation growth is the Dirichlet series

$$\zeta_G(s) := \sum_{n=1}^{\infty} r_n(G) n^{-s},$$

where s is a complex variable. The *abscissa of convergence* $\alpha(G)$ of $\zeta_G(s)$, i.e. the infimum of all $\alpha \in \mathbb{R}$ such that $\zeta_G(s)$ converges on the complex right half-plane $\{s \in \mathbb{C} \mid \mathrm{Re}(s) > \alpha\}$, gives the precise degree of polynomial growth: $\alpha(G)$ is the smallest α such that $R_N(G) = O(1 + N^{\alpha+\varepsilon})$ for every $\varepsilon \in \mathbb{R}_{>0}$. Representation zeta functions of groups are studied in a variety of contexts, e.g. in the setting of arithmetic and p-adic analytic groups, wreath products of finite groups and finitely generated nilpotent groups; cf. [**12, 16, 2, 3, 4**], [**5**] and [**9, 21, 20**]. In the case of

2000 *Mathematics Subject Classification.* Primary 22E50, 20F69, 11M41, 20C15, 20G25.
Avni was supported by NSF grant DMS-0901638.

finitely generated nilpotent groups, one enumerates representations up to 'twisting' by one-dimensional representations; see, for instance, [**17**, Theorem 6.6].

We state our main results, deferring the precise definitions of some of the technical terms involved as well as further corollaries and remarks to later sections. In our first theorem we provide general estimates for the abscissae of convergence of representation zeta functions of FAb compact p-adic analytic groups. These are expressed in terms of invariants ρ and σ which may be defined by minimal and maximal centraliser dimensions in the corresponding Lie algebras; see (2.4). The term 'permissible' is defined in Section 2.1; here it suffices to note that, for a given Lie lattice $\mathfrak{g}$, all sufficiently large $m \in \mathbb{N}$ are permissible for $\mathfrak{g}$.

THEOREM 1.1. *Let $\mathfrak{o}$ be a compact discrete valuation ring of characteristic* 0, *with maximal ideal $\mathfrak{p}$ and field of fractions $\mathfrak{k}$. Let $\mathfrak{g}$ be an $\mathfrak{o}$-Lie lattice such that $\mathfrak{k}\otimes_{\mathfrak{o}}\mathfrak{g}$ is a perfect Lie algebra. Let $d := \dim_{\mathfrak{k}}(\mathfrak{k}\otimes_{\mathfrak{o}}\mathfrak{g})$, and let $\sigma := \sigma(\mathfrak{g})$ and $\rho := \rho(\mathfrak{g})$, as defined in* (2.4). *Let $m \in \mathbb{N}$ be permissible for $\mathfrak{g}$, and let $\mathsf{G}^m := \exp(\mathfrak{p}^m\mathfrak{g})$.*

Then lower and upper bounds for the abscissa of convergence of $\zeta_{\mathsf{G}^m}(s)$ are given by

$$(d-2\rho)\rho^{-1} \leq \alpha(\mathsf{G}^m) \leq (d-2\sigma)\sigma^{-1}.$$

Our other main results provide explicit formulae for the representation zeta functions of members of specific families of 'simple' compact p-adic analytic groups. The groups can be realised as matrix groups over a compact discrete valuation ring $\mathfrak{o}$ of characteristic 0 and residue field characteristic p, with absolute ramification index $e(\mathfrak{o},\mathbb{Z}_p)$. In the unramified case $e(\mathfrak{o},\mathbb{Z}_p)=1$, all $m \in \mathbb{N}$ are permissible for a given $\mathfrak{o}$-Lie lattice $\mathfrak{g}$.

THEOREM 1.2. *Let $\mathfrak{o}$ be a compact discrete valuation ring of characteristic* 0. *Then for all $m \in \mathbb{N}$ which are permissible for the Lie lattice $\mathfrak{sl}_2(\mathfrak{o})$ the representation zeta function of the principal congruence subgroup $\mathrm{SL}_2^m(\mathfrak{o})$ is*

$$\zeta_{\mathrm{SL}_2^m(\mathfrak{o})}(s) = \begin{cases} q^{3m}(1-q^{-2-s})/(1-q^{1-s}) & \text{if } p>2, \\ q^{3m}(q^2-q^{-s})/(1-q^{1-s}) & \text{if } p=2 \text{ and } e(\mathfrak{o},\mathbb{Z}_2)=1. \end{cases}$$

In fact, our proof also yields an explicit formula in the case that $p=2$ and $e(\mathfrak{o},\mathbb{Z}_2)>1$.

THEOREM 1.3. *Let $\mathfrak{D}$ be a non-split quaternion algebra over a $\mathfrak{p}$-adic field $\mathfrak{k}$. Let $\mathfrak{R}$ denote the maximal compact subring of $\mathfrak{D}$, and suppose that $2e(\mathfrak{k},\mathbb{Q}_p)<p-1$, where p denotes the residue field characteristic of $\mathfrak{k}$.*

Then the representation zeta functions of $\mathrm{SL}_1(\mathfrak{D})=\mathrm{SL}_1(\mathfrak{R})$ and its principal congruence subgroups $\mathrm{SL}_1^m(\mathfrak{R})$, $m \in \mathbb{N}$, are

$$\zeta_{\mathrm{SL}_1(\mathfrak{R})}(s) = \frac{(q+1)(1-q^{-s})+4(q-1)((q+1)/2)^{-s}}{1-q^{1-s}},$$

$$\zeta_{\mathrm{SL}_1^m(\mathfrak{R})}(s) = q^{3m}\frac{q-q^{-1-s}}{1-q^{1-s}}.$$

THEOREM 1.4. *Let $\mathfrak{o}$ be a compact discrete valuation ring of characteristic* 0, *with residue field of cardinality q and characteristic* 3. *Suppose that $\mathfrak{o}$ is unramified* *over $\mathbb{Z}_3$. Then, for all $m \in \mathbb{N}$, one has*

$$\zeta_{\mathrm{SL}_3^m(\mathfrak{o})}(s) = \\ q^{8m-4}\frac{(q^2-q^{-s})(q^2+q^{-s}+(q^4-1)q^{-2s}-q^{1-3s}+(q^4-q^2-q)q^{-4s})}{(1-q^{1-2s})(1-q^{2-3s})}.$$

The four theorems are proved and discussed in Sections 2.3, 3.1, 4.2 and 5 respectively. Whilst all of the results in the current paper are of a local nature, they might be best appreciated in the 'global' context of representation zeta functions of arithmetic groups, as explained to some extent below and exposed, in much greater detail, in [**2, 3, 4**]. The main technical tools of these papers are the Kirillov orbit method, novel techniques from $\mathfrak{p}$-adic integration and Clifford theory. The current paper is closely related to these works, complements them in parts and provides concrete examples of the general methods developed in these papers. We tried, however, to keep it reasonably self-contained, and hope that it might help the reader appreciate these articles and their interconnections.

Let Γ be an arithmetic subgroup of a connected, simply connected semisimple algebraic group $\mathbf{G}$ defined over a number field k, and assume that Γ has the Congruence Subgroup Property. Relevant examples are groups of the form $\Gamma = \mathrm{SL}_n(\mathcal{O})$, where $\mathcal{O}$ is the ring of integers in a number field k, and $n \geq 3$. The Congruence Subgroup Property and Margulis super-rigidity imply that the 'global' representation zeta function $\zeta_\Gamma(s)$ of Γ is an Euler product of 'local' representation zeta functions, indexed by places of k; see [**16**, Proposition 1.3]. For example, if $\Gamma = \mathrm{SL}_n(\mathcal{O})$ with $n \geq 3$, we have

$$\zeta_{\mathrm{SL}_n(\mathcal{O})}(s) = \zeta_{\mathrm{SL}_n(\mathbb{C})}(s)^{|k:\mathbb{Q}|} \cdot \prod_v \zeta_{\mathrm{SL}_n(\mathcal{O}_v)}(s), \tag{1.1}$$

where each archimedean factor $\zeta_{\mathrm{SL}_n(\mathbb{C})}(s)$ enumerates the finite-dimensional, irreducible rational representations of the algebraic group $\mathrm{SL}_n(\mathbb{C})$ and, for every non-archimedean place v of k, we denote by $\mathcal{O}_v$ the completion of $\mathcal{O}$ at v, which is a finite extension of the p-adic integers $\mathbb{Z}_p$ if v prolongs p. In general, the local factors indexed by non-archimedean places are representation zeta functions of FAb compact p-adic analytic groups. In [**4**] we provide explicit, uniform formulae for the zeta functions of special linear groups $\mathrm{SL}_3(\mathfrak{o})$ and special unitary groups $\mathrm{SU}_3(\mathfrak{o})$, in the case that $p \geq 3e+4$, where $e = e(\mathfrak{o}, \mathbb{Z}_p)$ denotes the absolute ramification index of $\mathfrak{o}$. In the Euler product (1.1) this condition is satisfied by all but finitely many of the rings $\mathfrak{o} = \mathcal{O}_v$. It is a natural, interesting question to describe the local factors at the finitely many 'exceptional' places v of k, and in particular in the non-generic case $p = 3$.

In [**3**] we develop general methods to describe representation zeta functions of certain 'globally defined' FAb compact p-adic analytic pro-p groups. These include the principal congruence subgroups of compact p-adic analytic groups featuring in Euler products of representation zeta functions of 'semisimple' arithmetic groups, such as (1.1). In particular, we obtain there formulae for the zeta functions of principal congruence subgroups of the form $\mathrm{SL}_3^m(\mathfrak{o})$, provided the residue field characteristic of $\mathfrak{o}$ is different from 3 and $m \in \mathbb{N}$ is permissible for the $\mathfrak{o}$-Lie lattice $\mathfrak{sl}_3(\mathfrak{o})$, cf. [**3**, Theorem E].

In Theorem 1.4 of the current paper we complement [**3**, Theorem E] (or, equivalently, the analogous result in [**4**]) by providing a formula for the representation zeta functions of groups of the form $\mathrm{SL}_3^m(\mathfrak{o})$, where $\mathfrak{o}$ is an unramified extension of $\mathbb{Z}_3$ and $m \in \mathbb{N}$. The formula differs from the 'generic' formula in [**3**, Theorem E], valid for residue field characteristic $p \neq 3$, and is obtained by computations akin to the algebraic approach developed in [**4**]. It is noteworthy that the 'generic' formula only depends on the residue field of the local ring $\mathfrak{o}$, irrespective of ramification. Whilst it is clear that this does not hold in the case of residue field characteristic 3, we do not know how sensitive the zeta function is to ramification in this case. It is, for instance, an interesting open problem whether, for rings $\mathfrak{o}_1$ and $\mathfrak{o}_2$ of residue field characteristic 3 which have the same inertia degree and ramification index over $\mathbb{Z}_3$, the representation zeta functions of the groups $\mathrm{SL}_3(\mathfrak{o}_1)$ and $\mathrm{SL}_3(\mathfrak{o}_2)$ are the same, or at least those of $\mathrm{SL}_3^m(\mathfrak{o}_1)$ and $\mathrm{SL}_3^m(\mathfrak{o}_2)$ for permissible $m \in \mathbb{N}$.

A phenomenon of the latter kind is exhibited in Theorem 1.2, in which we record formulae for zeta functions of groups of the form $\mathrm{SL}_2^m(\mathfrak{o})$, where $m \in \mathbb{N}$ is permissible for the Lie lattice $\mathfrak{sl}_2(\mathfrak{o})$. The formula for $\zeta_{\mathrm{SL}_2^m(\mathfrak{o})}(s)$ in the generic case $p > 2$ only depends on m and on q, the residue field cardinality of $\mathfrak{o}$. The expression in the case $p = 2$, on the other hand, is sensitive to ramification, albeit only to the absolute ramification index $e = e(\mathfrak{o}, \mathbb{Z}_2)$ of $\mathfrak{o}$. In Section 3.3 we employ Clifford theory to compute the representation zeta function $\zeta_{\mathrm{SL}_2(\mathfrak{o})}(s)$ in the case that $p \geq e - 2$. This reproduces, in the given case, a formula first computed in [**12**, Theorem 7.5] for the zeta functions of groups of the form $\mathrm{SL}_2(R)$, where R is an arbitrary complete discrete valuation ring with finite residue field of odd characteristic. We record it here as it illustrates our broader and conceptually different approach. We note that this formula, too, only depends on the residue field cardinality and not, for instance, on ramification. It is a challenge to establish analogous formulae in residue field characteristic 2. We record a formula for the representation zeta functions of $\mathrm{SL}_2(\mathbb{Z}_2)$, based on [**18**], and a conjectural formula for its first principal congruence subgroup.

Our methods also provide a tool for calculations of representation zeta functions of norm one groups $\mathrm{SL}_1(\mathfrak{D})$ of central division algebras $\mathfrak{D}$ of Schur index ℓ, say, over $\mathfrak{p}$-adic fields $\mathfrak{k}$, and their principal congruence subgroups. Results obtained by Larsen and Lubotzky in [**16**] suggest that such 'anisotropic' groups may actually be more tractable than their 'isotropic' counterparts. In the case that ℓ is prime, we compute the abscissa of convergence of $\zeta_{\mathrm{SL}_1(\mathfrak{D})}(s)$ in terms of Lie-theoretic data associated to the Lie algebra $\mathfrak{sl}_1(\mathfrak{D})$; cf. Corollary 4.1. This result, which was first proved in [**16**] for general ℓ, is an easy application of some general estimates for the abscissae of convergence of representation zeta functions of groups to which our methods are applicable; cf. Theorem 1.1. Of special interest is the case $\ell = 2$, where $\mathfrak{D}$ is a non-split quaternion algebra over $\mathfrak{k}$ with maximal compact subring $\mathfrak{R}$, say. In this situation we give, in Theorem 1.3, formulae for the representation zeta functions of the groups $\mathrm{SL}_1(\mathfrak{D}) = \mathrm{SL}_1(\mathfrak{R})$ and $\mathrm{SL}_1^m(\mathfrak{R})$, $m \in \mathbb{N}$, which hold if p is large compared to the ramification index $e(\mathfrak{k}, \mathbb{Q}_p)$.

Organisation. The paper is organised as follows. In Section 2 we briefly recall the geometric method, developed in [**3**], to describe representation zeta functions of certain compact p-adic analytic pro-p groups. We show how it yields lower and upper bounds for abscissae of convergence as in Theorem 1.1. In Section 3 we compute representation zeta functions associated to groups of the form $\mathrm{SL}_2(\mathfrak{o})$ and its principal congruence subgroups, thereby proving Theorem 1.2. Results on representation zeta functions of subgroups of norm one groups of central division algebras and, in particular, Theorem 1.3 are obtained in Section 4. The computations for principal subgroups of groups of the form $\mathrm{SL}_3(\mathfrak{o})$ in residue field characteristic $p = 3$, resulting in Theorem 1.4, are carried out in Section 5.

Notation. Our notation is the same as the one used in [**3**]. Non-standard terms are briefly defined at their first occurrence in the text. Zeta function will always refer to representation zeta function. Throughout this paper, $\mathfrak{o}$ denotes a compact discrete valuation ring of characteristic 0 and residue field cardinality q, a power of a prime p. We write F^* to denote the multiplicative group of a field F and extend this notation as follows. For a non-trivial $\mathfrak{o}$-module M we write $M^* := M \smallsetminus \mathfrak{p}M$ and set $\{0\}^* = \{0\}$.

2. Zeta functions as $\mathfrak{p}$-adic integrals

Let $\mathfrak{o}$ be a compact discrete valuation ring of characteristic 0, with maximal ideal $\mathfrak{p}$. The residue field $\mathfrak{o}/\mathfrak{p}$ is a finite field of characteristic p and cardinality q, say. Let $\mathfrak{k}$ be the field of fractions of $\mathfrak{o}$.

2.1. Integral formula. Let $\mathfrak{g}$ be an $\mathfrak{o}$-Lie lattice such that $\mathfrak{k} \otimes_{\mathfrak{o}} \mathfrak{g}$ is perfect. In accordance with [**3**, Section 2.1], we call $m \in \mathbb{N}_0$ permissible for $\mathfrak{g}$ if the principal congruence Lie sublattice $\mathfrak{g}^m = \mathfrak{p}^m\mathfrak{g}$ is potent and saturable. Almost all non-negative integers m are permissible for $\mathfrak{g}$; see [**3**, Proposition 2.3]. A key property of a potent and saturable $\mathfrak{o}$-Lie lattice $\mathfrak{h}$ is that the Kirillov orbit method can be used to study the set $\mathrm{Irr}(H)$ of irreducible complex characters of the p-adic analytic pro-p group $H = \exp(\mathfrak{h})$, which is associated to $\mathfrak{h}$ via the Hausdorff series; see [**8**].

Let $m \in \mathbb{N}_0$ be permissible for $\mathfrak{g}$ and consider $\mathsf{G}^m := \exp(\mathfrak{g}^m)$. Then the orbit method provides a correspondence between the elements of $\mathrm{Irr}(\mathsf{G}^m)$ and the co-adjoint orbits of G^m on the Pontryagin dual $\mathrm{Irr}(\mathfrak{g}^m) = \mathrm{Hom}_{\mathbb{Z}}^{\mathrm{cts}}(\mathfrak{g}^m, \mathbb{C}^*)$ of the compact abelian group $\mathfrak{g}^m$. The radical of $\omega \in \mathrm{Irr}(\mathfrak{g}^m)$ is

$$\mathrm{Rad}(\omega) := \{x \in \mathfrak{g}^m \mid \forall y \in \mathfrak{g}^m : \omega([x, y]_{\mathrm{Lie}}) = 1\}.$$

The degree of the irreducible complex character represented by the co-adjoint orbit of ω is equal to $|\mathfrak{g}^m : \mathrm{Rad}(\omega)|^{1/2}$, and the size of the co-adjoint orbit of ω is equal to $|\mathfrak{g}^m : \mathrm{Rad}(\omega)|$. This shows that the zeta function of G^m satisfies

$$\zeta_{\mathsf{G}^m}(s) = \sum_{\omega \in \mathrm{Irr}(\mathfrak{g}^m)} |\mathfrak{g}^m : \mathrm{Rad}(\omega)|^{-(s+2)/2}; \tag{2.1}$$

cf. [**12**, Corollary 2.13] and [**8**, Theorem 5.2]. According to [**3**, Lemma 2.4], the Pontryagin dual of the $\mathfrak{o}$-Lie lattice $\mathfrak{g}^m$ admits a natural decomposition

$$\mathrm{Irr}(\mathfrak{g}^m) = \dot{\bigcup}_{n \in \mathbb{N}_0} \mathrm{Irr}_n(\mathfrak{g}^m), \qquad \text{where } \mathrm{Irr}_n(\mathfrak{g}^m) \cong \mathrm{Hom}_{\mathfrak{o}}(\mathfrak{g}^m, \mathfrak{o}/\mathfrak{p}^n)^*.$$

Moreover, for each $n \in \mathbb{N}_0$ there is a natural projection of $\mathfrak{o}$-modules $\mathrm{Hom}_{\mathfrak{o}}(\mathfrak{g}^m, \mathfrak{o}) \to \mathrm{Hom}_{\mathfrak{o}}(\mathfrak{g}^m, \mathfrak{o}/\mathfrak{p}^n)$, mapping $\mathrm{Hom}_{\mathfrak{o}}(\mathfrak{g}^m, \mathfrak{o})^*$ onto $\mathrm{Hom}_{\mathfrak{o}}(\mathfrak{g}^m, \mathfrak{o}/\mathfrak{p}^n)^*$. We say that $\omega \in \mathrm{Irr}_n(\mathfrak{g}^m)$ has *level* n and that $w \in \mathrm{Hom}_{\mathfrak{o}}(\mathfrak{g}^m, \mathfrak{o})^*$ is a representative of ω if w maps onto the appropriate element of $\mathrm{Hom}_{\mathfrak{o}}(\mathfrak{g}^m, \mathfrak{o}/\mathfrak{p}^n)^*$.

Let $\mathbf{b} := (b_1, \ldots, b_d)$ be an $\mathfrak{o}$-basis for the $\mathfrak{o}$-Lie lattice $\mathfrak{g}$, where $d = \dim_{\mathfrak{k}}(\mathfrak{k} \otimes_{\mathfrak{o}} \mathfrak{g})$. The structure constants λ_{ij}^h of the $\mathfrak{o}$-Lie lattice $\mathfrak{g}$ with respect to $\mathbf{b}$ are encoded in the commutator matrix

$$\mathcal{R}(\mathbf{Y}) := \mathcal{R}_{\mathfrak{g},\mathbf{b}}(\mathbf{Y}) = \left(\sum_{h=1}^{d} \lambda_{ij}^h Y_h \right)_{ij} \in \mathrm{Mat}_d(\mathfrak{o}[\mathbf{Y}]), \tag{2.2}$$

whose entries are linear forms in independent variables $Y_1, \ldots, Y_d$. We put $W(\mathfrak{o}) := (\mathfrak{o}^d)^* \cong \mathrm{Hom}_{\mathfrak{o}}(\mathfrak{g}^m, \mathfrak{o})^*$ and set

$$\sigma(\mathfrak{g}) := \min\left\{\tfrac{1}{2}\,\mathrm{rk}_{\mathfrak{k}}\,\mathcal{R}(\mathbf{y}) \mid \mathbf{y} \in W(\mathfrak{o})\right\}, \quad \rho(\mathfrak{g}) := \max\left\{\tfrac{1}{2}\,\mathrm{rk}_{\mathfrak{k}}\,\mathcal{R}(\mathbf{y}) \mid \mathbf{y} \in W(\mathfrak{o})\right\}.$$

Note that this definition is independent of the choice of basis for $\mathfrak{g}$, and that both $\sigma(\mathfrak{g})$ and $\rho(\mathfrak{g})$ are integers, because $\mathcal{R}(\mathbf{Y})$ is anti-symmetric.

The relevance of the commutator matrix in connection with (2.1) stems from [**3**, Lemma 3.3], which we record here as follows.

LEMMA 2.1. *Let m be permissible for $\mathfrak{g}$, let $n \in \mathbb{N}$ and suppose that $\omega \in \mathrm{Irr}_n(\mathfrak{g}^m)$ is represented by $w \in \mathrm{Hom}_{\mathfrak{o}}(\mathfrak{g}^m, \mathfrak{o})^*$. Let π denote a uniformiser for $\mathfrak{o}$.*

Then for every $z \in \mathfrak{g}^m$ we have

$$z \in \mathrm{Rad}(\omega) \quad \Longleftrightarrow \quad \underline{z} \cdot \mathcal{R}(\underline{w}) \equiv_{\mathfrak{p}^{n-m}} 0,$$

where $\underline{z}$ and $\underline{w}$ denote the coordinate tuples of z and w with respect to the shifted $\mathfrak{o}$-basis $\pi^m\mathbf{b}$ for $\mathfrak{g}^m$ and its dual $\pi^{-m}\mathbf{b}^\vee$ for $\mathrm{Hom}_{\mathfrak{o}}(\mathfrak{g}^m, \mathfrak{o})$.

By this lemma, the index $|\mathfrak{g}^m : \mathrm{Rad}(\omega)|$ can be expressed in terms of the elementary divisors of the matrix $\mathcal{R}(\underline{w})$ which in turn one computes from its minors. The zeta function of the group G^m, associated to the principal congruence Lie sublattice $\mathfrak{g}^m$, can thus be regarded as a Poincaré series encoding the numbers of solutions of a certain system of equations modulo $\mathfrak{p}^n$ for all $n \in \mathbb{N}_0$. Such Poincaré series can be expressed as generalised Igusa zeta functions, which are certain types of $\mathfrak{p}$-adic integrals over the compact space $\mathfrak{p} \times W(\mathfrak{o})$; cf. [**3**] and [**7, 11**].

For $j \in \{1, \ldots, \rho(\mathfrak{g})\}$ and $\mathbf{y} \in W(\mathfrak{o})$ we define

$$F_j(\mathbf{Y}) = \{f \mid f \text{ a } 2j \times 2j \text{ minor of } \mathcal{R}(\mathbf{Y})\},$$
$$\|F(\mathbf{y})\|_{\mathfrak{p}} = \max\{|f(\mathbf{y})|_{\mathfrak{p}} \mid f \in F\}.$$

It is worth pointing out that the sets $F_j(\mathbf{Y})$ may be replaced by sets of polynomials defining the same polynomial ideals. Specifically, one could define $F_j(\mathbf{Y})$ to be the set of all principal $2j \times 2j$ minors; see [**3**, Remark 3.6]. It is the geometry of the varieties defined by the polynomials in $F_j(\mathbf{Y})$ which largely determines the zeta function of G^m. Of particular interest are 'effective' resolutions of their singularities; cf. [**3**]. If $\mathfrak{k} \otimes_{\mathfrak{o}} \mathfrak{g}$ is a semisimple Lie algebra, then the varieties defined by the polynomials in $F_j(\mathbf{Y})$ admit a Lie-theoretic interpretation: they yield a stratification of the Lie algebra defined in terms of centraliser dimensions; cf. [**3**, Section 5]. We state the integral formula derived in [**3**, Section 3.2].

PROPOSITION 2.2. *Let $\mathfrak{g}$ be an $\mathfrak{o}$-Lie lattice such that $\mathfrak{k} \otimes_{\mathfrak{o}} \mathfrak{g}$ is a perfect $\mathfrak{k}$-Lie algebra of dimension d. Then for every $m \in \mathbb{N}_0$ which is permissible for $\mathfrak{g}$ one has*

$$\zeta_{\mathsf{G}^m}(s) = q^{dm}\left(1 + (1 - q^{-1})^{-1}\mathcal{Z}_{\mathfrak{o}}(-s/2 - 1, \rho(s+2) - d - 1)\right),$$

with $\rho = \rho(\mathfrak{g})$ and

$$\mathcal{Z}_{\mathfrak{o}}(r,t) = \int_{(x,\mathbf{y}) \in \mathfrak{p} \times W(\mathfrak{o})} |x|_{\mathfrak{p}}^t \prod_{j=1}^{\rho} \frac{\|F_j(\mathbf{y}) \cup F_{j-1}(\mathbf{y})x^2\|_{\mathfrak{p}}^r}{\|F_{j-1}(\mathbf{y})\|_{\mathfrak{p}}^r}\, d\mu(x,\mathbf{y}), \tag{2.3}$$

where $\mathfrak{p} \times W(\mathfrak{o}) \subseteq \mathfrak{o}^{d+1}$ and the additive Haar measure μ is normalised so that $\mu(\mathfrak{o}^{d+1}) = 1$.

In studying the integral (2.3), it is useful to distinguish between regular and irregular points of $W(\mathfrak{o})$. Let $\mathcal{U}_1$ denote the subvariety of $\mathbb{A}^d$ defined by the set of polynomials $F_\rho(\mathbf{Y})$ over $\mathfrak{o}$. We write $\mathbb{F}_q$ for the residue class field $\mathfrak{o}/\mathfrak{p}$. The reduction of $\mathcal{U}_1$ modulo $\mathfrak{p}$ is denoted by $\overline{\mathcal{U}_1}$. We call a point $\mathbf{a} \in (\mathbb{F}_q^d)^*$, and any $\mathbf{y} \in W(\mathfrak{o})$ mapping onto $\mathbf{a}$, *regular* if $\mathbf{a}$ is not an $\mathbb{F}_q$-rational point of $\overline{\mathcal{U}_1}$. A functional $w \in \mathrm{Hom}_\mathfrak{o}(\mathfrak{g}, \mathfrak{o})^*$ and the representations associated to the Kirillov orbits of the images of w in $\mathrm{Hom}_\mathfrak{o}(\mathfrak{g}, \mathfrak{o}/\mathfrak{p}^n)^*$, $n \in \mathbb{N}$, are said to be *regular*, if the co-ordinate vector $\mathbf{y} \in W(\mathfrak{o})$ corresponding to w is regular. Points, functionals and representations which are not regular are called *irregular*.

2.2. Adjoint versus co-adjoint action. In the special case where $\mathfrak{g}$ is an $\mathfrak{o}$-Lie lattice such that $\mathfrak{k} \otimes_\mathfrak{o} \mathfrak{g}$ is semisimple, one can use the Killing form, or a scaled version of it, to translate between co-adjoint orbits and adjoint orbits. This has some technical benefits when using the orbit method, as illustrated in Sections 4 and 5.

Let $\mathfrak{g}$ be an $\mathfrak{o}$-lattice such that $\mathfrak{k} \otimes_\mathfrak{o} \mathfrak{g}$ is semisimple, and suppose that m is permissible for $\mathfrak{g}$ so that the mth principal congruence sublattice $\mathfrak{g}^m = \mathfrak{p}^m\mathfrak{g}$ is saturable and potent. At the level of the Lie algebra $\mathfrak{k} \otimes_\mathfrak{o} \mathfrak{g}$, the Killing form κ is non-degenerate and thus provides an isomorphism ι of $\mathfrak{k}$-vector spaces between $\mathfrak{k}\otimes_\mathfrak{o}\mathfrak{g}$ and its dual space $\mathrm{Hom}_\mathfrak{k}(\mathfrak{k}\otimes_\mathfrak{o}\mathfrak{g}, \mathfrak{k})$. Moreover, this isomorphism is G-equivariant for any $G \leq \mathrm{Aut}(\mathfrak{k}\otimes_\mathfrak{o}\mathfrak{g})$.

At the level of the $\mathfrak{o}$-Lie lattices $\mathfrak{g}$ and $\mathfrak{g}^m$, the situation is more intricate, because the restriction of κ, or a scaled version κ_0 of it, may not be non-degenerate over $\mathfrak{o}$. Typically, the pre-image of $\mathrm{Hom}_\mathfrak{o}(\mathfrak{g}, \mathfrak{o}) \hookrightarrow \mathrm{Hom}_\mathfrak{k}(\mathfrak{k}\otimes_\mathfrak{o}\mathfrak{g}, \mathfrak{k})$ under the $\mathfrak{k}$-isomorphism $\iota_0 : \mathfrak{k}\otimes_\mathfrak{o}\mathfrak{g} \to \mathrm{Hom}_\mathfrak{k}(\mathfrak{k}\otimes_\mathfrak{o}\mathfrak{g}, \mathfrak{k})$ induced by κ_0 is an $\mathfrak{o}$-sublattice of $\mathfrak{k}\otimes_\mathfrak{o}\mathfrak{g}$ containing $\mathfrak{g}$ as a sublattice of finite index. For instance, if $\mathfrak{g}$ is a simple Lie algebra of Chevalley type, then it is natural to work with the normalised Killing form κ_0 which is related to the ordinary Killing form κ by the equation $2h^\vee\kappa_0 = \kappa$. Here $h^\vee$ denotes the dual Coxeter number; e.g., the dual Coxeter number for $\mathfrak{sl}_n$ is $h^\vee = n$.

Irrespective of the detailed analysis required to translate carefully between adjoint and co-adjoint orbits, we obtain from the general discussion in [**3**, Section 5] a useful description of the parameters $\sigma(\mathfrak{g})$ and $\rho(\mathfrak{g})$, which were introduced in Section 2.1. Indeed, they can be computed in terms of centraliser dimensions as follows:

$$\begin{aligned} \dim_\mathfrak{k}(\mathfrak{k}\otimes_\mathfrak{o}\mathfrak{g}) - 2\sigma(\mathfrak{g}) &= \max\{\dim_\mathfrak{k} \mathrm{C}_{\mathfrak{k}\otimes_\mathfrak{o}\mathfrak{g}}(x) \mid x \in (\mathfrak{k}\otimes_\mathfrak{o}\mathfrak{g}) \smallsetminus \{0\}\}, \\ \dim_\mathfrak{k}(\mathfrak{k}\otimes_\mathfrak{o}\mathfrak{g}) - 2\rho(\mathfrak{g}) &= \min\{\dim_\mathfrak{k} \mathrm{C}_{\mathfrak{k}\otimes_\mathfrak{o}\mathfrak{g}}(x) \mid x \in (\mathfrak{k}\otimes_\mathfrak{o}\mathfrak{g}) \smallsetminus \{0\}\}. \end{aligned} \tag{2.4}$$

2.3. General bounds for the abscissa of convergence. In this section we derive general bounds for the abscissae of convergence of zeta functions of compact p-adic analytic groups. We start by proving Theorem 1.1 which was stated in the introduction.

PROOF OF THEOREM 1.1. Roughly speaking, the idea is that systematically overestimating the size of orbits in the co-adjoint action leads to a Dirichlet series $\psi_{\mathrm{low}}(s)$ which converges at least as well as $\zeta_{\mathsf{G}^m}(s)$ and hence provides a lower bound for $\alpha(\mathsf{G}^m)$. Similarly, consistently underestimating the size of orbits leads to a Dirichlet series which converges no better than $\zeta_{\mathsf{G}^m}(s)$ and hence provides an upper bound for $\alpha(\mathsf{G}^m)$. For this we use the description of $\zeta_{\mathsf{G}^m}(s)$ in (2.1).

First we derive the lower bound for $\alpha(\mathsf{G}^m)$. Lemma 2.1, in conjunction with the definition of ρ, implies that $|\mathfrak{g}^m : \mathrm{Rad}(\omega)| \le q^{2\rho n}$ for all $n \in \mathbb{N}_0$ and $\omega \in \mathrm{Irr}_n(\mathfrak{g}^m)$. Clearly, the Dirichlet series

$$\psi_{\mathrm{low}}(s) := \sum_{n\in\mathbb{N}_0} \sum_{\omega\in\mathrm{Irr}_n(\mathfrak{g}^m)} q^{-\rho n(s+2)}$$

converges better than $\zeta_{\mathsf{G}^m}(s)$ and it suffices to show that the abscissa of convergence of $\psi_{\mathrm{low}}(s)$ is equal to $(d-2\rho)\rho^{-1}$. Indeed, this can easily be read off from the precise formula

$$\begin{aligned}\psi_{\mathrm{low}}(s) &= 1 + \sum_{n\in\mathbb{N}}(1-q^{-d})q^{dn}q^{-\rho n(s+2)}\\ &= 1 + (1-q^{-d})q^{(d-2\rho)-\rho s}(1-q^{(d-2\rho)-\rho s})^{-1}\\ &= (1-q^{-2\rho-\rho s})(1-q^{(d-2\rho)-\rho s})^{-1}.\end{aligned}$$

The argument for deriving the upper bound is essentially the same, but with a little extra twist. Recall that $W(\mathfrak{o}) = (\mathfrak{o}^d)^*$. Similarly as in [**3**, Section 3.1] we consider a map $\nu : W(\mathfrak{o}) \to (\mathbb{N}_0 \cup \{\infty\})^{\lfloor d/2\rfloor}$ which maps $\mathbf{y} \in W(\mathfrak{o})$ to the tuple $\mathbf{a} = (a_1, \ldots, a_{\lfloor d/2\rfloor})$ such that

(i) $a_1 \le \ldots \le a_{\lfloor d/2\rfloor}$ and
(ii) the elementary divisors of the anti-symmetric matrix $\mathcal{R}(\mathbf{y})$ are precisely $\mathfrak{p}^{a_1}, \ldots, \mathfrak{p}^{a_{\lfloor d/2\rfloor}}$, each counted with multiplicity 2, and one further divisor $\mathfrak{p}^\infty$ if d is odd.

The definition of σ ensures that by forming the composition of ν with the projection $(a_1, \ldots, a_{\lfloor d/2\rfloor}) \mapsto (a_1, \ldots, a_\sigma)$ we obtain a map $\nu_{\mathrm{res}} : W(\mathfrak{o}) \to \mathbb{N}_0^\sigma$. Clearly, ν_{res} is continuous and hence locally constant. Since $W(\mathfrak{o})$ is compact, this implies that the image of ν_{res} is finite. From Lemma 2.1 we deduce that there is a constant $c \in \mathbb{N}_0$ such that for all $n \in \mathbb{N}_0$ and $\omega \in \mathrm{Irr}_n(\mathfrak{g}^m)$ we have $|\mathfrak{g}^m : \mathrm{Rad}(\omega)| \ge q^{2\sigma n - c}$. Now a similar calculation as above gives the desired upper bound for $\alpha(\mathsf{G}^m)$. □

REMARK. (1) According to [**16**, Corollary 4.5], the abscissa of convergence $\alpha(G)$ is an invariant of the commensurability class of G. Thus Theorem 1.1 provides a tool for bounding the abscissa of convergence of the zeta function of any FAb compact p-adic analytic group; see Corollary 2.3.

(2) The algebraic argument given in the proof of Theorem 1.1 admits a geometric interpretation based on the integral formula in Proposition 2.2. To obtain the lower bound one assumes that all points are 'regular', to obtain the upper bound that all points are as 'irregular' as possible.

For 'semisimple' compact p-adic analytic groups, the lower bound in Theorem 1.1 specialises to a result first proved by Larsen and Lubotzky; see [**16**, Proposition 6.6]. We formulate our more general result in this setting. Recall that to any compact p-adic analytic group G one associates a $\mathbb{Q}_p$-Lie algebra, namely $L(G) := \mathbb{Q}_p \otimes_{\mathbb{Z}_p} \mathfrak{h}$ where $\mathfrak{h}$ is the $\mathbb{Z}_p$-Lie lattice associated to any saturable open pro-p subgroup H of G. This Lie algebra is an invariant of the commensurability class of G. Suppose that $L(G)$ is semisimple. Then it decomposes as a sum $L(G) = S_1 \oplus \ldots \oplus S_r$ of simple $\mathbb{Q}_p$-Lie algebras S_i. For each $i \in \{1, \ldots, r\}$ the centroid $\mathfrak{k}_i$ of S_i, viz. the ring of S_i-endomorphisms of S_i with respect to the adjoint action, is a finite extension field of $\mathbb{Q}_p$, and S_i is an absolutely simple $\mathfrak{k}_i$-Lie algebra. The fields $\mathfrak{k}_i$ embed into the completion $\mathbb{C}_p$ of an algebraic closure of $\mathbb{Q}_p$. The field $\mathbb{C}_p$ is, algebraically, isomorphic to the field $\mathbb{C}$ of complex numbers. Indeed, $\mathbb{C}_p$ and $\mathbb{C}$ are algebraically closed and have the same uncountable transcendence degree over $\mathbb{Q}$. Choosing an isomorphism between $\mathbb{C}_p$ and $\mathbb{C}$, we define $\mathfrak{L} := \mathfrak{L}(G) := \mathbb{C} \otimes_{\mathbb{Q}_p} L(G)$ and $\mathfrak{S}_i := \mathbb{C} \otimes_{\mathfrak{k}_i} S_i$ for $i \in \{1, \ldots, r\}$. We define $r_{\mathrm{abs}}(G)$ and $\Phi_{\mathrm{abs}}(G)$ to be the absolute rank and the absolute root system of $L(G)$; they are equal to the rank and the root system of the semisimple complex Lie algebra $\mathfrak{L}$. We denote by $h^\vee(\mathfrak{S})$ the dual Coxeter number of a simple complex Lie algebra $\mathfrak{S}$.

COROLLARY 2.3. *Let G be a compact p-adic analytic group such that its associated $\mathbb{Q}_p$-Lie algebra $L(G)$ is semisimple and decomposes as described above. Then the abscissa of convergence of $\zeta_G(s)$ satisfies*

$$\frac{2r_{\mathrm{abs}}(G)}{|\Phi_{\mathrm{abs}}(G)|} \leq \alpha(G) \leq \min_{i \in \{1,\ldots,r\}} \frac{\dim_{\mathbb{C}}(\mathfrak{S}_i)}{h^\vee(\mathfrak{S}_i) - 1} - 2.$$

PROOF. By our remark, we may assume without loss of generality that $G = \exp(\mathfrak{g})$ is associated to a potent and saturable $\mathbb{Z}_p$-Lie lattice $\mathfrak{g}$. The lower bound for $\alpha(G)$ follows immediately from Theorem 1.1 and the equations (2.4) on noting that $r_{\mathrm{abs}}(G) = \dim(G) - 2\rho(\mathfrak{g})$ and $|\Phi_{\mathrm{abs}}(G)| = 2\rho(\mathfrak{g})$.

It remains to establish the upper bound. Replacing G by an open subgroup, if necessary, we may assume that $\mathfrak{g} = \mathfrak{s}_1 \oplus \ldots \oplus \mathfrak{s}_r$ and $G = G_1 \times \ldots \times G_r$, where for each $i \in \{1, \ldots, r\}$ the summand $\mathfrak{s}_i$ is a potent and saturable $\mathbb{Z}_p$-Lie lattice such that $\mathbb{Q}_p \otimes_{\mathbb{Z}_p} \mathfrak{s}_i$ is simple and $G_i = \exp(\mathfrak{s}_i)$. As $\zeta_G(s) = \prod_{i=1}^r \zeta_{G_i}(s)$, we have $\alpha(G) = \min_{i \in \{1,\ldots,r\}} \alpha(G_i)$ and it is enough to bound $\alpha(G_i)$ for each $i \in \{1, \ldots, r\}$.

Fix $i \in \{1, \ldots, r\}$ and write $\mathfrak{s} := \mathfrak{s}_i$. As before, the centroid $\mathfrak{k}$ of $\mathbb{Q}_p \otimes_{\mathbb{Z}_p} \mathfrak{s}$ is a finite extension of $\mathbb{Q}_p$, and $\mathbb{Q}_p \otimes_{\mathbb{Z}_p} \mathfrak{s}$ is an absolutely simple $\mathfrak{k}$-Lie algebra. Without loss of generality we may regard $\mathfrak{s}$ as an $\mathfrak{o}$-Lie lattice, where $\mathfrak{o}$ is the ring of integers of $\mathfrak{k}$. Writing $\mathfrak{S} = \mathbb{C} \otimes_{\mathfrak{o}} \mathfrak{s}$, we deduce from Theorem 1.1 that it suffices to show: $\sigma(\mathfrak{s}) \geq h^\vee(\mathfrak{S}) - 1$. It is clear that $2\sigma(\mathfrak{s})$ is greater or equal to the dimension of a non-zero co-adjoint orbit of $\mathfrak{S}$ of minimal dimension. According to [**6**, Section 5.8], every sheet of $\mathfrak{S}$ contains a unique nilpotent orbit, and the dimension of a minimal nilpotent orbit in $\mathfrak{S}$ is equal to $2h^\vee(\mathfrak{S}) - 2$; see [**22**, Theorem 1]. It follows that $\sigma(\mathfrak{s}) \geq h^\vee(\mathfrak{S}) - 1$. □

It is worth pointing out that the absolute rank and the size of the absolute root system of a semisimple Lie algebra grow proportionally at the same rate under restriction of scalars; hence, if G is defined over an extension $\mathfrak{o}$ of $\mathbb{Z}_p$, then it is natural to work directly with the invariants of the Lie algebra over $\mathfrak{k}$, without descending to $\mathbb{Q}_p$. A similar remark applies to the upper bound in Corollary 2.3. For instance, for the family of special linear groups $\mathrm{SL}_n(\mathfrak{o})$, $n \in \mathbb{N}$, we obtain the estimates

$$2/n \leq \alpha(\mathrm{SL}_n(\mathfrak{o})) \leq n - 1,$$

reflecting the fact that $\mathfrak{sl}_n(\mathbb{C})$ has rank $n-1$, a root system of size $n^2 - n$, dimension $n^2 - 1$ and dual Coxeter number $h^\vee(\mathfrak{sl}_n(\mathbb{C})) = n$. More generally, we note that Corollary 2.3 provides upper bounds for the abscissae of convergence of zeta functions of groups corresponding to classical Lie algebras which are linear in the rank. We further remark that for 'isotropic simple' compact p-adic analytic groups the abscissa of convergence is actually bounded from below by $1/15$; see [**16**, Theorem 8.1].

Another consequence of Theorem 1.1 is recorded as Corollary 4.1 in Section 4.1.

3. Explicit formulae for $\mathrm{SL}_2(\mathfrak{o})$ and its principal congruence subgroups

In this section we use the setup from Sections 2.1, 2.2 and [**3**, Section 5] to compute explicitly the zeta functions of 'permissible' principal congruence subgroups of the compact p-adic analytic group $\mathrm{SL}_2(\mathfrak{o})$, where $\mathfrak{o}$ denotes a compact discrete valuation ring of characteristic 0. As before, we write $\mathfrak{p}$ for the maximal ideal of $\mathfrak{o}$; the characteristic and cardinality of the residue field $\mathfrak{o}/\mathfrak{p}$ are denoted by p and q. We write $e(\mathfrak{o}, \mathbb{Z}_p)$ for the absolute ramification index of $\mathfrak{o}$.

3.1. Principal congruence subgroups of $\mathrm{SL}_2(\mathfrak{o})$. Our aim in this section is to prove Theorem 1.2 which was stated in the introduction. It provides explicit formulae for the zeta functions of principal congruence subgroups $\mathrm{SL}_2^m(\mathfrak{o})$ for permissible m, with no restrictions if $p > 2$ and for unramified $\mathfrak{o}$ if $p = 2$.

REMARK. In fact, our proof also supplies an explicit formula, if $p = 2$ and $e(\mathfrak{o}, \mathbb{Z}_2) > 1$, but this formula is not as concise as the ones stated in Theorem 1.2. It is noteworthy that in this special case the formula only depends on the ramification index $e(\mathfrak{o}, \mathbb{Z}_2)$, but not on the more specific isomorphism type of the ring $\mathfrak{o}$. It would be interesting to investigate what happens for 'semisimple' groups of higher dimensions; already for $\mathrm{SL}_3(\mathfrak{o})$ the matter remains to be resolved; cf. Section 5.

PROOF OF THEOREM 1.2. Let $m \in \mathbb{N}$ be permissible for $\mathfrak{sl}_2(\mathfrak{o})$. We need to compute the integral (2.3) over $\mathfrak{p} \times W(\mathfrak{o})$, where $W(\mathfrak{o}) = (\mathfrak{o}^3)^*$. It is easy to write down the commutator matrix $\mathcal{R}(\mathbf{Y})$ for the $\mathfrak{o}$-Lie lattice $\mathfrak{sl}_2(\mathfrak{o})$, and one verifies immediately that $\rho = 1$. Indeed, working with the standard $\mathfrak{o}$-basis $\mathbf{e} = \left(\begin{smallmatrix} 0 & 1 \\ 0 & 0 \end{smallmatrix}\right)$, $\mathbf{f} = \left(\begin{smallmatrix} 0 & 0 \\ 1 & 0 \end{smallmatrix}\right)$, $\mathbf{h} = \left(\begin{smallmatrix} 1 & 0 \\ 0 & -1 \end{smallmatrix}\right)$ of $\mathfrak{sl}_2(\mathfrak{o})$ one obtains

$$\mathcal{R}(\mathbf{Y}) = \begin{pmatrix} 0 & Y_3 & -2Y_1 \\ -Y_3 & 0 & 2Y_2 \\ 2Y_1 & -2Y_2 & 0 \end{pmatrix}. \tag{3.1}$$

In view of (3.1) we distinguish two cases.

First suppose that $p > 2$. In this case it is easily seen that

$$\max\left(\{|f(\mathbf{y})|_{\mathfrak{p}} \mid f \in F_1(\mathbf{Y})\} \cup \{|x^2|_{\mathfrak{p}}\}\right) = 1 \quad \text{for all } x \in \mathfrak{p} \text{ and } \mathbf{y} \in W(\mathfrak{o}).$$

Thus the integral (2.3) takes the form

$$\mathcal{Z}_{\mathfrak{o}}(r,t) = \int_{(x,\mathbf{y}) \in \mathfrak{p} \times W(\mathfrak{o})} |x|_{\mathfrak{p}}^t \, d\mu(x,\mathbf{y}).$$

As

$$\int_{x \in \mathfrak{p}} |x|_{\mathfrak{p}}^s \, d\mu(x) = \frac{(1-q^{-1})q^{-1-s}}{1-q^{-1-s}}$$

and $\mu(W(\mathfrak{o})) = 1 - q^{-3}$ we obtain

$$\mathcal{Z}_{\mathfrak{o}}(r,t) = \frac{(1-q^{-1})q^{-1-t}(1-q^{-3})}{1-q^{-1-t}}$$

so that, by Proposition 2.2, we have

$$\zeta_{\mathrm{SL}_2^m(\mathfrak{o})}(s) = q^{3m}\left(1 + (1-q^{-1})^{-1}\mathcal{Z}_{\mathfrak{o}}(-s/2-1, s-2)\right) = \frac{q^{3m}(1-q^{-2-s})}{1-q^{1-s}}.$$

Now consider the exceptional case $p = 2$. Put $e := e(\mathfrak{o}, \mathbb{Z}_2)$. Defining

$$W(\mathfrak{o})^{[j]} := \begin{cases} \{\mathbf{y} \in W(\mathfrak{o}) \mid y_3 \in \mathfrak{p}^j \smallsetminus \mathfrak{p}^{j+1}\} & \text{if } 0 \le j \le e-1, \\ \{\mathbf{y} \in W(\mathfrak{o}) \mid y_3 \in 2\mathfrak{o}\} & \text{if } j = e, \end{cases}$$

we write $W(\mathfrak{o})$ as a disjoint union $W(\mathfrak{o}) = W(\mathfrak{o})^{[0]} \,\dot\cup\, W(\mathfrak{o})^{[1]} \,\dot\cup \ldots \dot\cup\, W(\mathfrak{o})^{[e]}$. From (3.1) we see that, for all $x \in \mathfrak{p}^i \smallsetminus \mathfrak{p}^{i+1}$ and $\mathbf{y} \in W(\mathfrak{o})^{[j]}$, where $i \in \mathbb{N}$ and $1 \le j \le e$,

$$\max\left(\{|f(\mathbf{y})|_\mathfrak{p} \mid f \in F_1(\mathbf{Y})\} \cup \{|x^2|_\mathfrak{p}\}\right) = q^{-2\min\{i,j,e\}}.$$

Furthermore, we note that $\mu(W(\mathfrak{o})^{[0]}) = (1-q^{-1})$, $\mu(W(\mathfrak{o})^{[j]}) = (1-q^{-1})(1-q^{-2})q^{-j}$ for $1 \le j \le e-1$, and $\mu(W(\mathfrak{o})^{[e]}) = (1-q^{-2})q^{-e}$. Thus the integral (2.3) takes the form

$$\begin{aligned} \mathcal{Z}_\mathfrak{o}(r,t) &= \mu(W(\mathfrak{o})^{[0]}) \int_{x\in\mathfrak{p}} |x|_\mathfrak{p}^t \, d\mu(x) \\ &+ \sum_{j=1}^{e} \mu(W(\mathfrak{o})^{[j]}) \left(\sum_{i=1}^{j-1} q^{-2ir} \int_{x\in\mathfrak{p}^i\smallsetminus\mathfrak{p}^{i+1}} |x|_\mathfrak{p}^t \, d\mu(x) + q^{-2jr} \int_{x\in\mathfrak{p}^j} |x|_\mathfrak{p}^t \, d\mu(x) \right), \end{aligned}$$

which in turn yields an explicit formula for $\zeta_{\mathrm{SL}_2^m(\mathfrak{o})}(s)$, again by Proposition 2.2.

In the special case $e = e(\mathfrak{o}, \mathbb{Z}_2) = 1$ the resulting formula is as concise as for $p > 2$: indeed, we have

$$\mathcal{Z}_\mathfrak{o}(r,t) = \left((1-q^{-1}) + (1-q^{-2})q^{-1-2r}\right) \frac{(1-q^{-1})q^{-1-t}}{1-q^{-1-t}}$$

and consequently

$$\zeta_{\mathrm{SL}_2^m(\mathfrak{o})}(s) = q^{3m}\left(1 + (1-q^{-1})^{-1}\mathcal{Z}_\mathfrak{o}(-s/2-1, s-2)\right) = \frac{q^{3m}(q^2-q^{-s})}{1-q^{1-s}}.$$

□

Computer-aided calculations suggest the following conjecture.

CONJECTURE 3.1. The zeta function of $\mathrm{SL}_2^1(\mathbb{Z}_2)$ is given by

$$\zeta_{\mathrm{SL}_2^1(\mathbb{Z}_2)}(s) = \frac{2^5(2^2-2^{-s})}{1-2^{1-s}}.$$

3.2. Clifford theory. We briefly recall some applications of basic Clifford theory to representation zeta functions. For more details we refer to [**3**, Section 7.2]. Let G be a group, and $N \trianglelefteq G$ with $|G : N| < \infty$. For $\vartheta \in \mathrm{Irr}(N)$, let $I_G(\vartheta)$ denote the inertia group of ϑ in G, and $\mathrm{Irr}(G, \vartheta)$ the set of all irreducible characters ρ of G such that ϑ occurs as an irreducible constituent of the restricted character $\mathrm{res}_N^G(\rho)$. One shows that, if N admits only finitely many irreducible characters of any given degree, then so does G and

$$\zeta_G(s) = \sum_{\vartheta\in\mathrm{Irr}(N)} \vartheta(1)^{-s} \cdot |G : I_G(\vartheta)|^{-1-s} \zeta_{G,\vartheta}(s), \tag{3.2}$$

where

$$\zeta_{G,\vartheta}(s) := \vartheta(1)^s |G : I_G(\vartheta)|^s \sum_{\rho\in\mathrm{Irr}(G,\vartheta)} \rho(1)^{-s}.$$

In the special case where ϑ extends to an irreducible character $\hat\vartheta$ of $I_G(\vartheta)$, there is an effective description of the elements $\rho \in \mathrm{Irr}(G, \vartheta)$, and $\zeta_{G,\vartheta}(s) = \zeta_{I_G(\vartheta)/N}(s)$. There are several basic sufficient criteria for the extendability of ϑ; cf. [**10**, Chapter 19].

3.3. The group $\mathrm{SL}_2(\mathfrak{o})$. In this section we combine the Kirillov orbit method and basic Clifford theory to compute explicitly the zeta function of the compact p-adic analytic group $\mathrm{SL}_2(\mathfrak{o})$. The zeta function of the group $\mathrm{SL}_2(R)$, where R is an arbitrary compact discrete valuation ring of odd residue characteristic, was first computed by Jaikin-Zapirain by means of a different approach; see [**12**, Theorem 7.5].

For our approach we assume that $p - 2 \geq e$ where $e := e(\mathfrak{o}, \mathbb{Z}_p)$; in particular, this implies $p > 2$. Then the $\mathfrak{o}$-Lie lattice $\mathfrak{sl}_2^1(\mathfrak{o})$ and the corresponding pro-p group $\mathrm{SL}_2^1(\mathfrak{o})$ are potent and saturable; see [**3**, Proposition 2.3]. This means that the orbit method can be applied to describe the irreducible characters of $\mathrm{SL}_2^1(\mathfrak{o})$.

Write $G := \mathrm{SL}_2(\mathfrak{o})$ and $N := \mathrm{SL}_2^1(\mathfrak{o})$. Clifford theory, as indicated in Section 3.2, provides a framework to link $\mathrm{Irr}(G)$ and $\mathrm{Irr}(N)$. Put $\mathfrak{g} := \mathfrak{sl}_2(\mathfrak{o})$ and $\mathfrak{n} := \mathfrak{sl}_2^1(\mathfrak{o}) = \mathfrak{p}\mathfrak{g}$. The Kirillov orbit method links characters $\vartheta \in \mathrm{Irr}(N)$ to co-adjoint orbits of N on $\mathrm{Hom}_\mathfrak{o}(\mathfrak{n}, \mathfrak{o})$. Choose a uniformiser π of $\mathfrak{o}$. Via the G-equivariant isomorphism of $\mathfrak{o}$-modules $\mathfrak{g} \to \mathfrak{n}$, $x \mapsto \pi x$, we can link co-adjoint orbits on $\mathrm{Hom}_\mathfrak{o}(\mathfrak{n}, \mathfrak{o})$ to co-adjoint orbits on $\mathrm{Hom}_\mathfrak{o}(\mathfrak{g}, \mathfrak{o})$.

We follow closely the approach outlined in [**3**, Section 5], which uses the normalised Killing form to translate between the adjoint action of G on $\mathfrak{g}$ and the co-adjoint action of G on $\mathrm{Hom}_\mathfrak{o}(\mathfrak{g}, \mathfrak{o})$. The dual Coxeter number of $\mathfrak{sl}_2$ is $h^\vee = 2$ so that the normalised Killing form

$$\kappa_0 : \mathfrak{g} \times \mathfrak{g} \to \mathfrak{o}, \quad \kappa_0(x, y) = (2h^\vee)^{-1}\,\mathrm{Tr}(\mathrm{ad}(x)\,\mathrm{ad}(y))$$

has the structure matrix

$$[\kappa_0(\cdot,\cdot)]_{(\mathbf{h},\mathbf{e},\mathbf{f})} = \begin{pmatrix} 2 & 0 & 0 \\ 0 & 0 & 1 \\ 0 & 1 & 0 \end{pmatrix}$$

with respect to the basis

$$\mathbf{h} = \begin{pmatrix} 1 & 0 \\ 0 & -1 \end{pmatrix}, \quad \mathbf{e} = \begin{pmatrix} 0 & 1 \\ 0 & 0 \end{pmatrix}, \quad \mathbf{f} = \begin{pmatrix} 0 & 0 \\ 1 & 0 \end{pmatrix}.$$

As $p > 2$, the form κ_0 is 'non-degenerate' over $\mathfrak{o}$ and induces a G-equivariant isomorphism of $\mathfrak{o}$-modules $\mathfrak{g} \to \mathrm{Hom}_\mathfrak{o}(\mathfrak{g}, \mathfrak{o})$, $x \mapsto \kappa_0(x, \cdot)$. We obtain a G-equivariant commutative diagram

$$(3.3)\qquad \begin{array}{ccccc} \mathfrak{g}^* & \xrightarrow{\cong} & \mathrm{Hom}_\mathfrak{o}(\mathfrak{g}, \mathfrak{o})^* & & \\ \downarrow & & \downarrow & & \\ (\mathfrak{g}/\mathfrak{p}^n\mathfrak{g})^* & \xrightarrow{\cong} & \mathrm{Hom}_\mathfrak{o}(\mathfrak{g}/\mathfrak{p}^n\mathfrak{g}, \mathfrak{o}/\mathfrak{p}^n)^* & \xrightarrow{\cong} & \mathrm{Irr}_n(\mathfrak{n}) \\ \downarrow & & \downarrow & & \\ \mathfrak{sl}_2(\mathbb{F}_q)^* & \xrightarrow{\cong} & \mathrm{Hom}_{\mathbb{F}_q}(\mathfrak{sl}_2(\mathbb{F}_q), \mathbb{F}_q)^* & & \end{array}$$

where the last row is obtained by reduction modulo $\mathfrak{p}$ and we have used the isomorphism $\mathfrak{o}/\mathfrak{p} \cong \mathbb{F}_q$. Following the approach taken in [**3**, Section 7], we are interested in the orbits and centralisers of elements $x \in \mathfrak{g}$ and their reductions $\overline{x}$ modulo $\mathfrak{p}$ under the adjoint action of G.

In order to apply Clifford theory, we require an overview of the elements in $\mathfrak{sl}_2(\mathbb{F}_q)$ up to conjugacy under the group $\mathrm{GL}_2(\mathbb{F}_q)$. We distinguish four different types, labelled 0, 1, 2a, 2b. The total number of elements of each type and the isomorphism types of their centralisers in $\mathrm{SL}_2(\mathbb{F}_q)$ are summarised in Tables 3.1 and 3.2; see Appendix A for a short discussion. We remark that in this particular case all elements are regular.

type		# of orbits	size of each orbit	total number
0	–	1	1	1
1	regular	1	q^2-1	q^2-1
2a	regular	$(q-1)/2$	q^2+q	$(q^2-1)q/2$
2b	regular	$(q-1)/2$	q^2-q	$(q-1)^2q/2$

TABLE 3.1. Orbits in $\mathfrak{sl}_2(\mathbb{F}_q)$ under conjugacy by $\mathrm{GL}_2(\mathbb{F}_q)$ where $q=p^r$

type		centraliser in $\mathrm{SL}_2(\mathbb{F}_q)$
0	–	$\mathrm{SL}_2(\mathbb{F}_q)$
1	regular	$\mu_2(\mathbb{F}_q)\times\mathbb{F}_q^+\cong C_2\times C_p^r$
2a	regular	$\mathbb{F}_q^*\cong C_{q-1}$
2b	regular	$\ker(N_{\mathbb{F}_{q^2}\mid\mathbb{F}_q})\cong C_{q+1}$

TABLE 3.2. Centralisers in $\mathrm{SL}_2(\mathbb{F}_q)$ of elements of $\mathfrak{sl}_2(\mathbb{F}_q)$ where $q=p^r$

Corollary 7.6 in [**3**] provides

LEMMA 3.2. *Let $x\in\mathfrak{sl}_2(\mathfrak{o})^*$, and let $\overline{x}\in\mathfrak{sl}_2(\mathbb{F}_q)^*$ denote the reduction of x modulo $\mathfrak{p}$. Then*

$$\mathrm{C}_{\mathrm{SL}_2(\mathfrak{o})}(\overline{x})=\mathrm{C}_{\mathrm{SL}_2(\mathfrak{o})}(x)\,\mathrm{SL}_2^1(\mathfrak{o}).$$

REMARK. Alternatively, a direct argument shows that under the conjugation action by $\mathrm{GL}_2(\mathfrak{o})$ the elements of $\mathfrak{sl}_2(\mathfrak{o})^*$ fall into orbits represented by matrices of the form

$$\left(\begin{smallmatrix}0&1\\\pi^n\nu&0\end{smallmatrix}\right)\text{ with }n\in\mathbb{N}\cup\{\infty\}\text{ and }\nu\in\mathfrak{o}\smallsetminus\mathfrak{p},$$
$$\left(\begin{smallmatrix}\lambda&0\\0&-\lambda\end{smallmatrix}\right)\text{ with }\lambda\in\mathfrak{o}\smallsetminus\mathfrak{p},$$
$$\left(\begin{smallmatrix}0&1\\\mu&0\end{smallmatrix}\right)\text{ with }\mu\in\mathfrak{o}\smallsetminus\mathfrak{p}\text{, not a square modulo }\mathfrak{p}.$$

A short computation shows that the centralisers of these matrices in $\mathrm{SL}_2(\mathfrak{o})$ are, respectively,

$$\left\{\left(\begin{smallmatrix}a&b\\\pi^n\nu b&a\end{smallmatrix}\right)\mid a,b\in\mathfrak{o}\text{ with }a^2-\pi^n\nu b^2=1\right\},$$
$$\left\{\left(\begin{smallmatrix}a&0\\0&b\end{smallmatrix}\right)\mid a,b\in\mathfrak{o}\text{ with }ab=1\right\},$$
$$\left\{\left(\begin{smallmatrix}a&b\\\mu b&a\end{smallmatrix}\right)\mid a,b\in\mathfrak{o}\text{ with }a^2-\mu b^2=1\right\}.$$

We may assume that x is one of the listed representatives. The centraliser of $\overline{x}$ in $\mathrm{SL}_2(\mathbb{F}_q)$ has a similar form as that of x; cf. our discussion above. Aided by Hensel's lemma, one successfully lifts any given element $g_0\in\mathrm{C}_{\mathrm{SL}_2(\mathfrak{o})}(\overline{x})$ to an element $g\in\mathrm{C}_{\mathrm{SL}_2(\mathfrak{o})}(x)$.

In any case, if $\vartheta\in\mathrm{Irr}(N)$ is represented by the adjoint orbit of $x|_n:=x+\mathfrak{p}^n\mathfrak{g}\in(\mathfrak{g}/\mathfrak{p}^n\mathfrak{g})^*$, we deduce that $\mathrm{C}_G(x)N=\mathrm{C}_G(x|_n)N=\mathrm{C}_G(\overline{x})$, hence

$$I_G(\vartheta)=\mathrm{C}_G(\overline{x})\qquad\text{and}\qquad I_G(\vartheta)/N\cong\mathrm{C}_{\mathrm{SL}_2(\mathbb{F}_q)}(\overline{x}).$$

The isomorphism types of these groups are given by Table 3.2, where $\mu_2(\mathbb{F}_q)$ is the group of square roots of unity in $\mathbb{F}_q^*$, we denote by $\mathbb{F}_q^+$ the additive group of the field $\mathbb{F}_q$, and $\ker(N_{\mathbb{F}_{q^2}|\mathbb{F}_q})$ is the multiplicative group of norm-1 elements in $\mathbb{F}_{q^2}|\mathbb{F}_q$.

Note that, if ϑ is of type 2a or 2b, the inertia group quotient $I_G(\vartheta)/N$ has order coprime to p. This implies that ϑ can be extended to a character $\hat{\vartheta}$ of $I_G(\vartheta)$. If ϑ is of type 1, the inertia group is a Sylow pro-p subgroup of $\mathrm{SL}_2(\mathfrak{o})$, and we can draw the same conclusion based on [**1**, Theorem 2.3]: being an algebra group, the character degrees of $I_G(\vartheta)$ are powers of q, and since $|I_G(\vartheta) : N| = q$, Lemma 7.4 in [**3**] shows that ϑ extends to a character $\hat{\vartheta}$ of $I_G(\vartheta)$. (If $2e < p - 1$ then the Sylow pro-p subgroup of $\mathrm{SL}_2(\mathfrak{o})$ corresponds to a potent and saturable $\mathfrak{o}$-Lie lattice; cf. [**14**]. Hence one can apply [**3**, Corollary 3.2] to show that its character degrees are powers of q.) Since the inertia group quotients are abelian, it follows that in all cases $\zeta_{G,\vartheta}(s) = \zeta_{I_G(\vartheta)/N}(s) = |I_G(\vartheta) : N|$.

The final task is to connect the formula

$$\zeta_G(s) = \sum_{\vartheta \in \mathrm{Irr}(N)} \vartheta(1)^{-s} \cdot |G : I_G(\vartheta)|^{-1-s} |I_G(\vartheta) : N|, \tag{3.4}$$

resulting from (3.2), and the explicit formula for $\zeta_N(s) = \sum_{\vartheta \in \mathrm{Irr}(N)} \vartheta(1)^{-s}$ which we obtained in Section 3.1. We parameterise the non-trivial characters in $\mathrm{Irr}(N)$ by means of the affine cone $W(\mathfrak{o}) = \left(\mathfrak{o}^3\right)^*$, uniformly for all levels, as in the integral formula (2.3). The affine cone $W(\mathfrak{o})$ decomposes as a disjoint union of subsets $W(\mathfrak{o})^{[1]}$, $W(\mathfrak{o})^{[2a]}$ and $W(\mathfrak{o})^{[2b]}$ corresponding to representations of types 1, $2a$ and $2b$. According to our explicit element count modulo $\mathfrak{p}$ (see Table 3.1), we have $\mu(W(\mathfrak{o})^{[1]}) = q^{-1}(1 - q^{-2})$, $\mu(W(\mathfrak{o})^{[2a]}) = (1 - q^{-2})/2$ and $\mu(W(\mathfrak{o})^{[2b]}) = (1 - q^{-1})^2/2$. With this preparation we may write

$$\zeta_N(s) = 1 + q^3 \left(\mu(W(\mathfrak{o})^{[1]}) + \mu(W(\mathfrak{o})^{[2a]}) + \mu(W(\mathfrak{o})^{[2b]})\right)(1 - q^{1-s})^{-1},$$

and this yields

$$\begin{aligned} \zeta_{\mathrm{SL}_2(\mathfrak{o})}(s) = {} & \zeta_{\mathrm{SL}_2(\mathbb{F}_q)}(s) + q^3 \Big(\mu(W(\mathfrak{o})^{[1]})((q^2-1)/2)^{-1-s} 2q + \\ & \mu(W(\mathfrak{o})^{[2a]})(q^2+q)^{-1-s}(q-1) + \\ & \mu(W(\mathfrak{o})^{[2b]})(q^2-q)^{-1-s}(q+1)\Big)(1-q^{1-s})^{-1} \\ = {} & 1 + X_1 + \tfrac{q-3}{2} X_2 + 2X_3 + \tfrac{q-1}{2} X_4 + 2X_5 + \\ & (1-qX_1)^{-1}\left(4qX_2X_5 + \tfrac{q^2-1}{2} X_1X_4 + \tfrac{(q-1)^2}{2} X_1X_2\right) \end{aligned} \tag{3.5}$$

where $X_1 = q^{-s}$, $X_2 = (q+1)^{-s}$, $X_3 = ((q+1)/2)^{-s}$, $X_4 = (q-1)^{-s}$ and $X_5 = ((q-1)/2)^{-s}$. The last formula is in agreement with [**12**, Theorem 7.5].

It is worth pointing out that $\zeta_{\mathrm{SL}_2(\mathfrak{o})}(s)$, unlike $\zeta_{\mathrm{SL}_2^1(\mathfrak{o})}(s)$, cannot be written as a rational function in q^{-s}. In [**3**, Theorem A], we establish in a rather general context local functional equations for the zeta functions associated to families of pro-p groups, such as $\mathrm{SL}_2^1(\mathfrak{o})$; cf. Theorem 1.2. It would be very interesting if these could be meaningfully extended to the zeta functions of larger compact p-adic analytic groups, such as $\mathrm{SL}_2(\mathfrak{o})$.

REMARK. It would be interesting to see if the use of Clifford theory, as explored in the current section, can also be employed to establish the conjectural formula for the zeta function of $\mathrm{SL}_2^1(\mathbb{Z}_2)$ stated in Conjecture 3.1. For this one would start from our analysis of the zeta function $\zeta_{\mathrm{SL}_2^2(\mathbb{Z}_2)}(s)$; cf. Theorem 1.2.

3.4. The group $\mathrm{SL}_2(\mathbb{Z}_2)$**.** The paper [**18**] contains an explicit construction of the irreducible representations of the group $\mathrm{SL}_2(\mathbb{Z}_2)$. This is achieved by decomposing Weil representations associated to binary quadratic forms and by considering tensor products of certain components of such Weil representations. To complement (3.5) we record the following immediate consequence of the work in [**18**].

THEOREM 3.3. *We have*

$$\zeta_{\mathrm{SL}_2(\mathbb{Z}_2)}(s) = \frac{(4-2^{1-s}-5\cdot 2^{1-2s}+2^{1-3s})+3^{-s}(28+2^{1-s}-5\cdot 2^{1-2s}+2^{1-3s})}{1-2^{1-s}}.$$

PROOF. It follows by inspection of the classification results in [**18**, pp. 522–524] that the continuous irreducible characters of $G := \mathrm{SL}_2(\mathbb{Z}_2)$ all have degrees of the form 2^i or $3\cdot 2^i$, for $i\in\mathbb{N}_0$. Concerning 2-power-degree characters, one has

$$r_1(G)=4,\quad r_2(G)=6,\quad r_{2^2}(G)=2,\quad \text{and}\quad r_{2^i}(G)=3\cdot 2^{i-2} \text{ for } i\geq 3.$$

This yields

$$\sum_{i=0}^{\infty} r_{2^i}(G)(2^{-s})^i = \frac{4-2\cdot(2^{-s})-10\cdot 2^{-2s}+2\cdot 2^{-3s}}{1-2\cdot 2^{-s}}. \tag{3.6}$$

The numbers of characters of degree $3\cdot 2^i$, for $i\in\mathbb{N}_0$, are given by

$$r_3(G)=28,\quad r_{3\cdot 2}(G)=58,\quad r_{3\cdot 2^2}(G)=106,\ \text{and}\ r_{3\cdot 2^i}(G)=107\cdot 2^{i-2} \text{ for } i\geq 3.$$

Indeed, for $i\geq 3$ the characters of degree $3\cdot 2^i$ come from levels $i+1, i+2, i+3$ and $i+4$, with contributions from these levels of 2^{i-2}, 2^{i-1}, $5\cdot 2^{i+1}$ and 2^{i+4} characters, respectively. One checks that $2^{i-2}+2^{i-1}+5\cdot 2^{i+1}+2^{i+4}=107\cdot 2^{i-2}$, as claimed. This yields

$$\sum_{i=0}^{\infty} r_{3\cdot 2^i}(G)(3^{-s})(2^{-s})^i = 3^{-s}\left(\frac{28+2\cdot 2^{-s}-10\cdot 2^{-2s}+2\cdot 2^{-3s}}{1-2\cdot 2^{-s}}\right). \tag{3.7}$$

Combining (3.6) and (3.7) yields the claimed expression. □

The results in [**18**] do not indicate how to compute the zeta function of $\mathrm{SL}_2(\mathfrak{o})$ for extension rings $\mathfrak{o}$ of $\mathbb{Z}_2$. In view of our earlier computations it would be particularly interesting to consider the case where $\mathfrak{o}$ is an unramified extension of $\mathbb{Z}_2$.

4. Explicit formulae for subgroups of quaternion groups $\mathrm{SL}_1(\mathfrak{D})$

The aim of this section is to provide a setup for computing the zeta function of the norm one group $\mathrm{SL}_1(\mathfrak{D})$ of a central division algebra $\mathfrak{D}$ over a $\mathfrak{p}$-adic field $\mathfrak{k}$. Our approach leads to immediate consequences in the special case where the Schur index ℓ of $\mathfrak{D}$ over $\mathfrak{k}$ is a prime number. Explicit formulae are given for the zeta functions of norm one groups of non-split quaternion algebras; see Theorem 1.3 below.

4.1. General division algebras. Let $\mathfrak{D}$ denote a central division algebra of Schur index $\ell \geq 2$ over a $\mathfrak{p}$-adic field $\mathfrak{k}$. Let $\mathfrak{o}$ denote the valuation ring in $\mathfrak{k}$, with maximal ideal $\mathfrak{p}$, and let $\mathfrak{R}$ denote the maximal compact subring of $\mathfrak{D}$, with maximal ideal $\mathfrak{P}$. Write q and p for the cardinality and characteristic of the residue field of $\mathfrak{o}$.

We consider the compact p-adic analytic group $G := \mathrm{SL}_1(\mathfrak{D}) = \mathrm{SL}_1(\mathfrak{R})$ of norm-1 elements in $\mathfrak{D}$ and its principal congruence subgroups $G_m := \mathrm{SL}_1^m(\mathfrak{D}) = \mathrm{SL}_1(\mathfrak{D}) \cap (1 + \mathfrak{P}^m)$, $m \in \mathbb{N}$. We remark that the resulting congruence filtration of G is a refinement of the filtration that one would get from restriction of scalars to $\mathfrak{o}$ and defining congruence subgroups in terms of $\mathfrak{p}$; this justifies the slight difference in notation from Section 2. The group G/G_1 is isomorphic to the multiplicative group of norm-1 elements of $\mathfrak{R}/\mathfrak{P} \cong \mathbb{F}_{q^\ell}$ over $\mathfrak{o}/\mathfrak{p} \cong \mathbb{F}_q$, and is hence cyclic of order $(q^\ell - 1)/(q-1)$. Each of the quotients G_m/G_{m+1}, $m \in \mathbb{N}$, embeds into the additive group $\mathfrak{R}/\mathfrak{P} \cong \mathbb{F}_{q^\ell}$ and is thus an elementary abelian p-group. It follows that G_1 is the unique Sylow pro-p subgroup of G. The $\mathfrak{k}$-Lie algebra associated to the group $\mathrm{SL}_1(\mathfrak{D})$ is $\mathfrak{sl}_1(\mathfrak{D})$, consisting of all trace-0 elements of $\mathfrak{D}$.

To begin with, we derive the following consequence of Theorem 1.1 and its Corollary 2.3, which is based on the extra assumption that ℓ is prime.

COROLLARY 4.1. *Let $\mathfrak{D}$ be a central division algebra of prime Schur index ℓ over $\mathfrak{k}$. Then the abscissa of convergence of $\zeta_{\mathrm{SL}_1(\mathfrak{D})}(s)$ is equal to $2/\ell = 2r_{\mathrm{abs}}/|\Phi_{\mathrm{abs}}|$, where r_{abs} is the absolute rank of $\mathfrak{sl}_1(\mathfrak{D})$ and Φ_{abs} denotes the absolute root system associated to $\mathfrak{sl}_1(\mathfrak{D})$.*

We remark that a more complex argument shows that the conclusion of the corollary remains true even if the Schur index ℓ is not prime; see [**16**, Theorem 7.1].

PROOF OF COROLLARY 4.1. The absolute rank of $\mathfrak{sl}_1(\mathfrak{D})$ is $r_{\mathrm{abs}} = \ell - 1$, and the size of the absolute root system associated to $\mathfrak{sl}_1(\mathfrak{D})$ is $|\Phi_{\mathrm{abs}}| = \ell^2 - \ell$. Hence $2/\ell = 2r_{\mathrm{abs}}/|\Phi_{\mathrm{abs}}|$. Clearly, $\dim_{\mathfrak{k}}(\mathfrak{sl}_1(\mathfrak{D})) = \ell^2 - 1$. Recall that the abscissa of convergence $\alpha(\mathrm{SL}_1(\mathfrak{D}))$ is a commensurability invariant of the group $\mathrm{SL}_1(\mathfrak{D})$. Thus, in view of Theorem 1.1 and the equations (2.4), it suffices to prove that $\dim_{\mathfrak{k}} \mathrm{C}_{\mathfrak{sl}_1(\mathfrak{D})}(x) = \ell - 1$ for all $x \in \mathfrak{sl}_1(\mathfrak{D}) \smallsetminus \{0\}$. Indeed, this will imply $\sigma(\mathfrak{sl}_1(\mathfrak{R})) = \rho(\mathfrak{sl}_1(\mathfrak{R})) = \ell(\ell-1)/2$ and the result follows.

Let $x \in \mathfrak{sl}_1(\mathfrak{D}) \smallsetminus \{0\}$. Then $\mathfrak{k}(x)|\mathfrak{k}$ is a non-trivial field extension, and the Centraliser Theorem for central simple algebras yields $\ell^2 = |\mathfrak{D} : \mathfrak{k}| = |\mathfrak{k}(x) : \mathfrak{k}| \cdot |\mathrm{C}_{\mathfrak{D}}(x) : \mathfrak{k}|$. Since ℓ is assumed to be prime, this implies $|\mathrm{C}_{\mathfrak{D}}(x) : \mathfrak{k}| = \ell$, and hence $\dim_{\mathfrak{k}} \mathrm{C}_{\mathfrak{sl}_1(\mathfrak{D})}(x) = \ell - 1$. □

As a step toward the explicit computation of the zeta functions of G_1 and G, we describe sufficient conditions for applying the Kirillov orbit method to capture the irreducible complex characters of the group G_1.

PROPOSITION 4.2. *Let $\mathfrak{D}$ be a central division algebra of Schur index ℓ over the $\mathfrak{p}$-adic field $\mathfrak{k}$. Let $\mathfrak{R}$ be the maximal compact subring of $\mathfrak{D}$, and suppose that $p \nmid \ell$, where p denotes the residue field characteristic of $\mathfrak{k}$.*

Then $G_1 = \mathrm{SL}_1^1(\mathfrak{R})$ is an insoluble maximal p-adic analytic just infinite pro-p group. Furthermore, if $e(\mathfrak{k}, \mathbb{Q}_p)\ell < p - 1$, then G_1 is potent and saturable.

PROOF. It is known that $\mathrm{Aut}(\mathfrak{sl}_1(\mathfrak{D})) \cong \mathrm{PGL}_1(\mathfrak{D})$; see [**13**, § XI e]. The groups $\mathrm{SL}_1(\mathfrak{D})$ and $\mathrm{PGL}_1(\mathfrak{D})$ are, of course, closely related. Indeed, $Z := \mathrm{Z}(\mathrm{GL}_1(\mathfrak{D})) = \mathfrak{k}^*$

and the norm map induces an isomorphism

$$\mathrm{GL}_1(\mathfrak{D})/Z \cdot \mathrm{SL}_1(\mathfrak{D}) \cong \mathfrak{k}^*/(\mathfrak{k}^*)^\ell.$$

Denote by $\mu_\ell(\mathfrak{k}^*)$ the finite subgroup of $\mathfrak{k}^*$ consisting of all elements whose order divides ℓ. Since $p \nmid \ell$, reduction modulo $\mathfrak{p}$ maps $\mu_\ell(\mathfrak{k}^*) \leq \mathfrak{o}^*$ injectively onto the cyclic subgroup of order $\gcd(\ell, q-1)$ of $\mathbb{F}_q^*$. Moreover, $G_1 \cap Z = G_1 \cap \mu_\ell(\mathfrak{k}^*) = 1$, and the order of $\mathfrak{k}^*/(\mathfrak{k}^*)^\ell$ is not divisible by p. Hence we see from the exact sequence

$$1 \to \mu_\ell(\mathfrak{k}^*) \to \mathrm{SL}_1(\mathfrak{D}) \to \mathrm{PGL}_1(\mathfrak{D}) \to \mathfrak{k}^*/(\mathfrak{k}^*)^\ell \to 1$$

that G_1 is isomorphic to a Sylow pro-p subgroup of the compact group $\mathrm{PGL}_1(\mathfrak{D})$. It follows that G_1 is an insoluble maximal p-adic analytic just infinite pro-p group; cf. [**13**, § III e].

Let $\mathfrak{K}$ be a splitting subfield of $\mathfrak{D}$, unramified and of degree ℓ over $\mathfrak{k}$. Then the $\mathfrak{K}$-algebra isomorphism $\mathfrak{K} \otimes_{\mathfrak{k}} \mathfrak{D} \cong \mathrm{Mat}_\ell(\mathfrak{K})$ provides an embedding of G_1 into a Sylow pro-p subgroup S of $\mathrm{SL}_\ell(\mathfrak{O})$, where $\mathfrak{O}$ denotes the valuation ring of $\mathfrak{K}$. Suppose that $e\ell < p-1$, where $e = e(\mathfrak{k}, \mathbb{Q}_p) = e(\mathfrak{O}, \mathbb{Z}_p)$. From [**15**, III (3.2.7)] we conclude that S is saturable. Now we conclude as in [**14**, Proof of Theorem 1.3] that G_1 is saturable. Moreover, $\gamma_{p-1}(G_1) \subseteq \gamma_{e\ell+1}(G_1) = G_1^p$ (cf. [**19**, § 1]) so that G_1 is potent. $\square$

4.2. The quaternion case. In this section we consider the special case that $\ell = 2$, where $\mathfrak{D}$ is a non-split quaternion algebra over the $\mathfrak{p}$-adic field $\mathfrak{k}$. Some of the arguments below, however, are equally relevant in the more general situation where $p \nmid \ell$. Concretely, we compute explicit formulae for the zeta functions of norm one quaternion groups $\mathrm{SL}_1(\mathfrak{D}) = \mathrm{SL}_1(\mathfrak{R})$ and their principal congruence subgroups $\mathrm{SL}_1^m(\mathfrak{R})$, as stated in Theorem 1.3 in the introduction.

REMARK. Similar formulae as the ones provided in Theorem 1.3 can be obtained for higher principal congruence subgroups $\mathrm{SL}_1^m(\mathfrak{R})$, even if the condition $e(\mathfrak{k}, \mathbb{Q}_p)\ell < p-1$ imposed in the theorem is not satisfied. Our computation can be carried out in a similar fashion whenever the orbit method can be applied.

It is a natural and interesting problem to compute explicit formulae for the zeta functions of norm one groups $\mathrm{SL}_1(\mathfrak{D})$ and their principal congruence groups, where the Schur index of $\mathfrak{D}$ over $\mathfrak{k}$ is greater than 2. Another interesting group to consider would be $\mathrm{SL}_2(\mathfrak{R})$, where $\mathfrak{R}$ is the maximal compact subring of a non-split quaternion algebra $\mathfrak{D}$ over $\mathfrak{k}$.

PROOF OF THEOREM 1.3. According to Proposition 4.2, the pro-p group G_1 is saturable and potent. Our first aim is to compute an explicit formula for the zeta functions of the principal congruence subgroups $G_m = \mathrm{SL}_1^m(\mathfrak{R})$, $m \in \mathbb{N}$. Put $\mathfrak{g} := \mathfrak{sl}_1(\mathfrak{R})$, and for $m \in \mathbb{N}$ let $\mathfrak{g}_m := \mathfrak{sl}_1^m(\mathfrak{R}) = \mathfrak{sl}_1(\mathfrak{R}) \cap \mathfrak{P}^m$ denote the mth principal congruence Lie sublattice so that $G_m = \exp(\mathfrak{g}_m)$. Here $\mathfrak{o}$ denotes, as usual, the valuation ring of $\mathfrak{k}$, with maximal ideal $\mathfrak{p} = \pi\mathfrak{o}$ generated by a uniformiser π.

Since $p \neq 2$, we may choose $1, \mathbf{u}, \mathbf{v}, \mathbf{uv}$ as a standard basis for $\mathfrak{D}$ over $\mathfrak{k}$, where $\mathbf{u}^2 = a \in \mathfrak{o}$ is not a square modulo $\mathfrak{p}$, $\mathbf{v}^2 = \pi$ and $\mathbf{uv} = -\mathbf{vu}$. Writing $\mathbf{i} := \frac{1}{2}\mathbf{u}$, $\mathbf{j} := \frac{1}{2}\mathbf{v}$ and $\mathbf{k} := \frac{1}{2}\mathbf{uv}$ we have

$$[\mathbf{i}, \mathbf{j}] = \mathbf{k}, \quad [\mathbf{i}, \mathbf{k}] = a\mathbf{j}, \quad [\mathbf{j}, \mathbf{k}] = -\pi\mathbf{i}. \tag{4.1}$$

An $\mathfrak{o}$-basis for $\mathfrak{g}_1$ is then given by $\pi\mathbf{i}$, $\mathbf{j}$, $\mathbf{k}$, with corresponding commutator matrix

$$\mathcal{R}(\mathbf{Y}) = \begin{pmatrix} 0 & \pi Y_3 & a\pi Y_2 \\ -\pi Y_3 & 0 & -Y_1 \\ -a\pi Y_2 & Y_1 & 0 \end{pmatrix}. \tag{4.2}$$

In view of Proposition 2.2, an argument similar as for the case $p = 2$ and $e = 1$ in Section 3.1 shows that

$$\zeta_{\mathrm{SL}_1^m(\mathfrak{R})}(s) = q^{3(m-1)} \frac{q^2 - q^{-s}}{1 - q^{1-s}} \quad \text{for } m \in \mathbb{N}.$$

Our next aim is to deduce a formula for the zeta function of the group $G = \mathrm{SL}_1(\mathfrak{R})$, using Clifford theory, similarly as in Section 3.3. For this it is useful to record the following intermediate formula which comes from a similar argument as in Section 3.1:

$$\zeta_{\mathrm{SL}_1^1(\mathfrak{R})}(s) = 1 + \left(\mu(W(\mathfrak{o})^{[1]})q^{1-s} + \mu(W(\mathfrak{o})^{[2]})q^3\right) \frac{1}{1 - q^{1-s}}, \tag{4.3}$$

where $W(\mathfrak{o})^{[1]} = \{\mathbf{y} \in W(\mathfrak{o}) \mid y_1 \in \mathfrak{o}^*\}$ and $W(\mathfrak{o})^{[2]} = \{\mathbf{y} \in W(\mathfrak{o}) \mid y_1 \in \mathfrak{p}\}$ have Haar measure $\mu(W(\mathfrak{o})^{[1]}) = 1 - q^{-1}$ and $\mu(W(\mathfrak{o})^{[2]}) = q^{-1}(1 - q^{-2})$ respectively.

We continue to write $G = \mathrm{SL}_1(\mathfrak{R})$ and put $N := G_1 = \mathrm{SL}_1^1(\mathfrak{R})$. Since G/N is cyclic, any irreducible character ϑ of N can a priori be extended to an irreducible character $\hat{\vartheta}$ of its inertia group $I_G(\vartheta)$. Thus, similarly as for the group $\mathrm{SL}_2(\mathfrak{o})$, the central task consists in describing the inertia groups in order to apply (3.4). We show below that

(i) $I_G(\vartheta) = G$, and hence $I_G(\vartheta)/N$ is cyclic of order $q+1$, if $\vartheta \in \mathrm{Irr}(G)$ corresponds to a co-adjoint orbit of a functional represented by an element of $W(\mathfrak{o})^{[1]}$,

(ii) $I_G(\vartheta) = \{1, -1\}N$, and hence $I_G(\vartheta)/N$ is cyclic of order 2, if $\vartheta \in \mathrm{Irr}(G)$ corresponds to a co-adjoint orbit of a functional represented by an element of $W(\mathfrak{o})^{[2]}$.

We conclude the proof by applying Clifford theory as in Section 3.3. Putting together the general formula (3.4), the specific formula (4.3) and statements (i) and (ii), we deduce that

$$\begin{aligned} \zeta_{\mathrm{SL}_1(\mathfrak{R})}(s) &= \zeta_{C_{q+1}}(s) + \Big(\mu(W(\mathfrak{o})^{[1]})q^{1-s}(q+1) + \\ &\qquad \mu(W(\mathfrak{o})^{(2)})q^3((q+1)/2)^{-1-s}2)\Big) \frac{1}{1-q^{1-s}} \\ &= \frac{(q+1)(1-q^{-s}) + 4(q-1)((q+1)/2)^{-s}}{1-q^{1-s}}. \end{aligned}$$

It remains to justify the assertions (i) and (ii). For this it is convenient to translate between the adjoint action of G on $\mathfrak{g}$ and the co-adjoint action of G on $\mathrm{Hom}_{\mathfrak{o}}(\mathfrak{g}, \mathfrak{o})$. From (4.1) one easily sees that the normalised Killing form $\kappa_0 : \mathfrak{g} \times \mathfrak{g} \to \mathfrak{o}$ has the structure matrix

$$[\kappa_0(\cdot,\cdot)]_{(\mathbf{i},\mathbf{j},\mathbf{k})} = \left(\begin{smallmatrix} a & 0 & 0 \\ 0 & \pi & 0 \\ 0 & 0 & -a\pi \end{smallmatrix}\right)$$

with respect to the basis $\mathbf{i}$, $\mathbf{j}$, $\mathbf{k}$. While κ_0 is degenerate over $\mathfrak{o}$, the form is non-degenerate over $\mathfrak{k}$ and induces a bijective linear map $\iota_0 : \mathfrak{sl}_1(\mathfrak{D}) \to \mathrm{Hom}_{\mathfrak{k}}(\mathfrak{sl}_1(\mathfrak{D}), \mathfrak{k})$. We have $\mathfrak{g}_{-1} := \mathfrak{sl}_1(\mathfrak{R}) \cap \mathfrak{P}^{-1} = \iota_0^{-1}(\mathrm{Hom}_{\mathfrak{o}}(\mathfrak{g}, \mathfrak{o}))$ and $\mathfrak{g}_{-2} := \mathfrak{sl}_1(\mathfrak{R}) \cap \mathfrak{P}^{-2} =$

$\iota_0^{-1}(\mathrm{Hom}_{\mathfrak{o}}(\mathfrak{g}_1, \mathfrak{o}))$. The decomposition $\mathrm{Hom}_{\mathfrak{o}}(\mathfrak{g}_1, \mathfrak{o})^* = W(\mathfrak{o})^{[1]} \cup W(\mathfrak{o})^{[2]}$ corresponds to the decomposition

$$\mathfrak{g}_{-2}^* = (\mathfrak{g}_{-2} \smallsetminus \mathfrak{g}_{-1}) \cup (\mathfrak{g}_{-1} \smallsetminus \mathfrak{g}). \tag{4.4}$$

The action of G/N on quotients $\mathfrak{g}_m/\mathfrak{g}_{m+1}$ of successive terms in the congruence filtration of $\mathfrak{g}$ is described in [**19**, § 1] and we will use a compatible notation as far as practical. The division algebra $\mathfrak{D}$ contains an unramified extension $\mathfrak{K} = \mathfrak{k}(\mathbf{i})$ of degree 2 over $\mathfrak{k}$ which is normalised by the uniformiser $\Pi := \mathbf{j}$. Let F denote the residue field of $\mathfrak{K}$, and let Φ denote the group of roots of unity in $\mathfrak{K}$. Thus $F \cong \mathbb{F}_{q^2}$ and $\Phi \cup \{0\}$ is a set of representatives for the elements of F. Observe that N is complemented in G by the subgroup H of Φ consisting of all roots of unity which are of norm 1 over $\mathfrak{k}$: we have $G = H \ltimes N$. Accordingly, we will think of $G/N \cong H$ as the group of elements in the finite field F which have norm 1 over $f := \mathfrak{o}/\mathfrak{p} \cong \mathbb{F}_q$. Every element of $\mathfrak{R}$ has a unique power series expansion in Π with coefficients in $\Phi \cup \{0\}$. For each $m \in \mathbb{N}$ this induces an embedding $\eta_m : \mathfrak{g}_m/\mathfrak{g}_{m+1} \hookrightarrow F$; we denote the image of η_m by $F(m)$. If $2 \nmid m$ then $F(m) = F$, and if $2 \mid m$ then $F(m) = \{x \in F \mid \mathrm{Tr}_{F|f}(x) = 0\}$.

Clearly, for each $m \in \mathbb{N}$ the action of G on $F(m)$ by conjugation factors through N and is therefore determined by the action of H. The latter is given by the explicit formula

$$x^h = h^{1-q^m} \cdot x, \qquad \text{for } x \in F(m) \text{ and } h \in H.$$

This finishes our preparations and we turn to the proof of assertions (i) and (ii) above.

First we consider a character $\vartheta \in \mathrm{Irr}(N)$ corresponding to the co-adjoint orbit of $\omega \in \mathrm{Irr}(\mathfrak{g}_1)$, where ω is represented by an element of $W(\mathfrak{o})^{[1]}$ and has level n, say; cf. Section 2.1. Then the inertia group $I_G(\vartheta)$ is equal to CN, where $C := \mathrm{C}_G(x+\mathfrak{g}_{2n-2})$ for a suitable $x \in \mathfrak{g}_{-2} \smallsetminus \mathfrak{g}_{-1}$; see (4.4). Here $\mathfrak{g}_0 := \mathfrak{g}$ if $n = 1$. We claim that $CN = G$, justifying (i). For this it is enough to prove that H centralises a suitable N-conjugate of x. Since H is a subgroup of the multiplicative group of the field $\mathfrak{K}$, it suffices to show that x is N-conjugate to an element of $\mathfrak{K}$. For this we construct recursively a sequence $x_0, x_1, \ldots$ of N-conjugates of x such that $x_i \equiv \lambda_i \pi^{-1}\mathbf{i}$ modulo $\mathfrak{g}_{i-1}$ with $\lambda_i \in \mathfrak{o} \smallsetminus \mathfrak{p}$ for each index i. From the construction one sees that the sequence converges and its limit is an N-conjugate of x in $\mathfrak{K}$. Since $x \in \mathfrak{g}_{-2} \smallsetminus \mathfrak{g}_{-1}$, we can take $x_0 := x$. Now suppose that $i \in \mathbb{N}_0$ and that

$$x_i = \lambda_i \pi^{-1}\mathbf{i} + \mu\mathbf{j} + \nu\mathbf{k} \qquad \text{with } \lambda_i, \mu, \nu \in \mathfrak{o},\ \lambda_i \notin \mathfrak{p} \text{ and } \mu\mathbf{j} + \nu\mathbf{k} \in \mathfrak{P}^{i-1}$$

is an N-conjugate of x. Then $x_{i+1} := z^{-1}x_i z$ with $z := 1 - \lambda_i^{-1}\pi(\nu\mathbf{j} - a^{-1}\mu\mathbf{k}) \in 1+\mathfrak{P}^{i+1} \subseteq N$ is an N-conjugate of x and satisfies the desired congruence, modulo $\mathfrak{g}_i$,

$$\begin{aligned} x_{i+1} &\equiv (1 + \lambda_i^{-1}\pi(\nu\mathbf{j} + a^{-1}\mu\mathbf{k}))(\lambda_i\pi^{-1}\mathbf{i} + (\mu\mathbf{j} + \nu\mathbf{k}))(1 - \lambda_i^{-1}\pi(\nu\mathbf{j} + a^{-1}\mu\mathbf{k})) \\ &\equiv \lambda_i\pi^{-1}\mathbf{i} + (\mu\mathbf{j} + \nu\mathbf{k}) + [\nu\mathbf{j} + a^{-1}\mu\mathbf{k}, \mathbf{i}] \\ &\equiv \lambda_i\pi^{-1}\mathbf{i}. \end{aligned}$$

Finally we consider a character $\vartheta \in \mathrm{Irr}(N)$ corresponding to the co-adjoint orbit of $\omega \in \mathrm{Irr}(\mathfrak{g}_1)$, where ω is represented by an element of $W(\mathfrak{o})^{[2]}$ and has level n, say. Then the inertia group $I_G(\vartheta)$ is equal to CN, where $C := \mathrm{C}_G(x + \mathfrak{g}_{2n-2})$ for a suitable $x \in \mathfrak{g}_{-1} \smallsetminus \mathfrak{g}$; see (4.4). We claim that $CN = \{1, -1\}N$, justifying (ii). Since $x \in \mathfrak{g}_{-1}$, we have $\{1, -1\}N \subseteq CN \subseteq \mathrm{C}_G(x + \mathfrak{g}) = \mathrm{C}_H(x + \mathfrak{g})N$. Hence it suffices to prove that $\mathrm{C}_H(x+\mathfrak{g}) = \{1, -1\}$. Indeed, multiplication by π provides an

H-equivariant isomorphism $\mathfrak{g}_{-1}/\mathfrak{g} \to \mathfrak{g}_1/\mathfrak{g}_2$. The action of $h \in H$ by conjugation on $\mathfrak{g}_1/\mathfrak{g}_2$ corresponds to multiplication by h^{1-q} on $F(1)$; to carry out the multiplication h is considered as a element of the residue field F with norm 1 in f, as described above. The group H has order $q+1$, hence the kernel of $H \to H$, $h \mapsto h^{1-q}$ is equal to $\{1, -1\}$. It follows that $\mathrm{C}_H(\tilde{x}) = \{1, -1\}$ for any non-zero element $\tilde{x} \in F(1)$. □

5. Principal congruence subgroups of $\mathrm{SL}_3(\mathfrak{o})$ for unramified $\mathfrak{o}$ of residue field characteristic 3

Let $\mathfrak{o}$ be a compact discrete valuation ring of characteristic 0 and residue field characteristic p. Except for the special case $p = 3$, Theorem E in [**3**] provides an explicit universal formula for the zeta functions of principal congruence subgroups of $\mathrm{SL}_3(\mathfrak{o})$. By a different approach, the same formula and indeed a formula for the group $\mathrm{SL}_3(\mathfrak{o})$ itself are derived in [**4**]. In this section we complement the generic formulae by proving Theorem 1.4 which was stated in the introduction. This theorem provides explicit formulae for the zeta functions of principal congruence subgroups of $\mathrm{SL}_3(\mathfrak{o})$, where $\mathfrak{o}$ has residue characteristic 3 and is unramified over $\mathbb{Z}_3$.

Residue field characteristic 3 was excluded from [**3**, Theorem E], whose proof is based on a geometric description of the variety of irregular elements in the 8-dimensional Lie algebra $\mathfrak{sl}_3(\mathfrak{k})$. This description breaks down when the map

$$\beta : \mathfrak{gl}_3 \to \mathfrak{sl}_3, \quad \mathbf{x} \mapsto \mathbf{x} - \mathrm{Tr}(\mathbf{x})/3$$

used in [**3**, Section 6.1] displays bad reduction modulo $\mathfrak{p}$. Moreover, the translation of the relevant $\mathfrak{p}$-adic integral via the normalised Killing form becomes more technical. In the present paper we restrict our attention to unramified extensions $\mathfrak{o}$ of $\mathbb{Z}_3$ for simplicity. Indeed, the results in Section 3.1 suggest that analogous formulae which are to cover the general case, including ramification, are likely to become rather cumbersome to write down.

The method we employ is algebraic and somewhat closer to the approach taken in [**4**]. In fact, the arguments which we shall supply can be employed *mutatis mutandis* in the generic case $p \neq 3$ and hence give an alternative, less geometric derivation of the formula provided in [**3**, Theorem E]. Moreover, our calculations illustrate the algebraic meaning of the $\mathfrak{p}$-adic formalism developed and applied in [**3**].

5.1. Let $\mathfrak{o}$ be unramified over $\mathbb{Z}_p$ with residue field $\mathfrak{o}/\mathfrak{p} \cong \mathbb{F}_q$ of characteristic $p = 3$. Throughout the section we will continue to write p as far as convenient, while keeping the concrete value $p = 3$ in mind. We remark that p is also a uniformiser for $\mathfrak{o}$, because $\mathfrak{o}$ is unramified over $\mathbb{Z}_p$, and we will use p instead of the symbol π. Let $T_\mathfrak{o} = \{0\} \cup \mu_{q-1}(\mathfrak{o})$ denote the set of Teichmüller representatives for $\mathfrak{o}$, which projects bijectively onto the residue field of $\mathfrak{o}$. More generally, for any $l \in \mathbb{N}$ we fix

$$T_\mathfrak{o}(l) := \left\{ \sum\nolimits_{j=0}^{l-1} t_j p^j \mid t_j \in T_\mathfrak{o} \text{ for } 0 \leq j < l \right\} \tag{5.1}$$

as a set of representatives for $\mathfrak{o}/\mathfrak{p}^l$. As usual, $\mathfrak{k}$ denotes the field of fractions of $\mathfrak{o}$.

Let $m \in \mathbb{N}$. Proposition 2.3 in [**3**] shows that m is permissible for $\mathfrak{sl}_3(\mathfrak{o})$. Thus the zeta function of the mth principal congruence subgroup $\mathrm{SL}_3^m(\mathfrak{o})$ is given by the formula

$$\zeta_{\mathrm{SL}_3^m(\mathfrak{o})}(s) = q^{8m}\left(1 + (1-q^{-1})^{-1} \mathcal{Z}_\mathfrak{o}(-s/2-1, 3s-3)\right), \tag{5.2}$$

which results from Proposition 2.2 on setting $d = \dim_{\mathfrak{k}} \mathfrak{sl}_3(\mathfrak{k}) = 8$ and

$$\rho = 2^{-1}(\dim_{\mathfrak{k}} \mathfrak{sl}_3(\mathfrak{k}) - r_{\mathrm{abs}}(\mathfrak{sl}_3(\mathfrak{k}))) = 3;$$

see (2.4). As indicated in Section 2.1, the integral $\mathcal{Z}_{\mathfrak{o}}(r,t)$ is intimately linked to the elementary divisors of the commutator matrix $\mathcal{R}(\mathbf{y})$ for $\mathfrak{sl}_3(\mathfrak{o})$, evaluated at points $\mathbf{y} \in W(\mathfrak{o}) = (\mathfrak{o}^8)^*$. We observe that the commutator matrix $\mathcal{R}(\mathbf{y})$, at $\mathbf{y} \in W(\mathfrak{o})$, has Witt normal form

$$\begin{pmatrix} \begin{smallmatrix} 0 & 1 \\ -1 & 0 \end{smallmatrix} & & & \\ & \begin{smallmatrix} 0 & 1 \\ -1 & 0 \end{smallmatrix} & & \\ & & \begin{smallmatrix} 0 & p^a \\ -p^a & 0 \end{smallmatrix} & \\ & & & \begin{smallmatrix} 0 & 0 \\ 0 & 0 \end{smallmatrix} \end{pmatrix}$$

so that all the information is condensed in a single parameter $a = a(\mathbf{y}) \in \mathbb{N}_0 \cup \{\infty\}$.

The integral $\mathcal{Z}_{\mathfrak{o}}(r,t)$ in (2.3) is defined so that it performs integration over the space $\mathfrak{p} \times \mathrm{Hom}_{\mathfrak{o}}(\mathfrak{sl}_3(\mathfrak{o}), \mathfrak{o})^*$ with respect to a particular choice of coordinate system $(x, \mathbf{y}) \in \mathfrak{p} \times W(\mathfrak{o})$. The normalised Killing form κ_0 of $\mathfrak{sl}_3(\mathfrak{k})$ is related to the ordinary Killing form $\kappa : \mathfrak{sl}_3(\mathfrak{k}) \times \mathfrak{sl}_3(\mathfrak{k}) \to \mathfrak{k}$ by the equation $\kappa = 2h^\vee \kappa_0 = 6\kappa_0$; see Section 2.2. In [**3**, Section 6], we provided the structure matrix of the normalised Killing form κ_0 with respect to the basis

$$\mathbf{h}_{12} = \begin{pmatrix} 1 & & \\ & -1 & \\ & & 0 \end{pmatrix}, \quad \mathbf{h}_{23} = \begin{pmatrix} 0 & & \\ & 1 & \\ & & -1 \end{pmatrix},$$
$$\mathbf{e}_{12} = \begin{pmatrix} 0 & 1 & \\ & 0 & \\ & & 0 \end{pmatrix}, \quad \mathbf{e}_{23} = \begin{pmatrix} 0 & & \\ & 0 & 1 \\ & & 0 \end{pmatrix}, \quad \mathbf{e}_{13} = \begin{pmatrix} 0 & 0 & 1 \\ & 0 & 0 \\ & & 0 \end{pmatrix},$$
$$\mathbf{f}_{21} = \begin{pmatrix} 0 & & \\ 1 & 0 & \\ & & 0 \end{pmatrix}, \quad \mathbf{f}_{23} = \begin{pmatrix} 0 & & \\ & 0 & \\ & 1 & 0 \end{pmatrix}, \quad \mathbf{f}_{13} = \begin{pmatrix} 0 & & \\ 0 & 0 & \\ 1 & 0 & 0 \end{pmatrix}.$$

This matrix has determinant 3. Thus the form κ_0 induces a bijective linear map $\iota_0 : \mathfrak{sl}_3(\mathfrak{k}) \to \mathrm{Hom}_{\mathfrak{k}}(\mathfrak{sl}_3(\mathfrak{k}), \mathfrak{k})$, but becomes more intricate at the level of $\mathfrak{o}$-lattices, due to the residue field characteristic 3. Indeed, the pre-image of $\mathrm{Hom}_{\mathfrak{o}}(\mathfrak{sl}_3(\mathfrak{o}), \mathfrak{o})$ under ι_0 is the $\mathfrak{o}$-lattice

$$\Lambda := \iota_0^{-1}(\mathrm{Hom}_{\mathfrak{o}}(\mathfrak{sl}_3(\mathfrak{o}), \mathfrak{o})) = \bigcup_{u \in T_{\mathfrak{o}}(1)} \left(u(\tfrac{2}{3}\mathbf{h}_{12} + \tfrac{1}{3}\mathbf{h}_{23}) + \mathfrak{sl}_3(\mathfrak{o}) \right).$$

Thus we have $p\Lambda \leq \mathfrak{sl}_3(\mathfrak{o}) \leq \Lambda$ with $|\mathfrak{sl}_3(\mathfrak{o}) : p\Lambda| = q^7$ and $|\Lambda : \mathfrak{sl}_3(\mathfrak{o})| = q$. We pull back the integral $\mathcal{Z}_{\mathfrak{o}}(r,t)$ over $\mathfrak{p} \times (\mathrm{Hom}_{\mathfrak{o}}(\mathfrak{sl}_3(\mathfrak{o}), \mathfrak{o}))^*$ to an integral over $\mathfrak{p} \times \Lambda^*$, taking into account the Jacobi factor $|3|_{\mathfrak{p}} = q^{-1}$. Dividing the new region of integration with respect to the second factor into cosets modulo $p\Lambda$, we write

$$\mathcal{Z}_{\mathfrak{o}}(r,t) = \mathcal{S}_1(r,t) + \mathcal{S}_2(r,t), \tag{5.3}$$

where the two summands correspond to the complementary subregions of integration $\mathfrak{p} \times (\mathfrak{sl}_3(\mathfrak{o}) \smallsetminus p\Lambda)$ and $\mathfrak{p} \times (\Lambda \smallsetminus \mathfrak{sl}_3(\mathfrak{o}))$ respectively. We show in Sections 5.2 and 5.3 that these summands are given by the formulae

$$\begin{aligned} \mathcal{S}_1(r,t) &= (q^4 + q^3 - q - 1)\mathcal{Z}_{\mathfrak{o}}^{[3]}(r,t) + (q^6 - q^4 - q^3 + q)\mathcal{Z}_{\mathfrak{o}}^{[2]}(r,t) \\ &\quad + (q^7 - q^6)\mathcal{Z}_{\mathfrak{o}}^{[0]}(r,t), \qquad (5.4) \\ \mathcal{S}_2(r,t) &= (q-1)((q^2+q+1)q^2\mathcal{Z}_{\mathfrak{o}}^{[1]}(r,t) + (q^7 - (q^2+q+1)q^2)\mathcal{Z}_{\mathfrak{o}}^{[0]}(r,t)), \end{aligned}$$

where

$$\begin{aligned}
\mathcal{Z}_{\mathfrak{o}}^{[0]}(r,t) &= \int_{(x,\mathbf{y})\in\mathfrak{p}\times\mathfrak{p}^{(8)}} |x|^t \, d\mu(x,\mathbf{y}),\\
\mathcal{Z}_{\mathfrak{o}}^{[1]}(r,t) &= \int_{(x,\mathbf{y})\in\mathfrak{p}\times\mathfrak{p}^{(8)}} |x|^t \|\{y_1,y_2,y_3,x\}\|_{\mathfrak{p}}^{2r} \, d\mu(x,\mathbf{y}),\\
\mathcal{Z}_{\mathfrak{o}}^{[2]}(r,t) &= q^{-2r}\mathcal{Z}_{\mathfrak{o}}^{[0]}(r,t),\\
\mathcal{Z}_{\mathfrak{o}}^{[3]}(r,t) &= \int_{(x,\mathbf{y})\in\mathfrak{p}\times\mathfrak{p}^{(8)}} |x|^t \|\{py_1,py_2,y_3,x\}\|_{\mathfrak{p}}^{2r} \, d\mu(x,\mathbf{y}).
\end{aligned}$$

In [**3**, Section 6.1] it is shown that

$$\mathcal{Z}_{\mathfrak{o}}^{[0]}(r,t) = \frac{q^{-9-t}(1-q^{-1})}{1-q^{-1-t}} \tag{5.5}$$

and

$$\mathcal{Z}_{\mathfrak{o}}^{[1]}(r,t) = \frac{q^{-9-2r-t}(1-q^{-4-t})(1-q^{-1})}{(1-q^{-4-2r-t})(1-q^{-1-t})}. \tag{5.6}$$

The latter may be obtained from the formula

$$\mathcal{Z}_{\mathfrak{o}}^{[1]}(r,t) = \sum_{(l,n)\in\mathbb{N}^2} (1-q^{-1})q^{-n} m_l^{[1]} \, q^{-nt-2\min\{l,n\}r},$$

with

$$m_l^{[1]} := \mu\left(\left\{\mathbf{y}\in\mathfrak{p}^{(8)} \mid \max\{|y_1|_{\mathfrak{p}},|y_2|_{\mathfrak{p}},|y_3|_{\mathfrak{p}}\} = q^{-l}\right\}\right) = (1-q^{-3})q^{-3l-5}, \tag{5.7}$$

using the fact that

$$\sum_{(l,n)\in\mathbb{N}^2} X_1^l X_2^n X_3^{\min\{l,n\}} = \frac{X_1X_2X_3(1-X_1X_2)}{(1-X_1X_2X_3)(1-X_1)(1-X_2)}. \tag{5.8}$$

Clearly, (5.5) implies that

$$\mathcal{Z}_{\mathfrak{o}}^{[2]}(r,t) = \frac{q^{-9-2r-t}(1-q^{-1})}{1-q^{-1-t}}. \tag{5.9}$$

This formula, too, may be written as a sum

$$\mathcal{Z}_{\mathfrak{o}}^{[2]}(r,t) = \sum_{(l,n)\in\mathbb{N}^2} (1-q^{-1})q^{-n} m_l^{[2]} \, q^{-nt-2\min\{l,n\}r},$$

with

$$m_l^{[2]} := \mu\left(\left\{\mathbf{y}\in\mathfrak{p}^{(8)} \mid |p|_{\mathfrak{p}} = q^{-l}\right\}\right) = \begin{cases} q^{-8} & \text{if } l=1,\\ 0 & \text{if } l\geq 2.\end{cases} \tag{5.10}$$

It remains to compute $\mathcal{Z}_{\mathfrak{o}}^{[3]}(r,t)$. We have

$$\mathcal{Z}_{\mathfrak{o}}^{[3]}(r,t) = \sum_{(l,n)\in\mathbb{N}^2} (1-q^{-1})q^{-n} m_l^{[3]} \, q^{-nt-2\min\{l,n\}r},$$

type		# of orbits	size of each orbit	total number modulo $\mathbb{F}_q\bar{\mathbf{z}}$
A		q	1	1
B	irreg.	q	$(q^3-1)(q^2-1)q$	$(q^3-1)(q^2-1)q$
C	irreg.	q	$(q^3-1)(q+1)$	$(q^3-1)(q+1)$
D	reg.	$(q-1)q/6$	$(q^2+q+1)(q+1)q^3$	$(q^3-1)(q+1)q^3/6$
E	reg.	$(q-1)q/2$	$(q^3-1)q^3$	$(q^3-1)(q-1)q^3/2$
F	reg.	$(q-1)q/3$	$(q^2-1)(q-1)q^3$	$(q^2-1)(q-1)^2q^3/3$

TABLE 5.1. Adjoint orbits in $\mathfrak{sl}_3(\mathbb{F}_q)$ under the action of $\mathrm{GL}_3(\mathbb{F}_q)$, $q \equiv_3 0$

where

$$m_l^{[3]} := \mu\left(\left\{\mathbf{y} \in \mathfrak{p}^{(8)} \mid \max\{|py_1|_{\mathfrak{p}}, |py_2|_{\mathfrak{p}}, |y_3|_{\mathfrak{p}}\} = q^{-l}\right\}\right)$$

$$= \begin{cases} (1-q^{-1})q^{-8} & \text{if } l = 1, \\ (1-q^{-3})q^{-3-3l} & \text{if } l \geq 2. \end{cases} \tag{5.11}$$

Using (5.8) this gives

$$\begin{aligned} \mathcal{Z}_{\mathfrak{o}}^{[3]}(r,t) &= (1-q^{-1})(m_1^{[3]} - (1-q^{-3})q^{-6})\sum_{n\in\mathbb{N}} q^{-(1+t)n-2r} \\ &\quad + (1-q^{-1})(1-q^{-3})q^{-3}\sum_{(l,n)\in\mathbb{N}^2} q^{-3l-(1+t)n-2\min\{l,n\}r} \\ &= -(1-q^{-1})(1-q^{-2})q^{-7-2r-t}(1-q^{-1-t})^{-1} \\ &\quad + (1-q^{-1})(1-q^{-4-t})q^{-7-t-2r}(1-q^{-1-t})^{-1}(1-q^{-4-t-2r})^{-1} \\ &= \frac{(1-q^{-1})q^{-9-t-2r}\left(1-q^{-2-t}+q^{-2-t-2r}-q^{-4-t-2r}\right)}{(1-q^{-1-t})(1-q^{-4-t-2r})}. \end{aligned}$$

A short computation, based on (5.2), (5.3) and (5.4), now yields the explicit formula for the zeta function of $\mathrm{SL}_3^m(\mathfrak{o})$, stated in Theorem 1.4.

The remainder of this section is devoted to an algebraic justification of the equations (5.4). It would be interesting to derive a geometric explanation, more similar to the argument in [**3**, Section 6.1] treating the generic case.

5.2. First we will derive the formula given for the summand $\mathcal{S}_1(r,t)$ in (5.4). For this we decompose $\mathfrak{sl}_3(\mathfrak{o}) \smallsetminus p\Lambda$ into cosets modulo $p\Lambda$, or equivalently the finite Lie algebra $\mathfrak{sl}_3(\mathbb{F}_q)$ into cosets modulo its centre. As $p = 3$, the centre of $\mathfrak{sl}_3(\mathbb{F}_q)$ is the 1-dimensional subalgebra $\mathbb{F}_q\bar{\mathbf{z}}$, spanned by the reduction modulo $\mathfrak{p}$ of $\mathbf{z} := \mathbf{h}_{12} - \mathbf{h}_{23}$, viz. the subalgebra of scalar matrices over $\mathbb{F}_q$. An overview of the orbits in $\mathfrak{sl}_3(\mathbb{F}_q)$ under the adjoint action of $\mathrm{GL}_3(\mathbb{F}_q)$ is provided in Table 5.1; see Appendix C for a short discussion. The second column indicates whether the corresponding elements are regular or irregular, as defined at the end of Section 2.1. The corresponding total number of cosets modulo $\mathbb{F}_q\bar{\mathbf{z}}$, given in the last column of the table, is obtained upon division by q. A short calculation confirms that the sizes of the orbits listed in Table 5.1 add up to $q^7 = |\mathfrak{sl}_3(\mathbb{F}_q) : \mathbb{F}_q\bar{\mathbf{z}}|$, as wanted.

The equation for $\mathcal{S}_1(r,t)$ in (5.4) indicates that $\mathcal{Z}_{\mathfrak{o}}^{[0]}(r,t)$ is the correct integral for the types D, E, F, which cover the regular elements modulo p. It remains to link the contributions to the summand $\mathcal{S}_1(r,t)$ by irregular elements belonging to cosets of types B and C to the integrals $\mathcal{Z}_{\mathfrak{o}}^{[2]}(r,t)$ and $\mathcal{Z}_{\mathfrak{o}}^{[3]}(r,t)$, respectively.

5.2.1. Let us consider first elements belonging to cosets modulo $p\Lambda$ of type C and work out the integral around such elements which results from pulling the original integral $\mathcal{Z}_{\mathfrak{o}}(r,t)$ over $\mathfrak{p} \times (\mathrm{Hom}_{\mathfrak{o}}(\mathfrak{sl}_3(\mathfrak{o}),\mathfrak{o}))^*$ back to $\mathfrak{p} \times \Lambda^*$. A typical coset of type C is $\mathbf{a}_0 + p\Lambda$, where $\mathbf{a}_0 := \left(\begin{smallmatrix} 0&1&0\\0&0&0\\0&0&0\end{smallmatrix}\right)$, and each coset has measure $\mu(p\Lambda) = q^{-7}$. As indicated earlier, the determinant of the Jacobi matrix associated to $\iota_0 : \Lambda \to \mathrm{Hom}_{\mathfrak{o}}(\mathfrak{sl}_3(\mathfrak{o}),\mathfrak{o})$ is 3 and thus contributes another factor $|3|_{\mathfrak{p}} = q^{-1}$. The integral over $\mathfrak{p} \times (\mathbf{a} + p\Lambda)$ with Jacobi factor q^{-1} can thus be described as an integral $\mathcal{I}(r,t)$ over $\mathfrak{p} \times \mathfrak{p}^{(8)}$. We argue that it is equal to $\mathcal{Z}_{\mathfrak{o}}^{[3]}(r,t)$, which may be computed from the integer sequence $a_n^{[3]}$, $n \in \mathbb{N}_0$, defined by

$$\begin{aligned}
a_n^{[3]} &:= \# \left\{ \mathbf{y} + (\mathfrak{p}^{n+1})^{(8)} \mid \mathbf{y} \in \mathfrak{p}^{(8)} \text{ such that } \|\{py_1, py_2, y_3, p^{n+1}\}\|_{\mathfrak{p}} = q^{-n-1} \right\} \\
&= \# \left\{ \mathbf{y} + (\mathfrak{p}^{n+1})^{(8)} \mid \mathbf{y} \in \mathfrak{p}^{(8)} \text{ such that } y_1, y_2 \in \mathfrak{p}^n \text{ and } y_3 \in \mathfrak{p}^{n+1} \right\} \\
&= |\mathfrak{p}^{\max\{1,n\}} : \mathfrak{p}^{n+1}|^2 \cdot |\mathfrak{p}^{(5)} : (\mathfrak{p}^{n+1})^{(5)}| \\
&= \begin{cases} 1 & \text{if } n = 0, \\ q^{5n+2} & \text{if } n \geq 1, \end{cases}
\end{aligned}$$

describing the lifting behaviour of points modulo $\mathfrak{p}^{n+1}$ on the variety defined by the integrand of $\mathcal{Z}_{\mathfrak{o}}^{[3]}(r,t)$. Indeed, we observe that the numbers $m_l^{[3]}$ defined in (5.11) satisfy

$$m_l^{[3]} = q^{-8l} a_{l-1}^{[3]} - q^{-8(l+1)} a_l^{[3]}.$$

The following proposition shows that the integral $\mathcal{I}(r,t)$ over $\mathfrak{p} \times \mathfrak{p}^{(8)}$ is equal to $\mathcal{Z}_{\mathfrak{o}}^{[3]}(r,t)$.

PROPOSITION 5.1. *For $n \in \mathbb{N}_0$ the set*

$$A_n^{[3]} := \Big\{ \mathbf{a} + p^{n+1}\Lambda \in \Lambda/p^{n+1}\Lambda \mid \mathbf{a} \equiv \left(\begin{smallmatrix} 0&1&0\\0&0&0\\0&0&0\end{smallmatrix}\right) \text{ modulo } p\Lambda \\ \text{and } |\mathfrak{sl}_3(\mathfrak{o}) : \mathrm{C}_{\mathfrak{sl}_3(\mathfrak{o})}(\mathbf{a} + p^{n+1}\Lambda)| = q^{4(n+1)} \Big\}$$

has cardinality $a_n^{[3]}$.

PROOF. The case $n = 0$ is a simple computation. Indeed, the only candidate for an element of $A_0^{[3]}$ is $\mathbf{a}_0 + p\Lambda$, where $\mathbf{a}_0 := \left(\begin{smallmatrix} 0&1&0\\0&0&0\\0&0&0\end{smallmatrix}\right)$, and a short computation reveals that, indeed,

$$|\mathfrak{sl}_3(\mathfrak{o}) : \mathrm{C}_{\mathfrak{sl}_3(\mathfrak{o})}(\mathbf{a}_0 + p\Lambda)| = |\mathfrak{sl}_3(\mathbb{F}_q) : \mathrm{C}_{\mathfrak{sl}_3(\mathbb{F}_q)}(\overline{\mathbf{a}_0})| = q^4,$$

where $\overline{\mathbf{a}_0}$ denotes the image of $\mathbf{a}_0$ in $\mathfrak{sl}_3(\mathbb{F}_q)$; cf. (B.2). Thus $|A_0^{[3]}| = 1$, as claimed.

Now suppose that $n \geq 1$. Arguing by induction on n, we prove in fact a little more than stated in the proposition. For any $l \in \mathbb{N}$, let $T_{\mathfrak{o}}(l)$ denote representatives for $\mathfrak{o}/\mathfrak{p}^l$ derived from the Teichmüller representatives for $\mathfrak{o}/\mathfrak{p}$; see (5.1).

Claim. Every matrix $\mathbf{a} \in \Lambda$ with $\mathbf{a} + p^{n+1}\Lambda \in A_n^{[3]}$ can be conjugated by elements of $\mathrm{GL}_3^1(\mathfrak{o})$ to the 'normal' form

$$\begin{pmatrix} 0 & 1 & 0 \\ 2p^2c^2 & pc & 0 \\ 0 & 0 & -pc \end{pmatrix} + p^n \begin{pmatrix} 0 & 0 & 0 \\ 0 & 0 & y_3 \\ y_2 & 0 & 0 \end{pmatrix} \qquad \text{modulo } p^{n+1}\Lambda,$$

where $c \in T_{\mathfrak{o}}(n)$ and $y_2, y_3 \in T_{\mathfrak{o}}(1)$. These matrices modulo $p^{n+1}\Lambda$ form a complete set of representatives for the $\mathrm{GL}_3^1(\mathfrak{o})$-orbits comprising $A_n^{[3]}$. Moreover, the index in $\mathrm{GL}_3^1(\mathfrak{o})$ of the centraliser of any such matrix modulo $p^{n+1}\Lambda$ is q^{4n}, and thus $|A_n^{[3]}| = |T_{\mathfrak{o}}(n)||T_{\mathfrak{o}}(1)|^2 q^{4n} = q^{5n+2}$, as wanted.

Finally, every matrix $\mathbf{a} \in \Lambda$ with $\mathbf{a} + p^{n+2}\Lambda \in A_{n+1}^{[3]}$ can be conjugated by elements of $\mathrm{GL}_3^1(\mathfrak{o})$ to a matrix which is, modulo $p^{n+1}\Lambda$, of the normal form above and satisfies the extra condition $y_2 = y_3 = 0$.

As indicated we use induction on n. Let $c \in T_{\mathfrak{o}}(n-1)$ and put

$$\mathbf{a}_c := \begin{pmatrix} 0 & 1 & 0 \\ 2p^2c^2 & pc & 0 \\ 0 & 0 & -pc \end{pmatrix}.$$

The eigenvalues of $\mathbf{a}_c$ are $-pc$ and $2pc$ with multiplicities 2 and 1 respectively. Hence c is an invariant of the $\mathrm{GL}_3^1(\mathfrak{o})$-orbit of $\mathbf{a}_c$ modulo $\mathfrak{sl}_3^n(\mathfrak{o})$. In view of our discussion of the case $n = 0$, if $n = 1$, or the induction hypothesis, if $n > 1$, it suffices to work out representatives of the elements of $A_n^{[3]}$ within the set $\mathbf{a}_c + \mathfrak{sl}_3^n(\mathfrak{o})$ modulo $\mathfrak{sl}_3^{n+1}(\mathfrak{o})$. We consider the set $\mathbf{a}_c + \mathfrak{sl}_3^n(\mathfrak{o})$ modulo $\mathfrak{sl}_3^{n+1}(\mathfrak{o})$, up to conjugation by $\mathrm{GL}_3^n(\mathfrak{o})$. Let

$$\mathbf{x} := \begin{pmatrix} x_1 & x_2 & x_3 \\ x_4 & x_5 & x_6 \\ x_7 & x_8 & x_9 \end{pmatrix} \in \mathrm{Mat}_3(\mathfrak{o}). \tag{5.12}$$

Then

$$[\mathbf{a}_c, \mathbf{x}] = \begin{pmatrix} x_4 - 2p^2c^2x_2 & (x_5 - x_1) - pcx_2 & x_6 + pcx_3 \\ pc(x_4 + 2pc(x_1 - x_5)) & -x_4 + 2p^2c^2x_2 & 2pc(x_6 + pcx_3) \\ -pc(x_7 + 2pcx_8) & -x_7 - 2pcx_8 & 0 \end{pmatrix} \tag{5.13}$$

$$\equiv_{\mathfrak{p}} \begin{pmatrix} x_4 & x_5 - x_1 & x_6 \\ 0 & -x_4 & 0 \\ 0 & -x_7 & 0 \end{pmatrix}. \tag{5.14}$$

If $\mathbf{b} = \mathbf{a}_c + p^n\mathbf{y} \in \mathbf{a}_c + \mathfrak{sl}_3^n(\mathfrak{o})$ and $g = 1 + p^n\mathbf{x} \in \mathrm{GL}_3^n(\mathfrak{o})$, then

$$\begin{aligned} g^{-1}\mathbf{b}g &= (1 - p^n\mathbf{x})(\mathbf{a}_c + p^n\mathbf{y})(1 + p^n\mathbf{x}) \\ &\equiv \mathbf{a}_c + p^n(\mathbf{y} + [\mathbf{a}_c, \mathbf{x}]) \end{aligned} \tag{5.15}$$

modulo $\mathfrak{sl}_3^{n+1}(\mathfrak{o})$. In view of (5.14) this shows that the elements

$$\mathbf{b}_c(y_1, y_2, y_3, y_4) := \mathbf{a}_c + p^n \begin{pmatrix} 0 & 0 & 0 \\ y_1 & y_4 & y_3 \\ y_2 & 0 & -y_4 \end{pmatrix}$$

with $y_1, y_2, y_3, y_4 \in T_{\mathfrak{o}}(1)$ form a complete set of representatives for the $\mathrm{GL}_3^n(\mathfrak{o})$-orbits of $\mathbf{a}_c + \mathfrak{sl}_3^n(\mathfrak{o})$ modulo $\mathfrak{sl}_3^{n+1}(\mathfrak{o})$, and indeed modulo $p^{n+1}\Lambda$.

Consider one of these lifts, $\mathbf{b} = \mathbf{b}_c(y_1, y_2, y_3, y_4)$. In order to simplify the notation, it is convenient to use the fact that $\mathbf{b} = \mathbf{b}_{\tilde{c}}(y_1, y_2, y_3, 0)$, where $\tilde{c} := c + p^{n-1}y_4 \in T_{\mathfrak{o}}(n)$, and to work with

$$\mathbf{a}_{\tilde{c}} := \begin{pmatrix} 0 & 1 & 0 \\ 2p^2\tilde{c}^2 & p\tilde{c} & 0 \\ 0 & 0 & -p\tilde{c} \end{pmatrix}$$

instead of $\mathbf{a}_c$. In order to describe the centraliser index of $\mathbf{b} + p^{n+1}\Lambda$ in $\mathfrak{sl}_3(\mathfrak{o})$, we consider again a generic matrix $\mathbf{x}$ as in (5.12), now with the additional restriction that $\mathbf{x} \in \mathfrak{sl}_3(\mathfrak{o})$, viz. $x_1 + x_5 + x_9 = 0$. One computes

$$[\mathbf{b}, \mathbf{x}] = [\mathbf{a}_{\tilde{c}}, \mathbf{x}] + p^n \begin{pmatrix} -y_1x_2 - y_2x_3 & 0 & -y_3x_2 \\ y_1(x_5 - x_1) - y_2x_6 + y_3x_7 & y_1x_2 + y_3x_8 & y_1x_3 + y_3(x_9 - x_5) \\ y_2(x_1 - x_9) - y_1x_8 & y_2x_2 & y_2x_3 - y_3x_8 \end{pmatrix}.$$

Taking into account (5.13), the condition $[\mathbf{b}, \mathbf{x}] \equiv 0$ modulo $p^{n+1}\Lambda$ can be expressed in terms of the following list of restrictions on the entries of $\mathbf{x}$, involving the parameters $y_1, y_2, y_3 \in T_{\mathfrak{o}}(1)$:

(i) $x_4 - 2p^2\tilde{c}^2x_2 - p^n(y_1x_2 + y_2x_3) \equiv_{p^{n+1}} -x_4 + 2p^2\tilde{c}^2x_2 + p^n(y_1x_2 + y_3x_8)$ from the $(1,1)$- and $(2,2)$-entries, equivalently $2x_4 \equiv_{p^{n+1}} 4p^2\tilde{c}^2x_2 + p^n(2y_1x_2 + y_2x_3 + y_3x_8)$,
(ii) $x_5 \equiv_{p^{n+1}} x_1 + p\tilde{c}x_2$ from the $(1,2)$-entry,
(iii) $x_6 \equiv_{p^{n+1}} -p\tilde{c}x_3 + p^n y_3x_2$ from the $(1,3)$-entry,
(iv) $x_7 \equiv_{p^{n+1}} -2p\tilde{c}x_8 + p^n y_2x_2$ from the $(3,2)$-entry,
(v) $0 \equiv_p -y_2x_6 + y_3x_7$ from the $(2,1)$-entry, but this condition becomes redundant if $x_6, x_7 \equiv_p 0$,
(vi) $0 \equiv_p y_2(x_1 - x_9) - y_1x_8 \equiv_p y_2(x_1 + x_5 + x_9) - y_1x_8 \equiv_p -y_1x_8$ from the $(3,1)$-entry and (ii),
(vii) $0 \equiv_p y_1x_3 + y_3(x_9 - x_5) \equiv_p y_1x_3 + y_3(x_1 + x_5 + x_9) \equiv_p y_1x_3$ from the $(2,3)$-entry and (ii),
(viii) $-x_4 + 2p^2\tilde{c}^2x_2 + p^n(y_1x_2 + y_3x_8) \equiv_{p^{n+1}} p^n(y_2x_3 - y_3x_8)$ from the $(2,2)$- and $(3,3)$-entries, equivalently $x_4 \equiv_{p^{n+1}} 2p^2\tilde{c}^2x_2 + p^n(y_1x_2 - y_2x_3 + 2y_3x_8)$.

If these conditions are to hold for $\mathbf{x} \in \mathfrak{sl}_3(\mathfrak{o})$, then the congruences (i)–(iv) show that the entries x_4, x_5, x_6, x_7 are determined completely modulo p^{n+1} by the remaining entries x_1, x_2, x_3, x_8, x_9. Because $x_1 + x_5 + x_9 = \mathrm{Tr}(\mathbf{x}) = 0$, we can also think of x_9 as being determined by x_1. We observe that $|\mathfrak{sl}_3(\mathfrak{o}) : \mathrm{C}_{\mathfrak{sl}_3(\mathfrak{o})}(\mathbf{b} + p^{n+1}\Lambda)| = q^{4(n+1)}$

(and not larger) if and only if one can choose x_1, x_2, x_3, x_8 freely, i.e. if the remaining conditions (v)–(viii) do not impose extra restrictions.

In fact, the congruence (v) will be automatically satisfied, as indicated, because $x_6, x_7 \equiv_p 0$. As $x_1 + x_5 + x_9 = \mathrm{Tr}(\mathbf{x}) = 0$, the conditions (vi) and (vii) will give rise to restrictions on x_8 or x_3, unless $y_1 \equiv_p 0$. Finally, the last condition (viii) is equivalent to the condition (i), since $p = 3$.

The discussion so far shows that, with respect to the action of $\mathrm{GL}_3^n(\mathfrak{o})$, the intersection of $A_n^{[3]}$ and $(\mathbf{a}_c + \mathfrak{sl}_3^n(\mathfrak{o}))/p^{n+1}\Lambda$ consists of q^3 orbits, represented by matrices

$$\mathbf{b}_c(0, y_2, y_3, y_4)$$

and each of size q^4. By induction, there are $q^{5(n-1)}$ matrices modulo $p^{n-1}\Lambda$, represented by matrices such as $\mathbf{a}_c$, which lift to elements of $A_n^{[3]}$. Hence $|A_n^{[3]}| = q^{5(n-1)}q^{3+4} = q^{5n+2}$. In order to show that the $|T_\mathfrak{o}(n-1)|q^3 = q^{n+2}$ matrices $\mathbf{b}_c(0, y_2, y_3, y_4)$ form a complete set of representatives for the $\mathrm{GL}_3^1(\mathfrak{o})$-orbits comprising $A_n^{[3]}$, it suffices to show that for any one of them, $\mathbf{b} = \mathbf{b}_c(0, y_2, y_3, y_4)$ say, one has

$$|\mathrm{GL}_3^1(\mathfrak{o}) : \mathrm{C}_{\mathrm{GL}_3^1(\mathfrak{o})}(\mathbf{b} + p^{n+1}\Lambda)| = q^{4n}. \tag{5.16}$$

We make three observations. Firstly, a straightforward translation between $\mathrm{GL}_3^1(\mathfrak{o})$ and its Lie lattice $\mathfrak{gl}_3^1(\mathfrak{o})$ yields

$$|\mathrm{GL}_3^1(\mathfrak{o}) : \mathrm{C}_{\mathrm{GL}_3^1(\mathfrak{o})}(\mathbf{b} + p^{n+1}\Lambda)| = |\mathfrak{gl}_3^1(\mathfrak{o}) : \mathrm{C}_{\mathfrak{gl}_3^1(\mathfrak{o})}(\mathbf{b} + p^{n+1}\Lambda)|.$$

Secondly, we note that $\mathbf{c} := \begin{pmatrix} 0 & & \\ & 0 & \\ & & p \end{pmatrix} \in \mathrm{C}_{\mathfrak{gl}_3^1(\mathfrak{o})}(\mathbf{b}+p^{n+1}\Lambda)$. Since $\mathfrak{gl}_3^1(\mathfrak{o}) = \mathfrak{o}\mathbf{c} \oplus \mathfrak{sl}_3^1(\mathfrak{o})$, this implies that $\mathfrak{gl}_3^1(\mathfrak{o}) = \mathrm{C}_{\mathfrak{gl}_3^1(\mathfrak{o})}(\mathbf{b} + p^{n+1}\Lambda) + \mathfrak{sl}_3^1(\mathfrak{o})$, and consequently

$$|\mathfrak{gl}_3^1(\mathfrak{o}) : \mathrm{C}_{\mathfrak{gl}_3^1(\mathfrak{o})}(\mathbf{b} + p^{n+1}\Lambda)| = |\mathfrak{sl}_3^1(\mathfrak{o}) : \mathrm{C}_{\mathfrak{sl}_3^1(\mathfrak{o})}(\mathbf{b} + p^{n+1}\Lambda)|.$$

Finally, we observe that the property $|\mathfrak{sl}_3(\mathfrak{o}) : \mathrm{C}_{\mathfrak{sl}_3(\mathfrak{o})}(\mathbf{b} + p^{n+1}\Lambda)| = q^{4(n+1)}$ is, in fact, equivalent to $|\mathfrak{sl}_3(\mathfrak{o}) : \mathrm{C}_{\mathfrak{sl}_3(\mathfrak{o})}(\mathbf{b} + p\Lambda)| = q^4$ so that

$$|\mathfrak{sl}_3^1(\mathfrak{o}) : \mathrm{C}_{\mathfrak{sl}_3^1(\mathfrak{o})}(\mathbf{b} + p^{n+1}\Lambda)| = q^{-4}q^{4(n+1)} = q^{4n}.$$

These three observations yield (5.16), as wanted.

To establish the last part of the induction claim, consider the relevance of the values of y_2, y_3 for lifting one of these matrices one step further. Consider $\mathbf{b} = \mathbf{b}_c(y_1, y_2, y_3, y_4) = \mathbf{b}_{\tilde{c}}(y_1, y_2, y_3, 0)$, where $\tilde{c} := c + p^{n-1}y_4 \in T_\mathfrak{o}(n+1)$, similarly as above, but allowing $y_1, \ldots, y_4 \in T_\mathfrak{o}(2)$. Then the centraliser condition $[\mathbf{b}, \mathbf{x}] \equiv 0$ modulo $p^{n+2}\Lambda$ leads, on the diagonal entries (cf. conditions (i) and (viii) in the list above), to the restriction

$$\begin{aligned} 0 &\equiv_{p^{n+2}} \left(4p^2\tilde{c}^2x_2 + p^n(2y_1x_2 + y_2x_3 + y_3x_8)\right) \\ &\quad - 2 \cdot \left(p^2\tilde{c}^2x_2 - p^n(y_1x_2 - y_2x_3 + 2y_3x_8)\right) \\ &= p^n(3y_2x_3 - 3y_3x_8) \end{aligned}$$

which is equivalent to $0 \equiv_p y_2x_3 - y_3x_8$, as $p = 3$. Unless both y_2 and y_3 are congruent to 0 modulo p, this leads to an unwanted restriction on x_3 and x_8. This shows that it is enough to look for representatives of the elements of $A_{n+1}^{[3]}$ within the sets $\mathbf{a}_{\tilde{c}} + \mathfrak{sl}_3^{n+1}(\mathfrak{o})$ modulo $\mathfrak{sl}_3^{n+2}(\mathfrak{o})$, where $\mathbf{a}_{\tilde{c}}$ arises from $\tilde{c} \equiv c$ modulo p^{n-1}. □

5.2.2. Next we consider elements belonging to cosets modulo $p\Lambda$ of type B. Similarly as for type C, we claim that the relevant integral for type B is equal to the $\mathcal{Z}_{\mathfrak{o}}^{[2]}(r,t)$. The latter may be computed easily from the integer sequence $a_n^{[2]}$, $n \in \mathbb{N}_0$, defined by

$$a_n^{[2]} := \#\{\mathbf{y} + (\mathfrak{p}^{n+1})^{(8)} \mid \mathbf{y} \in \mathfrak{p}^{(8)} \text{ such that } |p|_{\mathfrak{p}} = q^{-n-1}\} = \begin{cases} 1 & \text{if } n = 0, \\ 0 & \text{if } n \geq 1, \end{cases}$$

observing that the numbers $m_l^{[2]}$ defined in (5.10) satisfy

$$m_l^{[2]} = q^{-8l} a_{l-1}^{[2]} - q^{-8(l+1)} a_l^{[2]}.$$

The claim follows from the following proposition.

PROPOSITION 5.2. *For $n \in \mathbb{N}_0$ the set*

$$A_n^{[2]} := \Big\{\mathbf{a} + p^{n+1}\Lambda \in \Lambda/p^{n+1}\Lambda \mid \mathbf{a} \equiv \left(\begin{smallmatrix} 0&1&0\\0&0&1\\0&0&0 \end{smallmatrix}\right) \text{ modulo } p\Lambda \\ \text{and } |\mathfrak{sl}_3(\mathfrak{o}) : \mathrm{C}_{\mathfrak{sl}_3(\mathfrak{o})}(\mathbf{a} + p^{n+1}\Lambda)| = q^{4(n+1)}\Big\}$$

has cardinality $a_n^{[2]}$.

PROOF. The case $n = 0$ is a simple computation. The only candidate for an element of $A_0^{[2]}$ is $\mathbf{a}_0 + p\Lambda$, where $\mathbf{a}_0 := \left(\begin{smallmatrix} 0&1&0\\0&0&1\\0&0&0 \end{smallmatrix}\right)$, and a short computation reveals that

$$|\mathfrak{sl}_3(\mathfrak{o}) : \mathrm{C}_{\mathfrak{sl}_3(\mathfrak{o})}(\mathbf{a}_0 + p\Lambda)| = q^4.$$

Indeed, for

$$\mathbf{x} = \begin{pmatrix} x_1 & x_2 & x_3 \\ x_4 & x_5 & x_6 \\ x_7 & x_8 & x_9 \end{pmatrix} \in \mathfrak{sl}_3(\mathfrak{o}) \tag{5.17}$$

the commutator identity

$$[\mathbf{a}_0, \mathbf{x}] = \begin{pmatrix} x_4 & x_5 - x_1 & x_6 - x_2 \\ x_7 & x_8 - x_4 & x_9 - x_5 \\ 0 & -x_7 & -x_8 \end{pmatrix} \tag{5.18}$$

shows that $\mathbf{x} \in \mathrm{C}_{\mathfrak{sl}_3(\mathfrak{o})}(\mathbf{a}_0 + p\Lambda)$ if and only if the following congruences are satisfied:

$$x_4 \equiv_p -x_8, \quad x_5 \equiv_p x_9 \equiv_p x_1, \quad x_6 \equiv_p x_2, \quad x_7 \equiv_p 0,$$

where x_1, x_2, x_3, x_8 can be chosen freely. Thus $|A_0^{[2]}| = 1$, as claimed.

Next suppose that $n \geq 1$, and consider a lift $\mathbf{b} \in \mathbf{a}_0 + \mathfrak{sl}_3^1(\mathfrak{o})$ modulo $\mathfrak{sl}_3^2(\mathfrak{o})$. Replacing $\mathbf{b}$ by a conjugate under $\mathrm{GL}_3^1(\mathfrak{o})$ if necessary, we may assume, by a computation similar to (5.15), that $\mathbf{b}$ is of the form

$$\mathbf{b} = \mathfrak{b}(y_1, y_2) = \mathbf{a}_0 + \begin{pmatrix} 0 & 0 & 0 \\ 0 & 0 & 0 \\ py_1 & py_2 & 0 \end{pmatrix} = \begin{pmatrix} 0 & 1 & 0 \\ 0 & 0 & 1 \\ py_1 & py_2 & 0 \end{pmatrix},$$

type		# of orbits	size of each orbit	total number modulo $\mathbb{F}_q\overline{\mathbf{z}}$
G	irreg.	q	$(q^2+q+1)q^2$	$(q^2+q+1)q^2$
H	reg.	$(q-3)q/6$	$(q^2+q+1)(q+1)q^3$	$(q-3)(q^2+q+1)(q+1)q^3/6$
I	reg.	$(q-1)q/2$	$(q^3-1)q^3$	$(q^3-1)(q-1)q^3/2$
J	reg.	$q^2/3$	$(q^2-1)(q-1)q^3$	$(q^2-1)(q-1)q^4/3$
K	reg.	q	$(q^3-1)(q+1)q^2$	$(q^3-1)(q+1)q^2$

TABLE 5.2. Adjoint orbits in $\{\mathbf{x} \in \mathfrak{gl}_3(\mathbb{F}_q) \mid \operatorname{Tr}(\mathbf{x}) = -1\}$ under the action of $\mathrm{GL}_3(\mathbb{F}_q)$, $q \equiv_3 0$

where $y_1, y_2 \in T_{\mathfrak{o}}(1)$ are Teichmüller representatives in $\mathfrak{o}$. Suppose that $\mathbf{x}$, as in (5.17), lies in $\mathrm{C}_{\mathfrak{sl}_3(\mathfrak{o})}(\mathbf{b}+p^2\Lambda)$. Then the commutator identity

$$[\mathbf{b},\mathbf{x}] = [\mathbf{a}_0,\mathbf{x}] + \begin{pmatrix} -py_1x_3 & -py_2x_3 & 0 \\ -py_1x_6 & -py_2x_6 & 0 \\ p(y_1(x_1-x_9)+y_2x_4) & p(y_1x_2+y_2(x_5-x_9)) & p(y_1x_3+y_2x_6) \end{pmatrix}$$

in conjunction with (5.18) reveals that

$$x_4 - py_1x_3 \equiv_{p^2} (x_8-x_4) - py_2x_6 \equiv_{p^2} -x_8 + p(y_1x_3+y_2x_6),$$

hence $3x_4 \equiv_{p^2} 3x_8 \equiv_{p^2} 0$ irrespective of the particular values of y_1, y_2. Furthermore, inspection of the $(1,2)$- and $(2,3)$-entries of the commutator identity shows that $x_1 \equiv_p x_5 \equiv_{p^2} x_9$. Since $x_1+x_5+x_9 \equiv_p 0$, this yields $x_1 \equiv_p x_5 \equiv_p x_9 \equiv_p 0$. As before, x_4, x_5, x_6, x_7, x_9 are determined by the values of x_1, x_2, x_3, x_8, but the latter satisfy the extra condition $x_1 \equiv_p x_8 \equiv_p 0$. Hence the relevant index $|\mathfrak{sl}_3(\mathfrak{o}) : \mathrm{C}_{\mathfrak{sl}_3(\mathfrak{o})}(\mathbf{b}+p^2\Lambda)| \geq q^{10}$. This shows that $A_1^{[2]} = \varnothing$ and consequently $A_n^{[2]} = \varnothing$ for all $n \geq 1$. □

5.3. In order to conclude the justification of the equations (5.4) we explain how one obtains the summand $\mathcal{S}_2(r,t)$. For this we decompose $\Lambda \smallsetminus \mathfrak{sl}_3(\mathfrak{o})$ into cosets modulo $p\Lambda$. As $p=3$, every element in this domain is of the form $u(p^{-1}\operatorname{Id}+\mathbf{x})$, where $u \in \mathfrak{o}^*$ and $\mathbf{x} \in \mathfrak{gl}_3(\mathfrak{o})$ has trace -1. We are thus led to decomposing the finite affine space of matrices $\{\mathbf{x} \in \mathfrak{gl}_3(\mathbb{F}_q) \mid \operatorname{Tr}(\mathbf{x}) = -1\}$ into cosets modulo the 1-dimensional subspace $\mathbb{F}_q\overline{\mathbf{z}}$, spanned by the reduction modulo $\mathfrak{p}$ of $\mathbf{z} := \mathbf{h}_{12} - \mathbf{h}_{23}$. Of course, the latter coincides with the space of scalar matrices over $\mathbb{F}_q$. An overview of the orbits in $\{\mathbf{x} \in \mathfrak{gl}_3(\mathbb{F}_q) \mid \operatorname{Tr}(\mathbf{x}) = -1\}$ under the adjoint action of $\mathrm{GL}_3(\mathbb{F}_q)$ is provided in Table 5.2; see Appendix C for a short discussion. The corresponding total number of cosets modulo $\mathbb{F}_q\overline{\mathbf{z}}$, given in the last column of the table, is obtained upon division by q. Indeed, a short calculation confirms that the sizes of the orbits listed in Table 5.2 add up to q^7 so that with $q-1$ choices for u modulo $\mathfrak{p}$ we obtain $(q-1)q^7 = |\Lambda : p\Lambda| - |\mathfrak{sl}_3(\mathfrak{o}) : p\Lambda|$, as wanted.

That $\mathcal{Z}_{\mathfrak{o}}^{[0]}(r,t)$ is the correct integral for the types H, I, J, K is clear, because they correspond to regular elements modulo p. We need to link the contributions by irregular elements belonging to cosets of type G to the integral $\mathcal{Z}_{\mathfrak{o}}^{[1]}(r,t)$, as shown within the summand $\mathcal{S}_2(r,t)$ in (5.4).

Hence let us consider elements belonging to cosets modulo $p\Lambda$ of type G. A typical coset of this type is $\mathbf{a}_0 + p\Lambda$, where $\mathbf{a}_0 := \left(\begin{smallmatrix} 0&0&0\\0&0&0\\0&0&-1\end{smallmatrix}\right)$, and each coset has measure $\mu(p\Lambda) = q^{-7}$. The determinant of the Jacobi matrix associated to $\iota_0 : \Lambda \to \mathrm{Hom}_{\mathfrak{o}}(\mathfrak{sl}_3(\mathfrak{o}), \mathfrak{o})$ contributes another factor $|3|_{\mathfrak{p}} = q^{-1}$. The integral over $\mathfrak{p} \times (\mathbf{a} + p\Lambda)$ with Jacobi factor q^{-1} can thus be described as an integral over $\mathfrak{p} \times \mathfrak{p}^{(8)}$. Similarly as in cases B and C, we claim that this integral equals $\mathcal{Z}_{\mathfrak{o}}^{[1]}(r,t)$. The latter may be computed from the integer sequence $a_n^{[1]}$, $n \in \mathbb{N}_0$, defined by

$$\begin{aligned} a_n^{[1]} &:= \#\{\mathbf{y} + (\mathfrak{p}^{n+1})^{(8)} \mid \mathbf{y} \in \mathfrak{p}^{(8)} \text{ such that } \|\{y_1, y_2, y_3, p^{n+1}\}\|_{\mathfrak{p}} = q^{-n-1}\} \\ &= \#\{\mathbf{y} + (\mathfrak{p}^{n+1})^{(8)} \mid \mathbf{y} \in \mathfrak{p}^{(8)} \text{ such that } y_1, y_2, y_3 \in \mathfrak{p}^{n+1}\} \\ &= |\mathfrak{p}^{(5)} / (\mathfrak{p}^{n+1})^{(5)}| \\ &= q^{5n}, \end{aligned}$$

describing the lifting behaviour of points modulo $\mathfrak{p}^{n+1}$ on the variety defined by the integrand of $\mathcal{Z}_{\mathfrak{o}}^{[1]}(r,t)$. Indeed, we observe that the numbers $m_l^{[1]}$ defined in (5.7) satisfy

$$m_l^{[1]} = q^{-8l} a_{l-1}^{[1]} - q^{-8(l+1)} a_l^{[1]}.$$

The following proposition establishes the claim and thereby concludes the overall proof of Theorem 1.4.

PROPOSITION 5.3. *For $n \in \mathbb{N}_0$ the set*

$$\begin{aligned} A_n^{[1]} := \Big\{ \mathbf{a} + p^{n+1}\Lambda \in \Lambda / p^{n+1}\Lambda \mid \mathbf{a} \equiv p^{-1}\,\mathrm{Id} + \left(\begin{smallmatrix} 0 & 0 & 0 \\ 0 & 0 & 0 \\ 0 & 0 & -1 \end{smallmatrix}\right) \text{ modulo } p\Lambda \\ \text{and } |\mathfrak{sl}_3(\mathfrak{o}) : \mathrm{C}_{\mathfrak{sl}_3(\mathfrak{o})}(\mathbf{a} + p^{n+1}\Lambda)| = q^{4(n+1)} \Big\} \end{aligned}$$

has cardinality $a_n^{[1]}$.

PROOF. We argue by induction on n. The case $n = 0$ is a simple computation. Indeed, the only candidate for an element of $A_0^{[1]}$ is $\mathbf{a}_0 + p\Lambda$, where $\mathbf{a}_0 := p^{-1}\,\mathrm{Id} + \left(\begin{smallmatrix} 0 & 0 & 0 \\ 0 & 0 & 0 \\ 0 & 0 & -1 \end{smallmatrix}\right)$, and a short computation reveals that, indeed,

$$|\mathfrak{sl}_3(\mathfrak{o}) : \mathrm{C}_{\mathfrak{sl}_3(\mathfrak{o})}(\mathbf{a}_0 + p\Lambda)| = |\mathfrak{sl}_3(\mathbb{F}_q) : \mathrm{C}_{\mathfrak{sl}_3(\mathbb{F}_q)}(\overline{\mathbf{a}_0})| = q^4,$$

where $\overline{\mathbf{a}_0}$ denotes the image of $\mathbf{a}_0$ in $\mathfrak{sl}_3(\mathbb{F}_q)$; cf. (B.3). Thus $|A_0^{[1]}| = 1$, as claimed.

Now suppose that $n \geq 1$. In fact, we will prove more than stated in the proposition. For any $l \in \mathbb{N}$, let $T_{\mathfrak{o}}(l)$ denote the representatives for $\mathfrak{o}/\mathfrak{p}^l$ derived from the Teichmüller representatives for $\mathfrak{o}/\mathfrak{p}$; see (5.1). By induction on n, the following assertions are proved below.

Claim. Every matrix $\mathbf{a} \in \Lambda$ with $\mathbf{a} + p^{n+1}\Lambda \in A_n^{[1]}$ can be conjugated by elements of $\mathrm{GL}_3^1(\mathfrak{o})$ to the 'normal' form

$$p^{-1}\,\mathrm{Id} + \begin{pmatrix} c & 0 & 0 \\ 0 & c & 0 \\ 0 & 0 & -1-2c \end{pmatrix} \qquad \text{modulo } p^{n+1}\Lambda,$$

where $c \in T_{\mathfrak{o}}(n)$. These matrices modulo $p^{n+1}\Lambda$ form a complete set of representatives for the $\mathrm{GL}_3^1(\mathfrak{o})$-orbits comprising $A_n^{[1]}$. Moreover, the index in $\mathrm{GL}_3^1(\mathfrak{o})$ of the centraliser of any such matrix modulo $p^{n+1}\Lambda$ is q^{4n}, and $|A_n^{[1]}| = |T_{\mathfrak{o}}(n)| q^{4n} = q^{5n}$, as wanted.

Let $c \in T_{\mathfrak{o}}(n)$ and put

$$\mathbf{a} := \mathbf{a}_c := \begin{pmatrix} c & 0 & 0 \\ 0 & c & 0 \\ 0 & 0 & -1-2c \end{pmatrix}.$$

Clearly, the eigenvalues of $\mathbf{a}$ are c and $-1-2c$ with multiplicities 2 and 1 respectively. Hence c modulo p^{n-1} is an invariant of the $\mathrm{GL}_3^1(\mathfrak{o})$-orbit of $\mathbf{a}$ modulo $p^n\Lambda$. By induction, it suffices to look for representatives of the elements of $A_n^{[1]}$ within the set $\mathbf{a} + \mathfrak{sl}_3^n(\mathfrak{o})$ modulo $p^{n+1}\Lambda$. We consider the set $\mathbf{a} + \mathfrak{sl}_3^n(\mathfrak{o})$ modulo $\mathfrak{sl}_3^{n+1}(\mathfrak{o})$, up to conjugation by $\mathrm{GL}_3^n(\mathfrak{o})$. Let

$$\mathbf{x} := \begin{pmatrix} A & \underline{b}^{\mathrm{t}} \\ \underline{d} & a \end{pmatrix} \in \mathrm{Mat}_3(\mathfrak{o}), \qquad \text{where } A \in \mathrm{Mat}_2(\mathfrak{o}) \text{ and } \underline{b}, \underline{d} \in \mathfrak{o}^{(2)},\ a \in \mathfrak{o}.$$

Then, as $p = 3$,

$$[\mathbf{a}, \mathbf{x}] = \begin{pmatrix} 0 & (1+3c)\underline{b}^{\mathrm{t}} \\ -(1+3c)\underline{d} & 0 \end{pmatrix} \tag{5.19}$$

$$\equiv_{\mathfrak{p}} \begin{pmatrix} 0 & \underline{b}^{\mathrm{t}} \\ \underline{d} & 0 \end{pmatrix}. \tag{5.20}$$

As in the proof of Proposition 5.1, the congruence (5.20) shows that the elements

$$\mathbf{b}_c(Y) := \mathbf{a} + p^n \begin{pmatrix} Y & 0 \\ 0 & -\mathrm{Tr}(Y) \end{pmatrix}, \qquad \text{with } Y = \begin{pmatrix} y_1 & y_2 \\ y_3 & y_4 \end{pmatrix} \in \mathrm{Mat}_2(T_{\mathfrak{o}}(1)),$$

form a complete set of representatives for the $\mathrm{GL}_3^n(\mathfrak{o})$-orbits of $\mathbf{a} + \mathfrak{sl}_3^n(\mathfrak{o})$ modulo $\mathfrak{sl}_3^{n+1}(\mathfrak{o})$, and with the extra restriction $\mathrm{Tr}(Y) = 0$ a complete set of representatives modulo $p^{n+1}\Lambda$.

Consider one of these lifts, $\mathbf{b} = \mathbf{b}_c(Y)$ and put $z := \mathrm{Tr}(Y)$. In order to describe the centraliser index of $\mathbf{b} + p^{n+1}\Lambda$ in $\mathfrak{sl}_3(\mathfrak{o})$, we consider a generic matrix

$$\mathbf{x} := \begin{pmatrix} A & \underline{b}^{\mathrm{t}} \\ \underline{d} & -\mathrm{Tr}(A) \end{pmatrix} \in \mathfrak{sl}_3(\mathfrak{o}), \qquad \text{where } A \in \mathrm{Mat}_2(\mathfrak{o}) \text{ and } \underline{b}, \underline{d} \in \mathfrak{o}^{(2)}$$

and compute

$$[\mathbf{b}, \mathbf{x}] = [\mathbf{a}, \mathbf{x}] + p^n \begin{pmatrix} YA - AY & Y\underline{b}^{\mathrm{t}} - z\underline{b}^{\mathrm{t}} \\ z\underline{d} - \underline{d}Y & 0 \end{pmatrix}.$$

Taking into account (5.19), the condition $[\mathbf{b}, \mathbf{x}] \equiv 0$ modulo $p^{n+1}\Lambda$ can be expressed in terms of the following list of restrictions on the entries of $\mathbf{x}$, involving as parameters the matrix $Y \in \mathrm{Mat}_2(T_{\mathfrak{o}}(1))$ and $z = \mathrm{Tr}(Y)$:

(i) $YA - AY \equiv_p 0$,
(ii) $((1 + 3c - p^n z)\,\mathrm{Id} + p^n Y)\underline{b}^{\mathrm{t}} \equiv_{p^{n+1}} 0$,
(iii) $\underline{d}((1 + 3c - p^n z)\,\mathrm{Id} + p^n Y) \equiv_{p^{n+1}} 0$.

If these conditions are to hold then the last two congruences show that $\underline{b}$ and $\underline{d}$ are to be 0 modulo p^{n+1}. From this we observe that $|\mathfrak{sl}_3(\mathfrak{o}) : \mathrm{C}_{\mathfrak{sl}_3(\mathfrak{o})}(\mathbf{a}+p^{n+1}\Lambda)| = q^{4(n+1)}$ (and not larger) if and only if one can choose A freely, i.e. if the first condition does not impose extra restrictions. This implies that Y is scalar, and $\mathrm{Tr}(Y) = 0$ implies that $Y = 0$. The proof concludes in analogy to the proof of Proposition 5.1 $\square$

Appendix A. Adjoint action of $\mathrm{GL}_2(\mathbb{F}_q)$ on $\mathfrak{sl}_2(\mathbb{F}_q)$

In Section 3.3, we require an overview of the elements in $\mathfrak{sl}_2(\mathbb{F}_q)$ up to conjugacy under the group $\mathrm{GL}_2(\mathbb{F}_q)$. We distinguish four different types, labelled 0, 1, 2a, 2b. The total number of elements of each type and the isomorphism types of their centralisers in $\mathrm{SL}_2(\mathbb{F}_q)$ are summarised in Tables 3.1 and 3.2. We briefly discuss the four different types.

Type 0 consists of the zero matrix, which does not feature in our calculation but is shown for completeness. Its centraliser is the entire group $\mathrm{SL}_2(\mathbb{F}_q)$.

Type 1 consists of nilpotent matrices with minimal polynomial equal to X^2 over $\mathbb{F}_q$. The centraliser of a typical element is

$$\mathrm{C}_{\mathrm{SL}_2(\mathbb{F}_q)}\left(\left(\begin{smallmatrix} 0 & 1 \\ 0 & 0 \end{smallmatrix}\right)\right) = \left\{\left(\begin{smallmatrix} a & b \\ 0 & a \end{smallmatrix}\right) \mid a, b \in \mathbb{F}_q, a^2 = 1\right\}$$

and matrices of type 1 are regular.

Type 2a consists of semisimple matrices with distinct eigenvalues $\lambda, -\lambda \in \mathbb{F}_q^*$. The minimal polynomial of such elements over $\mathbb{F}_q$ is equal to $X^2 - \lambda^2$. The centraliser of a typical element is

$$\mathrm{C}_{\mathrm{SL}_2(\mathbb{F}_q)}\left(\left(\begin{smallmatrix} \lambda & 0 \\ 0 & -\lambda \end{smallmatrix}\right)\right) = \left\{\left(\begin{smallmatrix} a & 0 \\ 0 & b \end{smallmatrix}\right) \mid a, b \in \mathbb{F}_q, ab = 1\right\}$$

and matrices of type 2a are regular.

Type 2b consists of semisimple matrices with eigenvalues $\lambda, \lambda^q \in \mathbb{F}_{q^2} \setminus \mathbb{F}_q$. The minimal polynomial of such elements over $\mathbb{F}_q$ is equal to $(X - \lambda)(X - \lambda^q)$. The centraliser of a typical element is isomorphic to the group of elements of norm 1 in the field $\mathbb{F}_{q^2}$ and matrices of type 2b are regular.

Appendix B. Adjoint action of $\mathrm{GL}_3(\mathbb{F}_q)$ on $\mathfrak{sl}_3(\mathbb{F}_q)$

This appendix is almost identical to Appendix B in [**3**] and included for the reader's convenience. Let $\mathbb{F}_q$ be a finite field of characteristic not equal to 3. We give an overview of the elements in $\mathfrak{sl}_3(\mathbb{F}_q)$ up to conjugacy under the group $\mathrm{GL}_3(\mathbb{F}_q)$. For this we distinguish eight different types, labelled 0, 1, 2, 3, 4a, 4b, 4c, 5. The total number of elements of each type and the isomorphism types of their centralisers in $\mathrm{SL}_3(\mathbb{F}_q)$ are summarised in Tables 7.1 and 7.2 in [**3**]. We briefly discuss the eight different types.

Type 0 consists of the zero matrix, which does not feature in our calculation but is shown for completeness. Its centraliser is the entire group $\mathrm{SL}_3(\mathbb{F}_q)$.

Type 1 consists of nilpotent matrices with minimal polynomial equal to X^3 over $\mathbb{F}_q$. The centraliser of a typical element is

$$\mathrm{C}_{\mathrm{SL}_3(\mathbb{F}_q)}\left(\left(\begin{smallmatrix} 0 & 1 & 0 \\ 0 & 0 & 1 \\ 0 & 0 & 0 \end{smallmatrix}\right)\right) = \left\{\left(\begin{smallmatrix} a & b & c \\ 0 & a & b \\ 0 & 0 & a \end{smallmatrix}\right) \in \mathrm{GL}_3(\mathbb{F}_q) \mid a^3 = 1\right\} \tag{B.1}$$

and matrices of type 1 are regular.

Type 2 consists of nilpotent matrices with minimal polynomial equal to X^2 over $\mathbb{F}_q$. The centraliser of a typical element is

$$\mathrm{C}_{\mathrm{SL}_3(\mathbb{F}_q)}\left(\left(\begin{smallmatrix} 0 & 1 & 0 \\ 0 & 0 & 0 \\ 0 & 0 & 0 \end{smallmatrix}\right)\right) = \left\{\left(\begin{smallmatrix} a & b & c \\ 0 & a & 0 \\ 0 & d & e \end{smallmatrix}\right) \in \mathrm{GL}_3(\mathbb{F}_q) \mid a^2 e = 1\right\} \tag{B.2}$$

and matrices of type 2 are irregular.

Type 3 consists of semisimple matrices with eigenvalues $\lambda \in \mathbb{F}_q^*$ of multiplicity 2 and $\mu := -2\lambda$. The minimal polynomial of such elements over $\mathbb{F}_q$ is equal to $(X-\lambda)(X-\mu)$. The centraliser of a typical element is

$$\mathrm{C}_{\mathrm{SL}_3(\mathbb{F}_q)}\left(\left(\begin{smallmatrix}\lambda&0&0\\0&\lambda&0\\0&0&\mu\end{smallmatrix}\right)\right)=\left\{\left(\begin{smallmatrix}a&b&0\\c&d&0\\0&0&e\end{smallmatrix}\right)\in\mathrm{GL}_3(\mathbb{F}_q)\mid(ad-bc)e=1\right\} \tag{B.3}$$

and matrices of type 3 are irregular.

Type 4a consists of semisimple matrices with distinct eigenvalues $\lambda, \mu, \nu := -\lambda-\mu \in \mathbb{F}_q^*$. The minimal polynomial of such elements over $\mathbb{F}_q$ is equal to $(X-\lambda)(X-\mu)(X-\nu)$. The centraliser of a typical element is

$$\mathrm{C}_{\mathrm{SL}_3(\mathbb{F}_q)}\left(\left(\begin{smallmatrix}\lambda&0&0\\0&\mu&0\\0&0&\nu\end{smallmatrix}\right)\right)=\left\{\left(\begin{smallmatrix}a&0&0\\0&b&0\\0&0&c\end{smallmatrix}\right)\in\mathrm{GL}_3(\mathbb{F}_q)\mid abc=1\right\} \tag{B.4}$$

and matrices of type 4a are regular.

Type 4b consists of semisimple matrices with eigenvalues $\lambda, \mu := \lambda^q \in \mathbb{F}_{q^2} \smallsetminus \mathbb{F}_q$ and $\nu := -\lambda-\mu \in \mathbb{F}_q$. The minimal polynomial of such elements over $\mathbb{F}_q$ is equal to $(X-\lambda)(X-\mu)(X-\nu)$. The centraliser of a typical element is isomorphic to the multiplicative group of the field $\mathbb{F}_{q^2}$ and matrices of type 4b are regular.

Type 4c consists of semisimple matrices with eigenvalues $\lambda, \mu := \lambda^q, \nu := \lambda^{q^2} \in \mathbb{F}_{q^3} \smallsetminus \mathbb{F}_q$ with $\lambda+\mu+\nu=0$. The minimal polynomial of such elements over $\mathbb{F}_q$ is equal to $(X-\lambda)(X-\mu)(X-\nu)$. The centraliser of a typical element is isomorphic to the group of elements of norm 1 in the field $\mathbb{F}_{q^3}$ and matrices of type 4c are regular.

Type 5 consists of matrices with eigenvalues $\lambda \in \mathbb{F}_q^*$ of multiplicity 2 and $\mu := -2\lambda$. The minimal polynomial of such elements over $\mathbb{F}_q$ is equal to $(X-\lambda)^2(X-\mu)$. The centraliser of a typical element is

$$\mathrm{C}_{\mathrm{SL}_3(\mathbb{F}_q)}\left(\left(\begin{smallmatrix}\lambda&1&0\\0&\lambda&0\\0&0&\mu\end{smallmatrix}\right)\right)=\left\{\left(\begin{smallmatrix}a&b&0\\0&a&0\\0&0&c\end{smallmatrix}\right)\in\mathrm{GL}_3(\mathbb{F}_q)\mid a^2c=1\right\} \tag{B.5}$$

and matrices of type 5 are regular.

Appendix C. Auxiliary results regarding $\mathfrak{sl}_3(\mathbb{F}_q)$ and $\mathfrak{gl}_3(\mathbb{F}_q)$ for $q \equiv_3 0$

C.1. Let $\mathbb{F}_q$ be a finite field of characteristic equal to 3. In Section 5, we require an overview of the elements in $\mathfrak{sl}_3(\mathbb{F}_q)$ up to conjugacy under the group $\mathrm{GL}_3(\mathbb{F}_q)$. We distinguish six different types, labelled A, B, C, D, E and F, corresponding to the types 0, 1, 2, 4a, 4b and 4c in the generic case ($p \neq 3$); cf. Appendix B. There are no analogues to types 3 and 5 in characteristic 3. The total number of elements of each type, number of orbits and orbit sizes, are summarised in Table 5.1. We briefly discuss the six different types.

Type A (corresponding to type 0 in the generic case) consists of the scalar matrices. This type is listed for completeness and does not feature in our calculation.

Type B (corresponding to type 1) consists of matrices with minimal polynomial equal to $(X-\lambda)^3$ over $\mathbb{F}_q$. There are q possible values for $\lambda \in \mathbb{F}_q$, hence q orbits. The orbit sizes are as in the generic case; see (B.1). In contrast to the generic case, matrices of type B are irregular in characteristic 3.

Type C (corresponding to type 2) consists of matrices with minimal polynomial equal to $(X-\lambda)^2$ over $\mathbb{F}_q$. There are q possible values for $\lambda \in \mathbb{F}_q$, hence q orbits. The orbit sizes are as in the generic case; see (B.2). In contrast to the generic case, matrices of type B are irregular in characteristic 3.

Type D (corresponding to type 4a) consists of matrices with minimal polynomial equal to $(X-\lambda)(X-\mu)(X-\nu)$ over $\mathbb{F}_q$, with distinct λ, μ, ν such that $\lambda+\mu+\nu=0$. There are $(q-1)q/6$ possible choices for $\{\lambda, \mu, \nu\}$, hence the same number of orbits. The orbit sizes are as in the generic case; see (B.4). As in the generic case, matrices of type D are regular.

Type E (corresponding to type 4b) consists of semisimple matrices with eigenvalues $\lambda, \mu := \lambda^q \in \mathbb{F}_{q^2} \smallsetminus \mathbb{F}_q$ and $\nu := -\lambda-\mu \in \mathbb{F}_q$. The number of orbits and the orbit sizes are exactly as in the generic case; see Appendix B. Matrices of type E are regular.

Type F (corresponding to type 4c) consists of semisimple matrices with eigenvalues $\lambda, \mu := \lambda^q, \nu := \lambda^{q^2} \in \mathbb{F}_{q^3} \smallsetminus \mathbb{F}_q$ with $\lambda+\mu+\nu=0$. In characteristic 3, the number of elements in $\mathbb{F}_{q^3} \smallsetminus \mathbb{F}_q$ with trace 0 in $\mathbb{F}_q$ is q^2-q. Thus there are $(q-1)q/3$ orbits. The orbit sizes are as in the generic case; see Appendix B. Matrices of type F are regular.

C.2. Let $\mathbb{F}_q$ be a finite field of characteristic equal to 3. In Section 5, we also require an overview of the elements in $\{\mathbf{x}+\mathbb{F}_q\overline{\mathbf{z}} \mid \mathbf{x} \in \mathfrak{gl}_3(\mathbb{F}_q) \text{ with } \mathrm{Tr}(\mathbf{x})=-1\}$ up to conjugacy under the group $\mathrm{GL}_3(\mathbb{F}_q)$. We distinguish five different types, labelled G, H, I, J and K, corresponding to the types 3, 4a, 4b, 4c and 5 in the generic case ($p \neq 3$); cf. Appendix B. There are no analogues to types 1 and 2 in characteristic 3. The total number of elements of each type, number of orbits and orbit sizes, are summarised in Table 5.2. We briefly discuss the five different types.

Type G (analogous to type 3 in the generic case) consists of semisimple matrices with eigenvalues $\lambda \in \mathbb{F}_q$ of multiplicity 2 and $\mu := \lambda - 1$. The minimal polynomial of such an element over $\mathbb{F}_q$ is equal to $(X-\lambda)(X-\mu)$. There are q possible values for λ, hence q orbits. The orbit sizes are as in the generic case; see (B.3). As in the generic case, matrices of type G are irregular.

Type H (analogous to type 4a) consists of matrices with minimal polynomial equal to $(X-\lambda)(X-\mu)(X-\nu)$ over $\mathbb{F}_q$, with distinct λ, μ, ν such that $\lambda+\mu+\nu=-1$. One checks that there are $(q-3)q/6$ possible choices for $\{\lambda, \mu, \nu\}$, hence the same number of orbits. The orbit sizes are as in the generic case; see (B.4). As in the generic case, matrices of type H are regular.

Type I (analogous to type 4b) consists of semisimple matrices with eigenvalues $\lambda, \mu := \lambda^q \in \mathbb{F}_{q^2} \smallsetminus \mathbb{F}_q$ and $\nu := -\lambda-\mu-1 \in \mathbb{F}_q$. The number of orbits and the orbit sizes are exactly as in the generic case; see Appendix B. Matrices of type I are regular.

Type J (analogous to type 4c) consists of semisimple matrices with eigenvalues $\lambda, \mu := \lambda^q, \nu := \lambda^{q^2} \in \mathbb{F}_{q^3} \smallsetminus \mathbb{F}_q$ with $\lambda+\mu+\nu=-1$. In characteristic 3, the number of elements in $\mathbb{F}_{q^3} \smallsetminus \mathbb{F}_q$ with trace -1 in $\mathbb{F}_q$ is q^2. Thus there are $q^2/3$ orbits. The orbit sizes are as in the generic case; see Appendix B. Matrices of type J are regular.

Type K (analogous to type 5) consists of matrices with eigenvalues $\lambda \in \mathbb{F}_q$ of multiplicity 2 and $\mu := \lambda - 1$. The minimal polynomial of such an element over $\mathbb{F}_q$ is equal to $(X-\lambda)^2(X-\mu)$. There are q possible values for λ, hence q orbits. The orbit sizes are as in the generic case; see (B.5). As in the generic case, matrices of type K are irregular.

ACKNOWLEDGEMENTS. The authors would like to thank Alexander Lubotzky as well as the following institutions: the Batsheva de Rothschild Fund for the Advancement of Science, the EPSRC, the Mathematisches Forschungsinstitut Oberwolfach, the National Science Foundation and the Nuffield Foundation.

References

[1] C. A. M. André, A. P. Nicolás, *Supercharacters of the adjoint group of a finite radical ring*, J. Group Theory **11** (2008), 709–746. MR2446152 (2010m:16025)

[2] N. Avni, B. Klopsch, U. Onn, C. Voll, *On representation zeta functions of groups and a conjecture of Larsen-Lubotzky*, C. R. Math. Acad. Sci. Paris **348** (2010), 363–367. MR2607020 (2011c:11143)

[3] N. Avni, B. Klopsch, U. Onn, C. Voll, *Representation zeta functions of compact p-adic analytic groups and arithmetic groups*, arXiv:1007.2900v1 (2010). MR2607020 (2011c:11143)

[4] N. Avni, B. Klopsch, U. Onn, C. Voll, *Representation zeta functions for* SL_3, preprint.

[5] L. Bartholdi, P. de la Harpe, *Representation zeta functions of wreath products with finite groups*, Groups Geom. Dyn. **4** (2010), 209–249. MR2595090 (2011h:20010)

[6] W. Borho, H. Kraft, *Über Bahnen und deren Deformationen bei linearen Aktionen reduktiver Gruppen*, Comment. Math. Helvetici **54** (1979), 61–104. MR522032 (82m:14027)

[7] J. Denef, *Report on Igusa's local zeta function*, Séminaire Bourbaki, Vol. 1990/91. MR1157848 (93g:11119)

[8] J. González-Sánchez, *Kirillov's orbit method for p-groups and pro-p groups*, Comm. Algebra **37** (2009), 4476–4488. MR2588861 (2011a:20057)

[9] E. Hrushovski, B. Martin, *Zeta functions from definable equivalence relations*, arXiv:math/070111 (2007).

[10] B. Huppert, Character theory of finite groups, de Gruyter Expositions in Mathematics **25**, Walter de Gruyter & Co., Berlin, 1998. MR1645304 (99j:20011)

[11] J. Igusa, An introduction to the theory of local zeta functions, AMS/IP Studies in Advanced Mathematics **14**, American Mathematical Society, Providence, RI, International Press, Cambridge, MA, 2000. MR1743467 (2001j:11112)

[12] A. Jaikin-Zapirain, *Zeta functions of representations of compact p-adic analytic groups*, J. Amer. Math. Soc. **19** (2006), 91–118. MR2169043 (2006f:20029)

[13] G. Klaas, C. R. Leedham-Green, W. Plesken, Linear pro-p-groups of finite width, Lecture Notes in Mathematics **1674**, Springer-Verlag, Berlin, 1997. MR1483894 (98m:20028)

[14] B. Klopsch, *On the Lie theory of p-adic analytic groups*, Math. Z. **249** (2005), 713–730. MR2126210 (2005k:22012)

[15] M. Lazard, *Groupes analytiques p-adiques*, Publ. Math. IHÉS **26** (1965), 389–603. MR0209286 (35:188)

[16] M. Larsen, A. Lubotzky, *Representation growth of linear groups, J. Eur. Math. Soc. (JEMS)* **10** (2008), 351–390. MR2390327 (2009b:20080)

[17] A. Lubotzky, A. R. Magid, *Varieties of representations of finitely generated groups*, Mem. Amer. Math. Soc. **58** (1985), no. 336. MR818915 (87c:20021)

[18] A. Nobs, J. Wolfart, *Die irreduziblen Darstellungen der Gruppen* $\mathrm{SL}_2(\mathbb{Z}_p)$, *insbesondere* $\mathrm{SL}_2(\mathbb{Z}_2)$. *II.*, Comment. Math. Helv. **51** (1976), 491–526. MR0444788 (56:3136)

[19] G. Prasad, M. S. Raghunathan, *Topological central extensions of* $\mathrm{SL}_1(D)$, Invent. Math. **92** (1988), 645–689. MR939479 (89e:22018)

[20] A. Stasinski, C. Voll, *Representation zeta functions of nilpotent groups and generating functions for Weyl groups of type B*, arXiv:1104.1756 (2011).

[21] C. Voll, *Functional equations for zeta functions of groups and rings*, Ann. of Math. (2) **172** (2010), 1181–1218. MR2680489 (2011f:20057)

[22] W. Wang, *Dimension of a minimal nilpotent orbit*, Proc. Amer. Math. Soc. **127** (1999), 935–936. MR1610801 (99f:17010)

Department of Mathematics, Harvard University, One Oxford Street, Cambridge MA 02138, USA

E-mail address: avni.nir@gmail.com

Department of Mathematics, Royal Holloway, University of London, Egham TW20 0EX, United Kingdom

E-mail address: Benjamin.Klopsch@rhul.ac.uk

Department of Mathematics, Ben Gurion University of the Negev, Beer-Sheva 84105, Israel

E-mail address: urionn@math.bgu.ac.il

School of Mathematics, University of Southampton, University Road, Southampton SO17 1BJ, United Kingdom

Current address: Fakultät für Mathematik, Universität Bielefeld, D-33501 Bielefeld, Germany

E-mail address: C.Voll.98@cantab.net

Contemporary Mathematics
Volume **566**, 2012
http://dx.doi.org/10.1090/conm/566/11227

Applications of some zeta functions in group theory

Aner Shalev

ABSTRACT. We show that some Dirichlet series encoding maximal subgroup growth and representation growth have diverse applications in group theory, and can be used to solve various seemingly unrelated problems. These involve random generation, random walks, and properties of word maps and commutator maps in particular. This is a survey paper which also contains some new results and conjectures.

In fond memory of Fritz Grunewald

1. Introduction

Several growth functions were introduced in group theory, giving rise to Dirichlet series which sometimes can be regarded as natural generalizations of number theoretic zeta functions.

Let G be a group, and let n be a natural number. Denote by $a_n(G)$ the number of subgroups of index n in G, and by $m_n(G)$ the number of maximal subgroups of index n in G. We also let $r_n(G)$ denote the number of n-dimensional complex irreducible representations of G (up to equivalence).

In general these quantities need not be finite. If G is finitely generated then $a_n(G)$ and $m_n(G)$ are clearly finite. Groups G for which $r_n(G) < \infty$ for all n are called *rigid*, but in general no characterization of them is known.

For each of the above sequences we associate a Dirichlet series as follows.

$$\zeta_G^a(s) = \sum_{n\geq 1} a_n(G)n^{-s} = \sum_H |G:H|^{-s},$$

where the sum is over the finite index subgroups H of G,

$$\zeta_G^m(s) = \sum_{n>1} m_n(G)n^{-s} = \sum_M |G:M|^{-s},$$

where the sum is over the maximal subgroups of finite index M of G, and

$$\zeta_G^r(s) = \sum_{n\geq 1} r_n(G)n^{-s} = \sum_\chi \chi(1)^{-s},$$

2010 *Mathematics Subject Classification.* Primary 20D06, 20P05.

The author acknowledges the support of an Advanced ERC Grant 247034, an Israel Science Foundation Grant 754/08, a BSF grant 2008194, and the Miriam and Julius Vinik Chair in Mathematics which he holds.

where the sum is over the complex irreducible characters χ of G. The latter function was first studied in [**Wit**] for Lie groups, and is sometimes called the Witten zeta function.

We shall often assume our group G is finite, in which case there is no convergence problem. More generally, polynomial growth of the sequence b_n implies convergence of the associated Dirichlet series $\zeta(s) = \sum b_n n^{-s}$ in some half plane $Re(s) > \alpha$, and the minimal such α is called the abscissa of convergence of ζ.

The two seminal papers [**GSS**] and [**dSG**] by Fritz Grunewald and others explore important properties of the subgroup growth zeta function ζ_G^a for finitely generated nilpotent groups G. For more general background on subgroup growth, see the book [**LS**] by Lubotzky and Segal, which also presents results on the maximal subgroup growth function ζ_G^m. For background on the representation growth function ζ_G^r see Jaikin-Zapirain [**J**], Larsen and Lubotzky [**LL**] and the references therein. Properties and applications of these latter two functions are the main theme of this paper.

Recently there is also some interest in *density functions* associated with an infinite group G, counting the number of indices $\leq x$ of subgroups (or maximal subgroups) of G [**Sh1**], or the number of degrees $\leq x$ of irreducible representations of G [**LSSh**]. In particular it is shown in [**LSSh**] that a finitely generated linear group over a field of characteristic zero is either virtually abelian, or has many (at least x^α for some $\alpha > 0$ and all large x) irreducible representation degrees up to x.

For various applications it is natural to introduce a finitary analogue of the abscissa of convergence of Dirichlet series. Let $\mathcal{F}$ be an infinite family of finite groups. We shall be interested in real numbers s such that $\zeta_G^m(s) \to 0$, or $\zeta_G^r(s) \to 1$, for $G \in \mathcal{F}$ as $|G| \to \infty$, and we will try to minimize s. We shall see below that various asymptotic properties of the groups in $\mathcal{F}$ can be derived from these limit behaviors of the respective zeta functions, and the smaller s is the stronger the applications are.

We briefly mention some examples, starting with the maximal subgroup growth function ζ_G^m. If $\zeta_G^m(2) \to 0$ then it easily follows that the finite groups in $\mathcal{F}$ are generated with probability tending to 1 by two randomly chosen elements, and this applies for finite simple groups, proving a conjecture of Dixon (see [**Di, KL, LiSh1**]). The fact that $\zeta_G^m(7/5) \to 0$ for finite simple groups is used to prove that finite simple groups are generated with probability tending to 1 by a random involution and a random additional element, as conjectured by Kantor and Lubotzky (see [**LiSh2**]). The fact that $\zeta_G^m(66/65) \to 0$ for finite simple classical and alternating groups G implies random $(2,3)$-generation of these groups (with some exceptions) and helps studying the simple quotients of the modular group (see [**LiSh23**]).

As for the representation growth function ζ_G^r, if $\zeta_G^r(2) \to 1$ for $G \in \mathcal{F}$, then it follows that the commutator maps $f : G \times G \to G$ are almost measure preserving, and this applies for finite simple groups G, and was used to solve a conjecture of Guralnick and Pak [**GP**] on the product replacement algorithm (see [**GSh**]). The representation zeta function is also a key tool in many results on word maps in finite simple groups (see [**Sh2, Sh3, LaSh1, LaSh2, Se, ScSh, LST**]), and in the recent proof in [**LOST**] of Ore's conjecture [**O**] from 1951 that every element of a finite simple group is a commutator.

In a different direction, if $\zeta_G^r(2/3) \to 1$ for $G \in \mathcal{F}$, then results on mixing times of certain random walks on the groups G in $\mathcal{F}$ follow [**Sh2**]. The fact that

$\zeta_G^r(1/42) \to 1$ for alternating groups $G = A_n$ was used to derive various results on Fuchsian groups (see [**LiSh3, LiSh4**]) including a probabilistic proof of Higman's conjecture that every non-elementary Fuchsian group surjects onto all large enough alternating groups. More delicate properties of the characters of symmetric groups S_n established in [**LaSh1**] imply new results on the mixing times of random walks on these groups, including a proof of a conjecture of Lulov and Pak [**LP**].

In this paper we survey some of these applications. We also prove two new results. The first improves and generalizes previous bounds on the maximal subgroup growth of finite simple groups. Recall that an almost simple group is a group G satisfying $T \leq G \leq Aut(T)$ for some simple group T.

THEOREM 1.1. *There is an absolute constant c such that for any finite almost simple group G and a natural number n we have*

$$m_n(G) \leq cn(\log n)^3.$$

This implies $m_n(G) \leq n^{1+\epsilon}$ for every $\epsilon > 0$ and $n \geq N_\epsilon$, a result proved earlier for simple groups in [**LiSh23, LMSh**].

Our second new result deals with analogues of Ore's conjecture for some p-adic groups. While we cannot, at present, show that all elements of these groups are commutators, we can prove that the sets of commutators in some of these groups are very large.

THEOREM 1.2. *Fix $d = 2, 3$ and for a prime p let $G_p = SL_d(\mathbb{Z}_p)$. Let m_p denote the Haar measure of the set of commutators in G_p. Then $m_p \to 1$ as $p \to \infty$.*

The proof of this result relies heavily on computations and properties of the representation zeta functions $\zeta_{G_p}^r$, obtained in [**J**] for $d = 2$ and in the preprint [**AKOV**] for $d = 3$. More specifically, the result follows from the fact that $\zeta_{G_p}^r(2) \to 1$ as $p \to \infty$.

In fact [**J, AKOV**] contain computations of ζ_G^r for more general groups G, including $SL_2(O)$, $SL_3(O)$ and $SU_3(O)$, where O is a compact discrete valuation ring satisfying certain assumptions. Consequently our proof of Theorem 1.2 holds for these groups too; see Section 4 for more details.

It would be interesting to find out whether we actually have $m_p = 1$ in Theorem 1.2. We propose the following.

CONJECTURE 1.3. *Let $d \geq 2$ be an integer, p a prime, and if $d = 2$ suppose $p > 3$. Then every element of $SL_d(\mathbb{Z}_p)$ is a commutator.*

This conjecture may be strengthened, to include groups over more general discrete valuation rings.

Another challenging problem we propose is to study Ore's conjecture for arithmetic groups (say with the Congruence Subgroup Property). In particular we pose the following natural problem.

PROBLEM 1.4. *Is it true that all elements of $SL_d(\mathbb{Z})$ $(d \geq 3)$ are commutators?*

Clearly this is not the case for $SL_2(\mathbb{Z})$ (in fact this group is not even perfect). We note that by [**Th1, Th2, Th3**], if K is an arbitrary field, then (excluding the cases $d = 2$ and $|K| = 2, 3$), all elements of $SL_d(K)$ are commutators. However, special linear groups over rings are much harder to handle, and Conjecture 1.3 and Problem 1.4 seem highly non-trivial.

The structure of this paper is as follows. In Section 2 we discuss properties and applications of the maximal subgroup growth zeta function ζ_G^m. Most of the applications are related to generation and random generation. In particular we prove Theorem 1.1 there. In Section 3 we discuss properties of the representation zeta function ζ_G^r, and its applications to certain random walks. Finally, in Section 4 we present applications of ζ_G^r to word maps and commutator maps in particular. This is where Theorem 1.2 is proved.

I am grateful to Nir Avni for sending me the yet unposted preprint [**AKOV**], and to the anonymous referee of this paper for useful comments.

2. Maximal subgroup growth and random generation

Let G be a finite or profinite group, and let k be a positive integer. Denote by $P(G,k)$ the probability that k randomly chosen elements of G (with respect to the uniform distribution on G, or the normalized Haar measure if G is infinite). Since the probability that k random elements lie in a maximal subgroup M is $|G:M|^{-k}$, we easily obtain

$$1 - P(G,k) \leq \zeta_G^m(k).$$

It is shown in [**MSh**, Theorem 4] that, for G profinite, $P(G,k) > 0$ for some k if and only if $m_n(G) \leq n^c$ for some c and for all n, which is equivalent to $\zeta_G^m(s) < \infty$ for some s. Groups satisfying these equivalent conditions are termed *positively finitely generated*, and have been very recently characterized in [**JP**] (see also the references therein).

For finite simple groups the following was obtained in [**KL, LiSh1, LiSh23**].

THEOREM 2.1. *Let G be a finite simple group. Then $\zeta_G^m(2) \to 0$ as $|G| \to \infty$.*

Recall that Dixon [**Di**] proved that $P(A_n, 2) \to 1$ as $n \to \infty$, and conjectured a similar result for all finite simple groups G. Since $1 - P(G,2) \leq \zeta_G^m(2)$ which tends to 0 we immediately deduce:

THEOREM 2.2. *Dixon's conjecture holds: two randomly chosen elements of a finite simple group G generate G with probability tending to 1 as $|G| \to \infty$.*

A more delicate result is obtained in [**LiSh23**, Theorem 2.1] and [**LMSh**, Theorem 1.1].

THEOREM 2.3. *Fix $s > 1$, and let G be a finite simple group. Then $\zeta_G^m(s) \to 0$ as $|G| \to \infty$.*

In other words, for any $\epsilon > 0$ there exists $N = N_\epsilon$ such that $m_n(G) \leq n^{1+\epsilon}$ for all $n \geq N$ and finite simple group G. This confirms a conjecture from [**MSh**].

The above theorem has applications to more delicate results on random generation and to longstanding classical problems on the modular group $PSL_2(\mathbb{Z})$.

What are the finite simple quotients of the modular group? The study of this problem goes back to the beginning of the previous century, and possibly earlier. Since $PSL_2(\mathbb{Z})$ is isomorphic to the free product of groups of orders 2 and 3, a group G is a quotient of $PSL_2(\mathbb{Z})$ if and only if $G = \langle x, y \rangle$ with $x^2 = y^3 = 1$. Such groups are termed $(2,3)$*-generated.*

The study of $(2,3)$-generated finite groups was usually carried out with geometric and number-theoretic motivations, and was based on finding explicit generators of orders 2 and 3.

In [**LiSh23**] we introduce a different approach. Using probabilistic methods and properties of $\zeta_G^m(s)$ we show (in Theorem 1.5 of [**LiSh23**]) the following.

THEOREM 2.4. *All finite simple classical groups except $PSp_4(2^k)$, $PSp_4(3^k)$ and finitely many others are quotients of $PSL_2(\mathbb{Z})$.*

Let us now discuss the strategy of proof of Theorem 2.4. We need some notation. For $k \geq 1$ let x_k denote a randomly chosen element of order k in G. Set

$$P_{2,3}(G) = Prob(\langle x_2, x_3 \rangle = G).$$

We now formulate the probabilistic result behind Theorem 2.4 (see [**LiSh23**, Theorem 1.4]).

THEOREM 2.5. *Let $G \neq PSp_4(q)$ be a finite simple classical group. Then $P_{2,3}(G) \to 1$ as $|G| \to \infty$. If $G = PSp_4(p^k)$ $(p \geq 5)$ then $P_{2,3}(G) \to 1/2$ as $|G| \to \infty$.*

Clearly, Theorem 2.5 implies Theorem 2.4. In fact this is the classical way in which probabilistic methods are applied: prove existence theorems using probability estimates instead of explicit constructions.

We now briefly sketch the proof of Theorem 2.5. Let $i_k(G)$ denote the number of elements of order k in G. Note that

$$1 - P_{2,3}(G) \leq \sum_{M \max G} \frac{i_2(M) i_3(M)}{i_2(G) i_3(G)}.$$

Indeed, if $\langle x_2, x_3 \rangle \neq G$, then $x_2, x_3 \in M$ for some maximal subgroup $M < G$, and $\text{Prob}(x_k \in M) = i_k(M)/i_k(G)$.

Now, counting involutions and elements of order 3 in classical groups G and in their maximal subgroups M we deduce that, for some absolute constant c we have $i_2(M)/i_2(G) \leq c|G : M|^{-2/5}$, and (with few exceptions such as $PSp_4(q)$) $i_3(M)/i_3(G) \leq c|G : M|^{-8/13}$. We conclude that

$$1 - P_{2,3}(G) \leq c^2 \sum_M |G : M|^{-2/5} |G : M|^{-8/13}$$

$$= c^2 \zeta_G^m(2/5 + 8/13) = c^2 \zeta_G^m(66/65).$$

Using Theorem 2.3 above (proved in [**LiSh23**] for classical groups), we see that $\zeta_G^m(66/65) \to 0$ as $|G| \to \infty$, and this yields $P_{2,3}(G) \to 1$, completing the proof of the main part of Theorem 2.5.

Recall that the proof of Theorem 2.2 (Dixon's conjecture) was based on estimating $\zeta_G(2)$, while the proof of Theorems 2.4 and 2.5 is based on estimating $\zeta_G(66/65)$. It is intriguing that other generation results can be obtained by studying $\zeta_G(s)$ for other values of s. For instance, the probability that a randomly chosen involution and a randomly chosen additional element generate a finite simple classical group G is at least $1 - c\zeta_G(7/5)$, which tends to 1 by Theorem 2.3; and the probability that G is generated by three randomly chosen involutions is at least $1 - c\zeta_G(6/5)$, which also tends to 1. See [**LiSh23**, Corollary 2.5] for more details, as well as [**LiSh2**] for the case of exceptional groups of Lie type.

In [**LiShrs**] a similar method is applied to deal with random (r, s)-generation, where r, s are any primes (not both 2). We show there that if G is a finite simple classical group of dimension at least $f(r, s)$ (for a suitable function f), then the probability that random elements of G of orders r, s respectively generate G tends to 1 as $|G| \to \infty$. This again relies heavily on $\zeta_G^m(s) \to 1$ for $s > 1$.

We conclude this section with a proof of Theorem 1.1, which strengthens Theorem 2.3 above in two ways: first, the bound $m_n(G) \leq n^{1+\epsilon}$ is improved, and secondly, the result is also extended to almost simple groups.

Proof of Theorem 1.1

Our main tool is a new result of Guralnick, Larsen and Tiep [**GLT**], showing that an almost simple group G has at most $c(\log |G|)^3$ conjugacy classes of maximal subgroups.

Let G_0 be the simple socle of G. Then $G/G_0 \leq Out(G_0)$, and the structure of groups of outer automorphisms of simple groups is well known. In particular $Out(G_0)$ is an extension of a cyclic group of diagonal automorphisms by a cyclic group of field automorphisms, by a (small) group of graph automorphisms, and the orders of these groups can be found in [**KLi**], pp.170-171. Now, the maximal subgroups of G with non-trivial core correspond to maximal subgroups of G/G_0, and using the information on $Out(G_0)$ it is easy to verify that $m_n(G/G_0) \leq cn$ for some c and all n (in fact if G_0 is not of type PSL or PSU we even have $a_n(G/G_0) \leq cn$, since $Out(G_0)$ has a cyclic subgroup of bounded index).

It remains to count maximal subgroups M of G of index n with trivial core. Then M is self-normalizing, and the number of such subgroups is nt, where t is the number of conjugacy classes of such maximal subgroups. Hence it suffices to show that $t \leq c(\log n)^3$.

Suppose first that $G = S_k$ (the case of A_k being similar). We may assume $n \geq k$. The contribution to t of intransitive subgroups $M = S_l \times S_{k-l}$ is at most 1. The contribution to t of transitive imprimitive subgroups $M = S_l \wr S_{k/l}$ is at most the number of divisors of k, which is $o(k)$. Now, it is well known (see for instance [**Wi**, 14.2]) that the index of such subgroups is at least $2^{k/2}$, giving $k \leq 2\log n$, so $o(k) \leq o(\log n)$. Finally, if M is primitive then $|M| \leq 4^k$ by [**PS**], yielding $k \leq o(\log n)$. It is proved in [**LMSh**, Theorem 5.2] that S_k has $k^{o(1)}$ conjugacy classes of primitive maximal subgroups. Thus the contribution to t of primitive subgroups is $(\log n)^{o(1)}$. Altogether we obtain

$$t \leq 1 + o(\log n) + (\log n)^{o(1)} = o(\log n),$$

as required. Note that this implies $m_n(G) \leq o(n \log n)$ for $G = A_k, S_k$.

Now suppose G is an almost simple group of Lie type. Fix small $\epsilon > 0$. If $n \geq |G|^\epsilon$ then $\log |G| \leq \epsilon^{-1} \log n$. By Theorem 1.3 of [**GLT**] we have $t \leq c_1(\log |G|)^3 \leq c_2(\log n)^3$ where $c_2 = c_1\epsilon^{-3}$.

It remains to deal with the case $n < |G|^\epsilon$, assuming $m_n(G) > 0$. Write $G = X_r(q)$ where r is the Lie rank and q the field size. Choosing ϵ small enough it follows (using e.g. [**KLi**], p.175) that r is large, G is a classical group, and M is a reducible subgroup (when we regard $G_0 = Sp_{2m}(q)$ in characteristic 2 as $O_{2m+1}(q)$). The number of conjugacy classes of such subgroups M is at most $O(r)$, and it is known (by the previous reference) that $n \geq c_3 q^r$. Hence $r \leq c_4 \log n$ so $t \leq O(\log n)$ in this case.

This completes the proof. □

3. Representation growth and random walks

The abscissa of convergence of the representation zeta function was studied in [**LL**] for infinite linear groups. The finitary analogue in the case of finite simple groups was studied earlier in [**LiSh3, LiSh4, LiSh5**], where the following is obtained:

THEOREM 3.1. *Fix a real number s and let G be a finite simple group.*
(i) If $s > 1$ then $\zeta^{r}_{G}(s) \to 1$ as $|G| \to \infty$.
(ii) If $s > 0$ and G is alternating or classical of sufficiently large rank then $\zeta^{r}_{G}(s) \to 1$ as $|G| \to \infty$.
(iii) If G ranges over the groups of Lie type of fixed rank r and u positive roots, and $s > r/u$, then $\zeta^{r}_{G}(s) \to 1$ as $|G| \to \infty$.

Indeed, part (i) is included in Theorem 1.1 of [**LiSh4**], part (ii) for alternating groups is [**LiSh3**, Corollary 2.7] and for classical groups is [**LiSh5**, Theorem 1.2], and part (iii) is [**LiSh5**, Theorem 1.1]. See [**LiSh5**] for the exact definition of rank. It can be shown that this result is best possible, in the sense that the assumptions on s are necessary.

Theorem 3.1 has numerous applications, only some of which will be described here. In this section we focus on some applications to random walks, leaving further applications to the next section.

Let C be a conjugacy class of G. We wish to study the random walk on G based on C, namely we start with 1, multiply it by a random element $x_1 \in C$, then multiply again by a random element $x_2 \in C$ to get x_1x_2, and so on, so the resulting element after t steps of the random walk is $x_1 \cdots x_t$. (We do not assume at this stage that C generates G, so it might be that some elements of G are never reached).

Let U be the uniform distribution on G, P_C the uniform distribution on C, and P^t_C the distribution after t steps of the random walk. We are interested in the L_1-distance $||P^t_C - U|| = \sum_{g\in G} |P^t_C(g) - |G|^{-1}|$ between P^t_C and the uniform distribution. When this distance is smaller than $1/e$ we say that the *mixing time* $T(C, G)$ of the random walk is $\leq t$.

Let Irr$G$ denote the set of (complex irreducible) characters of G. Given $x \in G$ and a positive integer t define

$$d_t(x) = \sum_{1\neq\chi\in \mathrm{Irr}G} |\chi(x)|^{2t}/\chi(1)^{2t-2}.$$

Let $C = x^G$, the conjugacy class of x in G. By the upper bound Lemma of [**DS**] we have

$$||P^t_C - P||^2 \leq d_t(x).$$

The following somewhat surprising result obtained in [**Sh2**, Theorem 1.1] shows that the mixing time $T(C, G)$ is usually the smallest possible, namely 2, provided the representation growth of the finite group G is very slow.

THEOREM 3.2. *Let $\mathcal{F}$ be a family of finite groups such that $\zeta^{r}_{G}(2/3) \to 1$ as the order of $G \in \mathcal{F}$ tends to infinity. Let $x \in G$ be randomly chosen, and let $C = x^G$ be its conjugacy class. Then the probability that $T(C, G) = 2$ tends to 1 as $|G| \to \infty$.*

This means that the product of two random elements of a "typical" class C is almost uniformly distributed on G. In particular this shows that, for any $\epsilon > 0$ and for almost all $x \in G$ we have $|C^2| \geq (1-\epsilon)|G|$ where $C = x^G$ and $G \in \mathcal{F}$ is large enough. (Thus a random class C generates G in this case, and G is the normal closure of the cyclic subgroup generated by a random element). See [**Sh2**] for more details.

The idea behind the proof of Theorem 3.2 is to show that for almost all $x \in G$, $d_2(x) \to 0$, so the desired conclusion follows from the upper bound lemma mentioned above.

A main application of Theorem 3.2 is for random walks on finite simple groups with respect to a conjugacy class C as a generating set. These walks have been studied extensively in the past decades. See Diaconis and Shahshahani [**DS**] for transpositions in symmetric groups, Lulov [**Lu**], Vishne [**V**] and [**LiSh3**], [**LaSh1**] for more general classes in symmetric groups, as well as [**H**], [**Gl**], [**LiShdiam**], [**LiSh5**] for groups of Lie type. In many cases the mixing times $T(C, G)$ of these walks are still not known. For background see also [**D1**], [**D2**].

Combining Theorem 3.1 and 3.2 with some extra arguments we obtain the following.

COROLLARY 3.3. *Let G be a finite simple group, let $x \in G$ be randomly chosen, and let $C = x^G$ be its conjugacy class. Then the probability that $T(C, G) = 2$ tends to 1 as $|G| \to \infty$.*

Indeed, Theorem 3.1 implies that $\zeta_G^r(2/3) \to 1$ as $|G| \to \infty$ for all families of simple groups G except $PSL_2(q), PSL_3(q), PSU_3(q)$; these groups are dealt with directly using ad-hoc methods.

A longstanding conjecture of Thompson states that every finite simple group G has a conjugacy class C such that $C^2 = G$. This is known in various cases but is still open in general. See [**EG**] for background and results. Corollary 3.3 implies that the square of a class of a random element of G covers almost all of G. This provides positive evidence towards Thompson's conjecture, suggesting that C^2 might be equal to G for many classes C.

Other results on mixing times require information on character values. Bounds of the form $|\chi(x)| \leq \chi(1)^a$ for all $\chi \in \mathrm{Irr}G$ and some $x \in G$, where $a < 1$ depends on x, are particularly useful; indeed plugging them in the upper bound lemma enable us to bound $d_t(x)$, and hence $||P_C^t - U||^2$, by $\zeta_G^r(s)$ where s depends on a and t.

In [**LaSh1**, 1.6, 1.7, 1.8] we employ this strategy for the groups A_n and S_n, obtaining various new character bounds. These enable us to provide the sharpest bounds obtained so far on mixing times in symmetric and alternating groups.

THEOREM 3.4. *Let $\sigma \in A_n$, and $C = \sigma^{S_n}$, and let $T = T(C, A_n)$ denote the mixing time of the associated random walk on A_n.*

(i) The mixing time T is bounded if and only if σ has at most n^α fixed points, where $\alpha < 1$ is bounded away from 1.

(ii) If σ has n^α fixed points where $\alpha < 1$ then $\frac{1}{1-\alpha} \leq T \leq \frac{2}{1-\alpha} + 1$.

(iii) If σ is fixed-point-free or has $n^{o(1)}$ fixed points then $T \leq 3$.

(iv) If σ has at most $n^{o(1)}$ cycles of length 1 and 2 then $T = 2$.

Parts (iii) and (iv) are best possible, and extend Lulov's result [**Lu**] for permutations σ which consist of n/m m-cycles (where the mixing time is 3 if $m = 2$ and 2 if $m \geq 3$).

The main conjecture of Lulov and Pak in [**LP**] is the following. Let $C_n \subset S_n$ be a sequence of conjugacy classes of permutations with no fixed points. Then, as $n \to \infty$, the mixing time $T(C_n, S_n)$ is 2 or 3.

This means that in two or three steps we reach an almost uniform distribution on a suitable coset of A_n in S_n.

Part (iii) of Theorem 3.4 (with a similar variant when σ is an odd permutation) establishes this conjecture even when there are $n^{o(1)}$ fixed points. We therefore have

COROLLARY 3.5. *The Lulov-Pak conjecture holds.*

4. Representation growth, commutators and words

In this section we present further applications of the representation zeta function ζ_G^r and of Theorem 3.1 in particular. We first show, following [**GSh**], that the value of ζ_G^r at 2 is highly significant in analyzing the commutator structure of finite groups G.

Let $f : G \times G \to G$ be the commutator map, so that $f(x, y) = [x, y]$. Let P be the distribution of G induced by this map, namely $P(g) = |f^{-1}(g)|/|G|^2$.

Our first result, obtained in [**GSh**, Proposition 1.1], bounds the L_1-distance $||P - U||$ between the probability measure P above and the uniform distribution U on G.

PROPOSITION 4.1. *With the above notation we have*

$$||P - U|| \leq (\zeta_G^r(2) - 1)^{1/2}.$$

The proof uses a classical result of Frobenius showing that

$$P(g) = |G|^{-1} \sum_{\chi \in \mathrm{Irr} G} \chi(1)^{-1} \chi(g).$$

Proposition 4.1 shows that when $\zeta_G^r(2)$ is close to 1 the probability measure P is almost uniform.

We now deduce a general lower bound on the number of commutators in G (see [**GSh**, Corollary 1.2]).

COROLLARY 4.2. *A finite group G has at least $(1 - (\zeta_G^r(2) - 1)^{1/2})|G|$ commutators.*

By Theorem 3.1, if G is a finite simple group then $\zeta_G^r(2) \to 1$ as $|G| \to \infty$. Combining this with Proposition 4.1 we show that the commutator map is almost *measure preserving* on finite simple groups. More precisely we have:

THEOREM 4.3. *Let G be a finite simple group and let $f : G \times G \to G$ be the map sending (x, y) to $[x, y]$. Then*

(i) For every subset $Y \subseteq G$ we have

$$|f^{-1}(Y)|/|G|^2 = |Y|/|G| + o(1).$$

(ii) For every subset $X \subseteq G \times G$ we have

$$|f(X)|/|G| \geq |X|/|G|^2 - o(1).$$

(iii) In particular, if X is as above and $|X|/|G|^2 = 1 - o(1)$, then almost every element of G is a commutator of the form $[x, y]$ where $x, y \in X$.

Indeed, this is Corollary 1.6 of [**GSh**].

Applying this result for the set of generating pairs for G, which is of size $|G|^2(1 - o(1))$ (by Theorem 2.2 above) we obtain the following.

COROLLARY 4.4. *Almost every element of a finite simple group G can be expressed as a commutator $[x, y]$ where x, y generate G.*

The same result holds with the same proof for any family $\mathcal{F}$ of finite groups such that $\zeta_G^r(2) \to 1$ and $\zeta_G^m(2) \to 0$ for $G \in \mathcal{F}$.

Corollary 4.4 was used to prove a conjecture by Guralnick and Pak [**GP**] regarding the product replacement algorithm, see [**GSh**, 1.8, 1.9] for more details.

We shall now use a similar machinery to study some p-adic groups and prove Theorem 1.2 stated in the Introduction.

Proof of Theorem 1.2.

We first note that Corollary 4.2 is easily adjusted to profinite groups G. Here in ζ_G^r we restrict to representations which factor through finite quotients of G (when we factor out open subgroups), and we have

$$\mu(Comm(G)) \geq 1 - (\zeta_G^r(2) - 1)^{1/2},$$

where μ is the normalized Haar measure of G and $Comm(G)$ is the set of commutators in G.

Hence, to prove Theorem 1.2 it suffices to show that, if $G_p = SL_d(\mathbb{Z}_p)$ with $d = 2$ or 3, then $\zeta_{G_p}^r(2) \to 1$ as $p \to \infty$.

The functions $\zeta_{G_p}^r(s)$ are computed in [**J**] for $d = 2$ and $p > 2$, and in [**AKOV**] for $d = 3$ and p large. Substituting $s = 2$ in the explicit expressions yields the desired conclusion, proving the theorem. □

Some remarks are in order. First, the preprint [**AKOV**] has not yet appeared or posted. However, Theorem C of [**AKOV1**] implies that the abscissa of convergence of ζ_G^r for $G = SL_3(\mathbb{Z})$ is 1, and so $\zeta_G^r(s) < \infty$ for any $s > 1$. Now, Using the Euler factorization for this zeta function $\zeta_G^r(s) = \zeta_H^r(s) \prod_p \zeta_{G_p}^r(s)$ where $H = SL_3(\mathbb{C})$ and $G_p = SL_3(\mathbb{Z}_p)$ (see [**LL**]), it follows immediately that for each real number $s > 1$ we have

$$\zeta_{G_p}^r(s) \to 1 \text{ as } p \to \infty,$$

and the case $s = 2$ proves Theorem 1.2 for $d = 3$.

Secondly, the results in [**J**] hold for $SL_2(O)$ where O is any compact DVR (also in positive characteristic) provided its residue field has odd characteristic. Thus our proof of Theorem 1.2 extends to such groups $SL_2(O)$, showing that the measure of the set of commutators in them tends to 1 as the size of the residue field tends to infinity.

Thirdly, the results in [**AKOV**] hold for $SL_3(O)$ and $SU_3(O)$ where O is a compact DVR of characteristic zero, assuming the characteristic of the residue field is large enough. It follows that the set of commutators in these groups has measure tending to 1 as the size of the residue field tends to infinity.

It turns out that other values of $\zeta_G^r(s)$ yield additional information on the commutator structure of finite groups, and on commutator width in particular. Indeed we have

PROPOSITION 4.5. *Let G be a finite group, k a positive integer, and suppose $\zeta_G^r(2k-2) < 2$. Then every element of G can be expressed as a product of k commutators.*

This result follows immediately from Lemma 9.2 of [**Sh3**].

Commutators are a particular case of group words, namely elements $w(x_1, \ldots, x_d)$ of the free group F_d on $x_1, \ldots, x_d$. Given a word w and a group G one defines a *word map* $w = w_G : G^d \to G$ sending $(g_1, \ldots, g_d)$ to $w(g_1, \ldots, g_d)$. These maps were studied extensively in the past few years, with particular emphasis on their image, denoted by $w(G)$. See [**Bo, LiShdiam, La, LaSh1, LaSh2, Sh2, Sh3, Se, ScSh, LST, LST2**], as well as the recent preprint [**BGG**] by Fritz Grunewald et. al, dealing with some word maps on $PSL_2(q)$ and $SL_2(q)$.

Clearly primitive words (i.e. words which are part of a free generating set for F_d) are measure preserving on all finite groups G (namely $|w^{-1}(X)|/|G|^d = |X|/|G|$ for any subset $X \subseteq G$). It is conjectured that these are the only words which are measure preserving on all finite groups, and the case $d = 2$ was recently established by D. Puder [**Pu**].

However, we can show that various additional words are almost measure preserving on finite simple groups, in the sense of Theorem 4.3.

THEOREM 4.6. *(i) If w_1, w_2 are non-trivial words in different variables, then $w_1 w_2$ is almost measure preserving on alternating groups A_n.*

(ii) If w_1, w_2 are non-trivial words in different variables, then $w_1 w_2$ is almost measure preserving on finite simple groups of Lie type of bounded rank.

(iii) For any positive integers n, m, the word $x^n y^m$ is almost measure preserving on all finite simple groups.

(iv) Any admissible word is almost measure preserving on all finite simple groups.

Here a reduced word $w(x_1, \ldots, x_d) \neq 1$ is called *admissible* if each x_i occurs twice in w, once with exponent 1 and once with exponent -1. Thus commutators are admissible, as well more general words such as $x_1 x_2 \cdots x_d x_1^{-1} x_2^{-1} \cdots x_d^{-1}$, and so on.

Part (i) of Theorem 4.4 is obtained in [**LaSh1**, Theorem 1.18], and parts (ii), (iii) and (iv) are the main results [**LaSh3**]. Again character methods and the zeta function ζ_G^r play a crucial role in these proofs. These methods are also useful in studying random walks on finite simple groups G with respect to $w(G)$ as the generating set. Indeed we have:

THEOREM 4.7. *Let w be a non-identity word, and let G be a finite simple group. Then the mixing time $T(w(G), G)$ is equal to 2 if G is large enough.*

This result for alternating groups is obtained in [**LaSh1**, Theorem 1.17], and the result for groups of Lie type is [**ScSh**, Theorem 1.1]. Theorem 4.7 implies that $w(G)^2 \geq (1-\epsilon)|G|$ for any given $\epsilon > 0$ and a large enough finite simple group G. Is it true that we actually have $w(G)^2 = G$, namely every group element is a product of two values of w?

This was a major open problem for a few years. An affirmative answer is given in the following yet unpublished result which we announce here.

THEOREM 4.8. *For each non-identity word w there exists a number $N = N_w$ such that if G is a finite simple group of order at least N then $w(G)^2 = G$.*

This means that the word width of G is at most 2, improving [**Sh3**, Theorem 1.1], where it was shown that, under similar assumptions, the word width is at most 3. Clearly Theorem 4.8 is best possible, since some words, such as x^2, are never surjective on finite simple groups.

The proof of Theorem 4.8 for alternating groups and groups of Lie type of bounded rank appears in [**LaSh1, LaSh2**]. The proof for classical groups of unbounded rank (and hence for all finite simple groups) will appear in the joint work [**LST**] by Larsen, Tiep and myself. All these proofs employ character methods, and often some additional methods from geometry.

References

[AKOV] N. Avni, B. Klopsch, U. Onn and C. Voll, Representation zeta functions for SL_3, Preprint 2011.

[AKOV1] N. Avni, B. Klopsch, U. Onn and C. Voll, Representation zeta functions of compact p-adic analytic groups and arithmetic groups, arXiv:1007.2900v2.

[BGG] T. Bandamn, Sh. Garion and F. Grunewald, On the surjectivity of Engel words on PSL$(2, q)$, arXiv:1008.1397.

[Bo] A. Borel, On free subgroups of semisimple groups, *Enseign. Math.* **29** (1983), 151–164. MR702738 (85c:22009)

[D1] P. Diaconis, *Group Representations in Probability and Statistics*, Institute of Mathematical Statistics Lecture Notes - Monograph Series, Vol. 11, 1988. MR964069 (90a:60001)

[D2] P. Diaconis, Random walks on groups: characters and geometry, in *Groups St Andrews 2001 in Oxford, Vol.I*, 120–142, London Math. Soc. Lecture Note Series **304**, Cambridge Univ. Press, Cambridge, 2003. MR2051523 (2005c:20109)

[DS] P. Diaconis and M. Shahshahani, Generating a random permutation with random transpositions, *Z. Wahrscheinlichkeitstheorie Verw. Gebiete* **57** (1981), 159–179. MR626813 (82h:60024)

[Di] J.D. Dixon, The probability of generating the symmetric group, *Math. Z.* **110** (1969), 199–205. MR0251758 (40:4985)

[dSG] M.P.F. du Sautoy and F.J. Grunewald, Analytic properties of zeta functions and subgroup growth, *Ann. Math.* **152** (2000), 793–833. MR1815702 (2002h:11084)

[EG] E.W. Ellers and N. Gordeev, On conjectures of J. Thompson and O. Ore, *Trans. Amer. Math. Soc.* **350**, 3657–3671. MR1422600 (98k:20022)

[GSh] S. Garion and A. Shalev, Commutator maps, measure preservation, and T-systems, *Trans. Amer. Math. Soc.* **361** (2009), 4631–4651. MR2506422 (2010f:20019)

[Gl] D. Gluck, Characters and random walks on finite classical groups, *Adv. Math.* **129** (1997), 46–72. MR1458412 (98g:20027)

[GSS] F.J. Grunewald, D. Segal and G.C. Smith, Subgroups of finite index in nilpotent groups, *Invent. Math.* **93** (1988), 185–223. MR943928 (89m:11084)

[GLT] R.M. Guralnick, M. Larsen and Ph. Tiep, Representation growth in positive characteristic and conjugacy classes of maximal subgroups, Preprint 2010, arXiv:1009.2437v2.

[GP] R.M. Guralnick and I. Pak, On a question of B.H. Neumann, *Proc. Amer. Math. Soc.* **131** (2002), 2021–2025. MR1963745 (2004a:20029)

[H] M. Hildebrand, Generating random elements in $SL_n(F_q)$ by random transvections, *J. Alg. Combin.* **1** (1992), 133–150. MR1226348 (94k:60104)

[J] A. Jaikin-Zapirain, Zeta functions of representations of p-adic analytic groups, *J. Amer. Math. Soc.* **19** (2006), 91–118. MR2169043 (2006f:20029)

[JP] A. Jaikin-Zapirain and L. Pyber, Random generation of finite and profinite groups and group enumeration, *Ann. Math.* **173** (2011), 769–814. MR2776362

[KL] W.M. Kantor and A. Lubotzky, The probability of generating a finite classical group, *Geom. Ded.* **36** (1990), 67–87. MR1065213 (91j:20041)

[KLi] P. Kleidman and M.W. Liebeck, *The subgroup structure of the finite classical groups*, London Math. Soc. Lecture Note Series **129**, Cambridge Univ. Press, 1990. MR1057341 (91g:20001)

[La] M. Larsen, Word maps have large image, *Israel J. Math.* **139** (2004), 149–156. MR2041227 (2004k:20094)

[LL] M. Larsen and A. Lubotzky, Representation growth for linear groups, *J. Eur. Math. Soc.* **10** (2008), 351–390. MR2390327 (2009b:20080)

[LaSh1] M. Larsen and A. Shalev, Characters of symmetric groups: sharp bounds and applications, *Invent. Math.* **174** (2008), 645–687. MR2453603 (2010g:20022)

[LaSh2] M. Larsen and A. Shalev, Word maps and Waring type problems, *J. Amer. Math. Soc.* **22** (2009), 437–466. MR2476780 (2010d:20019)

[LaSh3] M. Larsen and A. Shalev, Words and measure preservation, Preprint 2011.

[LST] M. Larsen, A. Shalev and Ph. Tiep, The Waring problem for finite simple groups, *Ann. Math.* **174** (2011), 1885–1950. MR2846493

[LST2] M. Larsen, A. Shalev and Ph. Tiep, Waring problem for finite quasisimple groups, arXiv:1107.3341. To appear in *Int. Math. Res. Notices* (IMRN).

[LMSh] M.W. Liebeck, B.M.S. Martin and A. Shalev, On conjugacy classes of maximal subgroups of finite simple groups, and a related zeta function, *Duke Math. Journal* **128** (2005), 541–557. MR2145743 (2006b:20043)

[LOST] M.W. Liebeck, E.A. O'Brien, A. Shalev and Ph. Tiep, The Ore Conjecture, *J. Eur. Math. Soc.* **12** (2010), 939–1008. MR2654085 (2011e:20016)

[LSSh] M.W. Liebeck, D. Segal and A. Shalev, The density of representation degrees, to appear in *J. Eur. Math. Soc.*

[LiSh1] M.W. Liebeck and A. Shalev, The probability of generating a finite simple group, *Geom. Ded.* **56** (1995), 103–113. MR1338320 (96h:20116)

[LiSh2] M.W. Liebeck and A. Shalev, Simple groups, probabilistic methods, and a conjecture of Kantor and Lubotzky, *J. Algebra* **184** (1996), 31–57. MR1402569 (97e:20106b)

[LiSh23] M.W. Liebeck and A. Shalev, Classical groups, probabilistic methods and the $(2,3)$-generation problem, *Ann. Math.* **144** (1996), 77-125. MR1405944 (97e:20106a)

[LiShdiam] M.W. Liebeck and A. Shalev, Diameters of finite simple groups: sharp bounds and applications, *Ann. Math.* **154** (2001), 383–406. MR1865975 (2002m:20029)

[LiShrs] M.W. Liebeck and A. Shalev, Random (r,s)-generation of finite classical groups, *Bull. London Math. Soc.* **34** (2002), 185–188 MR1874245 (2002i:20100)

[LiSh3] M.W. Liebeck and A. Shalev, Fuchsian groups, coverings of Riemann surfaces, subgroup growth, random quotients and random walks, *J. Algebra* **276** (2004), 552–601. MR2058457 (2005e:20076)

[LiSh4] M.W. Liebeck and A. Shalev, Fuchsian groups, finite simple groups, and representation varieties, *Invent. Math.* **159** (2005), 317–367. MR2116277 (2005j:20065)

[LiSh5] M.W. Liebeck and A. Shalev, Character degrees and random walks in finite groups of Lie type, *Proc. London Math. Soc.* **90** (2005), 61–86. MR2107038 (2006h:20016)

[LS] Alexander Lubotzky and Dan Segal, *Subgroup growth*, Progress in Mathematics **212**, Birkhäuser, Basel, 2003. MR1978431 (2004k:20055)

[Lu] N. Lulov, Random walks on symmetric groups generated by conjugacy classes, Ph.D. Thesis, Harvard University, 1996. MR2695111

[LP] N. Lulov and I. Pak, Rapidly mixing random walks and bounds on characters of the symmetric groups, *J. Alg. Combin.* **16** (2002), 151–163. MR1943586 (2003i:60008)

[MSh] A. Mann and A. Shalev, Simple groups, maximal subgroups, and probabilistic aspects of profinite groups, *Israel. J. Math.* **96** (1997) (Amitsur memorial issue), 449–468. MR1433701 (97m:20038)

[O] O.O. Ore, Some remarks on commutators, *Proc. Amer. Soc.* **272** (1951), 307–314. MR0040298 (12:671e)

[PS] C.E. Praeger and J. Saxl, On the orders of primitive permutation groups, *Bull. London Math. Soc.* **12** (1980), 303–307. MR576980 (81g:20003)

[Pu] D. Puder, On Primitive Words II: Measure Preservation, arXiv:1104.3991.

[ScSh] G. Schul and A. Shalev, Words and mixing times in finite simple groups, *Groups, Geometry and Dynamics* (Fritz Grunewald Issue) **5** (2011), 509–527. MR2782183

[Se] D. Segal, *Words: notes on verbal width in groups*, London Math. Soc. Lecture Note Series **361**, Cambridge University Press, Cambridge, 2009. MR2547644 (2011a:20055)

[Sh1] A. Shalev, The density of subgroup indices, *J. Australian Math. Soc.* **85** (2008), 257–267. MR2470542 (2009i:20053)

[Sh2] A. Shalev, Mixing and generation in simple groups, *J. Algebra* **319** (2008), 3075–3086. MR2397424 (2009e:20152)

[Sh3] A. Shalev, Word maps, conjugacy classes, and a non-commutative Waring-type theorem, *Ann. Math.* **170** (2009), 1383–1416. MR2600876 (2011e:20017)

[Th1] R.C. Thompson, Commutators in the special and general linear groups, *Trans. Amer. Math. Soc.* **101** (1961), 16–33. MR0130917 (24:A771)

[Th2] R.C. Thompson, On matrix commutators, *Portugal. Math.* **21** (1962), 143–153. MR0147558 (26:5073)

[Th3] R.C. Thompson, Commutators of matrices with coefficients from the field of two elements, *Duke Math. J.* **29** (1962), 367–373. MR0140584 (25:4002)

[V] U. Vishne, Mixing and covering in symmetric groups, *J. Algebra* **205** (1998), 119–140. MR1631326 (99d:20005)

[Wi] H. Wielandt, *Finite Permutation Groups*, Academic Press, New York, 1964. MR0183775 (32:1252)

[Wit] E. Witten, On quantum gauge theories in two dimensions, *Comm. Math. Phys.* **141** (1991), 153–209. MR1133264 (93i:58164)

Institute of Mathematics, The Hebrew University, Jerusalem 91904, Israel
E-mail address: shalev@math.huji.ac.il